ANATOMY & PHYS

Laboratory Manual

Eric Wise

Santa Barbara City College

 **Higher Education**

Boston   Burr Ridge, IL   Dubuque, IA   New York   San Francisco   St. Louis
Bangkok   Bogotá   Caracas   Kuala Lumpur   Lisbon   London   Madrid   Mexico City
Milan   Montreal   New Delhi   Santiago   Seoul   Singapore   Sydney   Taipei   Toronto

fifth edition

LABORATORY MANUAL SALADIN ANATOMY & PHYSIOLOGY: THE UNITY OF FORM AND FUNCTION, FIFTH EDITION

Some ancillaries, including electronic and print components, may not be available to customers outside the United States.

This book is printed on recycled, acid-free paper containing 10% postconsumer waste.

1 2 3 4 5 6 7 8 9 0 QPD/QPD 0 9

ISBN 978–0–07–325095–3
MHID 0–07–325095–3

Publisher: *Michelle Watnick*
Senior Sponsoring Editor: *James F. Connely*
Director of Development: *Kristine Tibbetts*
Senior Developmental Editor: *Anne L. Winch*
Editorial Coordinator: *Ashley Zellmer*
Marketing Manager: *Lynn M. Breithaupt*
Project Coordinator: *Mary Jane Lampe*
Senior Production Supervisor: *Laura Fuller*
Senior Designer: *David W. Hash*
Cover/Interior Designer: *K. Wayne Harms*
Cover Illustration: *Paul Waldinger, Electronic Publishing Services Inc., NYC*
(USE) Cover Image: *©John Knill/Stockbyte/Getty Images*
Senior Photo Research Coordinator: *John C. Leland*
Compositor: *Electronic Publishing Services Inc., NYC*
Typeface: *10/12 Janson*
Printer: *Quebecor World Dubuque, IA*

The credits section for this book begins on page 615 and is considered an extension of the copyright page.

Some of the laboratory experiments included in this text may be hazardous if materials are handled improperly or if procedures are conducted incorrectly. Safety precautions are necessary when you are working with chemicals, glass test tubes, hot water baths, sharp instruments, and the like, or for any procedures that generally require caution. Your school may have set regulations regarding safety procedures that your instructor will explain to you. Should you have any problems with materials or procedures, please ask your instructor for help.

www.mhhe.com

Contents

Instructor Preface

Anatomy and physiology can be the jewel in our students' education or the bane of their college career. As an instructor of anatomy and physiology for many years, I decided to write a lab manual that was student-friendly and with a singular focus on the lab portion of the course. This lab manual was written for the undergraduate student of anatomy and physiology, and it consists of 47 exercises designed to help students learn basic human anatomy and the practical lab applications in physiology.

The diversity of interests in today's anatomy and physiology students is due, in part, to the number of majors that either require or recommend the subject. This lab manual provides a framework for understanding anatomy and physiology for students interested in nursing, radiology, physical or occupational therapy, physical education, dental hygiene, or other allied health majors.

This manual was written to be used with Saladin, *Anatomy & Physiology*, fifth edition. The illustrations are labeled; therefore, students do not need to bring their lecture text to the lab. The lab manual accompanies the lecture text and lecture portion of the course and can be used in either a one-term or full-year course. The illustrations are outstanding, and the balanced combination of line art and photographs provides effective coverage of material. The amount of lecture material in the manual is limited, so there is little material included that is not part of the lab experience.

Practical lab experience is an invaluable opportunity to reinforce lecture concepts, enrich students' understanding of anatomy and physiology, and allow them to explore new dimensions in the subject area. The educational benefit of reinforcing lecture material with hands-on experiments and acquiring knowledge with a learn-by-doing philosophy makes the anatomy and physiology lab a very special educational environment. Many of us use lab experiences to present conceptually difficult material in physiology and to provide students with different learning styles another avenue for learning.

The 47 exercises in this lab manual provide a comprehensive overview of the human body. Each exercise presents the core elements of the subject matter. You can tailor this manual to match your vision of the course or use it in its entirety. There are significant differences in the laboratories found around the country, and the advances in physiology equipment, especially computer modules, are numerous and continually evolving. The materials section in each lab is designed for a lab of 24 students and includes the amounts and types of reagents to be used. The labs generally take between 2 and 3 hours to complete.

This lab manual was written for three types of anatomy and physiology courses. For those courses that use the cat as the primary dissection animal, cat dissections or mammalian organ dissections follow the material on humans. For those courses that use models or charts, numerous cadaver photographs are included so that students can see the representative structures as they exist in the cadaver material. Finally, for those courses that use cadavers, this lab manual can be used by studying the human material and omitting the cat dissection sections.

Key Features

1. **Dynamic Art Program.** All the illustrations have been rendered by a state-of-the-art digital illustration company. They are extremely accurate, use bold and appealing colors, and offer a unique, three-dimensional look.
2. **Labels.** Illustrations are labeled for students to learn the names and terminology by looking at real-life examples or models and by referring to the illustrations in the manual.
3. **Instructional Photographs.** Numerous full-color photographs, including detailed histological light micrographs and cadaver and cat dissections, show detailed structures.
4. **Focus on the Laboratory.** This manual focuses primarily on the material necessary for the laboratory and does not repeat the material presented in the lecture text, with the expectation that students can look up material in the lecture text when necessary.

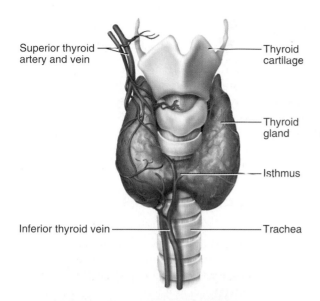

Superior thyroid artery and vein
Thyroid cartilage
Thyroid gland
Isthmus
Inferior thyroid vein
Trachea

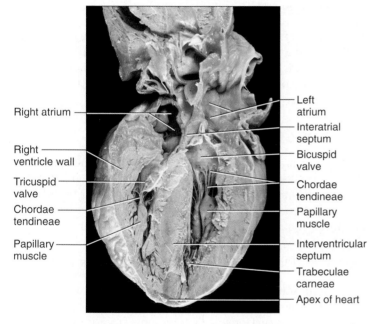

Right atrium
Right ventricle wall
Tricuspid valve
Chordae tendineae
Papillary muscle
Left atrium
Interatrial septum
Bicuspid valve
Chordae tendineae
Papillary muscle
Interventricular septum
Trabeculae carneae
Apex of heart

and special attention to when preparing for or performing the laboratory exercise.

8. **Clean Up.** At the end of many laboratory exercises an icon for clean up (🖐) reminds the student to clean up the laboratory. Special instructions are given where appropriate.

9. **Data Collection.** Collection of data is embedded within each exercise as opposed to in a separate table at the back of the manual.

10. **Summary Data.** These sections can be found at the end of appropriate exercises for ease of handing in material.

11. **User-Friendly Format.** Each exercise begins on a right-hand page, and the pages are perforated to allow students to more easily remove the exercises to turn them in and later store them.

12. **Study Hints.** Study hints are located in selected chapters where additional information is provided to help students comprehend and retain the more difficult material presented.

13. **Key Terms.** Current anatomical terminology is used throughout the laboratory manual. Key terms are boldfaced.

Saladin Website

The *Anatomy & Physiology* text website at www.mhhe.com/saladin5 offers an extensive array of learning and teaching tools.

New! Digital Images Full-color digital files of all illustrations and key photographs in the laboratory manual can be readily incorporated into lab presentations, exams, or custom-made classroom materials. All files are pre-inserted into blank PowerPoint slides for ease of preparation.

Instructor's Manual for the Laboratory Manual This helpful preparation guide includes suggestions for coordinating lab exercises with the textbook, set-up instructions, materials lists, and answers to the laboratory review questions at the end of each exercise.

Along with these outstanding online laboratory tools, the Saladin website features content for both students and instructors using the fifth edition of *Anatomy & Physiology* including quizzing and interactive study tools.

5. **Integrated Use of the Cat for Dissection Specimen.** The cat is used as the dissection animal; however, it is integrated with material on human anatomy, so that animals do not have to be relied upon as dissection specimens unless so desired.

6. **Virtual Lab.** An icon appears in some of the exercises after the materials section. This icon represents Ph.I.L.S. (Physiology Interactive Lab Simulations) and signals the reader that a supplemental laboratory exercise can be found. Ph.I.L.S. can be packaged with *Anatomy & Physiology Laboratory Manual.*

7. **Safety.** Safety guidelines appear in the inside front cover for reference. The international symbol for caution (⚠) is used throughout the manual to identify material that the reader should pay close

Anatomy & Physiology Revealed

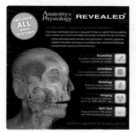

This amazing multimedia tool is designed to help students learn and review human anatomy using cadaver specimens. Detailed cadaver photographs blended together with a state-of-the-art layering

technique provide a uniquely interactive dissection experience. This easy-to-use program features the following sections:

- **Dissection**—Peel away layers of the human body to reveal structures beneath the surface. Structures can be pinned and labeled, just as in a real dissection lab. Each labeled structure is accompanied by detailed information and an audio pronunciation. Dissection images can be captured and saved.
- **Animation**—Compelling animations demonstrate muscle actions, clarify anatomical relationships, or explain difficult concepts.
- **Histology**—Study microscopic anatomy using labeled micrographs.
- **Imaging**—Labeled X-ray, MRI, and CT images familiarize students with the appearance of key anatomical structures as seen through different medical imaging techniques.
- **Self-Test**—Challenging exercises let students test their ability to identify anatomical structures in a timed practical exam format or traditional multiple choice. A results page provides analysis of test scores plus links back to all incorrectly identified structures for review.

Physiology Interactive Lab Simulations (Ph.I.L.S)

Ph.I.L.S 3.0 contains **37 lab simulations** that allow students to perform experiments without using expensive lab equipment or live animals. This easy-to-use software offers students the flexibility to change the parameters of every lab experiment, with no limit to the amount of times a student can repeat experiments or modify variables. This power to manipulate each experiment reinforces key physiology concepts by helping students view outcomes, make predictions, and draw conclusions.

MediaPhys 3.0

This interactive software tool offers detailed explanations, high-quality illustrations, and animations to provide students with a thorough introduction to the world of physiology—giving them a virtual tour of physiological processes. MediaPhys 3.0 is filled with interactive activities and quizzes to help reinforce physiology concepts that are often difficult to understand.

Cat Dissection Review for Human Anatomy and Physiology

This multimedia program, created by John Waters of Pennsylvania State University and Melissa Janssen and Donna White of Collin County Community College, contains vivid, high-quality, labeled cat dissection photographs. The program helps students easily identify and compare cat structures with the corresponding human structures.

Student Supplements

McGraw-Hill offers various tools and technology products to support the textbook. Students can order supplemental study materials by contacting their campus bookstore or online at www.shopmcgrawhill.com.

Instructor Supplements

Instructors can obtain teaching aids by calling the McGraw-Hill Customer Service Department at 1-800-338-3897, visiting our online catalog at www.mhhe.com, or by contacting their local McGraw-Hill sales representative.

Acknowledgments

Many people have been involved in the production of this lab manual, and I would like to thank the editorial staff at McGraw-Hill, including Marty Lange, Michelle Watnick, James Connely, Anne Winch, and Ashley Zellmer. Thanks also go to the McGraw-Hill production team—Mary Jane Lampe, John Leland, and David Hash—for their input and encouragement.

I would like to dedicate this book to my two daughters, Sarah and Caitlin, whom I love so very much.

Please feel free to write me or e-mail me with your comments, suggestions, and criticisms. I value your input and hope that your comments will lead to an even better sixth edition of the laboratory manual.

Eric Wise
Santa Barbara City College
721 Cliff Drive
Santa Barbara, CA 93109
wise@sbcc.edu

Reviewers

The writing of a laboratory manual is a collaborative effort, and I was impressed by the dedication and insight given to me by the reviewers of this edition and previous editions. I would like to gratefully acknowledge them as they made significant contributions regarding improving the content and clarity of the manual. I appreciate their time and energy in making this a better lab manual; however, I take responsibility for any remaining errors.

Richard R. Adler, Ph.D.
University of New Orleans

Karen A. Baskerville
Lincoln University

Janice Beaird
University of West Alabama

Ruth E. Birch, Ph.D.
St. Louis Community College–Forest Park

David N. Chambers
Concord University

Bertram W. Coltman, Ph.D.
Associate Professor of Biology, Concordia University

Gerard Cronin
Salem Community College

Joy Davis
Baton Rouge Community College

Abdeslem El Idrissi, Ph.D.
The College of Staten Island/CUNY

John Enz
Alderson-Broaddus College

Annette M. Gabaldon
Colorado State University–Pueblo

Virgil Gaddy

Shelley A. Kirkpatrick
Saint Francis University

William L'Amoreaux
CUNY College of Staten Island

Kent Lange
Purdue University North Central

Steven A. Leadon
Durham Technical Community College

Eric Lippincott
Lock Haven University

Kara Miller
Great Basin College

Lourdes P. Norman, Ph.D.
Florida Community College Jacksonville

Olumide O. Ogunmosin
Texas Southern University

Vanessa Passler
Wallace Community College

Andrew J. Petto, Ph.D.
University of Wisconsin, Milwaukee

Don Plantz
Mohave Community College

John M. Ripper
Butler County Community College

Mrs. Cynthia P. Robison
Wallace Community College

Donald Rodd
University of Evansville

Susan Safford
Lincoln University

Paul K. Small
Eureka College

James Stittsworth
Florida Community College at Jacksonville

Andrzej Wieraszko
The College of Staten Island/CUNY

Student Preface

This laboratory manual was written to help you gain experience in the lab as you learn human anatomy and physiology. The 47 exercises explore and explain the structure and function of the human body. You will be asked to study the structure of the body using the materials available in your lab, which may consist of models, charts, mammal study specimens such as cats, preserved or fresh internal organs of sheep or cows, and possibly cadaver specimens. You may also examine microscopic sections from various organs of the body. You should familiarize yourself with the microscopes in your lab very early, so that you can take the best advantage of the information they can provide.

The physiology portion of the course involves experiments that you will perform on yourself or your lab partner. They may also involve the mixing of various chemicals and the study of the functions of live specimens. Because the use of animals in experiments is of concern to many students, a significant attempt has been made to reduce (but, unfortunately, not eliminate) the number of live experiments in this manual. Until there is a sound replacement for live animals, their use will continue to be part of the college physiology lab. Your instructor may have alternatives to live animal experimentation exercises. It is important to get the most out of what live specimen experimentation there is. Coming to the lab unprepared and then sacrificing a lab animal while gaining little or no information is an unacceptable waste of life. Use the animals with care. Needless use or inhumane treatment of lab animals is not acceptable or tolerated.

As a student of anatomy and physiology you will be exposed to new and detailed information. The time it takes to learn the information will involve *more* than just time spent in the lab. You should maximize your time in the lab by reading the assigned lab exercises before you come to class. You will be doing complex experiments, and if you are not familiar with the procedure, equipment, and time involved you could ruin the experiment for yourself and/or your lab group. The exercises in physiology are written so you can fill in the data as you proceed with the experiment. At the end of each exercise are review sheets that your instructor may wish to collect to evaluate the data and conclusions of your experiments. The illustrations are labeled except on the review pages. All review materials can be used as study guides for lab exams or they may be handed in to the instructor.

The anatomy exercises are written for cat and human study, though these exercises can be used with or without cats or cadavers. Get *involved* in your lab experience. Don't let your lab partner do all the dissections or all the experimentation; likewise, don't insist on doing everything yourself. Share the responsibility and you will learn more.

Safety

Safety guidelines appear in the inside front cover for reference. The international symbol for caution (⬦) is used throughout the manual to identify material that you should pay close and special attention to when preparing for or performing laboratory exercises.

Clean Up

Special instructions are provided for clean up at the end of appropriate laboratory exercises and are identified by this unique icon (✋).

How to Study for This Course

Some people learn best by concentrating on the visual, some by repeating what they have learned, and others by writing what they know over and over again. In this course, you will have to adapt your learning style to different study methods. You may use one study method to learn the muscles of the body and a completely different method for understanding the function of the nervous system. Some students need only a few hours per week to succeed in this course, while others seem to study far longer.

You need to go to class. Go to lab on time. The beginning of the lab is when most instructors go over the material and point out what material to omit, what to change, and how to proceed. If you do not attend lab, you do not get the necessary information.

Read the material ahead of time. The subject matter is visual, and you will find an abundance of illustrations in this manual. Record on your calendar all the lab quizzes and exams listed on the syllabus provided by your instructor. Budget your time so you study accordingly.

Work hard! There is absolutely no substitute for hard work to achieve success in a class. Some people do math easier than others, some people remember things easier, and some people express themselves better. Most students succeed because they work hard at learning the material. It is a rare student who gets a bad grade because of a lack of intelligence. Working at your studies will get you much farther than worrying about your studies.

Be *actively* involved with the material and you will learn it better. Outline the material after you study it for a while. Read your notes, go over the material in your mind, and then make the information your own. There are several ways that you can get actively involved.

Draw and doodle a lot. Anatomy is a visual science, and drawing helps. You do not have to be a great illustrator. Visualize the material in the same way you would draw a map to your house for a friend. You do not draw every bush and tree but, rather, create a *schematic* illustration that your friend could use to get the *pertinent* information. As you know, there are differences in maps. Some people need more practice than others, but anyone can do it. The head can be a circle, which can be divided into pieces representing the bones of the skull. Draw and label the illustration after you have studied the material and without the use of your text! Check yourself against the text to see if you really know the material. Correct the illustration with a colored pen, so that you highlight the areas you need work on. Go back and do it again until you get it perfect. This does take some time, but not as much as you might think.

Write an outline of the material. Take the mass of information to be learned and go from the general to the specific. Let's use the skeletal system as an example. You may wish to use these categories:

1. Bone composition and general structure
2. Bone formation
3. Parts of the skeleton
   a. Appendicular skeleton
      (1) Pectoral girdle
      (2) Upper extremity
      (3) Pelvic girdle
      (4) Lower extremity
   b. Axial skeleton
      (1) Skull
      (2) Hyoid
      (3) Ribs
      (4) Vertebral column
      (5) Sternum

An outline helps you organize the material in your mind and lets you sort the information into areas of focus. If you do not have an organizational system, then this course is a jumble of terms with no interrelationships. The outline can get more detailed as you progress, so you eventually know that the specific nasal bone is one of the facial bones and the facial bones are skull bones, which are part of the axial skeleton, which belongs to the skeletal system.

Test yourself before the exam or quiz. If you have practiced answering questions about the material you have studied, then you should do better on the real exam. As you go over the material, jot down possible questions to be answered later, after you study. If you compile a list of questions as you review your notes, then you can answer them later to see if you have learned the material well. You can also enlist the help of friends, study partners, or family (if they are willing to do this for you). You can also study alone. Some people make flash cards for the anatomy portion of the course. It is a good idea to do this for the muscle section, but you may be able to get most of the information down by using the preceding technique. Flash cards take time to fill out, so use them carefully.

Use memory devices for complex material. A mnemonic device is a memory phrase that has some relationship to the study material. For example, there are two bones in the wrist right next to one another, the trapezium and the trapezoid. The mnemonic device used by one student was that trape*zium* rhymes with th*umb* and it is the one under the thumb.

Use your study group as a support group. A good study group is very effective in helping you do your best in class. Study with people who will push you to do your best. If you get discouraged, your study partners can be invaluable support people. A good group can help you improve your test scores, develop study hints, encourage you to do your best, and let you know that you are not the *only* one living, eating, and breathing anatomy and physiology.

Just as a good study group can really help, a bad group can drag you down farther than you might go on your own. If you are in a group that constantly complains about the instructor, that the class is too hard, that there is too much work, that the tests are not fair, that you don't really need to know this much anatomy and physiology for your field of study, and that this isn't medical school, then get yourself out of that group and into one excited by the information. Don't listen to people who complain constantly and make up excuses instead of studying. There is a tendency to start believing the complaints, and that begins a cycle of failure. Get out of a bad situation early and get with a group that will move forward.

Do well in the class and you will feel good about the experience. If you set up a study time with a group of people and they spend most of the time talking about parties, sports, or personal problems, then you aren't studying. There is nothing wrong with talking about parties, sports, or helping someone with personal problems, but you need to address the task at hand, learning anatomy and physiology. Don't feel bad if you must get out of your study group. It is *your* education, and if your partners don't want to study, then they don't really care about your academic well-being. A good study partner is one who pays attention in class, who is prepared ahead of the study time session, and who can explain information that you may have gotten wrong in your notes. You may want to get the phone number of two or three such classmates.

Test-Taking

Finally, you need to take quizzes and exams in a successful manner. By doing practice tests, you can develop confidence. Do well early in the semester. Study extra hard early (there is no such thing as overstudying!). If you fail the first test or quiz, then you must work yourself out of an emotional ditch. Study early and consistently, and then spend the evening before the exam going over the material in a general way and solving those last few problems. Some people do succeed under pressure and cram before exams; however, the information is stored in short-term memory and does not serve them well in their major field! If you study on a routine basis, then you can get up on the morning of a test, have a good breakfast, listen to some encouraging music, maybe review a bit, and be ready for the exam.

Your instructor is there to help you learn anatomy and physiology, and this laboratory manual was written with you in mind. Relate as much of the material as you can to your body and keep an optimistic attitude.

Please feel free to write me or e-mail me with your comments, suggestions, and criticisms. I value your input and hope that your comments will lead to an even better sixth edition of this laboratory manual.

Eric Wise
Santa Barbara City College
721 Cliff Drive
Santa Barbara, CA 93109
wise@sbcc.edu

LABORATORY

Introduction to Lab Science, Measurement, and Chemistry

INTRODUCTION

Science is the study of physical phenomena and follows specific guidelines that make it unique from other disciplines. The human body's structure and function fall within the realm of scientific investigation; for example, **human anatomy** involves the understanding of the structure of the human body, while **human physiology** is the study of the function of the body. There are many ways that people can understand anatomy and physiology. One of these involves the use of the **scientific method.**

Scientific understanding frequently begins with a question. An example of a question is how the body digests food. The next part in this understanding frequently involves the development of a **hypothesis,** which is a testable proposal that seeks to explain a scientific question. The testing done to prove or disprove the hypothesis is usually carried out in the form of an **experiment.** A more lengthy discussion of the scientific method is provided in the Saladin text in chapter 1, "Major Themes of Anatomy and Physiology."

Much of science involves measurement and the collection of **data.** Experimental data are the pieces of information, or "facts," obtained and later examined to support or reject the proposed hypothesis. In this exercise you collect data and examine how to graph data so that the information becomes more comprehensible. You will also look at what type of measurement is done in science.

OBJECTIVES

At the end of this exercise you should be able to

1. describe the advantages of the metric system over the U.S. customary system;
2. define independent variable and dependent variable;
3. list four base units of the metric system;
4. convert fractions into decimal equivalents;
5. collect and graph data taken in class;
6. determine some factors involved in experimental error;
7. discuss pH, acid, base, ionic bond, and covalent bond.

MATERIALS

Acid/Base

Five 10 mL test tubes

Test tube rack

10 mL graduated cylinder

Permanent marker

Distilled water in dropper bottle

0.1 M HCl in dropper bottle

0.1 M NaOH in dropper bottle

Baking soda (sodium bicarbonate)

Sodium chloride (table salt)

Wide-ranging pH paper (pH 1-14)

Parafilm®

Small metal spatula

Balance and weigh paper

Ionic and Covalent Molecules

18 gauge wire

Alligator clips

9 volt battery

6 volt flashlight bulb in holder

Two 50 mL beakers

15% sucrose solution in dropper bottle

15% sodium chloride solution in dropper bottle

Hydrogen Bonds

Graduated cylinder

Two 50 mL beakers

Small bottle of distilled water

Small bottle of ethanol (70% or greater)

Hot plate (do not use open flame)

Measurement Section

Meter stick

Small metric ruler (30 cm)

PROCEDURE

Measurements in Science

Members of the scientific community and people of many nations of the world use the **metric system** to record quantities such as length, volume, mass (weight), and time. This is because the metric system is based on units of 10, and conversion to higher or lower values is relatively easy when compared to using the U.S. customary system. For example, assume you are working on a bicycle and are using a ½-inch wrench. If you need a larger wrench you move to a 9/16-inch, then a 5/8-inch, then an 11/16-inch, or perhaps as large as a ¾-inch wrench. This requires a bit of computation as you move from one size to the next. On the other hand, if you are using the metric system and a 12-millimeter (mm) wrench is too small, you progressively move to a 13 mm, 14 mm, or 15 mm wrench.

The same idea can be applied to volume, temperature, or weight. In the case of volume, there are 8 ounces per cup, 128 ounces per gallon. The calculation for the number of ounces in 7 gallons is a little cumbersome (7 gallons × 128 ounces). In the metric system, there are 1,000 milliliters in 1 liter, so there are 7,000 milliliters in 7 liters. The conversions are much easier. Medical dosages are given frequently in milliliters or cubic centimeters (cc). One milliliter occupies 1 cubic centimeter, so these values are interchangeable. Examine table 1.1 and compare the quantity, base unit, and U.S. equivalent. Additional conversions can be found in Appendix A.

If the quantity measured is much larger or smaller than the base unit, then the base unit can be expressed in multiples of 10. For example, if you had 1,000 grams, you would have a **kilo**gram. If you had one-thousandth of a gram (1/1,000 gram), you would have a **milli**gram. Examine table 1.2 as you answer the following questions. You can also fill in the Chapter Summary Data section at the end of the exercise with your results if your instructor wants you to hand in your results.

? What is 1/100 gram? _____ 1

? What is 1,000 seconds? _____ 2

? What is 10 meters? _____ 3

? What is 1/1,000,000,000 liter? _____ 4

As you can see from table 1.2, some measurements in science are very small. For example, the amounts of hormones circulating in the blood are very minute. To provide a shortened notation of very large or small numbers we use scientific notation. A number such as 60,000 is written as 6×10^4. You move the decimal point four places to the left and thus the superscript above the 10 is a 4. Write 6,000 in scientific notation _____. For very small numbers, the superscript is written as a negative number. The number 0.00006 is written as 6×10^{-5}, as you move the decimal point five places to the right. Convert the following numbers into scientific notation:

? 4,300,000 _____ 5

? 0.000034 _____ 6

? 2,200 _____ 7

? 0.0019 _____ 8

Working in the science lab requires you to focus on the procedures and materials at hand. You may work as part of a group in some labs, and it is important that you read your lab manual *before* coming into the lab. Some of the materials you work with may be dangerous, and a thorough prior knowledge of the lab exercise will ensure a safer lab.

Pay attention to the experiment and what is to be done and when. Casual observation and carelessness may lead to incorrect results. Establish a procedure for conducting experiments. If you are working with one or more lab partners, divide responsibilities before the experiment begins. If you are responsible for a particular portion of the experiment, make sure your lab partners see the results. Make a careful record of the results of your experiment.

Be honest. Fudging data is not tolerated in the scientific community. Record your data as you measure them. If your results do not seem to be what they should, then discuss this with your instructor. Never record data that you think you *should* get but, rather, *record the observed data.*

Correlation Between Leg Length and Height

Scientists experiment by altering one variable and seeing what effect occurs. The variable that scientists change in an experiment is called the **independent variable.** The effect or result caused by this change is known as the **dependent variable.** For example, if we look at the amount of food a

TABLE 1.1	Metric System and Equivalents	
Quantity	**Base Unit**	**U.S. Equivalent**
Length	Meter (m)	1.09 yards (39.4 inches)
Volume	Liter (L)	1.06 quarts
Mass	Gram (g)	.036 ounce (1/454 of a pound)
Time	Second (s)	Second

TABLE 1.2	Decimals of the Metric System			
Name	**Description**	**Multiple/Fraction**		**Scientific Notation**
Kilo	One thousand times greater		1,000	1.0×10^3
Deca	Ten times greater		10	1.0×10^1
Base Unit				
Deci	One-tenth as much	$1/10$	0.1	1.0×10^{-1}
Centi	One-hundredth as much	$1/100$	.01	1.0×10^{-2}
Milli	One-thousandth as much	$1/1,000$	.001	1.0×10^{-3}
Micro	One-millionth as much	$1/1,000,000$	.000001	1.0×10^{-6}
Nano	One-billionth as much	$1/1,000,000,000$	.000000001	1.0×10^{-9}

person eats versus weight gain, the amount of food is the independent variable. Weight gain occurs due to the amount of food eaten; therefore, weight gain is the dependent variable.

In this part of the exercise you will look for a correlation between the length of the leg and the overall height of the individual.

Make a hypothesis about any relationship between leg length and height.

? Record your hypothesis in the following space.

_____ 9

_____

By graphing data you can more easily see potential correlations in a sample size. One problem in sampling is that you need to have a large enough number to have a valid sample. Let's suppose that one-half of the basketball team is enrolled in your lab section. This might have a rather unusual effect on your graph (see figure 1.1). On the other hand, if you are able to sample your entire school, the effects of the size of the basketball players in the sample would be minimized (see figure 1.2).

Your lab partner should measure the length of your leg from the back of your knee to the bottom of your heel. To do this, kneel on a chair with the thigh vertical and the back of the calf at a 90 degree angle to the thigh (see figure 1.3). Place a meter stick along the outside (lateral) edge of the back of your calf. You should feel a tendon, which is the tendon of the biceps femoris muscle. Make sure that the meter stick is just at the edge of the tendon,

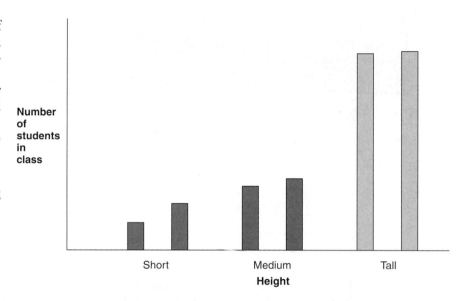

FIGURE 1.1 Distribution of Students in a Class

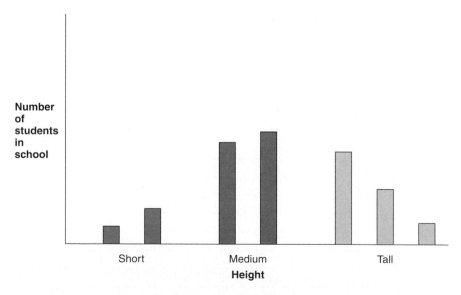

FIGURE 1.2 Distribution of Students in Entire School

FIGURE 1.3 Measurement of Leg Length

and record the distance, in centimeters, from this tendon to the bottom of your heel.

? Length (cm) between tendon and heel: _____ 10

Now record your overall height in centimeters. Determine this value directly. Have your lab partner measure your height against a wall in the lab, after removing your shoes.

? Overall height in centimeters: _____ 11

Determine the distribution of the entire class in terms of leg length. Give your data to other members of the class to record in chart 1. Plot the length of the leg on the x-axis (the independent variable is plotted on the horizontal x-axis) and the overall height of the individual on the y-axis (the dependent variable) to see if leg length determines overall height. Is there a correlation?

Plot the distribution of leg length and overall height on chart 2.

Chart 1 Leg Length and Height Data		
Student Number	Leg Length (cm)	Height (cm)
1		
2		
3		
4		
5		
6		
7		
8		
9		
10		
11		
12		
13		
14		
15		
16		
17		
18		
19		
20		
21		
22		
23		
24		

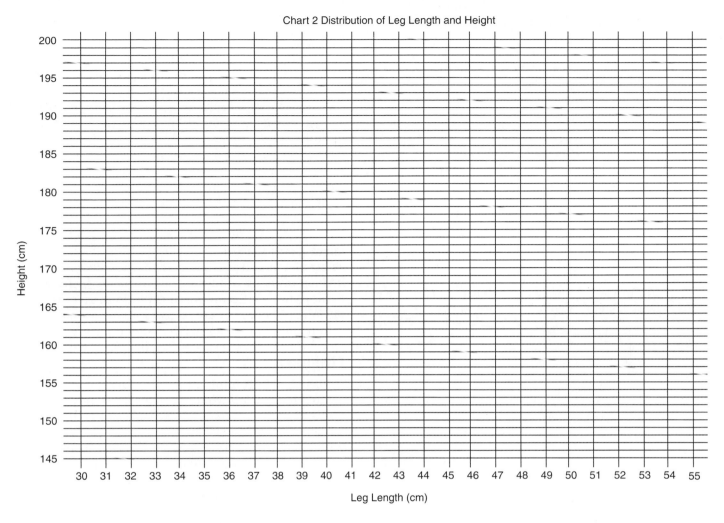

Chart 2 Distribution of Leg Length and Height

For every 5-centimeter change in leg length, what is the approximate change in height?

? _____ 12

Range and Mean in Measurements

The extremes of measurement, which represent the longest and shortest measurements, represent the **range** of the measurements. Using the values obtained by members of your lab class, record the range of the leg length in centimeters for the entire class.

? Longest value: _____ 13

? Shortest value: _____ 14

The **mean** leg length of the class is the average length for the group. Using the data obtained for the leg lengths, add all of the values together (the sum of the leg lengths) and divide that number by the number of individuals in the class. This is the mean leg length. Record the number in centimeters.

? Mean leg length: _____ 15

Measurement and Experimental Error

In this part of the lab all members of the lab will take a measurement of a single individual's hand. You should perform the next exercise by following these directions.

1. Select *one* person in your lab class who will serve as the experimental subject for this section of the lab.
2. Everyone in the lab should take a metric ruler from the lab table and measure the length of the hand of the selected subject. To obtain individual results, the measurements should be done alone and the results recorded without any discussion with other classmates.
3. With a ruler, measure the test subject's hand from the wrist to the tip of the middle finger and record this value in millimeters.

? Length of the hand in millimeters: _____ 16

4. After everyone has made a measurement and entered the data in his or her lab book, every individual in the class should write his or her results on the board.

In chart 3, record all of the data collected by members of the class.

Chart 3

Length of hand in millimeters:

————	————	————	————
————	————	————	————
————	————	————	————
————	————	————	————
————	————	————	————
————	————	————	————
————	————	————	————
————	————	————	————

? If everyone measured the same person, all of the values should be the same. Was this the case in your lab?

_____ 17

? If this was not the case, what are some reasons that the measurements were different?

_____ 18

_____

? If you were to do this experiment again, how would you change the procedure to try to get a more accurate recording of lengths among the people measuring the subject's hand?

_____ 19

_____

Lab Reports

Your instructor may ask you to write up your results for specific experiments. This process is valuable in that you evaluate your experimental data and form an understanding of the process that comes from the results of your experiment. Writing up scientific experiments generally follows a very specific process, and this is described in Appendix C at the back of the lab manual. You should refer to it before you begin your lab write-ups.

Scientific Words

In science, many terms are derived from Greek or Latin words. Remembering words are easier if their meanings are understood. A word may consist only of a **root word.** For example, a *gastric* ulcer refers to an ulcer of the stomach. The term *gastric* comes from *gastro,* which means stomach. In addition to the root word, **prefixes** or **suffixes** may be added. A prefix is added to the front of the root word. The word *epigastric* means on top of the stomach (*epi* = upon). *Hypogastric* means under the stomach (*hypo* = below). A suffix is added to the end of a root word. *Gastritis* is an inflammation of the stomach (*itis* = inflammation).

Sometimes you might feel like you are learning a new language and, in many respects, this is true. To make some of the words you learn more coherent there is a lexicon of many of these word elements in Appendix D. Refer to this section if you need to look up a word for its meaning. You may want to tear out the lexicon and have it in hand as you use the lab manual.

Chemistry

In order to fully understand the functions of the body, a fundamental knowledge of chemistry is essential. For those of you who have studied chemistry the following pages are a simplified review. For those of you who have never had chemistry a small chemistry book or online resources may be invaluable to you for the rest of the course. In the following section you will experiment with acids, bases, and chemical bonds.

Acid/Base Relationships   Many solutions can be **acidic** or **basic (alkaline).** Solutions such as hydrochloric acid and carbonic acid are common. Basic solutions, such as bicarbonate and ammonia, also play important roles in the body. The acidic or basic condition of a solution is measured in **pH** units. The abbreviation pH refers to the **power of hydrogen.** It is the negative logarithm of the hydrogen ion concentration. In simple terms the pH unit varies 10 times compared to the next pH value. Thus, a solution of pH 5 is 10 times more acidic than a solution of pH 6. A solution with a pH of 7 has a neutral pH because, in pure water, the number of hydrogen ions equals the number of hydroxyl ions (see figure 1.4). As the solution becomes more acidic the hydrogen ion concentration increases and the pH decreases. Adding more hydrogen ions to the solution, such as when the stomach wall adds hydrochloric acid (HCl) to the stomach cavity, causes the pH to drop to about 2. If you increase the alkalinity of the solution (such as adding ammonia to water), then the pH increases. Household ammonia has a pH of about 11. **Buffers** are materials that resist changes in pH. Any introductory chemistry book will give you a more detailed description of pH, and there are

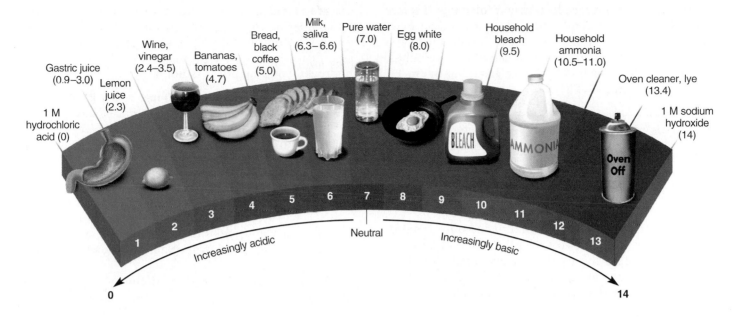

FIGURE 1.4 The pH Scale

many online tutorials that you can find if you search using the words *acid base*.

Caution! When using acids and bases, make sure to wear safety goggles and gloves. This material is potentially caustic and can cause skin burns. Do not drink any of the solutions.

For this experiment select five test tubes and label them 1–5 with a permanent marker. Place them in a test tube rack. Use pH paper to measure pH.

1. Use a graduated cylinder and put 6 ml of distilled water into each of the five test tubes.
2. Measure the pH of test tube 1 (distilled water). Record this pH value _____.
3. To test tube 2, add 10 drops of HCl. Place a small square of Parafilm® over the top and mix the contents. Measure the pH and record the value _____. Is this solution acidic, basic, or neither (circle one)?
4. To test tube 3, add 0.1 g sodium bicarbonate. Place a small square of Parafilm® over the top and mix the contents. Record the pH _____.
5. To test tube 3, add 10 drops of HCl. Place a small square of Parafilm® over the top and mix the contents. Record the pH of the solution _____. Does the pH drop to the same level as test tube 2? _____.
6. Add 10 more drops to test tube 3 and record the pH value _____. How might you describe the action of sodium bicarbonate with respect to HCl?

7. To test tube 4, add 10 drops of sodium hydroxide. Place a small square of Parafilm® over the top and mix the contents. Record the pH value _____. Is this solution acidic, basic, or neither (circle one)?
8. To test tube 5, add 0.2 g of sodium chloride (table salt). Place a small square of Parafilm® over the top and mix the contents. Record the pH value _____. Is this solution acidic, basic, or neither (circle one)?

Chemical Bonds Most of the material around us is held together by chemical bonds. These bonds can be strong, intermediate, or weak. There are several types of bonds but the three that we will examine are covalent bonds, ionic bonds, and hydrogen bonds. **Covalent bonds** are generally thought of as those where bonded atoms share electrons. These are strong bonds. The nucleus of the atom is positively charged due to the presence of protons. The positive charges of two separate atoms repel one another. When an electron is shared between these atoms the electrons are attracted to both of the positive nuclei and this holds them together.

In **ionic bonds** electrons from one atom are transferred to another atom and the result is an atom with a positive charge and an atom with a negative charge. These oppositely charged particles, called **ions** or **electrolytes**, attract one another and form ionic bonds. The bond is due to the electrostatic attraction between the two ions. These are also considered strong bonds.

You can examine the differences between covalent and ionic bonds with a simple experiment. Ionic bonds pull apart (dissociate) by the action of water. Electrolytes (ions) in solution conduct electricity. Molecules that are held together with covalent bonds do not conduct electricity so easily. This can be demonstrated with the use of a 6 volt flashlight bulb, appropriate cables and switches, and a DC generator, or a 9 volt battery. Examine figure 1.5 for the proper set up.

Caution! Be careful when working with electricity. You will be using low-voltage direct current. Do NOT use any voltage higher than 10 volts for this experiment and do not put any part of your body into the solutions being tested.

---

1. Connect a 9 volt battery terminal, using one alligator clip and wire, to the 6 volt flashlight bulb apparatus.
2. Connect the other terminal to an insulated wire that is placed in a glass beaker with 25 mL of 15% sucrose solution. The wire should have about an inch of insulation removed.
3. Connect the vacant terminal of the flashlight bulb setup to an insulated wire.
4. Do not touch the ends of the wires together but insert the ends of the insulated wire into the water.
5. Record the results (light on or light off).
6. Repeat the experiment but insert the wires into a beaker with 25 mL of 15% sodium chloride solution.
7. Record the results (light on or light off).

Which one of the solutions has solute particles composed of ions?

Which solution has solute particles with covalent bonds?

Hydrogen bonds arise from molecules, such as water, in which the electron from the hydrogen atom is strongly attracted to another atom (in water it is oxygen), giving it a negative charge. The hydrogen atoms becomes more positively charged. When two or more of these molecules are associated with one another, the postively charged hydrogen end of the molecule forms a weak bond with the negatively charged end of another molecule, as seen in figure 1.6. This bond is known as a hydrogen bond and it is considered a weak bond.

We can demonstrate the impact of hydrogen bonds by comparing the evaporation rates of water to those of ethanol (alcohol). Water is an example of a molecule with significant hydrogen bonding. Ethanol does have some hydrogen bonding but it is not as strong as in water.

1. Pour 5 mL of water into a 50 mL beaker labeled "water."
2. Pour 5 mL of ethanol into a 50 mL beaker labeled "alcohol."
3. Make sure that both liquids are at room temperature before you begin the experiment.
4. Warm the two beakers on a hot plate (NOT an open flame).

NOTE: You can use 70% lab alcohol but the remaining 30% is water; therefore, the results will not be as dramatic. Which fluid evaporates the slowest, alcohol or water (circle one)?

Slow evaporation is due to the greater number of hydrogen bonds "holding" the molecules together.

Caution! Despite the fact that pure water, or water with covalent molecules, does not conduct electricity as well as electrolytes it still can conduct electricity and, when the voltage is high enough, can kill you. This is why you do not swim outside during lightning storms or place electrical appliances on the edge of a bathtub when you are taking a bath.

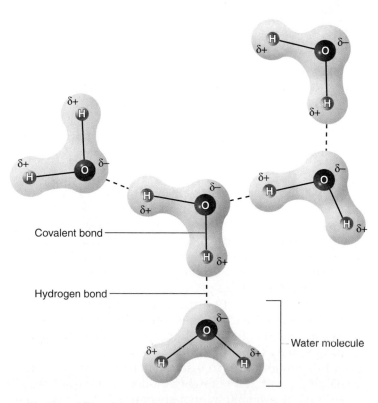

FIGURE 1.6 Hydrogen Bonding Positive ends of hydrogen atoms are attracted to negative ends of oxygen atoms.

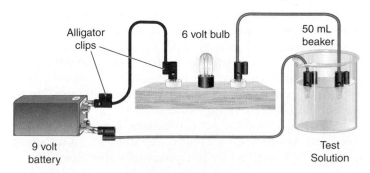

FIGURE 1.5 Electrolyte Conductivity Set-up

REVIEW SECTION

Introduction to Lab Science, Measurement, and Chemistry

Name _____ *Date* _____

Lab Section _____ *Time* _____

Review Questions

1. In terms of base units

   a. What is the base unit of length in the metric system?

   b. What is the base unit of volume in the metric system?

2. How many cubic centimeters are there in 200 milliliters?

3. Assume a pill has a dosage of 350 mg of medication. How much medication is this in grams?

4. How would you write 0.000345 liter in scientific notation?

5. How many milligrams are there in 4.5 kilograms?

6. How much of a meter is 250 millimeters?

7. If given a length of 1/10,000 of a meter

   a. Convert this number into a decimal:

   b. Convert it into scientific notation:

8. Use a word to describe

   a. One-thousandth of a second:

   b. One-thousand liters:

   c. One-hundredth of a meter:

9. According to your graph, predict how tall a person would be if his or her leg length were 37 cm. How tall do you predict this person would be if the individual had a leg length of 48 cm?

10. In the case of leg length and overall height, which is the independent variable and which is the dependent variable?

11. What was the range in values for the hand length measured in lab?

12. Define the term *buffer*.

13. What is the neutral pH?

14. Is a pH of 8 more basic or less basic than a pH of 6?

15. As a solution becomes more acidic what happens to the concentration of hydrogen ions?

16. The term *electrolyte* is derived from the words *electro* ("electricity") and *lyte* ("to separate"). How does this term correlate with your experiment?

17. In which bond are electrons significantly shared between atoms?

18. Which bond—covalent, ionic, or hydrogen—is a weak bond?

? Chapter Summary Data

Use this section to record your results from questions within the exercise.

1. _____

2. _____

3. _____

4. _____

5. _____

6. _____

7. _____

8. _____

9. _____

   _____

10. _____

11. _____

12. _____

13. _____

14. _____

15. _____

16. _____

17. _____

18. _____

19. _____

   _____

LABORATORY

Organs, Systems, and Organization of the Body

INTRODUCTION

The study of the human body requires a thorough understanding of the orientation of the body or organ of study and the detail of presentation and knowledge of body regions. Knowledge of the material in this exercise is vital to your study of anatomy and physiology. The information that you learn in this exercise will be used throughout the rest of the lab manual and provides the foundation for anatomy and physiology. In this exercise you examine the major organ systems of the body, learn directional terms, and describe the major regions of the body. These topics are covered in your lecture text in Atlas A, "General Orientation to Human Anatomy."

OBJECTIVES

At the end of this exercise you should be able to

1. list the 11 organ systems;
2. place major organs, such as the heart, lungs, and stomach, in the proper organ system;
3. determine from an illustration whether a section is in the frontal, transverse, or sagittal plane;
4. give directional terms equivalent to up, down, front, back, toward the midline or edge, and toward the core or surface of the body;
5. explain anatomical position;
6. identify the quadrants and nine abdominal regions;
7. list the major regions of the body and provide a common name, if known.

MATERIALS

Models of human torso

Charts of human torso

PROCEDURE

Organ Systems

Anatomy can be studied in many ways. **Regional anatomy** is the study of particular areas of the body, such as the head or leg. Most undergraduate college courses in anatomy and physiology (and the format of this lab manual)

involve **systemic anatomy,** the study of **organ systems,** such as the skeletal system and nervous system. Although organ systems are studied separately, it is important to realize the intimate connections between the systems. If the heart fails to pump blood as part of the cardiovascular system, then the lungs do not receive blood for oxygenation and the intestines do not transfer nutrients to the blood as fuel. The brain is no longer capable of functioning, and the result is death. From a clinical standpoint, the failure of one system has impacts on many other organ systems.

Examine the torso models and charts in the lab and locate various organs. Using figure 2.1, find the following organ systems:

Reproductive	Respiratory	Digestive
Urinary	Skeletal	Endocrine
Nervous	Lymphatic	Cardiovascular
Muscular	Integumentary	

A quick way to recall all 11 systems is to remember this phrase: "Run Mrs. Lidec." Each letter of the phrase represents the first letter of one of the names of the organ systems.

Reproductive The gonads (testes and ovaries) contain the sex-producing cells of the body, and the accessory organs, such as the uterus, vagina, penis, and seminal vesicles, play a part in the transport of the sex cells and the development of the fetus.

Urinary The kidneys serve as filters of the body, and the urinary bladder is a storage organ. The ureters connect the kidneys to the bladder and the urethra is the exit tube from the body. The urinary system plays an important role in ridding the body of nitrogenous wastes while maintaining the chemical balance of body fluids.

Nervous The brain and spinal cord, as well as the numerous nerves, make up the nervous system. The nervous system coordinates body regions, interprets environmental cues, and integrates information.

Muscular Individual muscles are the organs of this system. Muscles move and strengthen joints, generate heat, and serve other functions, such as abdominal compression.

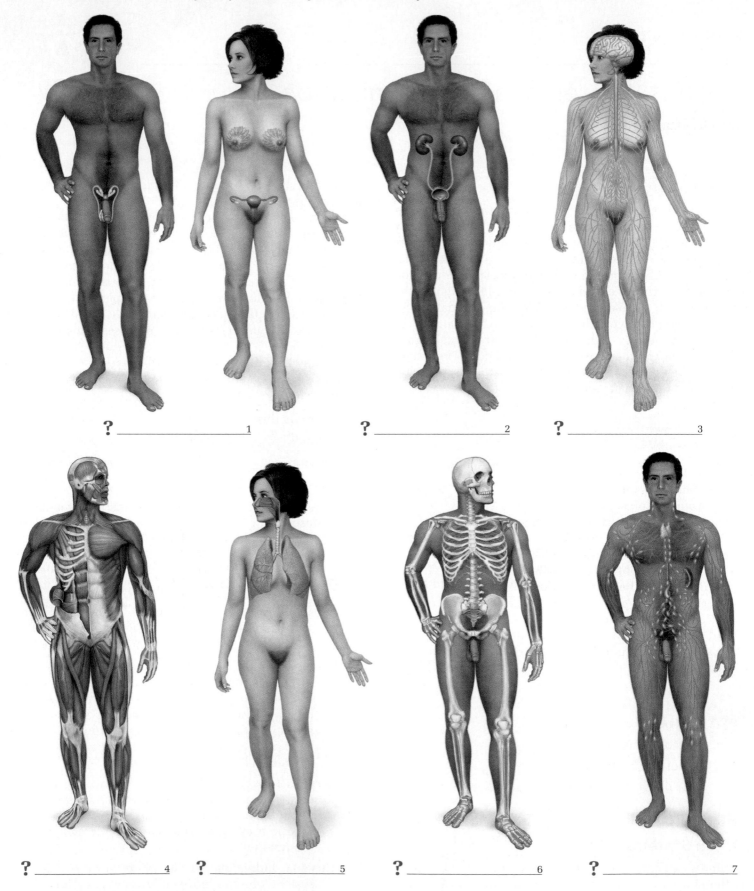

? _____ 1   ? _____ 2   ? _____ 3

? _____ 4   ? _____ 5   ? _____ 6   ? _____ 7

FIGURE 2.1 Organ Systems of the Human Body Fill in the terms for the organ systems here and in the Chapter Summary Data section at the end of the exercise.

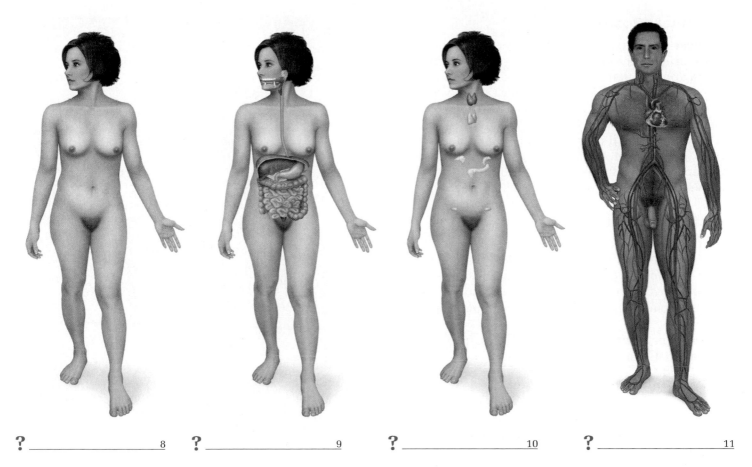

? _____ 8   ? _____ 9   ? _____ 10   ? _____ 11

FIGURE 2.1 *Continued*

Respiratory The nose, larynx, trachea, and lungs are part of the respiratory system. The function of the system is the exchange of gases (oxygen and carbon dioxide) between the blood and the air.

Skeletal Each bone is considered an organ, with blood vessels and nerves found in each bone. The skeletal system supports the body, protects delicate organs, and produces blood.

Lymphatic The lymph nodes, spleen, and tonsils are part of the lymphatic system. A major function of the lymphatic system is to protect the body from foreign particles, such as bacteria, viruses, and fungi. Cells from the lymphatic system make up the **immune system**. The immune system is not an organ system but a functional system consisting of an assemblage of cells that defend the body.

Integumentary The skin is the largest organ of the body and makes up most of the integumentary system. The system also contains associated structures, such as hair follicles, hair, nails, and the glands of the skin. The skin protects the body against microorganisms, keeps it from drying out, and produces vitamin D.

Digestive The mouth, esophagus, stomach, intestines, and liver are parts of the digestive system, which provides nutrients and water to the body.

Endocrine This system is composed of organs that produce hormones. Organs such as the thyroid gland and the adrenal glands are primarily endocrine glands (glands that secrete hormones without the use of ducts). Organs such as the pancreas and the gonads have dual functions; one is endocrine and the other is exocrine (secretion of material through ducts). Hormones are vital in regulating growth and development and maintaining a constant internal body condition.

Cardiovascular The heart, blood, and blood vessels make up this system, sometimes included with the lymphatic system as the circulatory system. The heart is the pump of the system and the blood vessels are the delivery and return portion of the system. The cardiovascular system is primarily involved in transporting oxygen, carbon dioxide, and other materials throughout the body.

You should examine models or charts in lab and locate the organs of the various organ systems.

Anatomical Position

In clinical settings it is vital to have a proper orientation when dealing with patients. If two physicians are operating on a patient and one tells the other to make an incision to the left, the physician making the cut does not have to ask, "My left or your left?" because the cut is always to the *patient's* left side. When referring to the human body you will orient the body in **anatomical position.** In this position, the body is upright, facing forward, arms and legs straight, palms facing forward, feet flat on the ground, and eyes open (see figure 2.2).

Directional Terms

With the body in anatomical position there are specific terms used to describe the location of one part with respect to another. These terms are somewhat different for four-footed animals and for humans, so both examples are given. Table 2.1 lists the directional terms used for humans.

Locate the terms in figure 2.3. In quadrupeds (four-footed animals) the directional terms are somewhat different. Quadrupeds do not have a superior/inferior designation. Dorsal is the back and ventral is on the belly side, while anterior (or craniad) is the front, or head-end, of the animal and posterior (or caudad) is the rear, or tail end, of the animal. There are other terms for location that have unique meanings. For the digestive system, **proximal** refers to regions closer to the mouth, while **distal** is in reference to regions closer to the anus. **Parietal** is in reference to the body wall when compared to **visceral,** which refers to areas closer to the internal organs (see figure 2.4). The heart, for example, has a visceral layer closer to the heart proper, while it also has a parietal layer farther from the heart. **Ipsilateral** refers to being on the same side of the body, and **contralateral** refers to being on the other half

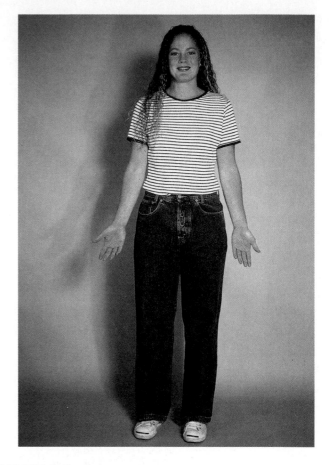

FIGURE 2.2 Anatomical Position

(left side/right side). The right hand and right arm are ipsilateral, while the ears are contralateral. Directional terms do not change if the body position changes. If you stand on your head, it is still superior to your feet because you reference the body as if it were in anatomical position.

TABLE 2.1	Directional Terms Used for Humans	
Term	**Meaning**	**Example**
Superior	Above	The nose is superior to the chin.
Inferior	Below	The stomach is inferior to the head.
Medial	Toward the midline	The sternum is medial to the shoulders.
Lateral	Toward the side	The ears are lateral to the nose.
Superficial	Toward the surface	The skin is superficial to the heart.
Deep	Toward the core	The lungs are deep to the ribs.
Ventral (or anterior)	To the front	The toes are ventral/anterior to the heel.
Dorsal (or posterior)	To the back	The spine is dorsal/posterior to the sternum.
Proximal	For extremities, meaning near the trunk	The elbow is proximal to the wrist.
Distal	For extremities, meaning away from the trunk	The toes are distal to the knee.

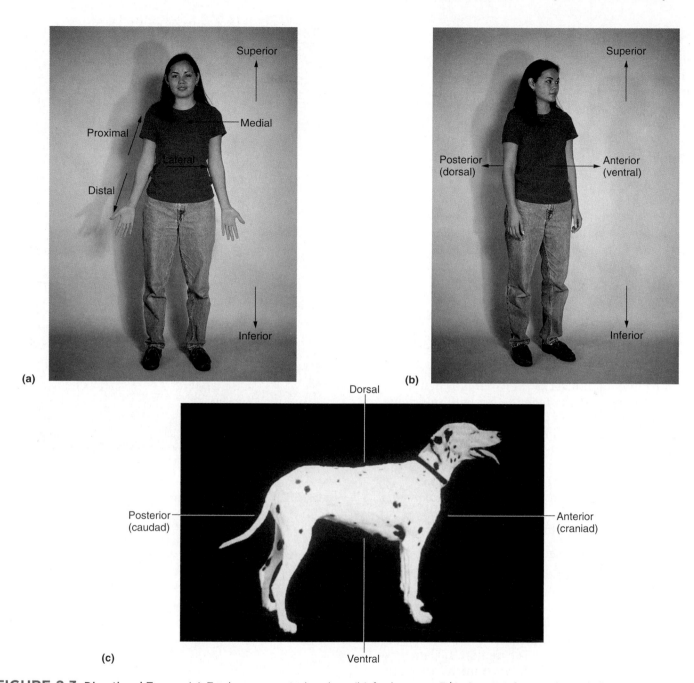

FIGURE 2.3 Directional Terms (a) For humans, anterior view; (b) for humans, 3/4 view; (c) for quadrupeds, lateral view.

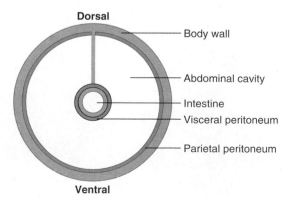

FIGURE 2.4 Idealized Cross Section of the Body with Relationships of Visceral and Parietal Terms

Planes of Sectioning

When viewing a picture of an organ that has been cut it is important to understand how the cut was made. Just as an apple looks different when cut crosswise as opposed to lengthwise, so do some organs. Examine figure 2.5 for the following sectioning planes. A cut that divides the body or organ into superior and inferior parts is in the **transverse,** or **horizontal, plane.** A cut that divides the body into anterior and posterior portions is in the **coronal,** or **frontal, plane.** A cut that divides the body into left and right portions is in the **sagittal plane.** A cut that divides the body equally into left and right halves is in the **midsagittal,** while one that divides the body into unequal left and right parts is in the **parasagittal plane.** A **median** cut occurs in the sagittal plane. Examine figure 2.6 for examples of sectioning planes.

Body Cavities

The body contains two major body cavities, the **dorsal cavity** and the **ventral cavity.** These can be further divided into smaller cavities. The dorsal cavity is composed of the **cranial cavity** and the **vertebral canal** (see figure 2.7). Compare the models and charts in the lab with this figure. The ventral cavity is divided into (1) the **thoracic cavity,** superior to the diaphragm, which is divided into the **mediastinum** and the **pleural** and **pericardial cavities,** and (2) the **abdominopelvic cavity,** inferior to the diaphragm, divided into the **abdominal** and **pelvic cavities.** The mediastinum is the region medial to the lungs. It contains, among other things, the heart, esophagus, and trachea. The pleural cavities are lateral to the mediastinum and contain the lungs.

Regions of the Body

Overview

Examine figure 2.8 for specific areas of the body. You will refer to these areas throughout this lab manual, so a complete study of these regions here is essential. Locate these regions on a torso model in lab. In anatomical usage, the **arm** is the region between the shoulder and the elbow, and the **leg** is the region between the knee and the ankle.

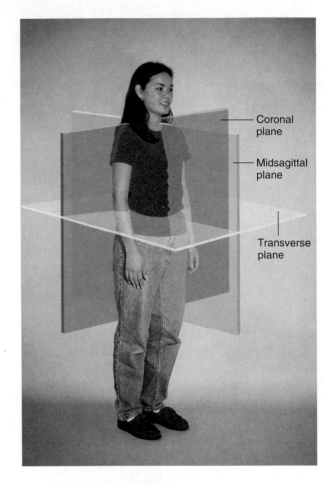

Coronal plane

Midsagittal plane

Transverse plane

FIGURE 2.5 Sectioning Planes

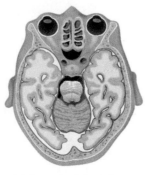 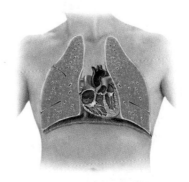 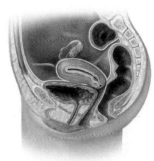

FIGURE 2.6 Planes of Sectioning   **(a) Transverse section**   **(b) Frontal section**   **(c) Sagittal section**

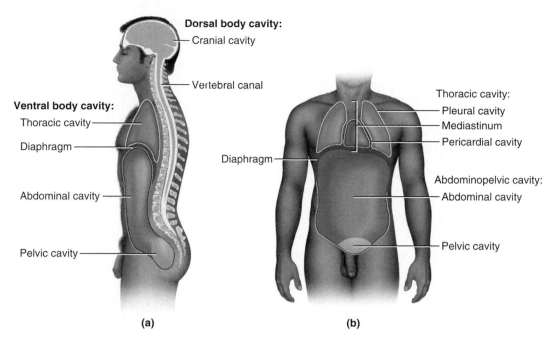

FIGURE 2.7 Body Cavities (a) Left lateral view; (b) anterior view.

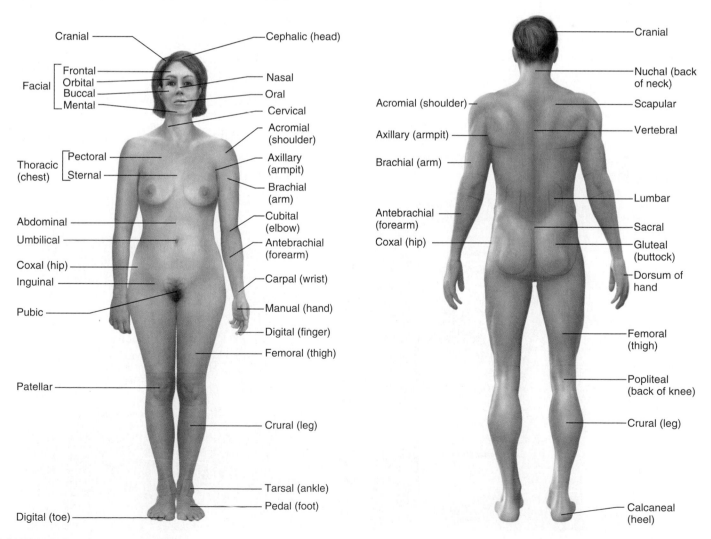

FIGURE 2.8 Regions of the Body

Examine a muscle model in lab, and name the region where the following muscles are found. Fill these in at the end of the exercise as well.

? Pectoralis major _____ 12

? Trapezius _____ 13

? External oblique (front) _____ 14

? Rectus femoris _____ 15

? Gastrocnemius _____ 16

? Gluteus maximus _____ 17

? Flexor carpi radialis _____ 18

? Triceps brachii _____ 19

? Latissimus dorsi _____ 20

Abdominal Regions

The abdomen can be further divided into either four quadrants or nine regions (see figure 2.9). Clinicians typically use four quadrants terminology, while anatomists generally use nine regions.

Right hypochondriac

Left hypochondriac

Epigastric

Right lumbar (lateral abdominal)

Left lumbar (lateral abdominal)

Umbilical

Hypogastric

Right iliac (inguinal)

Left iliac (inguinal)

Right upper quadrant

Left upper quadrant

Right lower quadrant

Left lower quadrant

Examine a torso model in lab. For each of the following organs write what abdominal region(s) contain(s) that organ. Fill in the following spaces and at the end of the exercise.

? Liver _____ 21

? Urinary bladder _____ 22

? Small intestine _____ 23

? Descending colon _____ 24

? Left kidney _____ 25

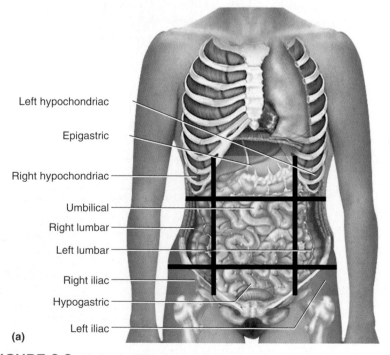

Left hypochondriac

Epigastric

Right hypochondriac

Umbilical

Right lumbar

Left lumbar

Right iliac

Hypogastric

Left iliac

(a)

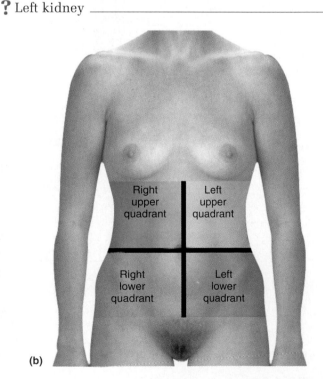

Right upper quadrant

Left upper quadrant

Right lower quadrant

Left lower quadrant

(b)

FIGURE 2.9 Abdominal Regions (a) Nine regions; (b) four quadrants.

REVIEW SECTION

Organs, Systems, and Organization of the Body

Name _____ Date _____

Lab Section _____ Time _____

Review Questions

1. The kidneys belong to the _____ system.

2. The liver belongs to the _____ system.

3. The spleen belongs to the _____ system.

4. In anatomical position, the palms of the hand are facing _____.

5. In anatomical terms, referring to front and back, the pectoral region is _____ to the scapular region.

6. In terms of nearness to the trunk, the antebrachium is _____ to the carpal region.

7. In terms of front to back, the shoulder blades are _____ to the nipples.

8. What type of section plane is illustrated in figure 2.7(a)?

9. What type of section plane is illustrated in figure 2.7(b)?

10. The lungs are found in what main body cavity (dorsal/ventral)?

11. Name the body cavities in the illustration.

    a. _____

    b. _____

    c. _____

    d. _____

    e. _____

12. The lungs are found in what specific cavities?

13. The brain is located in what specific cavity?

14. If you were to sit on a horse's back you would be on the _____ aspect of the horse.

    a. anterior   b. ventral   c. posterior   d. dorsal

15. Pain in the appendix would be felt in what quadrant of the abdomen?

16. What is the difference between the abdomen and the abdominal cavity?

17. The region of the abdomen directly under the right side of the rib cage is the _____ region.

18. If a hairline fracture occurred in the proximal humerus (arm bone), would the injury be closer to the shoulder or to the elbow? Why?

19. In a clinical report there is a note of a laceration (cut) on the posterior crural region. Where does this occur in layperson's terms?

20. Name the regions of the body in the illustration.

a. _____

b. _____

c. _____

d. _____

e. _____

f. _____

g. _____

h. _____

i. _____

j. _____

k. _____

l. _____

m. _____

n. _____

? Chapter Summary Data

Use this section to record your results from questions within the exercise.

1. _____

2. _____

3. _____

4. _____

5. _____

6. _____

7. _____

8. _____

9. _____

10. _____

11. _____

12. _____

13. _____

14. _____

15. _____

16. _____

17. _____

18. _____

19. _____

20. _____

21. _____

22. _____

23. _____

24. _____

25. _____

LABORATORY

Microscopy

INTRODUCTION

Originally the study of anatomy and physiology was based on macroscopic, or gross, observation. With the invention and use of the light microscope much greater detail was seen, and thus began the study of cells and tissues. **Light microscopy** involves the use of visible light and glass lenses to magnify and observe a specimen. Further advances in microscopy led to the development of electron microscopy, which uses electrons passing through or bouncing off of material. Electron microscopy has revealed much greater detail than observable with the light microscope. These topics are covered in the Saladin text in chapter 1, "Major Themes of Anatomy and Physiology." This exercise involves the use of the light microscope, how to examine prepared slides under the microscope, and how to make slides of fresh material for study.

OBJECTIVES

At the end of this exercise you should be able to

1. list the rules for proper microscope use;
2. name the parts of the microscope presented in this exercise;
3. demonstrate the proper use of the light microscope;
4. place a microscope slide on the microscope and observe the material, in focus, under all magnifications of the microscope;
5. calculate the total magnification of a microscope based on the specific lenses used;
6. prepare a wet mount for observation.

MATERIALS

Light microscope

Prepared slide with the letter *e* (or newsprint and razor blades)

Transparent ruler or sections of overhead acetates of rulers

Glass microscope slides

Coverslips

Lens paper

Kimwipes or other cleaning paper

Lens cleaner

Small dropper bottle of water

1% methylene blue solution

Toothpicks

Histological slides of kidney, stomach, or liver

Slide with silk threads

PROCEDURE

Microscopes are very expensive pieces of equipment, and you should always take great care handling them.

1. Remove the microscope from the storage area. When you carry the microscope, place one hand under the base and the other hand on the arm of the microscope. Never tilt the microscope from an upright position, as lenses or filters may fall and break.
2. Take the microscope to your desk. Unwrap the electrical cord and familiarize yourself with the parts of the microscope. Compare the microscope that you have in lab to the one illustrated in figure 3.1. There may be differences between the microscope and the one in figure 3.1, but you should be able to locate the parts listed in the checklist. Use the checklist to make sure that you find all of the listed parts. Place a check mark next to the appropriate space when you locate each part of the microscope.

Microscope Parts Checklist

_____ Base	_____ Arm	_____ Objective lens
_____ Condenser	_____ Iris diaphragm lever	_____ Body tube
_____ Nosepiece	_____ Coarse-focus knob	_____ Fine-focus knob
_____ Stage	_____ Ocular (eyepiece) lens	_____ Light source

3. With the microscope in front of you, plug it in, making sure the ocular lens or lenses are facing you. Make sure the cord is not hanging over the counter or in the aisle where someone might trip on the cord

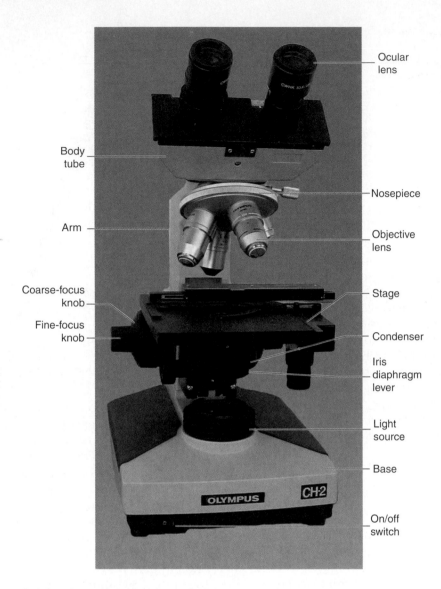

FIGURE 3.1 Compound Light Microscope Courtesy of Olympus America, Inc.

or pull the microscope off the counter. If the microscope has an illuminator (light source) dial, make sure that the setting is on the lowest level.

4. Turn the nosepiece until the low-power objective lens (the one with the lowest number or the shortest barrel) clicks into place. When you first look at a microscope slide, always use the low-power objective lens.

5. Take a piece of newsprint and, using a razor blade or scissors, cut out a single letter from the paper. Place the letter on a glass microscope slide and add a drop of water to the piece of paper on the slide.

6. Place a thin coverslip on the slide by touching one edge of the coverslip to the water and lowering it slowly over the newsprint specimen. If you drop the coverslip on top of the slide, you will probably trap several air bubbles, which may obscure some of your specimen.

7. Locate and turn on the light switch.

8. Place the slide on the microscope stage so that you can read the letter. If you cannot see the letter at first, move the slide so that you can see light coming through the specimen. This may require that you look at the specimen to see if it is centered in the middle of the microscope stage.

9. Examine the specimen under low power. Use the coarse-focus knob to bring the specimen into focus. Adjust the iris diaphragm for brightness.

10. Draw what the specimen looks like in the space provided.

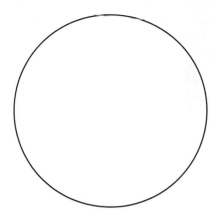

Does the letter appear right side up, or is the image inverted? _____

Is the letter oriented correctly, or is the image flipped horizontally? _____

The field of view is the circle you can see through the ocular lens. How much of the letter occupies the field of view? _____

11. Move the objective lens to the next higher-power lens. Examine the specimen again. You may need to adjust the fine-focus knob so that the image becomes clearer. As magnification increases, does the field of view increase or decrease in size?

If you are having a difficult time seeing anything or seeing things in focus, there could be several reasons. Use the following troubleshooting list to help you.

Problem	Solution
Nothing is visible in the lens.	Plug in the microscope. Turn the power supply on.
	Rotate the objective lens until you hear it click into place.
	The bulb is burned out; replace bulb.
You see a dark crescent.	The objective lens is not in proper position; click the lens into place.
All you see is a light circle.	The microscope is out of focus; adjust the coarse-focus knob.
	The light is up too high; turn down the light.
	The iris diaphragm is open too much; close it down.

Focusing the Microscope

Focusing the microscope requires a little bit of patience. When first looking at microscope slides always use the low-power objective lens. Make sure the specimen is on the stage and centered in the open circle on the stage. There should be light coming through the specimen. The coverslip should be close to the objective lens. Look at the microscope stage from the side; adjust the coarse-focus knob so that the coverslip is almost touching the objective lens. Look through the ocular lens and rotate the coarse-focus knob slowly so the objective lens and the slide begin to move away from each other. This should bring the object into focus in the field of view. The field of view is the circle that you see as you look into the microscope. Binocular microscopes have one fixed ocular and another that you can adjust. Focus the microscope so that the image in the nonadjustable ocular comes into focus. Once it is in focus, use the knurled ring on the adjustable ocular to bring that lens into focus for the other eye.

After you get the specimen in focus under low power, you should examine the material under higher powers. This is done by centering the image that you observe in the field of view and then switching the objective lens to the next higher power. The image should be pretty well in focus if your microscope is *parfocal*. Do not adjust the height of the stage. The next higher-power objective lens should clear the slide. Once you rotate the lens, adjust the focus by using the *fine-focus knob*. You can look at the subject under high power by the same procedure by turning the objective lens to the high-power lens. If you cannot focus on the high power or have lost the image that you were looking for, you should return to low power, make sure the specimen is centered in the field of view, and try the process again. If you still cannot find the object under high power, you should ask your instructor to help you.

Proper Lighting

Too much or too little light makes a specimen difficult to see. You can change the light level by either adjusting the light from the light source or adjusting the iris diaphragm lever. On some faintly stained specimens, the material may be difficult to see on low power. One trick is to locate the edge of the coverslip and turn the coarse-focus knob up and down until the edge is in sharp focus. This lets you know that you are in the approximate focal plane for examining the material on the slide. Move the slide to where the specimen should be, adjust the light, and re-examine it. The condenser lens will also help with lighting. Generally put the lens close to the specimen.

Magnification and Field of View

You can determine the size of the object under view if you know the diameter of the field of view. The field of view can be measured directly when under low magnification by using a clear ruler or a glass slide with a grid on it. If higher magnifications are used, rulers won't work and you have to calculate the field of view. Remember a millimeter is $^1/_{1,000}$th of a meter and a micrometer is $^1/_{1,000,000}$th of a meter. If you need to review these units, refer to Exercise 1. There is a relationship between the diameter of the field of view and the magnification used. You can first calculate the total magnification using the following procedure. Look at the barrels of your microscope and determine the magnification of each lens.

Eyepiece (ocular) magnification: _____

Low-power objective magnification: _____

Total magnification (= ocular magnification × objective magnification): _____

Place a transparent section of ruler, a glass slide with a grid, or a ruled section of acetate on the stage of the microscope. The space between each dark line that runs vertically on a ruler is 1 millimeter (mm). Count the number of millimeters at the broadest part of the field of view and enter this number as the diameter of the field of view in the following space.

Diameter of the field of view (mm): _____

You can calculate the length of an object by determining how much of the diameter of the field of view

it occupies. Let's say that the diameter of the field of view is 10 mm. If an object takes up one-half of the field of view, then you can estimate its size at 5 mm. If the object took up only one-third of the field of view, how large would it be? Record your answer in the following space.

Object size (mm): _____

As the magnification increases, the field of view decreases proportionally. Thus, if the diameter of the field of view is 10 mm at one magnification and you double the magnification by changing lenses, the field of view is reduced to a diameter of 5 mm. If you switch to a new lens and increase the magnification by 10 times, then the field of view is reduced to one-tenth of the original field of view. Look at figure 3.2 for a representation of this proportional change in the field of view.

Keep the clear ruler, a glass slide with a grid, or ruled acetate sheet under the microscope and increase the magnification to the next higher power by moving the next larger objective lens in place. Record the total magnification of your microscope with this objective lens.

Total magnification: _____

Examine the ruler or grid under the microscope and record the diameter of the field of view in millimeters.

Diameter of the field of view in millimeters: _____

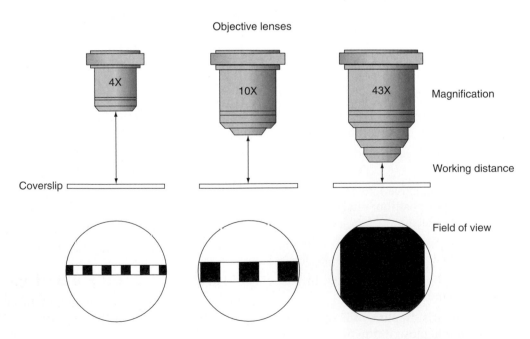

FIGURE 3.2 Increasing Magnification and Decreasing Field of View

Has the increase in magnification produced a decrease in the field of view? Is the decrease in the field of view proportional to the magnification?

Now calculate the total magnification of the microscope using the high-power objective lens.

Magnification with high-power objective lens: _____

You will not be able to measure the field of view accurately under high power with a ruler or an acetate sheet; however, you should be able to calculate the diameter of the field of view. For example, if the diameter of the field of view is 2.5 mm at 40 power ($40\times$), then it would be 0.25 mm at $400\times$ (10 times more magnified yet one-tenth the field of view). Calculate the diameter of the field of view under high power.

Diameter of the field of view under high power: _____

Preparation of a Wet Mount

You can make relatively quick and easy observations under the microscope as long as the material is thin enough and small enough. One technique for cell observation is to examine cells from the inside of the oral cavity. Ask your instructor for permission first. He or she may prefer that you select another type of cell, such as a scrape inside the abdominal cavity of a study animal. Do the following procedure.

1. Place a drop of methylene blue (a stain) on the slide.
2. With a toothpick, *gently* scrape the inside of your cheek.
3. Smear the cheek material from the toothpick on the slide.
4. Place a coverslip on the slide by holding the coverslip at a 45-degree angle and slowly lowering it on the slide to avoid trapping air bubbles. Examine it under the microscope (see figure 3.3). The small oval structures inside the cells are the nuclei.
5. Draw your observations in the space provided.
   Your illustration of a cheek cell:

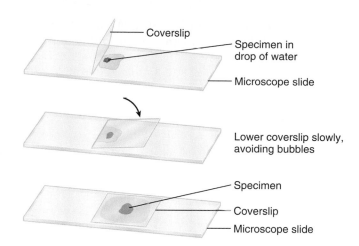

FIGURE 3.3 Preparation of a Wet Mount

Methylene blue is a stain that makes the nucleus of the cell more apparent.

For another observation, remove a hair from your head (preferably one with split ends) and examine it by making a wet mount. Place the hair in the center of the slide and add a drop of water. Place the coverslip on one edge of the drop and slowly lower it. Try to avoid trapping any air bubbles in the process.

Observation of a Prepared Slide

Examine a prepared slide of tissue provided by your instructor. Examine the entire sample using the low-power objective lens. You should scan the entire sample, looking for areas you want to observe closer. Move to the next higher power and adjust the focus using the fine-focus knob. Finally, examine the material with the high-power objective lens and draw what you see in the space provided.

Your illustration of material from a prepared slide:

Name of the sample drawn: _____

Observation of Silk Threads

Examine a slide that has crossed silk threads. Focus the slide on low power. How many of the threads can you see in focus without adjusting the focus knob further?

Move to medium and high power. What happens to the depth of field with increasing magnification?

Which thread is on top?

Which thread is on the bottom?

Oil Immersion Lens

The objective lenses you have used so far are called "dry" lenses. Your lab may be equipped with microscopes that have oil immersion lenses. Light bends differently through air than it does glass. Under high magnification, oil is used because it has the same level of refraction (it bends light the same) as glass. The techniques for using these lenses are somewhat different from those for dry lenses. Once you have examined the specimen using the high-power dry lens, find the spot you want to examine and center it in the field of view. Add a drop of immersion oil on top of the coverslip and carefully swing the oil immersion lens into place. Avoid getting oil on the other lenses of the microscope. Use the fine focus only, or you may drive the oil immersion lens through the slide and break it. Once you have examined the slide, swing the lens away and remove the slide, carefully wiping away the immersion oil with a clean piece of lens paper (do not use your shirt or a paper towel; these can scratch the lens). Use only lens paper to clean the oil from the oil immersion lens. Use another paper to remove any remaining oil.

Cleaning the Microscope

Smudges on the images you view through the microscope may be due to several things. There may be makeup or dirt on the ocular lens (or lenses). There may be dirt, oil, salt, stains, or other material on the objective lenses. To clean a lens, place a small amount of lens-cleaning fluid on a clean sheet of lens paper. Make one circular pass on the lens and throw away the paper. If you continue to clean the lens with the same lens paper, you can grind dirt or dust into the lens. Use a fresh piece of lens paper and repeat the procedure if further cleaning is needed.

You may want to clean the microscope slide before you examine it. Use a cleaning paper, such as a Kimwipe, to clean oil or dust from the slide.

Finally, dust may have collected inside the microscope over the years or the lenses may be scratched. There is nothing you can do about this, though you may want to bring this to your instructor's attention.

Care of the Microscope

There are a few rules concerning the care of a microscope.

1. When carrying the microscope, hold it securely with two hands—one hand under the base and one on the arm.
2. Keep the microscope upright at all times.
3. Keep microscope lenses clean with lens cleaner and softened lens paper. Do *not* use paper towels or your shirt.
4. Use only the fine-focus knob when using the high-power objective lens.
5. Remove slides from the microscope before putting it away.
6. Secure the cord with a rubber band or wrap the cord carefully around the base of the microscope.
7. Store the microscope with the *low-power* (scanning) objective lens in place.
8. Put the microscope away in its proper location.

REVIEW SECTION

Microscopy

Name _____ *Date* _____

Lab Section _____ _____ *Time* _____

Review Questions

1. If the ocular lens is 10×, what would be the total magnification for the following objective lenses?

   a. 5× _____

   b. 17× _____

   c. 35× _____

2. What is the function of the iris diaphragm of the microscope?

3. If the diameter of the field of view is 5.6 mm at 40×, what would the diameter be at 80×?

4. What is the name of the thin glass plate placed on top of a specimen?

5. The microscope that you use in lab is a(n)

   a. light microscope.   b. dissecting microscope.   c. electron microscope.

6. What is the name of the circle you see when you look through the ocular lens of the microscope?

7. When you switch from a low-power objective lens (for example, 4×) to a higher-power objective lens (for example, 10×), what happens to the working distance between the lens and the coverslip?

8. What happens to the field of view when you change from a low-power objective lens to a higher-power objective lens? Does it increase or decrease?

9. When should you use the scanning lens on the microscope?

10. How should you clean the lenses of a microscope?

11. What is the proper way to carry the microscope in lab?

12. Examine the following "field of view" and determine what the size of the object is.

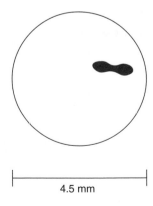

4.5 mm

13. Label the parts of the microscope illustrated, using the terms provided.

Light source   Body tube
Objective lens   Arm
Ocular lens   Stage
Coarse-focus knob   Base
Iris diaphragm lever

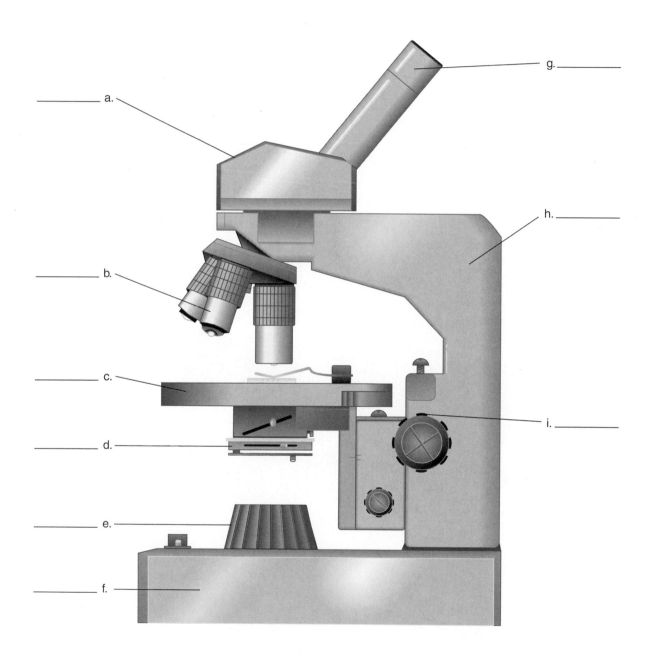

Notes

LABORATORY

Cell Structure and Function

INTRODUCTION

The cell is the structural and functional unit of the human body and is the fundamental unit of living organisms. Cytology is the scientific study of cells. Many diseases can be traced to some type of cellular change. Cells grow, divide, acquire nutrients, release wastes, and respond to local stimuli. Cells perform many functions, some unique to the particular organ where they are found. In this exercise you examine the structure of animal cells and learn how cells of the body divide to make new cells. These topics are discussed in chapter 3, "Cellular Form and Function," and chapter 4, "Genetics and Cellular Function," in the Saladin text.

OBJECTIVES

At the end of this exercise you should be able to

1. describe the importance of cells in the makeup of the body;
2. list the functions of the nucleus;
3. list all of the organelles and their functions;
4. describe the structure and functions of the mitochondria;
5. draw a representation of each of the organelles;
6. describe the three main events of the cell cycle;
7. name the four phases of mitosis and the events occurring in each phase.

MATERIALS

Models or charts of animal cells

Electron micrographs of cells or textbook with electron micrographs

Prepared slides of whitefish blastula

Microscopes

Modeling clay (Plasticine)—two colors

Marbles

PROCEDURE

Overview of the Cell

There are many different types of cells in the body. Some are long and thin, some are spherical, and some are flat. We will examine a representative cell as an example, but realize that there is tremendous diversity in cell shapes and functions.

Cells consist of two main parts, the **plasma** (or **cell**) **membrane** and the **cytoplasm.** The plasma membrane is the outer boundary of the cell, and though it cannot be seen using the light microscope, its location is determined by the difference in color between the cytoplasm and the surrounding liquid on the microscope slide. The cytoplasm is the portion of the cell in which water, dissolved materials, and small cellular **organelles** (*organelle* = small organ) are found. The fluid in which the organelles are suspended is called the **cytosol.** Small filaments and tubules make up the **cytoskeleton** (also considered part of the cytoplasm). The nucleus directs the cell's activities and stores its genetic information. It is visible with the light microscope and frequently appears as a spherical or oblong structure. Figure 4.1 shows the plasma membrane and cytoplasm. Locate these on the model or charts in the lab.

Plasma Membrane

The plasma membrane is composed of a **phospholipid bilayer,** proteins, cholesterol, and other molecules. Phospholipids consist of a hydrophilic phosphate group attached to hydrophobic lipid groups (see figure 4.2). Interspersed among the phospholipid molecules are **cholesterol molecules,** which provide stability to the membrane or, in larger concentrations, can make the membrane more fluid. Two major types of proteins are also found in the membrane. **Peripheral proteins** are found on the inner or outer surface of the membrane, while those passing through the membrane are known as **transmembrane proteins**. Transmembrane proteins may have carbohydrates or other molecules associated with them and frequently serve as cell markers. Some transmembrane proteins function as channels

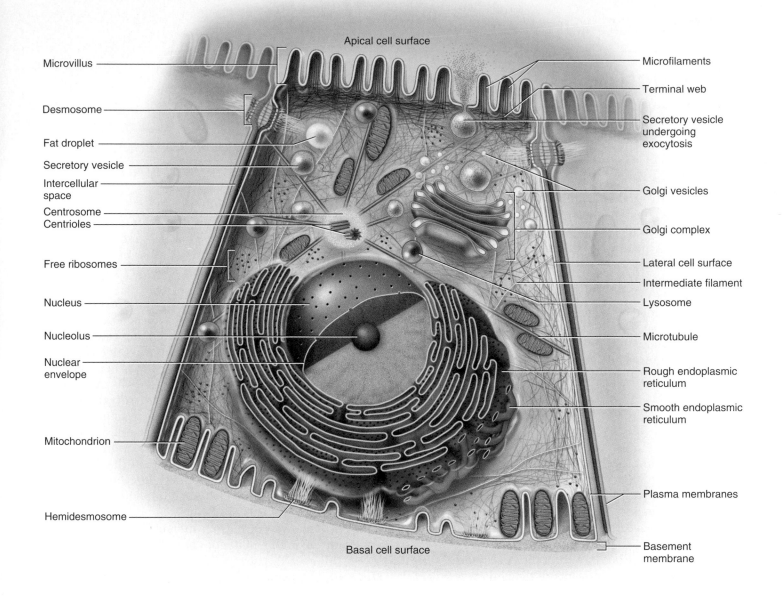

FIGURE 4.1 Overview of a Representative Cell

by which specific materials can pass through the membrane. The plasma membrane is important in establishing electro-chemical charge differentials, which allow for nerve impulse conduction and muscle contraction. It is also a selectively permeable membrane that provides an entrance to or exit from the cell for some materials while excluding other mate-rial from entering or exiting the cell's interior. Proteins may anchor one cell to another, provide a place for metabolic reactions to take place, act as cell markers that identify a par-ticular cell, or act as receptors or channels.

Cytoplasm

Most of the inside of the cell is cytoplasm, which consists of the inner fluid portion of the cell known as the cytosol,

the inner framework of the cell called the cytoskeleton, and the small specialized units of the cell called organelles. The cytosol is composed of water with dissolved materials, such as sugars, ions, proteins, and amino acids.

Figure 4.3 illustrates the cytoskeleton (not visible under the light microscope), which consists of micro-tubules, microfilaments, and intermediate filaments, all of which provide shape to the cell, a place to anchor organelles, and resistance to gravitational and other forces acting on the cell. **Microtubules** are made of the protein tubulin and are approximately 25 nanometers (nm) in diameter. **Intermediate filaments** are composed of fibrous proteins and are approximately 10 nm in diameter. **Microfilaments** are made of **actin** and are approximately 6 nm in diameter.

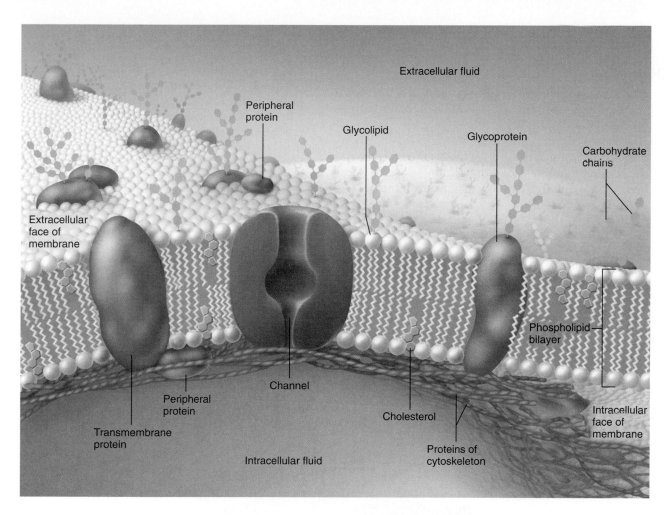

FIGURE 4.2 Plasma Membrane

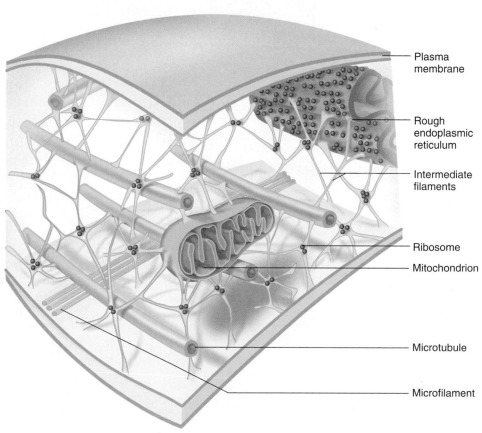

FIGURE 4.3 Cytoskeleton

Organelles

Organelle literally means "small organ." Each organelle serves a particular function in the cell. There are two types of organelles—membranous and nonmembranous. Organelles represent a wonderful example of **specialization** on a microscopic scale—individual organelles have structural characteristics reflecting their specific functions. Look at the illustrations of the various organelles as you read the following text.

Mitochondria

The major function of the mitochondrion (MY-toe-con-dree-un) (plural, *mitochondria*) is to convert the stored chemical energy in sugars, fatty acids, and amino acids to stored chemical energy in molecules of **adenosine triphosphate (ATP).** Mitochondria are elongated organelles approximately 0.2 to 5.0 micrometers (μm) in length. Mitochondria have two membranes similar in structure to the plasma membrane (phospholipid bilayers). Examine the illustration of a mitochondrion in figure 4.4. Note the separate outer and inner membranes. The inner membrane

has folds called **cristae.** These cristae increase the surface area of the mitochondrion. The inner region of the mitochondrion is known as the mitochondrial matrix (stroma), and it is the area where the citric acid cycle occurs. The citric acid cycle is involved in converting food molecules into energy molecules that the cell can use.

Ribosomes

Ribosomes (RYE-boh-somz) are the smallest of the organelles (about 25 nm in diameter), are nonmembranous, and produce proteins. Ribosomes are composed of two subunits that come together during protein synthesis.

Ribosomes come in two forms, **free ribosomes** (cytoplasmic ribosomes) and **bound ribosomes** (endoplasmic ribosomes). Free ribosomes make proteins for use inside the cell, while bound ribosomes make proteins for use outside the cell. An example of an externally used protein is mucus of the salivary gland. In this case, cells in the salivary gland have bound ribosomes that secrete the glycoprotein mucin. The cells of the gland produce mucus, secreted into the oral cavity. Figure 4.5 shows examples of ribosomes.

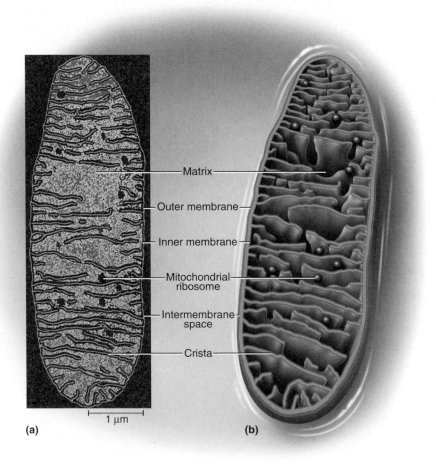

Matrix
Outer membrane
Inner membrane
Mitochondrial ribosome
Intermembrane space
Crista

1 μm

(a)   (b)

FIGURE 4.4 Mitochondrion

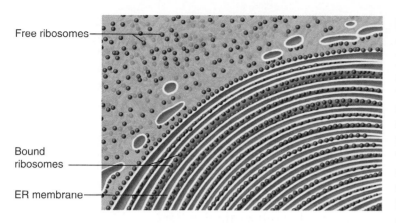

FIGURE 4.5 Ribosomes

Free ribosomes

Bound ribosomes

ER membrane

Endoplasmic Reticulum

The **endoplasmic reticulum** (EN-do-PLAZ-mic re-TIK-u-lum) is an organelle composed of a network of enclosed channels. The name *endoplasmic reticulum* means the "little net inside the cytoplasm." There are two types of endoplasmic reticulum (see figures 4.1 and 4.6): **rough endoplasmic reticulum (RER),** which contains bound ribosomes, and **smooth endoplasmic reticulum (SER),** which does not have ribosomes on its surface. The rough endoplasmic reticulum is associated with the nucleus and synthesizes proteins for transport and use outside the cell. The smooth endoplasmic reticulum is often a distal extension of the rough endoplasmic reticulum and produces lipid and steroid compounds and detoxifies

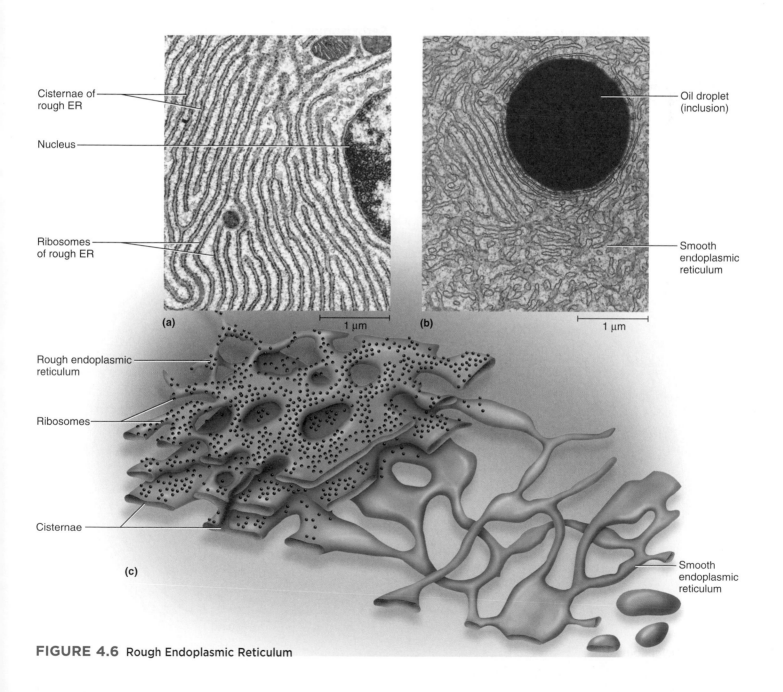

Cisternae of rough ER

Nucleus

Ribosomes of rough ER

(a)   1 µm

Oil droplet (inclusion)

Smooth endoplasmic reticulum

(b)   1 µm

Rough endoplasmic reticulum

Ribosomes

Cisternae

(c)

Smooth endoplasmic reticulum

FIGURE 4.6 Rough Endoplasmic Reticulum

material. **Cisternae** are enclosed spaces in the endoplasmic reticulum that receive proteins and secrete them in **vesicles** (fluid-filled sacs) for modification.

Golgi Complex

The **Golgi** (GOAL-jee) **complex** (or Golgi body) receives material from the endoplasmic reticulum and other parts of the cytoplasm and serves as an assembly and packaging organelle. Examine the Golgi complex in figure 4.7 and note the vesicles released from this organelle. Proteins from the endoplasmic reticulum are joined with carbohydrates, lipids, metals (such as the iron in hemoglobin), or other materials in the Golgi complex and then secreted in vesicles for transport from the cell.

Vesicles

There is a limit to the size of a molecule permitted to pass through the phospholipid bilayer or through the protein channels in the plasma membrane. Transportation of large molecular weight materials out of the cell is an important function of vesicles. They protect the integrity of the plasma membrane by maintaining the isolation of the cytoplasm from the external environment. If a substance were to exit the cell by opening a hole in the plasma membrane, the likely consequence would be that the cell would burst. Large molecular weight substances are enclosed in vesicles, and the vesicular membrane fuses with the plasma membrane, ejecting the material without disrupting the plasma membrane. Two common types of vesicles are discussed next.

Lysosomes

Lysosomes (LY-so-somz) are vesicles filled with digestive enzymes. Some cells, such as white blood cells, engulf foreign particles and release the contents of the lysosome near the foreign particles. In this way digestive enzymes hydrolyze the substance, which can later be removed from or incorporated into the body. Figure 4.8 illustrates lysosomes.

Peroxisomes

Peroxisomes are vesicles that contain peroxidase, an enzyme that converts hydrogen peroxide (H_2O_2) to water (H_2O) and oxygen ($\frac{1}{2}\ O_2$). Hydrogen peroxide is damaging to tissue, and removal of peroxide lessens potential damage to cells.

Go back to figure 4.1 and locate as many of the organelles listed in table 4.1 as you can. Be able to recognize the variances in structure and function of the organelles.

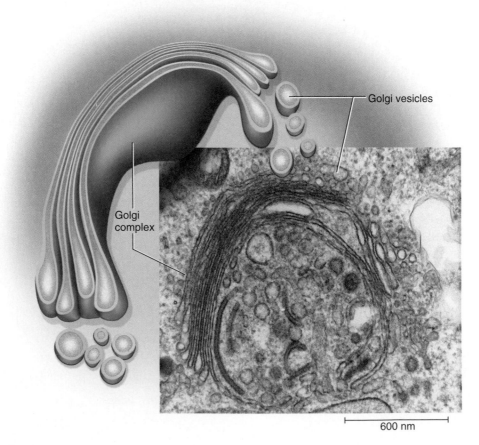

Golgi vesicles

Golgi complex

600 nm

FIGURE 4.7 Golgi Complex

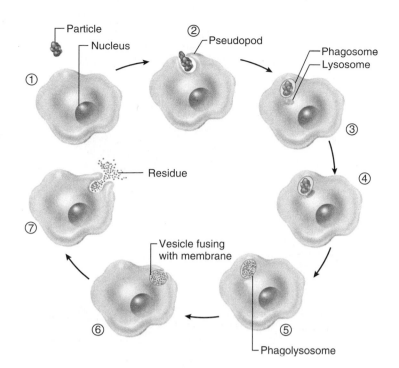

FIGURE 4.8 **Lysosomes** The sequence of digestion progresses from low numbers to high numbers.

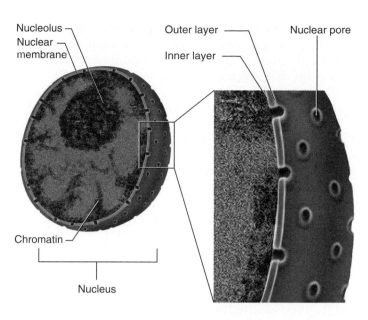

FIGURE 4.9 Nucleus of a Cell

Nucleus

The **nucleus** has two major functions: one is to house the genetic information of the cell and the other is to direct many cellular functions. These two tasks are carried out by **DNA (deoxyribonucleic acid),** which combines with proteins to form a material called **chromatin** in the nucleus. The nucleus is bounded by a **nuclear membrane** (nuclear envelope), a double membrane. Each layer of the nuclear membrane is a phospholipid bilayer. The nuclear membrane has holes in it called **nuclear pores,** which allow movement of proteins into and mRNA out of the nucleus.

One or more structures known as the **nucleoli** (singular, *nucleolus*) are inside the nucleus. The nucleoli consist of

RNA (ribonucleic acid) and protein and make ribosomes, the protein-producing organelles in the cytoplasm of the cell. Figure 4.9 shows the structures of the nucleus.

Other Cellular Components

Additional structures occur in the cell that do not fall neatly into the category of plasma membrane, cytoplasm, or nucleus. Cilia and flagella are two such structures. These structures are associated with the cytoskeleton and extend from the body of the cell. They consist of microtubules covered by the plasma membrane. Most cells do not have cilia, and flagella are found in humans only in the sperm cells. Cilia are found on cells involved in some type of movement, such as the movement of mucus along

TABLE 4.1	Organelles and Their Functions in Cells	
Organelle	**Membrane**	**Function**
Mitochondrion	Double	ATP production; fatty acid oxidation
Ribosome	None	Protein production
Rough endoplasmic reticulum	Single	Protein production for export
Smooth endoplasmic reticulum	Single	Lipid and steroid synthesis; detoxification
Golgi complex	Single	Assembly of macromolecules; transport from cell for secretion
Lysosome	Single	Digestion of material
Peroxisome	Single	Conversion of H_2O_2 to $H_2O + O_2$

the free edge of the respiratory passage (see figure 4.10). The cilia move mucus, which traps dust and keeps it from entering the lungs. The structure of cilia and flagella is similar but cilia are shorter than flagella.

Centrosomes are also unique cellular structures. Typically found close to the nucleus, they are involved in microtubule formation and in the formation of a structure known as the spindle apparatus, involved in cellular division.

Finally, microvilli are small extensions of the surface membrane of some cells. They increase the surface area and are involved in absorption of material (such as cells of the digestive and urinary systems).

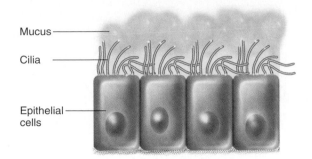

FIGURE 4.10 Cilia in the Respiratory Passage

The Cell Cycle

One of the great wonders of science is the mechanism by which a single cell, the result of the fusion of egg and sperm, develops into a complex, multicellular organism, such as a human. Various estimates put the number of cells in the human body in the trillions. All of these cells came from the first cell, or **zygote.** In this part of the lab exercise we examine the mechanism by which this occurs.

Most cells produce more cells by a process known as the cell cycle. The cell cycle can be divided into three events: **interphase, mitosis,** and **cytokinesis.** These events are illustrated in figure 4.11. For most of the time of the average life of a cell, it is in interphase.

Interphase

Interphase is the time when a cell undergoes growth and duplication of DNA in preparation for the next cell division. If a cell is not going to divide any further (such as brain cells and some muscle cells), then interphase is regarded as the time when a cell carries out normal cellular function.

Interphase has three separate phases, known as the **G_1 phase, S phase,** and **G_2 phase.** In the G_1 phase (G stands for *gap*), cells are in the process of growing in size and producing organelles. In the S phase (S stands for *synthesis*) the DNA of the cell is duplicated. The **double helix** of the DNA molecule unzips and two new, identical DNA molecules are produced. In the final phase of interphase, the G_2 phase, the cell continues to grow and prepares for the process of mitosis. Some cells do not undergo further division and are said to be in the G_0 (G-zero) phase.

Mitosis

Mitosis (my-TOE-sis) is a continuous event divided into four distinct phases. Mitosis is **nuclear division** and it involves the division of genetic information to produce two identical nuclei. For mitosis to occur, the chromatin in the nucleus of the cell must condense into compact units called **chromosomes.** Chromosomes consist of two

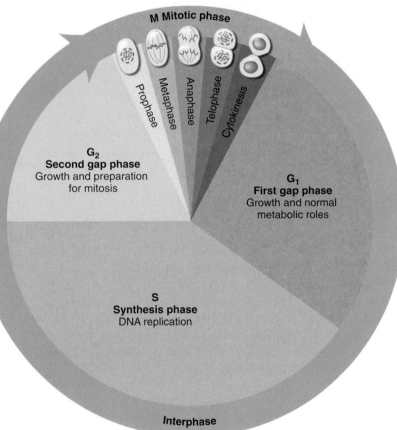

FIGURE 4.11 The Cell Cycle

chromatids held at the center by a **centromere.** The two chromatids are identical and result from the duplication of DNA. Examine figure 4.12 for the structure of a chromosome. You should also note the structure of chromosomes as you study the cells undergoing mitosis. The four phases of mitosis, **prophase, metaphase, anaphase,** and **telophase,** are described next. Refer to figure 4.13 as you read the descriptions.

Prophase

Cells in interphase have a distinct nuclear membrane, and the genetic information is dispersed in the nucleus as chromatin (see figure 4.13a). The first indication

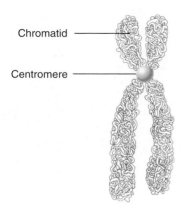

FIGURE 4.12 Structure of an Isolated Chromosome

that a cell is undergoing mitosis is the condensation of chromatin into visible chromosomes. Each chromosome consists of two elongated arms known as **chromatids,** connected to each other by a centromere.

In addition to the thickening of the chromosomes the nucleolus disappears and the nuclear membrane begins to disassemble. In order for the chromosomes to separate and move away from each other the nuclear membrane must not be present. Additionally, the **mitotic apparatus** becomes apparent. The mitotic apparatus consists of **spindle fibers,** which attach to the chromosomes at regions of the centromere known as the **kinetochores.** Two **asters** are points of radiating fibers at each end (pole) of the cell. In the center of the asters are two small structures known as **centrioles.**

Metaphase

In this phase the chromosomes align between the poles of the cell in a region known as the **metaphase plate.**

Anaphase

In anaphase the chromosomes separate at the centromere and each chromatid is now known as a **daughter chromosome.** The spindle fibers attach to the region of the centromere known as the kinetochore and pull the daughter chromosomes toward opposite poles of the cell. The centromere region moves first and the arms of the chromosomes follow.

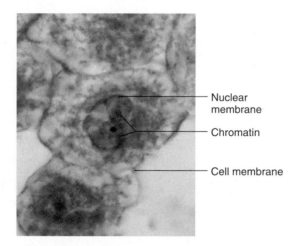

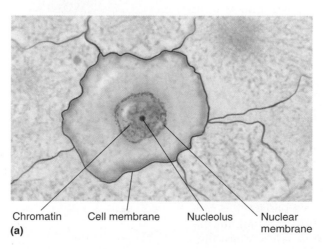

(a)

FIGURE 4.13 **The Cell Cycle (1,000×)** (a) Interphase, mitosis; (b) prophase; (c) metaphase; (d) anaphase; (e) telophase with cytokinesis.

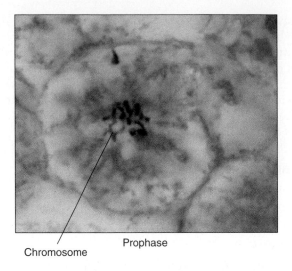

Chromosome   Prophase

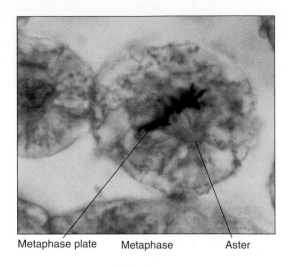

Metaphase plate   Metaphase   Aster

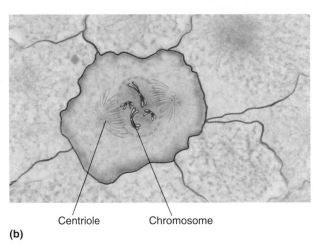

(b)

Centriole   Chromosome

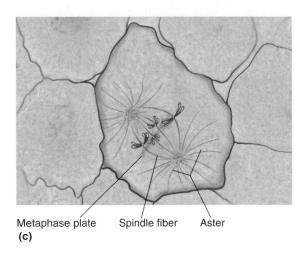

(c)

Metaphase plate   Spindle fiber   Aster

FIGURE 4.13 *Continued*

Telophase

Once the daughter chromosomes reach the poles, telophase (TEE-lo-faze) begins. The daughter chromosomes begin to unwind into chromatin, the nucleolus reappears, and the nuclear envelope begins to re-form. The mitotic apparatus disassembles, thus terminating mitosis.

Cytokinesis

The splitting of the cell's cytoplasm into two parts is known as **cytokinesis** (SY-toe-kih-NEE-sis). Although cytokinesis is a distinct process, it frequently begins during late anaphase or early telophase. In late anaphase, as the chromosomes are moving to the poles, the plasma membrane begins to constrict at a region known as the **cleavage furrow.** This begins the process of dividing the cytoplasm, ending as the cell splits into two daughter cells. The cytoplasm and the organelles are effectively divided into two parts.

Examine a slide of whitefish blastula and look for the cells in the slide. Most of the cells that you see are in a particular part of the cell cycle. What is this phase, and why are most of the cells in this phase?

Compare the slide with figure 4.13. Draw representative cells in each phase of mitosis in the following space.

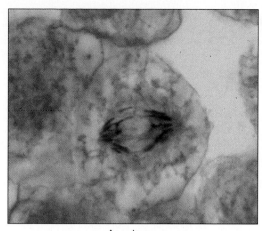

Anaphase

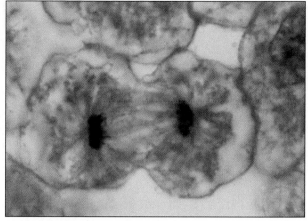

Telophase

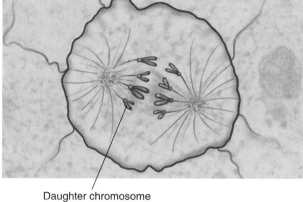

Daughter chromosome

(d)

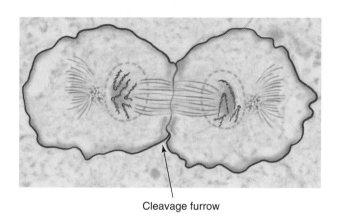

Cleavage furrow

(e)

FIGURE 4.13 *Continued*

Review the phases of mitosis in table 4.2. Take two colors of modeling clay and make chromosomes from the clay. You should have a long chromosome and a short chromosome of each color, for a total of four chromosomes. Each chromosome should have two chromatids, and the chromosomes should be joined by a marble, which represents the centromere. Draw a large circle on a sheet of paper to represent a cell. Manipulate the clay chromosomes to show how mitosis occurs. After you have done this, describe the process of mitosis. List each stage and what happens in that stage.

TABLE 4.2	Major Events of Mitosis
Prophase	Chromatin condenses to form visible chromosomes.
	Nuclear membrane disappears.
	Spindle apparatus forms.
	Nucleolus disappears.
Metaphase	Chromosomes align on the metaphase plate.
Anaphase	Chromosomes split and daughter chromosomes migrate to poles; cytokinesis often begins.
Telophase	Chromosomes reach poles; nuclear membrane re-forms.
	Chromosomes unwind to chromatin; cytokinesis divides the cytoplasm.
	Nucleolus reappears.

Mitosis Drawings

REVIEW SECTION

Cell Structure and Function

Name _____ *Date* _____

Lab Section _____ *Time* _____

Review Questions

1. Name the two major parts of the cell.

2. Which organelle is responsible for ATP production?

3. Which organelle in the cell produces lipids?

4. The DNA that controls the functions of the cell is located in what cellular structure?

5. What cellular structure is responsible for ribosome production?

6. Fill in the name, structure, and function of the organelles.

Organelle	Function	Structure
1. _____	Protein production for use inside of the cell	Not membrane-bound
2. Smooth ER	_____	No attached ribosomes
3. _____	Packages and transports material	Vesicles on organelle
4. Mitochondrion	Produces energy for the cell	_____
5. Rough ER	Produces protein for use outside of the cell	_____

7. How is the rough endoplasmic reticulum related to the Golgi complex in terms of function?

8. What two organelles produce proteins?

9. What is the function of the cytoskeleton?

10. The cytoplasm has a liquid portion. What is it called?

11. What structure in the cell is composed mostly of a phospholipid bilayer?

12. In what cellular object is cholesterol vital for structural integrity?

13. What are the three parts of the cell cycle?

14. Name the stages of interphase and describe what happens during those phases.

15. Of the three parts of the cell cycle, in which one is DNA duplicated?

16. In what part of the cell cycle do chromosomes first split apart?

17. What part of the cell cycle involves the division of the cytoplasm?

18. Describe the four phases of nuclear (mitotic) division and what occurs during those phases.

19. Name the cellular structures, using the terms provided.

Mitochondrion   Centriole
Nucleolus   Plasma membrane
Golgi complex   Ribosomes
Rough endoplasmic reticulum   Smooth endoplasmic reticulum

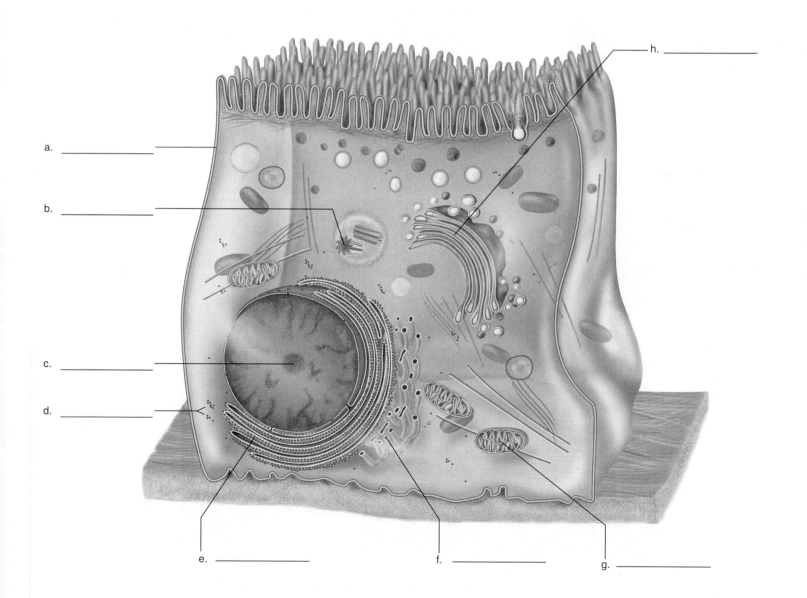

a. _____

b. _____

c. _____

d. _____

e. _____

f. _____

g. _____

h. _____

20. Name the phases of the cell cycle, as illustrated. Use the terms provided.

Interphase   Telophase

Prophase   Anaphase

Metaphase

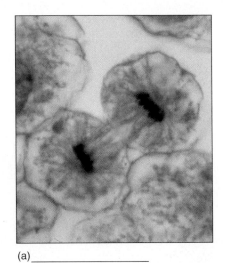

(a)_____

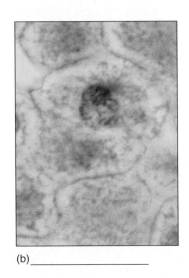

(b)_____

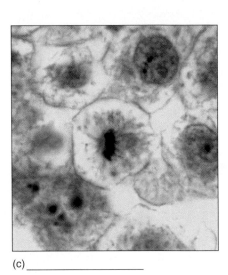

(c)_____

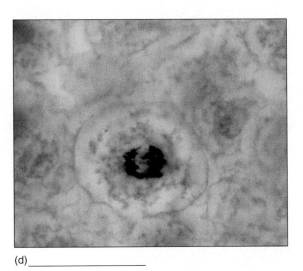

(d)_____

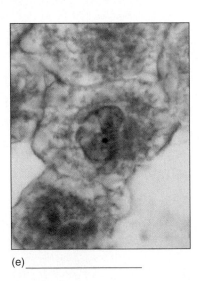

(e)_____

LABORATORY

Some Functions of Cell Membranes

INTRODUCTION

The membrane of the cell is the dynamic interface between the internal, living environment of the cell and the external environment. In humans most cells are bathed in a liquid medium called extracellular fluid (ECF), which provides nutrients, oxygen, hormones, water, and other materials to the cell. From the interior of the cell, the cell releases ammonia, carbon dioxide, and other metabolic products into this same liquid. The cell membrane is vitally important in the exchange of materials between the internal fluid of the cell and the environment surrounding the cell. The exchange of materials between the cell and the ECF maintains the homeostatic balance the cell must have to survive. Even small changes in the concentration of certain materials in the cell might lead to cellular death, so the constant adjustment of water, ions, and other metabolic products is extremely important. In large part, the cell membrane actively regulates what enters and what exits the cell. This is done passively in some cases and, in others, actively, requiring the use of ATP to drive certain processes. The membrane also has enzymes embedded in it that play a significant role in cellular function. In this exercise we look at the physical processes that influence cell membrane dynamics and study the nature of membrane transport. These topics are discussed in chapter 3, "Cellular Form and Function," in the Saladin text.

OBJECTIVES

At the end of this exercise you should be able to

1. describe the processes by which substances move across membranes;
2. define the terms *hypertonic, hypotonic,* and *isotonic;*
3. define the terms *diffusion, osmosis,* and *filtration;*
4. describe the movement of water across a selectively permeable membrane;
5. compare and contrast diffusion and osmosis.

MATERIALS

Brownian Motion

India ink in dropper bottles

Dropper bottle of water

Microscopes

Microscope slides

Coverslips

Hot plate

Diffusion Demonstration

Potassium permanganate crystals

100 mL beaker

Water

Small forceps or spatula

Diffusion

Agar plates (three per table)

0.01 M potassium permanganate solution

0.01 M methylene blue solution

Plastic drinking straws

Millimeter ruler

Dishpan filled with crushed ice

Fine probe or small forceps

Warming tray (35–40° C) or incubator

Osmosis Demonstration

Two thistle tubes

Dialysis tubing

Rubber band

Molasses or concentrated sucrose solution (20%)

1% starch solution

Ring stand and clamp

250 mL beaker

Distilled water

Permanent marker

Osmosis Experiment

Four strips of 20 cm long dialysis tubing (one set of four per table)

Four 200 mL beakers

String or "W" clamp with wooden applicator stick

Scissors

Four solutions (2 L each) of 0%, 5%, 15%, and 30% sucrose (color each solution with a different color of food coloring)

One liter of 15% sucrose solution

Balances

Towels

Pipettes (10 mL)

Pipette pumps

Osmosis and Living Cells

Clean glass microscope slides

Coverslips

5 mL of mammal blood

Distilled water in dropper bottle

0.9% sodium chloride solution in dropper bottle

5% sodium chloride solution in dropper bottles

Latex or plastic gloves

Filtration

Filter paper

Funnel

Ring stand with ring clamp

10 mL graduated cylinder

500 mL beaker

Iodine solution in dropper bottles

Filtration solution (500 mL of 1% starch, 1% charcoal, and 1% copper sulfate), consists of 5 g each of starch, charcoal, and copper sulfate in one bottle or flask

Stopwatch or clock with second hand

PROCEDURE

This lab is most efficiently done if the timing of the experiments overlaps. While you are waiting for the results of one experiment, begin another.

Brownian Motion

All matter has **kinetic energy** unless it is at absolute zero (−273°C). Kinetic energy is the energy of motion, and it is the driving force of the movement of atoms and molecules. Atoms and molecules are too small to be seen, even with the use of a light microscope, yet their movement can be inferred as they hit large particles that are visible under the microscope. Items such as dust or ink particles can be seen vibrating or jiggling, and we interpret this as the collective collisions of many molecules striking a larger structure and causing it to move. This is named **Brownian motion,** after the botanist Robert Brown, who first described this in the

nineteenth century. You will observe Brownian motion by examining ink particles in the following activity:

1. Place a drop of India ink on a clean glass microscope slide.
2. Carefully set a coverslip on the slide.
3. Examine the slide under high power and record the movement of the ink particles in the space provided, as well as in the Chapter Summary Data section at the end of the exercise.

? Movement of ink particles: _____ 1

1. Remove the slide from the microscope and place it on a warm surface (such as a hot plate on the low temperature setting) for a few seconds until the slide becomes warm.
2. Quickly return the slide to the microscope and note the movement of the particles, compared to the initial observation.
3. Record any difference in the space provided.

? Observation of warm slide: _____ 2

? How might kinetic energy play a part in the differences between the first observation and the second observation?

_____ 3

_____

Diffusion

Knowing that kinetic energy moves particles in solution (or in a gas, for that matter), we can see that a concentration of particles would be struck by chance collisions and that some of those particles would begin to disperse. This process is known as **diffusion,** which is defined as the movement of particles from regions of high concentration to regions of low concentration. If a small amount of perfume is poured into a dish in a room, the perfume molecules diffuse from the dish into the air of the room. The concentration of particles is higher in the dish and lower in the air, so the molecules move from regions of high concentration to regions of low concentration. The difference between the two concentrations is known as the **concentration gradient.** If the particles become uniformly dispersed in the air in the room, then the system has reached **equilibrium.**

Equilibrium can occur in solutions as well. A **solution** consists of a liquid portion, the **solvent,** and the dissolved portion, or **solutes.** If the solutes are concentrated in one area, they will diffuse in the solvent. There are many solvents but water is a vital solvent for the body, as it frequently contains sugars, ions, and amino acids.

Diffusion, driven by kinetic energy, is a process equally important to the cell. An example of the essential nature of diffusion is the movement of oxygen into the blood vessels in the lungs. Oxygen in the air is at a higher concentration than in the blood of the lung capillaries; consequently, oxygen moves from the air to the blood.

You can demonstrate diffusion by placing a crystal of potassium permanganate in a 100 mL beaker of water. Do this as a group at each table.

1. Fill a 100 mL beaker almost to the top with tap water.
2. Place the beaker on your table and drop a small crystal of potassium permanganate into it.
3. Leave the beaker undisturbed, but note the changes that occur during 1 hour.
4. Record your observations in the place provided. While you are waiting, continue with the other experiments.

? Description of potassium permanganate movement:

_____
                                                       4
_____

Many factors can affect the rate of diffusion, including changes in temperature, changes in concentration of the solute, the size or weight of the solute particles, and interactions between the solute and the solvent. In the following two experiments you will examine the effects of the weight of the particle and of temperature on diffusion rate.

Diffusion Rates and Particle Weight

You may want to do this experiment as a group of 3 or 4 students.

Agar, a liquid that forms a gel, can be used as a medium in which to measure diffusion rates of materials. Agar consists of water and algal polysaccharides. The liquid nature of agar is such that diffusion occurs in the gel at a slow rate.

1. Using a plastic drinking straw and a petri dish filled with agar, gently push the straw vertically into the dish making two holes that are approximately equidistant from one another (see figure 5.1). Do not twist the straw, or you will break the agar and leave a crack into which fluid will run.
2. Remove the small plug of agar with a fine probe or spatula, if needed, so a well is left in the agar.
3. Into one of the wells place three or four drops of 0.01 M potassium permanganate solution (molecular weight 158).

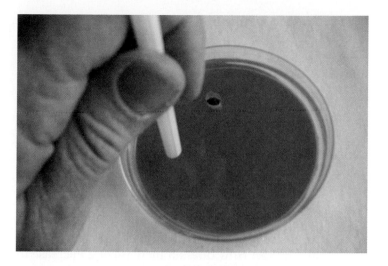

FIGURE 5.1 Placement of the Wells in Agar

4. Into another well place the same amount of 0.01 M methylene blue solution (molecular weight 320).

Be careful—these dyes stain clothes, skin, and lab notebooks!

Record the diameter of the diffusion in chart 1 and continue with the following experiments.

?

Diameter of Diffusion (in mm)				Chart 1

Fill in the data below and in the Chapter Summary Data section at the end of the exercises.

	20 min	40 min	60 min	80 min
Potassium permanganate	____	____	____	____ 5a
Methylene blue	____	____	____	____ 5b

Effects of Temperature on Diffusion Rates

1. Prepare two more petri dishes in the same way, except this time take one of the petri dishes from a refrigerator and another from a warming tray (such as an electric warming tray on the lowest setting or an incubator set at 37° C).
2. Make two wells in the cold petri dish and place three drops of the respective dyes, as you did previously.
3. This time, however, place the cold petri dish in a dishpan filled with crushed ice and examine after 80 minutes. You only need to record the diameter of diffusion at the end of the 80 minutes.
4. Remove a petri dish from the warming tray or incubator, make two wells in the dish, fill it with appropriate solutions, and return it to the warming tray or incubator. Examine after 80 minutes.

5. Record your results in chart 2.

? | **Diameter of Diffusion (in mm)** | **Chart 2**

Fill in the data below and in the Chapter Summary Data section at the end of the exercises.

	80 min (cold)	80 min (warm)	
Potassium permanganate	_____	_____	6a
Methylene blue	_____	_____	6b

? Compare the diffusion rates of the cold and warm temperatures with the dish left at room temperature for 80 minutes. How does an increase or a decrease in temperature affect the diffusion rate?

_____ 7

_____

? What can you say about temperature and the kinetic energy of the system?

_____ 8

_____

Osmosis

In the preceding diffusion experiments there was no barrier to the movement of the particles. Cell membranes are barriers to certain molecules, while they allow other molecules to pass through. This type of membrane is called a **selectively permeable membrane.** Water, some alcohols, oxygen, and carbon dioxide move easily across the cell membrane, while larger molecules, such as proteins, or charged particles (ions) are prevented from crossing the membrane.

The movement of water across a selectively permeable membrane from solutions of higher water concentration (water with less solutes) to lower water concentration (water with more solutes) is known as **osmosis.** Osmosis is a type of diffusion, and the process can be viewed from the perspective of the solvent or the perspective of the solute.

The Solvent Perspective of Osmosis

In diffusion, material moves from higher concentrations to lower concentrations. The same can be seen with osmosis. If you have two solutions separated by a selectively permeable membrane, one a 10% sugar solution (for example, 90% water) and one a 5% sugar solution (for example, 95% water), then the water will move from higher water concentration (95% water) to lower water concentration (90% water). The greater the difference between the two solutions, the greater the concentration gradient, which, in this case, is known as the **osmotic potential.**

The Solute Perspective of Osmosis

A 10% sugar solution has more solutes than a solution of 5% sugar. The 10% sugar solution is said to be **hypertonic** to the 5% sugar solution. The 5% sugar solution is said to be **hypotonic** to the 10% sugar solution. If these two solutions are separated by a selectively permeable membrane, then water flows *from* the hypotonic solution *to* the hypertonic solution. If enough water flows across the membrane and the two solutions reach the same concentration of sugar, then **equilibrium** is established and the net movement of water stops. If solutions have the same concentration of solutes, they are said to be **isotonic** to one another.

Demonstration of Osmosis

Observe osmosis with a thistle tube osmometer or construct an osmometer in which one end of a dialysis bag has been tied to a glass tube. The thistle tube consists of a hollow bell attached to a long tube. The tube is filled with molasses or sugar solution, and the opening of the bell is covered with dialysis tubing and secured with a rubber band. Sugar does not cross dialysis membranes, but water does, so the dialysis membrane is a selectively permeable membrane. The thistle tube osmometer is placed in a beaker of water and clamped to a ring stand (see figure 5.2). The level of the molasses or sugar solution is indicated with a wax mark from a china marker.

Examine the setup throughout the lab period. You may find that the liquid in the thistle tube eventually stops rising. This occurs when the gravitational pressure equals the force exerted by the process of osmosis. The amount of force needed to stop osmosis is the **osmotic pressure.**

Examine a thistle tube (or prepare one, if your lab instructor indicates for you to do so) that contains a starch solution. Does the liquid move up the tube?

? Why or why not?

_____ 9

_____

? What does this say about the osmotic activity of starch?

_____ 10

_____

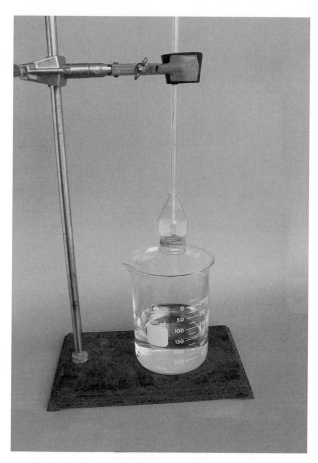

FIGURE 5.2 Osmosis Demonstration

Osmosis and the Concentration Gradient

As stated earlier, the osmotic potential varies, depending on the concentration gradient between the two solutions. In this experiment you determine the effects of various concentrations of a sucrose solution on the rate of osmosis.

1. Secure the bottom end of four dialysis bags that have been soaking in water.
2. Fill the dialysis bags with 10 mL of a 15% sucrose solution. You can do this by placing a pipette pump (or bulb) on the end of a 10 mL pipette, drawing liquid to reach the 10 mL mark, and filling the dialysis bag (see figure 5.3).
3. Tie or clamp the bags and then rinse each bag in distilled water and blot with a towel.
4. Weigh each bag to the nearest tenth of a gram and record the weights in chart 3. This weight is the **initial weight** of the bag.

(a)

Initial Weight of Dialysis Bags	**Chart 3**

Fill in the data below and in the Chapter Summary Data section at the end of the exercises.

Bag 1 _____ 11a grams

Bag 2 _____ 11b grams

Bag 3 _____ 11c grams

Bag 4 _____ 11d grams

(b)

FIGURE 5.3 Dialysis Bags (a) Filling; (b) tying.

5. Place bag 1 in a beaker filled about two-thirds to the top with a 0% sugar solution. Make sure the bag is covered with the solution, and leave it there for 20 minutes. Label the beakers with a wax marker.

6. Place bag 2 in a 5% sugar solution and leave it for 20 minutes.

7. Place bag 3 in a 15% sugar solution and leave it for 20 minutes.

8. Place bag 4 in a 30% sugar solution and leave it for 20 minutes.

9. Remove the bags from the beakers after 20 minutes, and blot and weigh each bag.

10. Remember to place each bag back in its proper solution! Record the weight of the bags each 20 minutes for a total of 80 minutes in chart 4.

Weight of Bags (in g) for Each Time Period			**Chart 4**

Fill in the data below and in the Chapter Summary Data section at the end of the exercises.

	Time			
	20 min	40 min	60 min	80 min
Bag				
1	12a	12e	12i	12m
2	12b	12f	12j	12n
3	12c	12g	12k	12o
4	12d	12h	12l	12p

Calculate the **change of weight** of each bag from the initial weight for each of the time periods. For example,

let's assume a bag weighed 20.5 grams as the initial weight and the recorded weights are as follows:

20 min	40 min	60 min	80 min
23.5 g	24.2 g	25.0 g	25.6 g

The change in weight is as follows:

3.0 g	3.7 g	4.5 g	5.1 g

Graph the change in weight for each bag of your experiment in chart 5 and connect the points with a line. Indicate which line represents which solution.

? Which of the bags (if any) gained weight? _____ 13

? Which of the bags (if any) lost weight? _____ 14

Determine the osmotic relationship (hypertonic, hypotonic, isotonic) of the beaker to the solution in the bag.

? Bag 1: The beaker solution is _____ to the bag solution. 15

? Bag 2: The beaker solution is _____ to the bag solution. 16

? Bag 3: The beaker solution is _____ to the bag solution. 17

? Bag 4: The beaker solution is _____ to the bag solution. 18

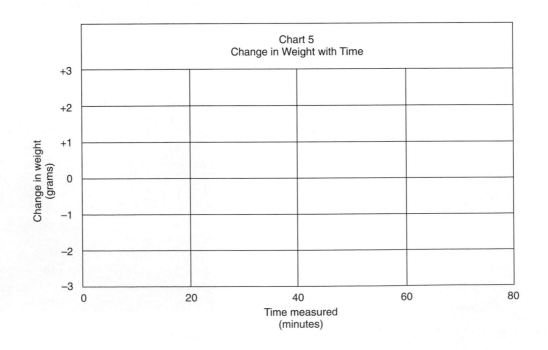

Chart 5
Change in Weight with Time

Change in weight (grams)

Time measured (minutes)

? Does the change in weight correlate to what you know about osmosis? If so, how does it correlate?

_____ 19

_____

_____

Osmosis and Living Cells

The importance of isotonic solutions can be demonstrated by the following procedure: Hold on to all slides for comparison. Mark each slide with a permanent marker.

Caution!   Make sure you wear protective gloves while conducting this experiment to avoid any potential transmission of disease.

---

1. Place a drop of fresh mammal blood on a slide.
2. To this slide add a drop of physiological saline (0.9% sodium chloride) and place a coverslip on the slide.
3. Observe the cells on high power under the microscope and note their shape.
4. Record this in the space provided.

Shape of the cells:

5. Place another drop of blood on a new slide.
6. To this slide add a drop of 5% sodium chloride (NaCl) solution.
7. Place a coverslip on the slide and record your observations.
8. Examine the slide for at least a few minutes or until a change of shape becomes obvious.

Shape of the cells in 5% NaCl:

Finally, with a third slide, repeat the previous procedure, except add a drop of distilled water instead of the sodium chloride. Immediately observe this slide and then continue to look at it for a few minutes. Record your observations.

? Cells in distilled water: _____ 20

? When red blood cells lose water they shrivel in a process known as **crenation.** Did any of the cells show crenation?

_____ 21

? Which solution might produce this? _____ 22

When water moves into a cell at a rapid rate the cell becomes inflated and sometimes bursts in a process known as **hemolysis.**

? Did any of the cells undergo hemolysis?_____ 23

? Which solution might produce this effect?_____ 24

Examine figure 5.4 for the various effects of solutions on red blood cells.

Note: In clinical settings, 0.9% saline and 5% dextrose in water (D5W) are isotonic to human red blood cells.

Filtration

The process of **filtration** is important in certain cells of the body and results as the pressure of a fluid forces particles through a filtering membrane. Filtration is a major component of kidney function. Hydrostatic pressure from the blood forces urea, ions, sugars, and other particles from the blood through the membranes of the kidney cells. In this way, small particles in the blood are forced into kidney tubules, while larger particles, such as proteins, remain in the blood. In this exercise you learn the basic principles of filtration, such as the **selectivity** of the filtration membrane and the **filtration rate** of a system.

1. Fold a piece of filter paper in half and then fold it in half again so that it forms a cone (see figure 5.5).
2. Place the cone in a funnel mounted on a ring stand over a beaker.
3. Into this cone filter add the filtration solution, a mixture of copper sulfate, powdered charcoal, and starch in water.
4. Fill the funnel to near the top of the brim and let the filtrate (the material passing through the filter) collect in the beaker.
5. When the funnel is approximately half full, place a 10 mL graduated cylinder under it and record the time it takes to fill the cylinder to the 2 mL mark.

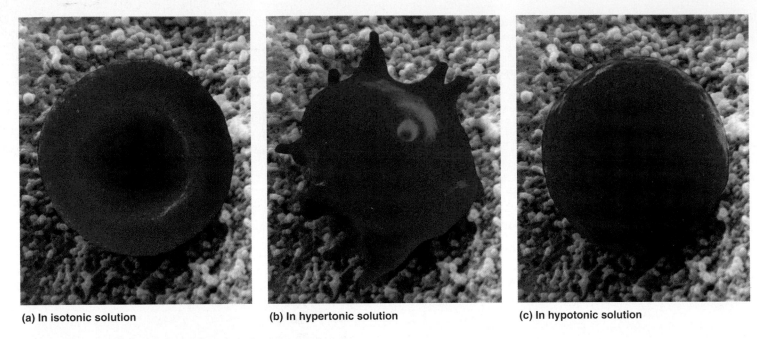

(a) In isotonic solution   (b) In hypertonic solution   (c) In hypotonic solution

FIGURE 5.4 **Stages of Red Blood Cells** (a) Normal; (b) crenoted; (c) inflated. David M. Phillips/Visuals Unlimited.

FIGURE 5.5 Filtration Apparatus

6. Divide the time by 2 and calculate the filtration rate expressed as mL/minute.
7. Record this value.

? Number of seconds to produce 2 mL: _____ 25

? Filtration rate: _____ mL/minute 26

You should be able to determine what material passes through the membrane by the following method. Remove the funnel from the beaker. If the solution in the beaker has a blue cast to it, then copper sulfate passed through the membrane. If black particles are found in the filtrate, then charcoal passed through the membrane. If you add a few drops of iodine to the beaker and the solution turns black, then starch passed through the membrane. Record what material passed through the membrane in the space provided.

? **Substance**	Yes	No
Copper sulfate	_____	_____ 27a
Activated charcoal	_____	_____ 27b
Uncooked starch	_____	_____ 27c

The force that drives filtration in the funnel is the force of gravity on the liquid. In the kidney the force that drives filtration is blood pressure.

REVIEW SECTION

Some Functions of Cell Membranes

Name _____ Date _____

Lab Section _____ Time _____

Review Questions

1. Based on your experimentation with Brownian motion, what do you predict would happen to the movement of the particles you examined if you cooled the slide? Why?

2. What process drives Brownian motion?

3. Particles in a gas or liquid moving from a region of higher concentration to a region of lower concentration represents what process?

4. Define solute.

5. What is a concentration gradient?

6. What process occurs as water moves from regions of higher water concentration to regions of lower water concentration across a semipermeable membrane?

7. Which osmosis bag in your experiment gained the most weight?

8. Was the bag that gained the most weight hypertonic, hypotonic, or isotonic to the solution in the beaker?

9. Which bag lost weight in your experiment? What caused the weight loss?

10. Why should you not inject a patient with a 10% saline solution? What osmotic effect would that cause?

11. What effect would lowering the temperature to absolute zero (–273° C) have on Brownian motion?

12. What particle diffused the farthest, potassium permanganate or methylene blue? What was the factor in the difference in diffusion rates?

13. The pressure needed to stop osmosis is called the _____.

14. Define hemolysis.

15. In terms of filtration, define selectivity.

16. What would happen to the filtration rate if you were to apply pressure to the filtration system?

What might happen to filtration if the water pressure damaged the membrane?

How might this relate to high blood pressure and kidney disease?

Chapter Summary Data

Use this section to record your results from questions within the exercise.

1. _____

2. _____

3. _____

4. _____

_____

5a. _____

5b. _____

6a. _____

6b. _____

7. _____

_____

8. _____

_____

9. _____

_____

10. _____

_____

11a. _____

11b. _____

11c. _____

11d. _____

12a. _____

12b. _____

12c. _____

12d. _____

12e. _____

12f. _____

12g. _____

12h. _____

12i. _____

12j. _____

12k. _____

12l. _____

12m. _____

12n. _____

12o. _____

12p. _____

13. _____

14. _____

15. _____

16. _____

17. _____

18. _____

19. _____

20. _____

21. _____

22. _____

23. _____

24. _____

25. _____

26. _____

27a. _____

27b. _____

27c. _____

LABORATORY

Tissues

INTRODUCTION

The study of tissues, called **histology,** is microscopic anatomy. Individual tissues consist of cells and extracellular material that have a particular function. Organs of the body are formed by two or more tissues. The study of histology is important because many organic dysfunctions of the human body are diagnosed at the tissue level. Surgical specimens are routinely sent to pathology labs so that accurate assessment of the health of the tissue, and consequently the health of the individual, can be made.

Stem cells are unspecialized, or **undifferentiated, cells** that have great potential for regeneration. Stem cells may divide to produce more stem cells, or they may **differentiate** to become other cells of the body, such as those in neural tissue or blood. Developing embryos contain many stem cells **(human embryonic stem cells),** which differentiate into embryonic tissues. Adult stem cells repair old or damaged tissue or regenerate new cells, such as blood cells. Therapeutic uses of stem cells may hold cures for diseases (such as Parkinson's disease), neural injury, or tissue repair in burn patients. Stem cells are currently used to regenerate white blood cells in patients who have been treated for leukemia. As you study the tissues in this exercise remember that they all arose from stem cells.

There are four main tissue types found in the human body, **epithelial tissue, connective tissue, muscular tissue,** and **nervous tissue.** These topics are discussed in chapter 5, "Histology," in the Saladin text. In this exercise you examine numerous slides of organ tissue and begin an introduction to histology. In later exercises you revisit histology as you examine various organ systems.

OBJECTIVES

At the end of this exercise you should be able to

1. recognize the various types of epithelium;
2. associate a particular tissue type with an organ, such as kidney or bone;
3. examine a slide under the microscope or a picture of a tissue and name the cell type or specific tissue represented;
4. distinguish between cartilage and other connective tissues;
5. list the three parts of a neuron;
6. describe the muscle cell types according to location and structure.

MATERIALS

Microscope

Colored pencils

Epithelial Tissue Slides

Simple squamous epithelium

Simple cuboidal epithelium

Simple columnar epithelium

Pseudostratified ciliated columnar epithelium

Stratified squamous epithelium

Transitional epithelium

Muscular Tissue Slides

Skeletal muscle

Cardiac muscle

Smooth muscle

All three muscle types

Nervous Tissue Slides

Spinal cord smear

Connective Tissue Slides

Dense regular connective tissue

Dense irregular connective tissue

Elastic connective tissue

Reticular connective tissue

Areolar connective tissue

Adipose tissue

Ground bone

Hyaline cartilage

Fibrocartilage

Elastic cartilage

Blood

PROCEDURE

Before you begin this exercise you should be thoroughly familiar with the microscopes in your lab. If you are not familiar with the use of the microscope, review Laboratory Exercise 3. Tissues consist of cells and extracellular material known as matrix. Tissues are usually stained so that the details become visible. The most common stain is hematoxylin and eosin (H&E) stain. Hematoxylin stains the nucleus purple, and eosin colors the cytoplasm pink.

Epithelial Tissue

Epithelial tissue is a highly cellular tissue, meaning that it is composed mostly of cells, with little matrix. It covers or lines parts of the body (such as the skin on the outside or the digestive tract on the inside) or is found in glandular tissue, such as the sweat glands or the pancreas. In most cases, epithelial tissue adheres to the underlying layers by way of a **basement membrane,** a noncellular adhesive layer. In the skin, the epidermis is made of epithelial tissue and the basement membrane connects the epidermis to the underlying dermis. Epithelial tissue is classified according to the shape of the cells and the number of layers present. The cell shapes are **squamous** (flattened), **cuboidal,** and **columnar.** The number of layers are **simple** (cells in a single layer) or **stratified** (cells stacked in more than one layer). Examine tables 6.1 and 6.2 for an overview of epithelium. Epithelial tissue is listed by cell type in the following discussion. The edge of the cell touching the basement membrane is the basal surface, and the upper edge is the apical surface.

TABLE 6.1	Simple Epithelia
Simple Squamous Epithelium	
Microscopic appearance: single layer of flat cells	
Significant locations: lungs, inside of heart and blood vessels	
Functions: diffusion and smooth lining, secretion of serous fluid	
Simple Cuboidal Epithelium	
Microscopic appearance: small cubes or wedge-shaped cells in single layer	
Significant locations: kidney tubules and liver	
Functions: absorption and secretion	
Simple Columnar Epithelium	
Microscopic appearance: tall cells in one layer with nuclei typically in basal part of cell	
Significant locations: from stomach to intestines	
Functions: absorption and secretion	
Pseudostratified Columnar Epithelium	
Microscopic appearance: looks stratified but all cells arise from basement membrane, often ciliated	
Significant locations: respiratory passages	
Functions: secretes mucus and traps dust particles, moving them away from lung	

TABLE 6.2	Stratified Epithelia
Stratified Squamous Epithelium	
Microscopic appearance: many layers, cells cuboidal but flatten toward surface	
Significant locations: epidermis, oral cavity, and vagina	
Functions: resists abrasion, prevents microbial infection, retards water loss in skin	
Transitional Epithelium	
Microscopic appearance: many layers with teardrop-shaped cells that do not flatten toward surface	
Significant locations: urinary bladder	
Functions: allows stretching of urinary bladder	

Simple Epithelia

Simple epithelium is only one cell layer thick. The cells are located on the basement membrane. Each epithelial cell type is further classified according to shape.

Simple Squamous (SKWAY-mus) ***Epithelium***   This cell type is composed of thin, flat cells that lie on the basement membrane like floor tiles. If these cells are seen from a side view, they look flat and their resemblance to floor tiles becomes apparent. Examine a prepared slide of simple squamous epithelium, usually seen as either a surface view or side view. Simple squamous epithelium occurs in the air sacs of lungs, lines blood vessels, and is called **endothelium.** It can be found as the surface layer of many membranes, where it is called **mesothelium.** It provides a smooth lining (as found on the inside of blood vessels),

STUDY HINTS

As you examine various tissues look for distinguishing features that will identify the tissue. It is a good idea to examine more than one slide of a particular tissue so that you see a range of samples of that tissue. Frequently, the material you see in the lab is from a slice of an organ and as such includes more than one tissue type. For example, a sample of cartilage taken from the trachea contains epithelial tissue, adipose tissue, and other connective tissues in addition to the cartilage you want to study. If you are looking for smooth muscle from the digestive tract, you may also find both epithelial tissue and connective tissue in the slide. Use the figures in this exercise to help you locate tissues on the prepared slides. Examine each slide by holding the slide up to the light and visually locating the sample. Then put the slide on the microscope and examine it on low power, scanning around the slide. Move to progressively higher powers after you have identified the tissue. If you cannot identify the tissue after some searching, then ask your lab partner or your instructor for help. When you look at epithelial tissue under the microscope, examine the edge of the sample because epithelial tissue is frequently found as a lining.

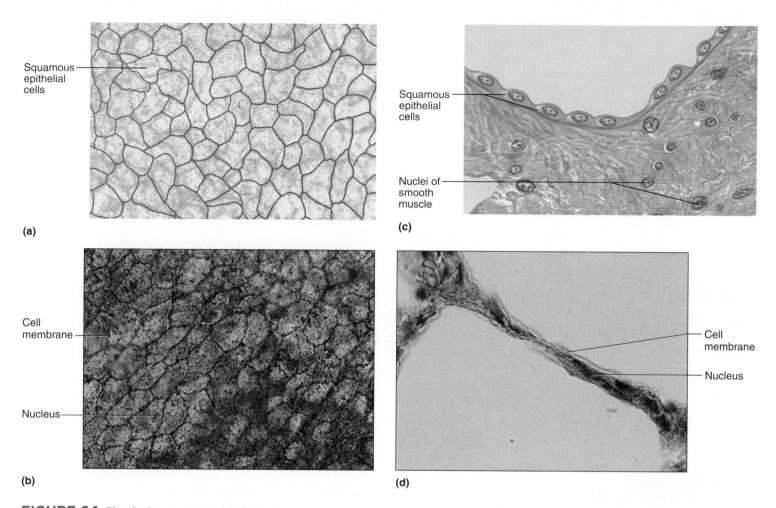

FIGURE 6.1 Simple Squamous Epithelium Top view of mesothelium, (a) diagram; (b) photograph (400x). Side view of human lung (c) diagram; (d) photograph (1,000x).

filters, or allows diffusion (as in the lungs). Compare your slide to figure 6.1. Draw what you see under the microscope in the space provided. Note whether your slide shows the cell as a surface or side view.

Illustration of simple squamous epithelium:

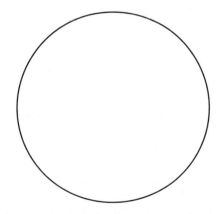

Simple Cuboidal Epithelium Simple cuboidal epithelium consists of cubelike or wedge-shaped cells mostly uniform in size. These cells form many of the major glands and glandular organs of the body and are the major

cell type of the kidney. Simple cuboidal epithelium is frequently found lining tubules, such as sweat ducts. Simple cuboidal epithelium often is involved in the secretion of fluids (sweat, oil) or in reabsorption (as in the kidneys). Examine a prepared slide of simple cuboidal epithelium and compare it to figure 6.2. In the space provided, draw what you see under the microscope and label the nucleus of the cell and the basement membrane.

Illustration of simple cuboidal epithelium:

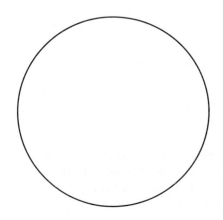

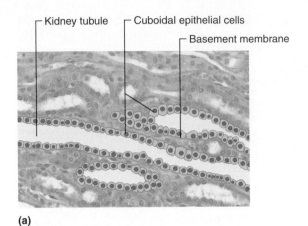

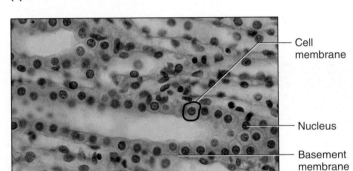

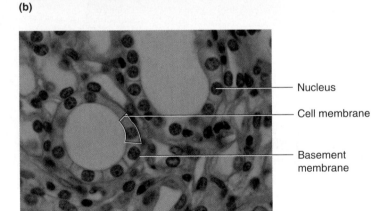

FIGURE 6.2 **Simple Cuboidal Epithelium—Kidney** Long section (a) diagram; (b) photograph (400×). Cross section (c) photograph (800×).

Simple Columnar Epithelium This cell type resembles tall columns anchored at one end to the basement membrane. The nuclei are frequently aligned in a row. Nonciliated simple columnar epithelium lines the inner portion of the digestive tract and provides an absorptive area for digested food. Mucus-secreting **goblet cells** are frequently found with it. Ciliated simple columnar epithelium lines the uterine tubes. Examine a prepared slide of simple columnar epithelium and compare it to figure 6.3. Draw a representation in the space provided.

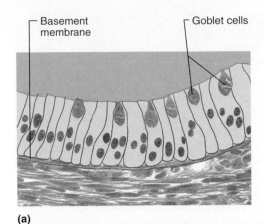

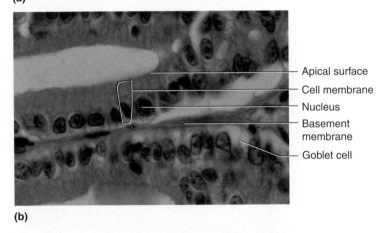

FIGURE 6.3 **Simple Columnar Epithelium—Stomach** (a) Diagram; (b) photograph (400×).

Illustration of simple columnar epithelium:

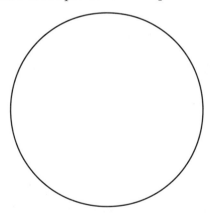

Pseudostratified Ciliated Columnar Epithelium These cells appear as if they occur in a few layers, but all of the cells rest on the basement membrane; thus, it is known as a simple epithelium. The nuclei are at different levels in the tissue. Pseudostratified ciliated columnar epithelium lines some portions of the respiratory passages, where it protects the lungs by trapping dust particles in a mucous sheet and by moving the particles away from them. Mucus is secreted by goblet cells. Examine a prepared slide of pseudostratified ciliated columnar epithelium and compare it to figure 6.4. Draw a representative sample in the space provided.

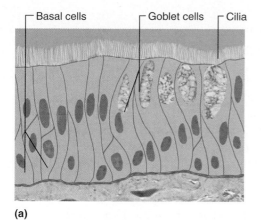

Basal cells   Goblet cells   Cilia

(a)

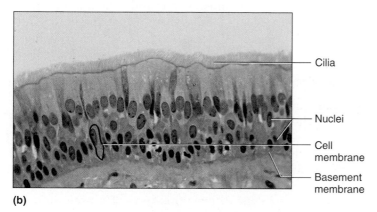

(b)

— Cilia

— Nuclei

— Cell membrane

— Basement membrane

FIGURE 6.4 **Pseudostratified Ciliated Columnar Epithelium—Trachea** (a) Diagram; (b) photograph (400×).

Illustration of pseudostratified ciliated columnar epithelium:

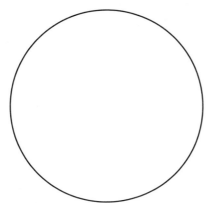

Stratified Epithelia

Stratified epithelium is so named because many epithelial cells occur in layers on a basement membrane. The term *stratified* comes from the word *strata* and refers to the layering of these cells. You will examine two common cell types belonging to this group.

Stratified Squamous Epithelium Stratified squamous epithelium is a cell type that covers the outermost layer of skin. It also lines the vaginal canal and mouth. The

multiple layers of this cell type protect the underlying tissue from mechanical abrasion. This epithelium may have cuboidal cells at the basement layer, but it derives its name from the cell shape at the free surface. Stratified squamous epithelium comes in two distinct types, keratinized and nonkeratinized (keratin is a tough protein that hardens cells in the outer layer of the skin). Examine a prepared slide of stratified squamous epithelium and locate the basement membrane. Compare your slide to figure 6.5 and draw what you see under the microscope in the space provided. Locate the basement membrane, and count the layers of cells that occur between it and the surface of this cell type.

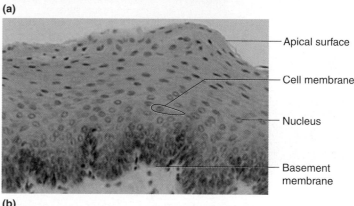

— Squamous epithelial cells

(a)

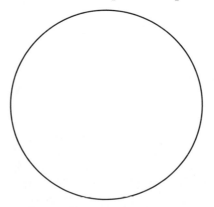

— Apical surface

— Cell membrane

— Nucleus

— Basement membrane

(b)

FIGURE 6.5 **Stratified Squamous Epithelium—Esophagus** (a) Diagram; (b) photograph (400×).

Illustration of stratified squamous epithelium:

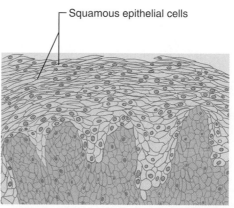

Record the numbers of layers of cells: _____

Transitional Epithelium Transitional epithelium is unusual in that it has some remarkable stretching capabilities. Transitional epithelium lines the ureter, urinary bladder, and proximal urethra and allows the bladder to expand as it fills with urine. Examine a prepared slide of transitional epithelium. Look at figure 6.6 and draw the cell type in the space provided.

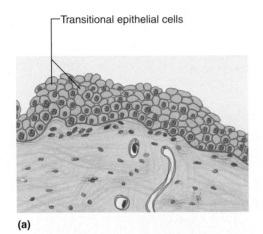

Transitional epithelial cells

(a)

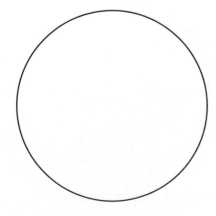

Apical surface

Cell membrane

Nucleus

Basement membrane

(b)

FIGURE 6.6 Transitional Epithelium—Bladder (a) Diagram; (b) photograph (400×).

Illustration of transitional epithelium:

Cell Connections

Cells are held together in many different ways. Epithelial cells are held to other structures by the basement membrane, as discussed previously. Cells are frequently held together at specific regions known as **desmosomes. Tight junctions** hold the cells even more closely together. **Gap junctions** connect one cell to another and allow ions and other solutes to pass from one cell to another. These are illustrated, along with other types of connections in figure 6.7.

Muscular Tissue

Like epithelial tissue, muscular tissue is a cellular tissue with the tissue having mostly cells and little matrix. These cells are contractile and shorten in length due to the sliding of protein filaments across one another. There are three types of muscle: skeletal, cardiac, and smooth muscle.

Skeletal Muscle

The cells of skeletal muscle are called **fibers.** When you refer to the muscles of your body you are referring to organs made mostly of skeletal muscle. Skeletal muscle is sometimes known as striated muscle because it has obvious **striations** (stry-A-shuns) (they look like stripes) in the fiber. Skeletal muscle is **voluntary** because you have conscious control over this type of muscle in your body. The individual muscle cells have many nuclei and are thus called **multinucleate** or **syncytial.** The nuclei are elongated and occur on the periphery of the cell. Examine a prepared slide of skeletal muscle under high power. Compare this slide to figure 6.8. In the space provided, draw a representation of the muscle. How many widths of muscle cells fit across the diameter of your microscope when viewed at high power?

Illustration of skeletal muscle:

How many muscle fiber widths do you count? _____

STUDY HINTS

By counting the cell widths in the slide you should be able to see yet another way that you can distinguish among skeletal muscle, cardiac muscle, and smooth muscle.

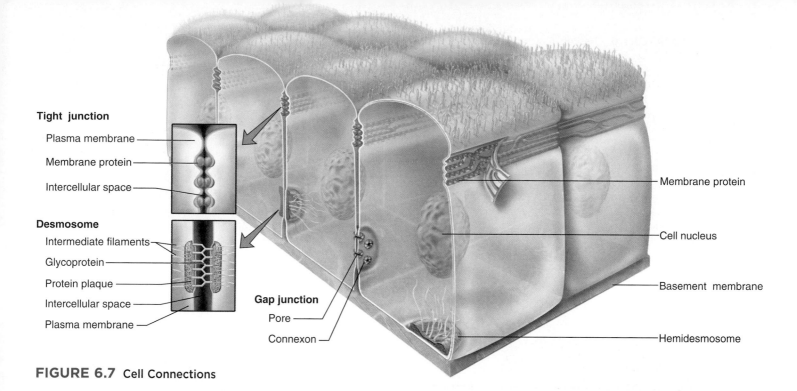

Tight junction
- Plasma membrane
- Membrane protein
- Intercellular space

Desmosome
- Intermediate filaments
- Glycoprotein
- Protein plaque
- Intercellular space
- Plasma membrane

Gap junction
- Pore
- Connexon

- Membrane protein
- Cell nucleus
- Basement membrane
- Hemidesmosome

FIGURE 6.7 Cell Connections

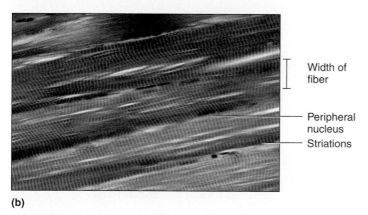

- Nuclei
- Striations
- Muscle fiber
- Connective tissue

(a)

- Width of fiber
- Peripheral nucleus
- Striations

(b)

FIGURE 6.8 Skeletal Muscle—Longitudinal Section
(a) Diagram; (b) photograph (400×).

Cardiac Muscle

Cardiac muscle is found only in the heart and is the main tissue making up that organ. Cardiac muscle is somewhat similar to skeletal muscle in that it is striated, but the striations are much less obvious and the individual cell diameter is less than that of skeletal muscle cells. Cardiac muscle cells are called **myocytes.** Another difference between cardiac muscle and skeletal muscle is that cardiac muscle is branched. Place a prepared slide of cardiac muscle under the microscope, and count the number of widths of cardiac muscle fibers across the diameter of the microscope under high power.

Record the number of cell widths that occupy the diameter of your field of view at high power: _____

Cardiac muscle contracts on its own and is therefore called **involuntary muscle.** The cells are mostly uninucleate with the nucleus centrally located and oval in shape. Cardiac muscle cells are joined together by **intercalated discs,** which facilitate the transmission of the electrical impulses in the heart. These intercalated discs have gap junctions. When one muscle cell receives an impulse it sends it on to the next cell. Look at a prepared slide of cardiac muscle and compare it to figure 6.9, noting the striations, intercalated discs, and nuclei of the cells. In the space provided, draw the cells.

Illustration of cardiac muscle:

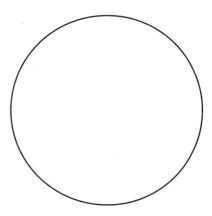

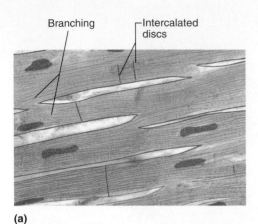

(a)

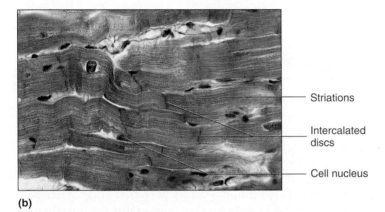

(b)

FIGURE 6.9 Cardiac Muscle—Longitudinal Section
(a) Diagram; (b) photograph (1,000×).

Smooth Muscle

Smooth muscle is **nonstriated** in that the fibers do not have crosshatchings perpendicular to the length of the fiber. Smooth muscle is **involuntary,** like cardiac muscle, and is found in the intestine, where it propels food along by a process known as peristalsis and segmentation. The uterine contractions during labor are smooth muscle contractions. Smooth muscle is also found in other areas involved in contraction. The cells of smooth muscle are spindle-shaped and uninucleate, and the nucleus is centrally located and elongated in shape when the muscle is relaxed. When the muscle is contracted the nucleus appears corkscrew-shaped. Examine a prepared slide of smooth muscle and note the narrow diameter of the fibers. How many fiber widths do you count across the diameter of the field of view of your microscope under high power? Compare your slide to figure 6.10, and in the space provided draw an illustration of smooth muscle.

Illustration of smooth muscle:

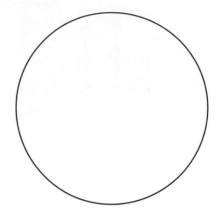

Number of cell widths under high power: _____

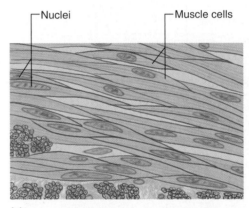

(a)

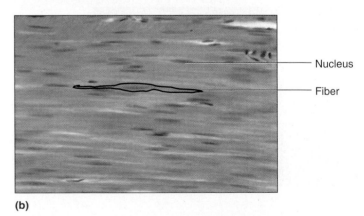

(b)

FIGURE 6.10 Smooth Muscle—Small Intestine, Longitudinal **Section** (a) Diagram; (b) photograph (400×).

Examine a prepared slide of all three muscle types, if available. Work with your lab partner and identify each muscle cell. Fill in the chart comparing the various muscle types and their characteristics. Review muscles in table 6.3.

Muscle Characteristics

Muscle Type	Number of Nuclei	Location Found	Voluntary/Involuntary
	Many		
	One		
	Mostly One		

TABLE 6.3	Muscular Tissue

Skeletal Muscle

Microscopic appearance: large cell with many nuclei and obvious striations

Significant locations: skeletal muscles (biceps brachii, rectus abdominis, etc.)

Functions: voluntary contractions

Cardiac Muscle

Microscopic appearance: smaller branched cell with one nucleus, intercalated discs, and less obvious striations

Significant locations: heart

Functions: rhythmic contraction of heart

Smooth Muscle

Microscopic appearance: small, slender cell with one central nucleus and no striations

Significant locations: digestive tract and uterus

Functions: sustained contractions, propulsion of food or delivery of infant

Nervous Tissue

Nervous tissue, as with epithelial and muscular tissues, is a cellular tissue. Nervous tissue is found in the brain, the spinal cord, the ganglia, and the peripheral nerves of the body. The conductive cell of this tissue is the **neuron,** which receives and transmits electrochemical impulses. A neuron is a specialized cell with three major regions, the **dendrites,** the **nerve cell body** (or **soma**), and the **axon.** Examine figure 6.11 for the regions of a typical neuron; also see table 6.4.

Examine the neurons of a smear of the spinal cord. Look for purple star-shaped structures, the nerve cell bodies. Compare them to figure 6.11. Draw what you see in the space provided.

Illustration of nervous tissue:

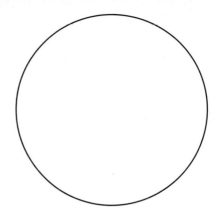

Special cells of the nervous system are called **glial cells** or **neuroglia.** These cells guide developing neurons to synapses, remove some neurotransmitters from the synapse, and perform numerous other functions. You will study glial cells in greater detail in Laboratory Exercise 20.

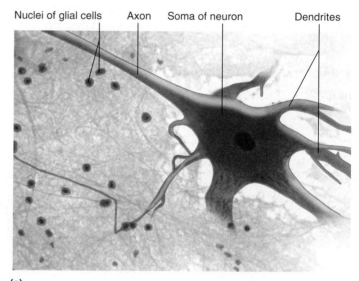

(a)

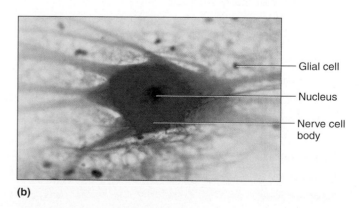

(b)

FIGURE 6.11 Typical Neuron (a) Diagram; (b) photograph (400×).

TABLE 6.4	Nervous Tissue

Microsopic appearance: large, star-shaped cells (neurons in brain and spinal cord) with smaller cells (glial cells) nearby

Significant locations: brain and spinal cord (nerves and ganglia)

Functions: transmission of information, assimilation

Connective Tissue

The tissues that you have seen so far, epithelial, muscular, and nervous, are composed mostly of cells with very little matrix. This is not the case with connective tissue. There is usually more matrix than cells in connective tissue. **Matrix** is extracellular material, and it usually consists of fibers and fluid, gel, or solid ground substance. Because of the abundance of matrix, connective tissue is classified by *specific tissue.*

Connective tissues have diverse appearances and functions, and specific tissues do not seem to have much in common with one another; however, all arise from an embryonic tissue called **mesenchyme.** Connective tissue can be divided into several subgroups for easier identification. These subgroups are **fibrous connective tissue, cartilage, bone,** and **blood.** They are described in more detail next and are outlined in table 6.5.

Fibrous Connective Tissue

Dense Connective Tissue   Dense connective tissue is composed of collagen fibers that either is parallel, known as **dense regular connective tissue,** or runs in many directions, known as **dense irregular connective tissue.** Dense regular connective tissue is found in tendons and ligaments. Dense irregular connective tissue is found in

TABLE 6.5	Connective Tissue

Fibrous connective tissue
  Dense connective tissue
    Dense regular connective tissue
    Dense irregular connective tissue
    Elastic connective tissue
  Loose connective tissue
    Reticular connective tissue
    Areolar connective tissue
    Adipose tissue
Cartilage
  Hyaline cartilage
  Fibrocartilage
  Elastic cartilage
Bone
Blood

the deep layers of the skin and the white of the eye. The fibers in these tissues are made of a protein called **collagen** and are called **collagenous fibers. Fibroblasts** are found in the tissue in addition to the collagenous fibers. Fibroblasts secrete fibers and then mature into fibrocytes. Examine a slide of dense connective tissue and compare it to figure 6.12. In the space provided, draw a section of dense regular or dense irregular connective tissue.

Illustration of dense connective tissue:

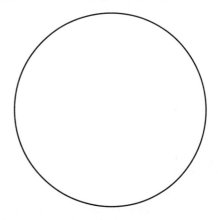

Elastic Connective Tissue   This tissue has a cellular component of fibroblasts. The fibers in this tissue are made of collagenous and **elastic fibers.** Elastic fibers are

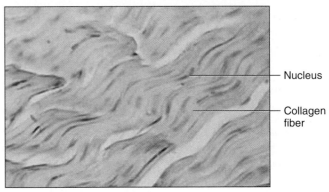

(a)

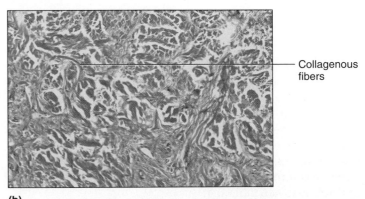

(b)

FIGURE 6.12 Dense Connective Tissue (a) Dense regular connective tissue—tendon (400×); (b) dense irregular connective tissue—skin (100×).

made of the protein **elastin.** Elastic tissue occurs in the walls of arteries and can be identified as dark, squiggly lines. Examine figure 6.13 as you look at a prepared slide of elastic tissue. Then, in the space provided, draw the specific tissue. Review these tissues in table 6.6.

Illustration of elastic connective tissue:

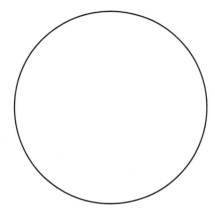

TABLE 6.6	Dense Connective Tissue

Dense Regular Connective Tissue

Microscopic appearance: closely packed, wavy collagen fibers (or elastic fibers, in the case of elastic connective tissue)

Significant locations: tendons and ligaments (vocal cords in elastic connective tissue)

Functions: binds bones together or muscle to bone (provides elastic structure)

Dense Irregular Connective Tissue

Microscopic appearance: randomly appearing collection of densely clustered collagen fibers

Significant locations: dermis and sheaths around cartilage and bone

Functions: provides strength and resists stress and strain against tearing

Reticular Connective Tissue Reticular connective tissue has fibroblasts with **reticular fibers.** Reticular connective tissue is found in soft internal organs, such as the liver, spleen, and lymph nodes. This tissue provides an internal framework for these organs. If your slide is stained with a silver stain, the reticular fibers appear black. If your slide is stained with Masson stain, they appear blue. In either case, look for fibers that appear branched among small, round cells. The other cells in the prepared slide are the cells of the organ that the reticular connective tissue is holding together. Examine a prepared slide of reticular tissue and compare it to figure 6.14. Compare what you see with table 6.7. In the space provided, draw a sample of what you see.

Illustration of reticular connective tissue:

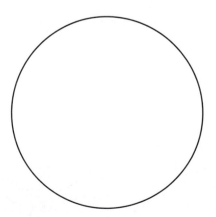

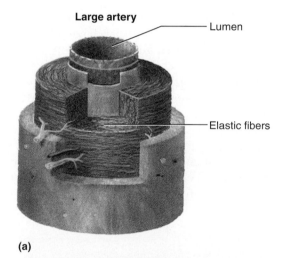

Large artery
— Lumen
— Elastic fibers

(a)

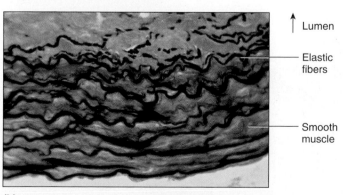

↑ Lumen
— Elastic fibers
— Smooth muscle

(b)

FIGURE 6.13 **Elastic Connective Tissue** (a) Overview of artery; (b) photomicrograph (400×).

Areolar Connective Tissue Areolar connective tissue is a complex collection of fibers and cells that has a distinctive look. Areolar connective tissue is found as a wrapping around organs, as sheets of tissue between muscles, and in many areas of the body where two different tissues meet (such as the boundary layer between fat and muscle).

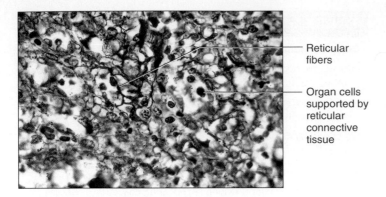

Reticular fibers

Organ cells supported by reticular connective tissue

FIGURE 6.14 Reticular Connective Tissue (400×)

TABLE 6.7	Loose Connective Tissue

Reticular Connective Tissue

Microscopic appearance: reticular fibers forming meshwork around organ cells

Significant locations: spleen, thymus, and lymph nodes

Functions: internal skeleton (framework) for soft organs

Areolar Connective Tissue

Microscopic appearance: scattered arrangement of collagenous fibers with elastic and reticular fibers along with many cells (many with protective immune functions)

Significant locations: attaches epithelia to lower layers and around many internal organs

Functions: binds epithelia to lower layers, insulates organs from infections

Adipose Tissue

Microscopic apperance: large, pale, open cells with nuclei near periphery of cell

Significant locations: under skin, breast tissue, and outside of heart and kidney

Functions: energy storage, physical protection

Areolar connective tissue consists of large, pink-stained **collagenous fibers;** smaller, dark **elastic** and **reticular fibers;** and a collection of cells that includes **fibroblasts, mast cells,** and **macrophages.** Examine a prepared slide of areolar connective tissue and compare it to figure 6.15. Make a drawing of it in the space provided.

Illustration of areolar connective tissue:

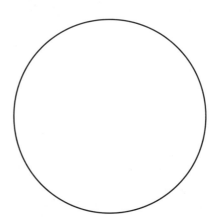

Adipose Tissue   Adipose tissue (fat) is unusual as a connective tissue in that it is highly cellular. The cells that make up adipose tissue are called **adipocytes.** Adipocytes store lipids in large vacuoles, while the nucleus and cytoplasm remain on the outer part of the cell near the cell membrane. In prepared slides of adipose tissue the fat typically has been dissolved in the preparation process. Locate the large, empty cells, with the nuclei on the edge of some of the cells. Compare what you see in the microscope to figure 6.16 and make a drawing in the space provided. You may need to close down the iris diaphragm or reduce the amount of light shining on the specimen to best see this tissue.

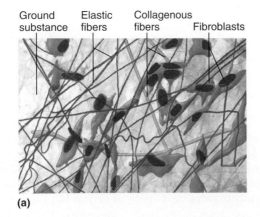

Ground substance   Elastic fibers   Collagenous fibers   Fibroblasts

(a)

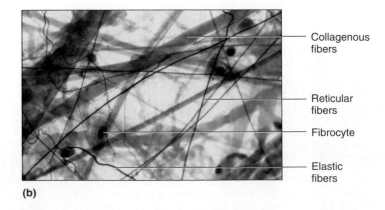

Collagenous fibers

Reticular fibers

Fibrocyte

Elastic fibers

(b)

FIGURE 6.15 Areolar Connective Tissue (a) Diagram; (b) photograph (400×).

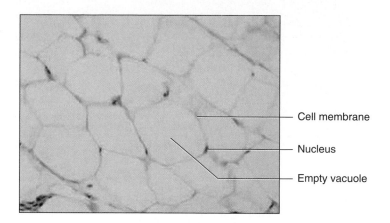

FIGURE 6.16 Adipose Tissue (400×)

Illustration of adipose tissue:

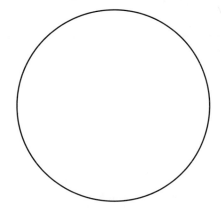

Cartilage (Cartilage and Perichondrium)

Cartilage is a type of connective tissue in which the matrix is composed of a pliable material that allows for some degree of movement. The matrix of cartilage has abundant amounts of chondroitin sulfate and forms a semisolid gel in which fibers and cells are found. In prepared slides, frequently the cells have come out of the matrix and left a cavity. This cavity is known as a **lacuna** and is reasonably diagnostic of cartilage tissue. The **perichondrium** is dense irregular connective tissue that is found as a thin layer on the surface of cartilage. There are three types of cartilage in the human body, and they vary by the number and type of protein fibers. These three types are hyaline cartilage, fibrocartilage, and elastic cartilage. Compare their features in table 6.8.

Hyaline Cartilage   The most common cartilage in the body is hyaline cartilage, found at the apex of the nose, at the ends of many long bones, between the ribs and the sternum, in the respiratory passages, and in other locations as well. Hyaline cartilage is clear and glassy in fresh tissue but will probably appear light pink in prepared slides. The cells that occur in hyaline cartilage are called **chondrocytes,** and the fibers are **collagenous fibers.** Find the chondrocytes in a prepared slide of hyaline cartilage. The fibers will not appear distinct because they blend

TABLE 6.8	Cartilage

Hyaline Cartilage

Microscopic appearance: usually light blue- or pink-stained matrix, frequently with pairs of cells appearing like "eyes"

Significant locations: ends of long bones, ribs, larynx, and trachea

Functions: reduces friction at joints, keeps air passages open

Fibrocartilage

Microscopic appearance: numerous collagen fibers; cells frequently in rows of four or five

Significant locations: symphysis pubis and intervertebral discs

Functions: protects from wear and tear at weight-bearing or stressed joints

Elastic Cartilage

Microscopic appearance: netlike pattern of fibers around chondrocytes

Significant locations: external ear and epiglottis

Functions: provides flexible framework

into the color of the matrix. The chondrocytes of hyaline cartilage frequently occur in pairs. Some people say that they look like pairs of eyes staring back at you as you look at them in the microscope. Look around the edge of the sample of hyaline cartilage and locate the perichondrium. Compare the material under the microscope to figure 6.17. In the space provided, draw what you see.

Illustration of hyaline cartilage:

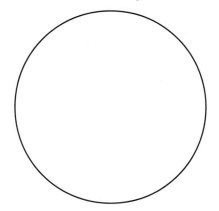

Fibrocartilage   Fibrocartilage is similar to hyaline cartilage in that it has chondrocytes and collagenous fibers, but it differs from hyaline cartilage in that it has many more collagenous fibers. These can be seen in the prepared slides. Fibrocartilage is found in areas where more stress is placed on the cartilage, such as in the intervertebral discs, the symphysis pubis, and the menisci of each knee. One characteristic of fibrocartilage is that the chondrocytes are frequently found in rows of four or five in the tissue. Examine a prepared slide of fibrocartilage, and locate the collagenous fibers and chondrocytes. Compare the slide to figure 6.18 and make a drawing of fibrocartilage in the space provided.

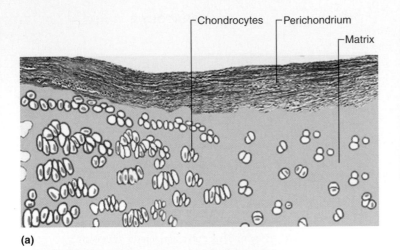

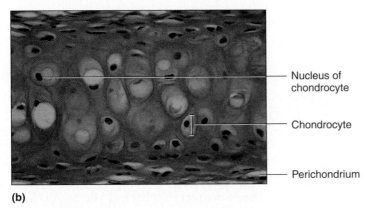

FIGURE 6.17 **Hyaline Cartilage—Trachea** (a) Overview; (b) photomicrograph (400×).

Illustration of fibrocartilage:

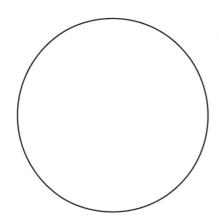

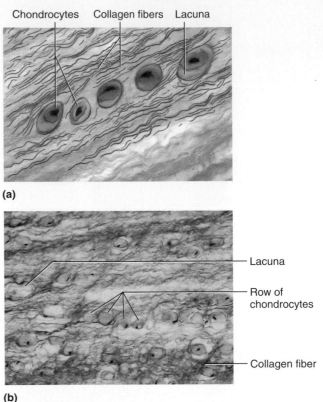

FIGURE 6.18 **Fibrocartilage** (a) Diagram; (b) photograph (400×).

Elastic Cartilage Elastic cartilage is unique in that it contains **elastic fibers,** which give the tissue its flexible nature. Elastic cartilage is found in the external ear and in the part of the larynx called the epiglottis. If the prepared slide is stained with a silver stain, the elastic fibers appear black. If the slide is not stained in this way, the fibers may be hard to distinguish. Examine a prepared slide of elastic tissue and look for the chondrocytes and elastic fibers. Compare your slide to figure 6.19 and make a drawing of elastic cartilage in the space provided.

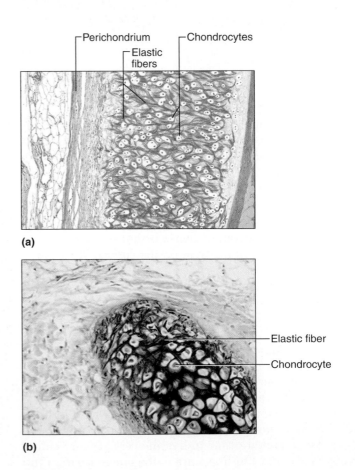

FIGURE 6.19 **Elastic Cartilage** (a) Diagram; (b) photograph (400×).

Illustration of elastic cartilage:

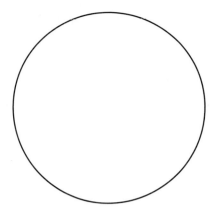

Illustration of ground bone:

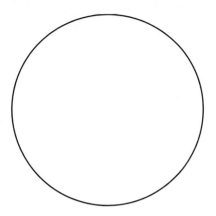

Bone

Bone is a type of connective tissue in which the matrix is made of a mineral component called **hydroxyapatite crystal,** or **calcium salt. Collagenous fibers** occur in the tissue along with osteocytes or bone cells. Look at a prepared slide of ground bone and compare it to figure 6.20. Locate the large, circular regions called **osteons** (which resemble a cross section of a tree trunk with the growth rings), the central canal, and the small spaces where osteocytes occur in the tissue. **Osteons** are long, thin units of bone with space in the middle of them known as the **central canal.** A more detailed examination of bone appears in Laboratory Exercise 8. Review the features of bone in table 6.9. After you study the slide, record your observations in the space provided.

Blood

Blood is a connective tissue in which the matrix is fluid and no fibers are present. The matrix of the blood is called **plasma,** and the cells of blood are either **erythrocytes** (eh-RITH-ro-sites) (red blood cells) or **leukocytes** (white blood cells). Small cellular fragments known as **platelets** are also present. Examine a prepared slide of blood and look for the numerous pink discs in the sample. These are the erythrocytes. You may find larger cells with lobed, purple nuclei. These are the leukocytes. Examine table 6.10. The details of blood are studied in Laboratory Exercise 29. Examine the blood slide under high power and compare it to figure 6.21. Make a drawing of blood in the space provided.

Illustration of blood:

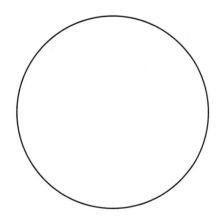

Review the types of connective tissue in table 6.5.

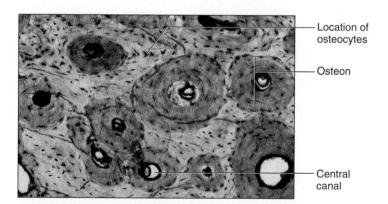

FIGURE 6.20 Ground Bone (100×)

Location of osteocytes

Osteon

Central canal

TABLE 6.9	Bone
Microscopic appearance: cells enclosed in calcium salts in concentric circles	
Significant locations: skeleton	
Functions: protection of soft organs, locomotion along with muscles, scaffold for body	

TABLE 6.10	Blood
Microscopic appearance: many cells, most with no nucleus, suspended in plasma	
Significant locations: heart and blood vessels	
Functions: transports gases, nutrients, hormones, water, and other material throughout the body	

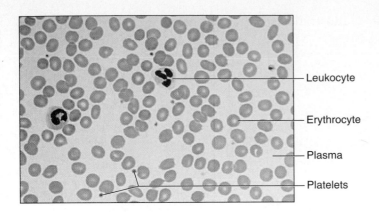

Leukocyte

Erythrocyte

Plasma

Platelets

FIGURE 6.21 Blood (400×)

STUDY HINTS

Examine as many different images of tissues as you can. Drawing the material observed in the lab is a good memory tool. Try to make associations between the abstract image of tissue and something common. For example, you could think of bone as looking like a series of tree rings. Adipose tissue may look like angular balloons. If you have difficulty distinguishing between two specimens, try to find a characteristic that separates the two. For example, smooth muscle and dense connective tissue may look similar in certain stained preparations. You will note that smooth muscle has nuclei inside the fibers, while dense connective tissue has nuclei of the fibrocytes between adjacent fibers.

After you have studied the slides in lab you can review the material by finding histology images online and looking for similarities and differences in the material that you have studied.

REVIEW SECTION

Tissues

Name _____ *Date* _____

Lab Section _____ *Time* _____

Review Questions

1. List the four main tissues of the body.

2. What is the name of the noncellular layer that attaches epithelial tissue to other layers?

3. How many layers of cells are in simple epithelium?

4. Name the three general cell shapes of epithelial tissue.

5. A single, flattened layer of cells represents what type of epithelium?

6. What functional problem could occur if the digestive tract were lined with stratified squamous epithelium instead of simple columnar epithelium?

7. A multiple layer of flattened epithelial cells represents what cell type?

8. What cell type lines the inside of the urinary bladder?

9. Can you make the hairs on your arm "stand on end?" Based on your results, what type of muscle is responsible for doing this?

10. What muscle cell types have the following features?

    a. striated and voluntary _____

    b. striated and involuntary _____

    c. nonstriated and involuntary _____

11. What muscle cell type has intercalated discs?

12. The intestine is lined with what kind of muscle?

13. The heart is composed primarily of what type of muscle?

14. The muscles of your arm are primarily composed of what cell type?

15. What cell type is responsible for the transmission of electrochemical impulses?

16. What is the matrix in blood called?

17. Name the three types of fibers in areolar connective tissue.

18. What type of connective tissue is springlike and found in the middle walls of arteries?

19. What is the cell type in adipose tissue?

20. Name the outer connective tissue layer that envelops cartilage.

21. What types of fibers are in fibrocartilage?

22. Describe the functional difference between fibrocartilage and hyaline cartilage.

23. What is another name for calcium salts in bone?

24. How do you distinguish generally between epithelium and connective tissue?

25. Label the following photomicrographs by tissue type and by using the terms provided.
    a. collagenous fibers, elastic fibers, reticular fibers

a. _____

b. _____

c. _____

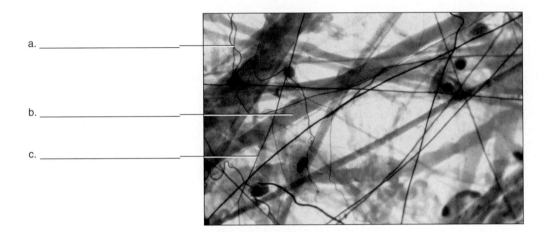

tissue name (400×) _____

b. cilia, basement membrane, nucleus

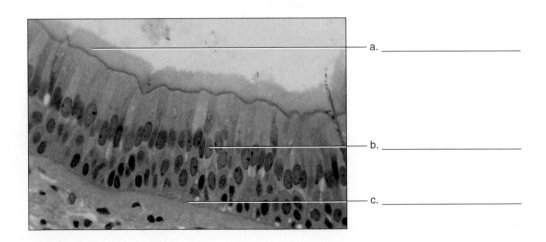

a. _____

b. _____

c. _____

tissue name (400×) _____

c. intercalated discs, striations, nucleus

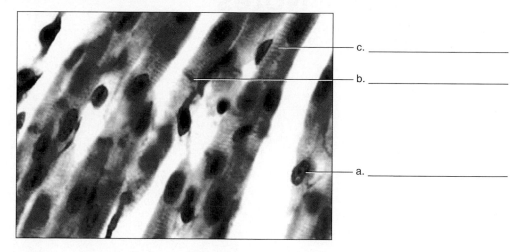

c. _____

b. _____

a. _____

tissue name (1,000×) _____

d. nerve cell body, glial cells

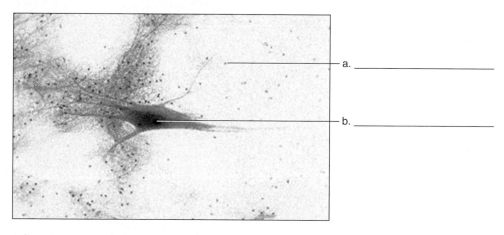

a. _____

b. _____

tissue name (100×) _____

Notes

LABORATORY

Integumentary System

INTRODUCTION

The integumentary system consists of the skin and associated structures, such as hair, nails, and various glands. The integument consists primarily of a cutaneous membrane composed of a superficial epidermis and a deeper dermis. Below these two layers is an underlying layer, called the hypodermis, which anchors the integument to deeper structures. The skin is the largest organ of the body and is vital for temperature regulation and for protection against water loss and invasion from microorganisms. These topics are discussed in chapter 6, "The Integumentary System," in the Saladin text.

OBJECTIVES

At the end of this exercise you should be able to

1. list the two layers of the integument;
2. list all the layers of the epidermis from superficial to deep;
3. describe the structure and function of sudoriferous glands and sebaceous glands;
4. draw a hair in a follicle in longitudinal and cross section;
5. discuss the protective nature of melanin;
6. describe the hypodermis and its relationship to the integument.

MATERIALS

Models and charts of the integumentary system

Prepared Microscope Slides

Thick skin (with Pacinian corpuscles)

Hair follicles

Pigmented and nonpigmented skin

Microscopes

PROCEDURE

Overview of the Integument

Examine models or charts of the **integumentary system** in the lab and locate the two major regions of the integument, the **epidermis** and the **dermis.** The dermis consists primarily of collagenous fibers. The epidermis is the most superficial layer and can be determined by the epithelial cells of that layer. You can look for cells that have obvious nuclei. Also locate the **hypodermis,** which is not a layer of the integument but anchors the integument to underlying bone or muscle. The hypodermis can be distinguished by the abundant adipose tissue. Compare the models in the lab to figure 7.1.

Locate the **hair follicles** in the dermis, **sebaceous (oil) glands,** and **sudoriferous (sweat) glands.** The sudoriferous glands are connected to the surface by **sweat ducts.**

Microscopic Examination of the Integument

Examine a prepared slide of thick skin under low power. Locate the hypodermis, dermis, and epidermis. The slide will probably show three layers—one relatively clear due to a significant amount of adipose tissue, a pink layer, and a multicolored layer. The clear layer is the hypodermis, the pink layer is the dermis, and the multicolored layer is the epidermis. The epidermis may be determined by the epithelial cells of that layer. Look for cells with obvious nuclei. The colors may vary in the slides used in your lab. Compare these layers in the prepared slide to figure 7.2.

Draw and label the overview of the integument in the following space, locating the epidermis, dermis, and hypodermis.

Illustration of integument:

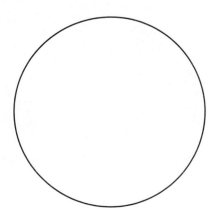

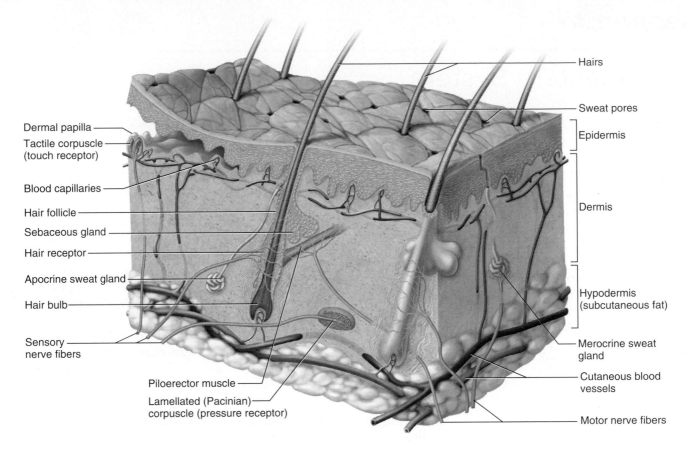

FIGURE 7.1 Diagram of the Integument

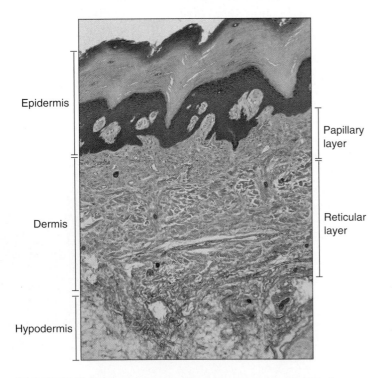

FIGURE 7.2 Photomicrograph of the Integument (40×)

Epidermis

The epidermis is composed of keratinized **stratified squamous epithelium** hardened with the protein **keratin.** Keratin is a tough material that protects the lower layers of the epidermis from wear and tear. The epidermis has a number of layers known as **strata** (singular, *stratum*). The epidermis does not have blood vessels between the cells but is nourished by the vessels in the dermis.

Strata of the Epidermis

In thick skin, the deepest layer of the epidermis is known as the **stratum basale.** This single layer of cells is attached to the dermis by the **basement membrane,** a noncellular layer. Cells in the stratum basale divide repeatedly and push up toward the superficial layers as the cells increase in age. This process is illustrated in figure 7.3.

In the stratum basale are specialized cells known as **melanocytes.** These cells produce a brown pigment called **melanin.** Melanin protects the stratum basale from the damaging effects of ultraviolet radiation. The number of melanocytes is relatively constant among people of all skin colors, but the amount of melanin produced by the melanocytes is variable. People of darker pigmentation produce more melanin, while people of lighter pigmentation have melanocytes that produce less melanin. People of lighter

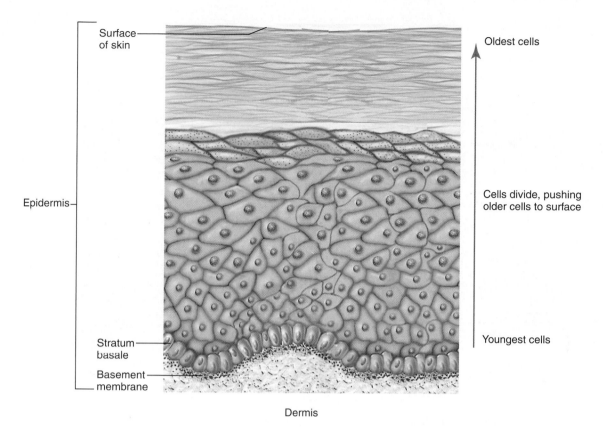

Surface of skin

Oldest cells

Epidermis

Cells divide, pushing older cells to surface

Youngest cells

Stratum basale

Basement membrane

Dermis

FIGURE 7.3 Proliferation of the Epidermis

pigmentation are more susceptible to the effects of ultraviolet radiation and skin cancer. Locate the melanocytes and the stratum basale and draw them in the following space.

Illustration of lower epidermis:

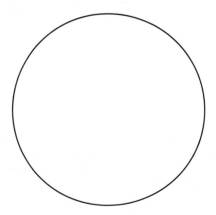

The layer of numerous cells superficial to the stratum basale is the **stratum spinosum.** In prepared slides the cells appear star-shaped because the cell membrane pulls away from the other cells, except in areas where the cells are joined at microscopic junctions called **desmosomes.** The stratum basale and stratum spinosum are frequently classified as the stratum germinativum.

A layer of granular cells is present in the next layer. This is known as the **stratum granulosum.** The cells have purple-staining keratohyalin granules in their cytoplasm. The granules are precursors of keratin found in the outermost layer of the epidermis.

Superficial to this layer is the **stratum lucidum.** It is found only in the palms of the hand and soles of the feet. It is so named (*lucid* = light) because this layer appears translucent in fresh specimens. This layer generally appears clear or pink in prepared slides stained with hematoxylin and eosin (H&E) stain. Draw a representation of the stratum spinosum, stratum granulosum, and stratum lucidum in the following space.

Illustration of midlayers of epidermis:

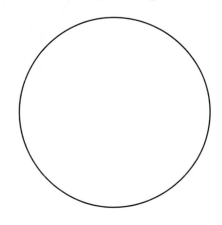

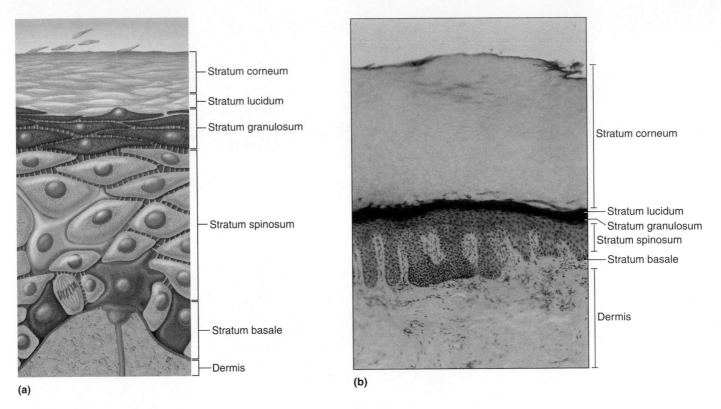

Stratum corneum
Stratum lucidum
Stratum granulosum
Stratum spinosum
Stratum basale
Dermis

(a)

Stratum corneum
Stratum lucidum
Stratum granulosum
Stratum spinosum
Stratum basale
Dermis

(b)

FIGURE 7.4 **Strata of the Epidermis** (a) Diagram; (b) photomicrograph (100×).

The most superficial layer of the integument is the **stratum corneum,** a tough layer consisting of dead cells flattened at the surface. Cells of the stratum corneum have been toughened by complete keratinization; thus, the epidermis is said to be made of **keratinized stratified squamous epithelium.** Examine a prepared slide of thick skin and locate the various strata of the epidermis, as illustrated in figure 7.4. Draw an example of the stratum corneum in the following space.

Illustration of stratum corneum:

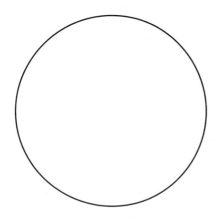

Cells in the epidermis go through three major phases. In the deepest layer, the stratum basale, the cells are involved in rapid cell division. More superficial to this layer, cells undergo a process of producing precursor molecules that leads to the general waterproofing of the cells. The final phase is the completion of the waterproofing process. Cells eventually die, providing a tough barrier. The most superficial cells of the epidermis are tough. They protect the skin from microorganisms and desiccation and eventually slough off. The epidermis renews itself about every 6 weeks.

Dermis

The dermis is responsible for the structural integrity of the integumentary system. It is composed primarily of closely apposed fibers along with blood vessels, nerves, sensory receptors, hair follicles, and glands. The dermis consists of two major regions, the superficial **papillary layer** and a deeper **reticular layer.** The papillary layer is so named for the **papillae** (bumps) that interdigitate with the epidermis, providing good adhesion between the two layers. The reticular layer is the largest layer of the dermis. Examine a slide of thick skin and locate the two layers of the dermis, as seen in figure 7.5.

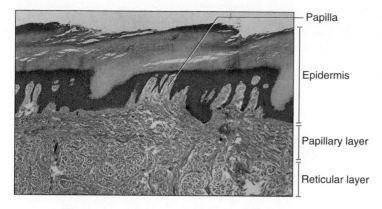

FIGURE 7.5 Layers of the Dermis (40×)

The majority of the fibers of the dermis are irregularly arranged **collagenous** along with lesser numbers of **elastic** and **reticular fibers.** The collagenous fibers provide strength and flexibility. The elastic fibers, as their name implies, provide elasticity to the skin. Draw a representation of the dermis in the following space.

Illustration of dermis:

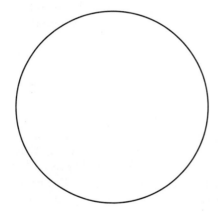

Blood vessels pass into the dermis, bringing nutrients to both the dermis and epidermis. The blood vessels do not enter into the epidermis, but the nutrients and oxygen that these vessels carry diffuse from the dermis into the nearby cells of the epidermis. Blood vessels are also important here in that they release heat to the external environment. As the internal temperature of the body core rises, vasodilation of the blood vessels of the dermis occurs and allows heat to radiate from the skin. In cold weather the vessels constrict, decreasing the amount of blood to the dermis, which conserves heat in the body core.

Nerves also receive sensory information from the dermis and transmit impulses to the central nervous system. Sensory information is received by numerous structures, such as the **light touch corpuscles** of the dermis. Two of these light touch corpuscles are the **Meissner's (tactile) corpuscles,** in the upper portion of the dermis, and the **Merkel (tactile) discs,** located in the upper dermis and lower epidermis. These two receptors allow for the perception of very slight touch stimuli (such as a fly lightly walking over your cheek). In addition, deep touch or pressure receptors are found in the dermis, farther away from the epidermis. **Pacinian** (lamellated) **corpuscles** sense pressure, such as when you lean against a wall or feel a vibration. Other receptors in the skin are **warm receptors** and **cool receptors.** When you are at a comfortable temperature both of these receptors are firing. If you increase or decrease the skin temperature beyond the range of these receptors, **pain receptors** are stimulated. Pain receptors are naked nerve endings in the dermis that respond to numerous environmental stimuli. Locate the Meissner's corpuscles and Pacinian corpuscles in a prepared slide of skin and compare them to figure 7.6. Draw an example of Meissner's corpuscle and Pacinian corpuscle in the following space.

Illustration of Meissner's corpuscle:

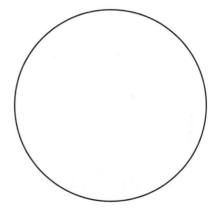

Illustration of Pacinian corpuscle:

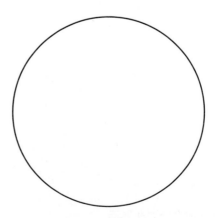

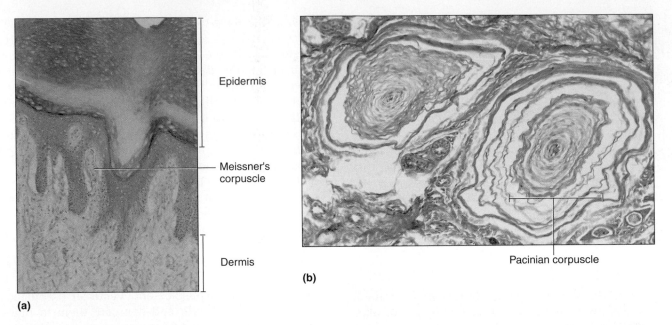

(a)

(b)

Epidermis

Meissner's corpuscle

Dermis

Pacinian corpuscle

FIGURE 7.6 **Skin Receptors (100×)** (a) Meissner's corpuscle; (b) Pacinian corpuscle.

The Hypodermis

The hypodermis (see figures 7.1 and 7.2) is also known as the subcutaneous layer. It is a layer deep to the dermis consisting of dense irregular connective tissue, adipose tissue, nerves, blood vessels, and sense receptors.

The hypodermis attaches to the dermis by collagenous and elastic fibers. In some locations, such as the front of the neck, the hypodermis is loosely attached to the dermis and underlying fascia. In other areas, such as the soles of the feet, it is dense and strongly attached to the adjacent tissues.

The adipocytes in the hypodermis are clustered in lobules and can be mobilized to the venous system when the body needs energy. Draw the hypodermis in the following space and label the adipocytes and collagenous fibers.

Illustration of hypodermis:

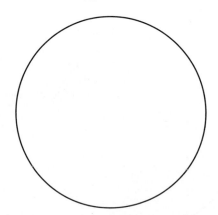

Integumentary Glands

A significant number of glands occur in the dermis and hypodermis. These include **sudoriferous (sweat) glands,** **lactiferous (milk) glands, sebaceous (oil) glands,** and **ceruminous (earwax) glands.** The sudoriferous glands are composed of simple cuboidal epithelium and come in two different types. **Eccrine (merocrine) glands** are the most common, are sensitive to temperature, and produce normal body perspiration. Perspiration reduces the temperature of the body by **evaporative cooling.** **Apocrine glands** secrete water and a higher concentration of organic acids than eccrine glands. The organic acids, along with bacterial action, produce the characteristic pungent body odor. These glands are concentrated in the region of the axilla and groin and are associated with hair follicles. Examine a prepared slide of skin and look for clusters of cuboidal epithelial cells with purple nuclei in the dermis or hypodermis. The glands are tubules, and when these are cut in section on your slide you are mostly seeing the tubules cut in cross section. These are the sudoriferous glands illustrated in figure 7.7. Draw a sudoriferous gland.

Illustration of sudoriferous gland:

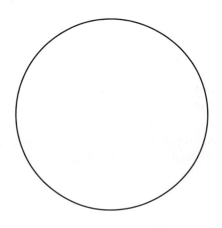

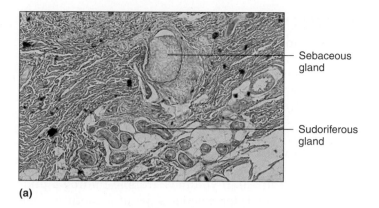

(a)

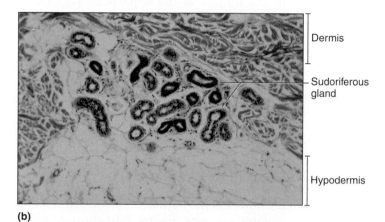

(b)

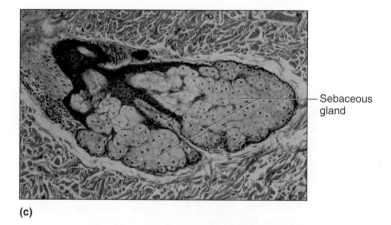

(c)

FIGURE 7.7 Integumentary Glands (a) Overview of glands (20×); (b) sudoriferous gland (100×); (c) sebaceous gland (100×).

Sebaceous glands are typically associated with hair follicles and are larger than sudoriferous glands. You may not see a hair follicle with the sebaceous gland if the section was made through the gland and not through the gland and follicle. These glands secrete an oily material called **sebum,** which lubricates the hair and decreases the wetting action of water on the skin. Examine a slide of hair follicles and compare the slide to figure 7.7. Draw a sebaceous gland.

Illustration of sebaceous gland:

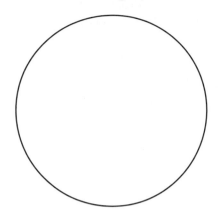

Hair

Hair, considered an accessory structure of the integumentary system, consists of keratinized cells produced in **hair follicles.** The follicle encloses the hair in the same way a bud vase encloses the stem of a rose. Hair follicles are projections of the epidermal layers into the dermis. Hair can thus be considered an epidermal derivative. The hair consists of the **shaft,** the portion that erupts from the skin surface; the **root,** a deeper portion that is enclosed by the follicle; and the **hair bulb,** the actively growing portion of the hair. There are two different types of hair. **Determinate hair** grows to a specific length and then stops. Determinate hair is found in many places in the body, including the axilla, groin, eyelashes, and eyebrows. **Indeterminate hair** continues to grow without regard to length. It is found on the scalp and in beard hair in men. Examine models and charts in the lab and compare them to figure 7.8.

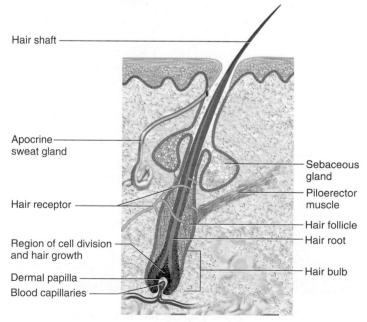

FIGURE 7.8 Diagram of Hair in a Follicle

Look at a prepared slide of hair and locate the root, follicle, and hair bulb. At the center of the bulb is a small structure known as the **dermal papilla,** a region with blood vessels and nerves that reach the hair bulb. The follicle is composed of an outer dermal layer of connective tissue and an inner epithelial layer. These form the **root sheath** of the hair. Locate these structures and compare them to figure 7.9. Draw a longitudinal section of hair.

Illustration of hair:

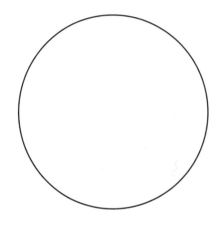

Piloerector

The hair "stands on end" by the contraction of **piloerector (arrector pili)** muscles. A piloerector muscle is a cluster of parallel smooth muscle fibers that connects the hair follicle to the upper regions of the dermis. These muscles are controlled by the autonomic nervous system. In response to cold conditions the piloerector muscles contract, producing goose bumps. The piloerector muscles also contract when a person is suddenly frightened. Find a piloerector muscle in a prepared slide and compare it to figures 7.8 and 7.9. If you have trouble finding a piloerector muscle, look for slender slips of smooth muscle that angle from the follicle to the superficial regions of the dermis. You may have to look at more than one slide to find them.

Cross Section of Hair

The central portion of the hair is known as the **medulla,** enclosed by an outer **cortex.** The cortex may contain a number of pigments, which give the hair its color. The brown and black pigments are due to varying types and amounts of **melanin.** Iron-containing melanin pigments produce red hair. Gray hair is the result of a lack of pigment in the cortex and air in the medulla. The layer superficial to the cortex is the **cuticle** of the hair and looks rough in microscopic sections. Figure 7.10 shows a cross section of hair.

Nails

Nails are keratinized cells of the stratum corneum of the skin that occur on the distal ends of the fingers and toes.

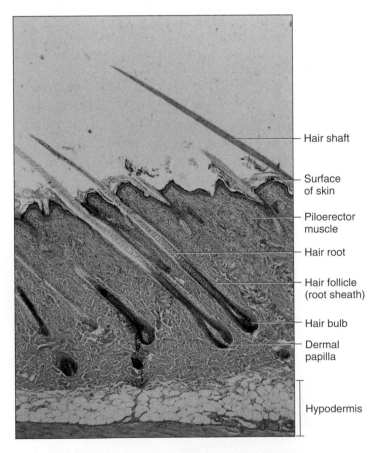

FIGURE 7.9 Photomicrograph of Hair in a Follicle (40×)

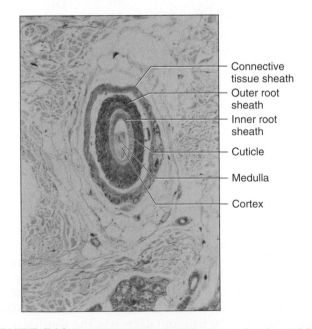

FIGURE 7.10 Hair Root and a Follicle, Cross Section (100×)

The nail is called the **nail body** and, as it grows, the **free edge** of the nail is the part you clip. The cuticle of the nail is known as the **eponychium,** and the **nail root** is underneath the eponychium and is the area that generates the nail body. The **nail bed** is the layer deep to the nail body, while the **lunule** is the small white crescent that occurs at the base of the nail. The **hyponychium** is the region under the free edge of the nail. Examine figure 7.11 and compare your fingernails to the diagram.

On each side of the nail is the **nail groove,** a depression between the nail body and the skin of the fingers. The ridge above the groove is the **nail fold.**

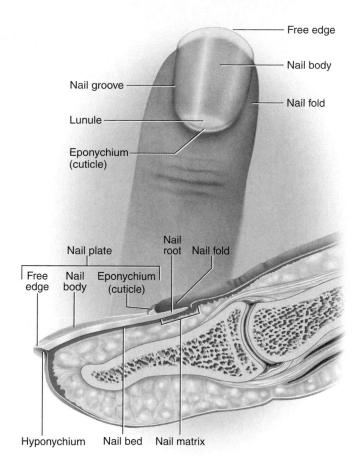

FIGURE 7.11 Fingernail, Top and Midsagittal View

Notes

REVIEW SECTION

Integumentary System

Name _____ *Date* _____

Lab Section _____ *Time* _____

Review Questions

1. The main organ of the integumentary system is the skin. Name three associated structures in the integumentary system.

2. The dermis has two main layers. Which one of these is the most superficial?

3. What is the most common connective tissue fiber found in the dermis?

4. The release of heat from the body by blood vessels occurs in what main layer of the integument?

5. The most superficial layer of skin, made of keratinocytes, is dead and provides a protective barrier. If the keratinocytes were alive at the surface, how could that compromise the protective function of the integument?

6. What material makes the epidermis tough?

7. Which type of sweat gland (merocrine or apocrine) is involved in evaporative cooling?

8. Distinguish among Pacinian corpuscles, Meissner's corpuscles, and pain receptors in the skin.

9. Hair of the axilla can be considered determinate/indeterminate (circle the correct answer).

10. Electrolysis is the process of hair removal by using electric current. Explain how this might destroy the process of hair growth in relation to the hair bulb.

11. Since hair color is determined by pigment in the cortex and the hair shaft is dead, explain the fallacy of a person's hair turning white overnight.

12. Tattoos consist of ink injected into the skin. Do you think that the ink is injected into the epidermis or the dermis? What support do you give for your answer?

13. Approximately how long does it take for the epidermis to renew itself?

14. What cell type produces a pigment that darkens the skin?

15. Which one of the three layers (epidermis, dermis, hypodermis) is not part of the integument?

16. A subcutaneous injection is inserted in the hypodermal or subcutaneous layer. Based on structures found in the hypodermis and not in the epidermis, why is this a preferred area for administering an injection?

17. What integumentary gland secretes an oil-like substance?

18. What part of the hair is found on the outside of the skin?

19. What does a piloerector muscle do?

20. The outermost portion of the hair is known as the

    a. cuticle.   b. lunula.   c. root.   d. medulla.

21. Label the following illustration using the terms provided.

hair root   hair shaft   hair follicle
piloerector muscle   epidermis   dermis
hypodermis   sweat gland   sebaceous gland
stratum corneum

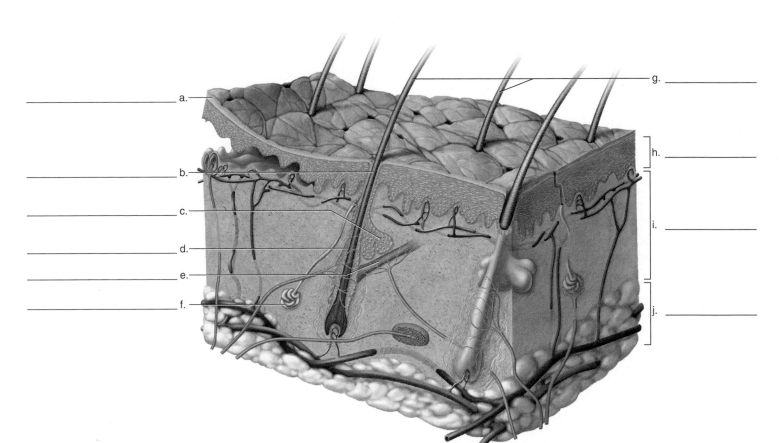

LABORATORY

Introduction to the Skeletal System

INTRODUCTION

The skeletal and muscular systems are intimately related. The skeletal system has numerous functions, including support of the body, protection of soft tissues (such as the brain and lungs), provision of a structure for locomotion, storage of calcium, and blood formation (hemopoiesis). These topics are covered further in the Saladin text in chapter 7, "Bone Tissue," and chapter 8, "The Skeletal System."

The interaction between the skeletal and muscular systems is one in which the skeletal system provides the lever system on which the muscles can work, resulting in movement. This will be discussed in greater detail in Exercise 12. In this exercise you examine the structure, composition, and development of the skeletal system.

OBJECTIVES

At the end of this exercise you should be able to

1. describe the composition of bone tissue;
2. discuss how the skeletal system provides for efficient movement;
3. list the two major types of bone material found in long bones;
4. draw or describe the microscopic structure of compact bone;
5. relate the anatomy of the bone to its growth.

MATERIALS

Chart of the skeletal system

Cut sections of bone showing compact and spongy bone

Articulated human skeleton

Articulated cat skeleton (if available)

Disarticulated human skeleton

Prepared slide of ground bone

Prepared slide of decalcified bone

Bone heated in oven at 350°F for 3 hours or more

Bone placed in vinegar for several weeks or 1 N nitric acid for a few days

Microscopes

PROCEDURE

Divisions of the Skeletal System

The skeletal system can be sorted into two main divisions according to location. The **axial skeleton** is named because it is in the vertical axis of the body. It consists of the skull, hyoid bone, vertebral column, ribs, and sternum. The **appendicular skeleton** is that part of the skeleton "attached to" the axial skeleton. It can be divided further into the **pectoral girdle,** the **upper extremity,** the **pelvic girdle,** and the **lower extremity.** Locate the major bones in lab and compare them to figure 8.1 and table 8.1.

Major Bones of the Body

Figure 8.1 shows the major bones or bone regions of the body. You should be familiar with them before you study the specifics of the individual bones in later exercises. It helps to take a bone from a disarticulated skeleton and compare it to an articulated skeleton. Locate the major bones in the figure as you examine the bones in the lab.

The typical young adult has around 206 bones, with more found in younger individuals and fewer in older individuals. Many bones fuse as a person ages, forming larger bones, thus reducing the overall number.

TABLE 8.1	Axial and Appendicular Skeletons
Axial Skeleton	
Skull	Ribs
Hyoid bone	Sternum
Vertebral column	
Appendicular Skeleton	
Pectoral girdle	Pelvic girdle
Clavicle	Hip bone
Scapula	Lower extremity
Upper extremity	Femur
Humerus	Patella
Radius	Tibia
Ulna	Fibula
Carpals	Tarsals
Metacarpals	Metatarsals
Phalanges	Phalanges

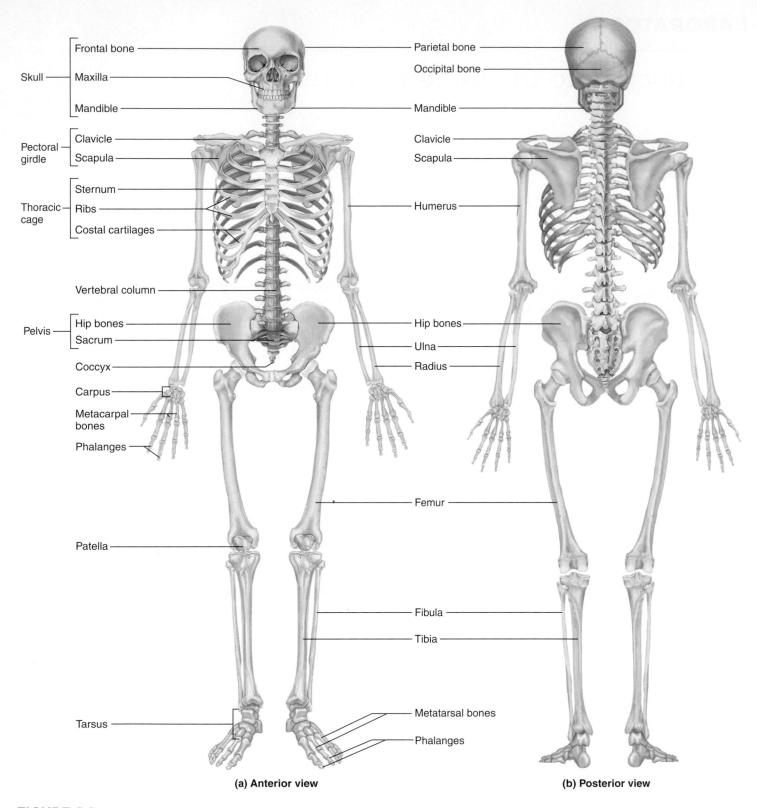

(a) Anterior view

(b) Posterior view

FIGURE 8.1 **Major Bones of the Body**   Axial skeleton in beige, appendicular skeleton in green.

Bone Features

Bones have many surface features that have specific names. These are grouped by projections, depressions, or holes in the bone. Projections on bones reflect different functions. Some projections, such as condyles and heads, are surfaces where the bone articulates with another. Other projections, such as tubercles, spines, and trochanters, are attachment points for muscles or ligaments. Depressions on bones, such as foramina, canals, and fissures, are often passageways for blood vessels and/or nerves. You should study the terms in table 8.2 and be able to find them on bones in lab.

Composition of Bone Tissue

Bone consists of **organic material** and **inorganic material.** The organic portion is composed mostly of cells and collagenous fibers and makes up about one-third of the bone by weight. The inorganic portion makes up the remaining two-thirds of the bone and mostly consists of **hydroxyapatite,** a complex salt of calcium phosphate. There is also a lesser amount of calcium carbonate in the inorganic portion of bone. If available in the lab, examine bones that have been soaked in acid. The acid has dissolved the mineral salts, leaving the collagenous fibers as a flexible framework. Now examine the bones baked in an oven for a few hours. In this preparation the structural characteristics of proteins have been destroyed, and though the bones remain hard because of the mineral salts, they are brittle due to the destruction of the protein fibers by the heat. Notice how easily these bones break.

Bone Morphology

The general structure of a long bone consists of the proximal and distal ends of the bone, called the **epiphyses** (eh-PIF-uh-seez), and the shaft of the bone, called the **diaphysis** (die-AH-fuh-sis) (see figure 8.2). The epiphyses of bones have **articular cartilage** at the ends of the bone. The articular cartilage is composed of hyaline cartilage and helps reduce friction as the joint moves.

In individuals who are growing in height, there is a plate of hyaline cartilage between the epiphysis and

TABLE 8.2	Bone Features	
Projections from the Bone Surface	**Example**
Process—a general term for a projection from the surface of the bone	Styloid process of ulna
Tubercle—a relatively small bump on a bone	Greater tubercle of humerus
Tuberosity—a relatively large, rough area on a bone	Deltoid tuberosity of humerus
Spine—a short, sharp projection	Vertebral spine
Condyle—an irregular, smooth surface that articulates with another bone	Lateral condyle of femur
Epicondyle—a bump on a condyle	Medial epicondyle of humerus
Head—a hemispheric projection that articulates with another bone	Head of femur
Neck—a constriction below the head	Neck of rib
Crest—an elevated ridge of bone	Crest of ilium
Line—a smaller elevation than a crest	Gluteal line of ilium
Facet—a smooth, flat face	Articular facets of vertebrae
Trochanter—a large bump (on femur)	Greater and lesser trochanter of femur
Ramus—a branch	Ramus of mandible
Depressions or Holes in the Bone Surface	**Example**
Foramen—a shallow hole	Foramen magnum of occipital bone
Sinus—a cavity	Maxillary sinus
Meatus or canal—a deep hole	External auditory meatus of temporal bone
Fossa—a shallow depression in a bone	Iliac fossa
Notch—a deep cut-out	Greater sciatic notch of ilium
Groove or sulcus—an elongated depression	Intertubercular groove of humerus
Fissure—a long, deep cleft	Inferior orbital fissure

the diaphysis. This is the **epiphyseal** (EP-ih-FIZZ-ee-ul) **plate,** which increases in thickness by division of the chondrocytes. As the cartilage grows, the individual increases in height. During growth, the cartilage away from the epiphyseal plate is replaced by bone. When a person reaches adult height all the cartilage is replaced by bone and the remnants of the plate are seen as the **epiphyseal line.** Examine long bones cut lengthwise for the epiphyseal line. In the cut sections of bone you can also see the hard, **compact** (cortical) **bone** on the outside of the bone and the inner **spongy** (or cancellous) **bone.** The spongy bone is made of **trabeculae,** thin rods or plates of bone that run in the same direction as the stress applied to the bone. Stress on the bone may be in the form of gravity, or it may occur due to a common force applied to the limb. Trabeculae make an internal framework that strengthens the bone.

The innermost section of bone is hollow and is called the **marrow** (or **medullary** (MED-you-lerr-ee)) **cavity.** Marrow is the material that occupies this cavity and can be of two types, a hemopoietic (blood-forming)

red marrow** or an adipose-containing **yellow marrow.** In flat bones there is an outer and an inner layer of compact bone with spongy bone sandwiched between them. In flat bones the spongy bone is called **diploe** (DIP-lo-ee). Compare the cut bones in the lab to figures 8.2 and 8.3 and note the features listed in the discussion.

Note how the walls of the compact bone in the middle of the long section of bone is thicker than the walls at the ends of the bone. Many long bones have this feature and resemble an archery bow in this regard. This reinforcement decreases breakage at the region of bone that would be most prone to break, the middle.

On the surface of the diaphysis, locate the **nutrient foramina** (for-AM-ih-nuh). These are small holes that allow for the passage of blood vessels into and out of the bone. The nutrient foramina lead to **perforating** (Volkmann's) **canals** that pass through compact bone.

The outer surface of the bone is covered with a dense connective tissue sheath called the **periosteum** (PEAR-ee-OS-tee-um), where nerves and blood vessels are located. The periosteum is an anchoring point for **tendons** and

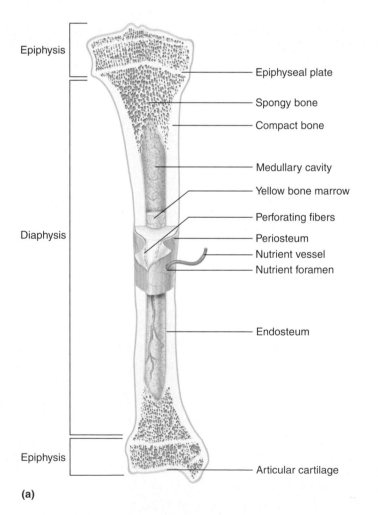

(a)

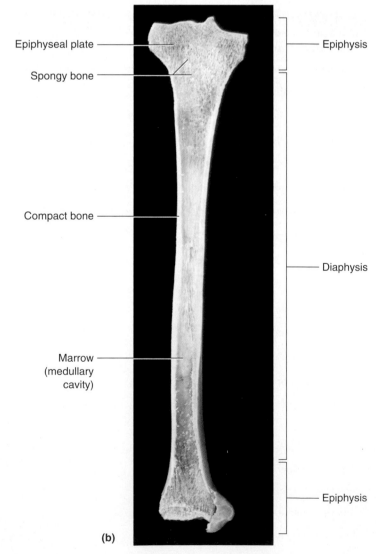

(b)

FIGURE 8.2 Long Bone, Longitudinal Section (a) Diagram; (b) photograph.

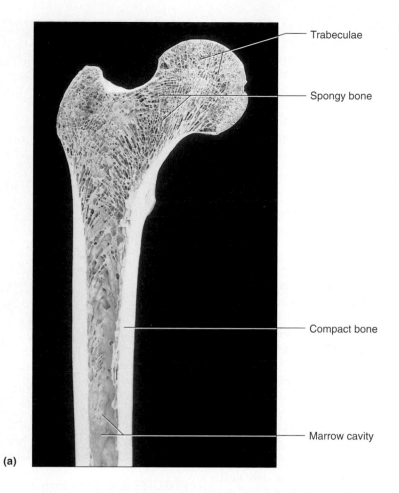

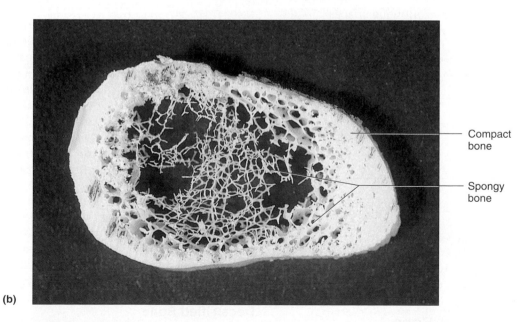

FIGURE 8.3 Cut Bone (a) Long section; (b) cross section.

ligaments. Tendons attach muscles to bone at the periosteum, and this attachment is strengthened by **perforating** (Sharpey's) **fibers** that penetrate the compact bone. **Ligaments** are parallel straps of connective tissue that connect one bone to another, and they are secured to the bone by the periosteum.

The inner surface of long bones, near the marrow cavity, is lined with a layer known as the **endosteum** (end-OS-tee-um).

Bone Classification

Bones may be classified in a number of ways. Anatomically, bones are classified according to shape and location.

Bone Shape

Bones are classified according to four main shapes: **long, short, flat,** and **irregular.** Long bones are more long than broad. Bones such as those of the arm, forearm, fingers, thigh, and leg are long bones. Short bones are more or less equal in all dimensions. Bones of the wrist and ankle are examples of short bones. Flat bones, such as bones of the cranium, pelvic girdle, pectoral girdle, and sternum, appear compressed in one dimension. Unique types of flat bones in the skull are small sutural or wormian bones. The ribs are also considered flat bones. Irregular bones are hard to place into any of the other categories. Bones such as those in the floor of the skull, facial bones, and vertebrae are considered irregular. Examine bones in the lab and compare them to figure 8.4. Look through the bones in lab, and find a bone that represents each of the bone shapes. Write the name of the bone in the following spaces.

Long bone _____

Short bone _____

Flat bone _____

Irregular bone _____

Bone Location

Bones can be classified not only as part of the axial and appendicular skeleton but also as part of the upper extremity, lower extremity, pelvic girdle, skull, and vertebral column.

Microscopic Structure of Bone

Ground Bone

Examine a prepared slide of ground bone and locate the modular units of bone called **osteons.** Each osteon has a hole in the middle called a **central (haversian** or **osteonic)**

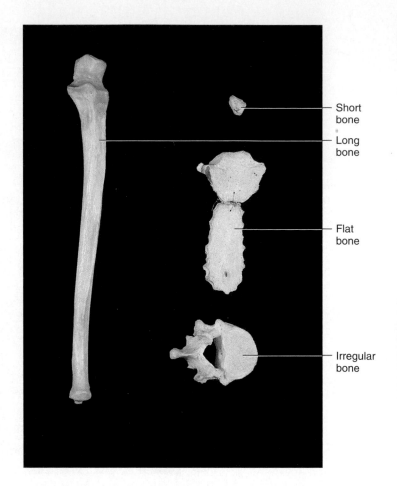

FIGURE 8.4 Bone Shapes

canal, which houses blood vessels and nerves in the dense bone tissue. The central canals typically run vertically in bones, while **perforating canals** carry nutrients horizontally. Around the central canal are rings of dark spots. These are the **lacunae** (singular, *lacuna*), spaces that the bone cells, or **osteocytes,** occupied in living tissue. In the prepared slide the lacunae have filled with bone dust from the grinding process. The lacunae are connected by thin tubes called **canaliculi** (CAN-uh-LIC-you-lie). The role of canaliculi is to provide a passageway through the dense bone material. Osteocytes are living and need to receive oxygen and nutrients and remove wastes. They do this by extending cellular strands through the canaliculi between osteocytes. The canaliculi connect to the central canal. The osteon has layers of dense mineral salts that form concentric rings between the lacunae. These are the **lamellae.** Compare what you see to figure 8.5.

Decalcified Bone

Examine a slide of decalcified bone. In this preparation the bone salts have been removed and the remaining thin section of bone has been sliced, mounted, and stained. The periosteum is visible on the surface of the bone, with the compact bone in the middle and the endosteum on

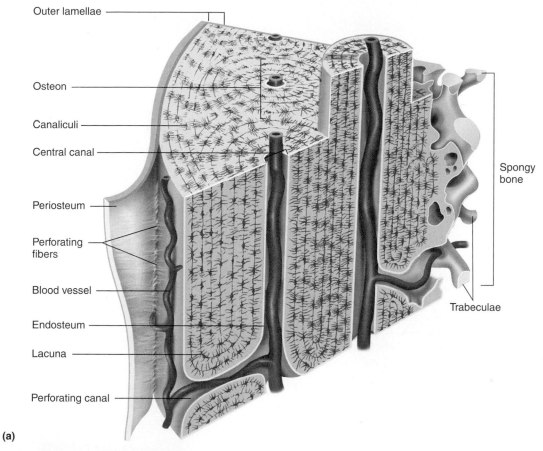

- Outer lamellae
- Osteon
- Canaliculi
- Central canal
- Periosteum
- Perforating fibers
- Blood vessel
- Endosteum
- Lacuna
- Perforating canal
- Spongy bone
- Trabeculae

(a)

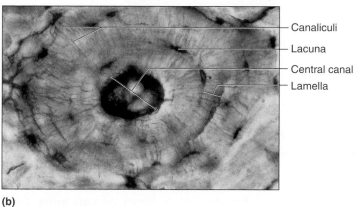

- Canaliculi
- Lacuna
- Central canal
- Lamella

(b)

FIGURE 8.5 Histology of Ground Bone (Osteon)
(a) Diagram; (b) cross section (400×).

the inside of the bone (see figure 8.6). Look for as many bone cells as you can find. These cells are described in the following section and illustrated in figure 8.7.

Bone Cells

There are three main types of bone cells: **osteoblasts, osteocytes,** and **osteoclasts.** Osteoblasts originate from stem cells called **osteogenic cells** (see figure 8.7), which occur in the periosteum, endosteum, and central canals of osteons. These osteogenic cells maintain mitotic activity and produce osteoblasts, which secrete collagen fibers

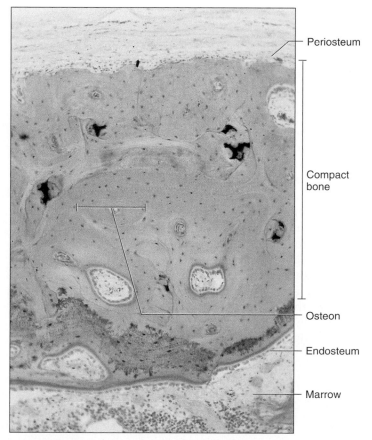

- Periosteum
- Compact bone
- Osteon
- Endosteum
- Marrow

FIGURE 8.6 Decalcified Bone (100×)

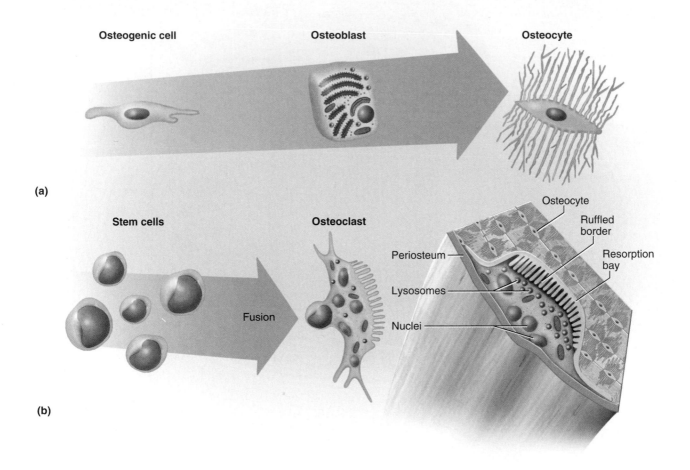

Osteogenic cell   **Osteoblast**   **Osteocyte**

(a)

Stem cells   **Osteoclast**

Osteocyte

Ruffled border

Periosteum

Resorption bay

Lysosomes

Fusion

Nuclei

(b)

FIGURE 8.7 Bone Cells (a) Osteogenic cells produce osteoblasts that develop into osteocytes. (b) Stem cells in bone marrow fuse to form osteoclasts.

and calcium salts. Osteoblasts are most numerous in the endosteum and periosteum.

Osteocytes are mature bone cells; they respond to stresses placed on bone and remodel the bone in response to those stresses. For example, an increase in weight-bearing activity increases the density of bone. This is seen in the bones of athletes, which are denser than those of people who do not exercise. Osteocytes deposit or reabsorb bone matrix, regulating bone density, and control calcium and phosphate balance. **Osteoclasts** are cells that remove collagen by acid phosphatase and minerals by HCl. They arise by the fusion of **stem cells.** Examine the inner portion of the same slide and locate the osteoclasts. Compare them to figure 8.7.

Skeletal Anatomy of the Cat

The cat skeleton differs from the human skeleton in several ways. Humans are bipedal, and this is reflected in the nature of the vertebral column. In humans the bodies of the vertebrae become much larger as you move from superior to inferior. Cats are quadrupeds, so the weight of the vertebral column is distributed more evenly and their vertebrae are more evenly matched in size. In humans the absence of a tail is reflected in a corresponding absence of tail vertebrae. In cats the tail vertebrae are present and fully functional.

Another difference is that humans walk on the entire inferior surface of the foot, while cats walk on their toes. Examine a cat skeleton in the lab and note the general features of the skeleton (see figure 8.8).

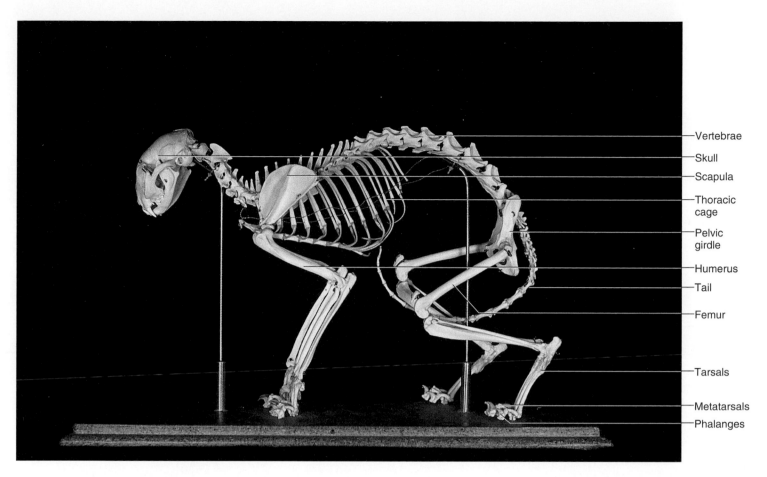

- Vertebrae
- Skull
- Scapula
- Thoracic cage
- Pelvic girdle
- Humerus
- Tail
- Femur
- Tarsals
- Metatarsals
- Phalanges

FIGURE 8.8 Cat Skeleton, Lateral View

Notes

REVIEW SECTION

Introduction to the Skeletal System

Name _____ Date _____

Lab Section _____ Time _____

Review Questions

1. The hyoid bone belongs to the

   a. appendicular skeleton.　b. axial skeleton.　　c. upper extremity.　　d. skull.

2. The clavicle belongs to the

   a. axial skeleton.　　　b. pectoral girdle.　　c. pelvic girdle.　　d. upper extremity.

3. In the disease osteoporosis there is a significant loss of spongy bone. Explain how the loss of this specific bone material can weaken a bone.

4. Label the following illustration using the terms provided.

   central canal

   lacuna

   osteon

   canaliculi

   lamella

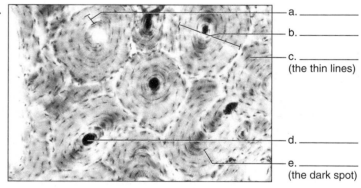

a. _____

b. _____

c. _____
(the thin lines)

d. _____

e. _____
(the dark spot)

5. The ends of a long bone are known as the _____.

6. A carpal bone is classified as a _____ bone in terms of shape.

7. Name two bones of the forearm.

8. The ribs are part of which skeletal division?

   a. axial   b. appendicular

9. Name the bone found between the femur and the tibia.

10. A young adult has how many bones, on average? _____

11. The inorganic portion of bone tissue is made of what complex mineral salt (consisting of calcium phosphate)?

12. The bones of the cranium are what shape?

   a. long   b. short   c. flat   d. irregular

13. What is the shape of the sternum?

   a. long   b. short   c. flat   d. irregular

14. What shape are the metacarpals?

   a. long   b. short   c. flat   d. irregular

15. What shape are the tarsals?

   a. long   b. short   c. flat   d. irregular

16. What is an osteon?

17. Synthetic bone material known as hydroxyapatite is frequently molded into the shape of bone. This procedure is done to replace diseased, damaged, or surgically removed bone. Cells in the body remodel the synthetic material and produce new bone. What bone cells do you predict to be involved in this remodeling, and what role do these cells have?

18. What role do osteocytes have in bone tissue?

19. How does the shape of a long bone resist breaking when put under stress?

20. Describe the nature of decalcified bone. What was removed in the process of decalcification, and what impact did this have on the bone structure?

21. How does the central canal differ from a lacuna in terms of location and the material found in each respective space?

22. What is another name for the shoulder blade, and what two bones attach to it?

23. Label the illustration of the skeletal system using the terms provided.

hip bone	rib	phalanges (feet)	ulna
mandible	phalanges (hand)	humerus	tibia
fibula	radius	vertebra	clavicle
sternum	carpals	metatarsals	scapula
femur	patella	metacarpals	

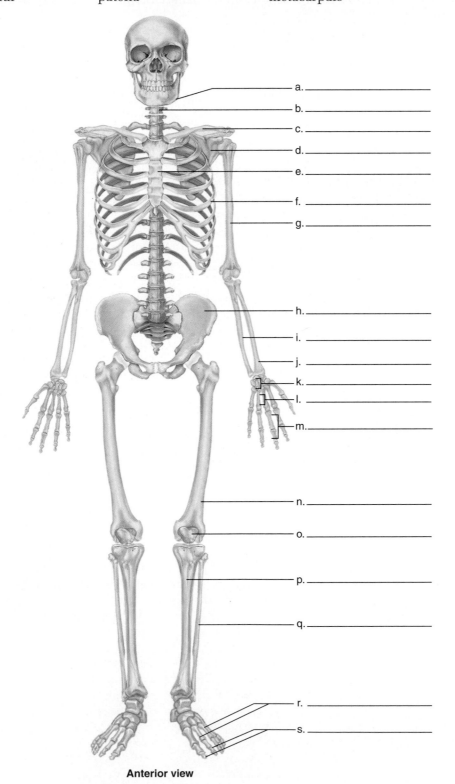

a. _____

b. _____

c. _____

d. _____

e. _____

f. _____

g. _____

h. _____

i. _____

j. _____

k. _____

l. _____

m. _____

n. _____

o. _____

p. _____

q. _____

r. _____

s. _____

Anterior view

LABORATORY

Appendicular Skeleton

INTRODUCTION

The appendicular skeleton consists of approximately 126 bones in four major groupings. These are the pectoral girdle, upper extremity, pelvic girdle, and lower extremity. The **pectoral girdle** is composed of the scapula and the clavicle on each side. The **upper extremity** consists of the humerus, radius, ulna, carpals of the wrist, metacarpals of the palm, and phalanges of the fingers. The lower portion of the appendicular skeleton is composed of the **pelvic girdle** (the hip bones), each of which consists of three fused bones, the ilium, the ischium, and the pubis. The **lower extremity** consists of the femur, patella, tibia, fibula, tarsals of the ankle, metatarsals (of the foot), and phalanges of the toes. These topics are discussed further in the Saladin text in chapter 8, "The Skeletal System."

OBJECTIVES

At the end of this exercise you should be able to

1. locate and name all the bones of the appendicular skeleton;
2. name the significant surface features of the major bones;
3. place a bone into one of the four major groupings;
4. name the individual carpal bones and tarsal bones in articulated hands and feet respectively;
5. determine whether a scapula or pelvic bone is from the left side or right side of the body;
6. determine whether a femur, a tibia, a humerus, a radius, or an ulna is from the left side or right side of the body;
7. place a clavicle with the sternal end pointing medially;
8. determine the sex of isolated or articulated hip bones.

MATERIALS

Articulated skeleton or plastic cast of articulated skeleton

Disarticulated skeleton or plastic casts of bones

Charts of the skeletal system

Plastic drinking straws cut on a bias or pipe cleaners for pointer tips

Foam pads of various sizes (to protect real bones from hard countertops)

PROCEDURE

General Considerations

Locate the bones of the appendicular skeleton on the material in the lab and on figure 9.1. When you study a bone, compare an isolated (or disarticulated) bone to the articulated skeleton (one that is joined together). Hold the individual bone up to the skeleton to see how it is positioned in relation to the other bones of the body. When examining bones in the lab, do not use your pen or pencil to locate a structure. They can leave marks on the bones. Use a cut plastic straw, wooden applicator stick, or pipe cleaner to point out structures. Your instructor may want you to place real bone material on foam pads to cushion the bone from the tabletop.

Pectoral Girdle

The pectoral, or shoulder, girdle consists of the right and left **scapulae** (singular, *scapula*) and the right and left **clavicles.** The pectoral girdle provides a movable yet stable support for the upper limb.

Scapula

The scapula, commonly known as the shoulder blade, is found on the posterior, superior portion of the thorax. The scapula is roughly triangular in shape and has three borders, a **superior border,** a **medial (vertebral) border,** and a **lateral (axillary) border.** On the superior border of the scapula is an indentation known as the **scapular notch,** which contains a nerve that innervates shoulder muscles. The anterior surface of the scapula is a smooth, hollowed depression known as the **subscapular fossa.** The posterior surface of the scapula is divided by the **scapular spine** into a **supraspinous fossa** and an **infraspinous fossa.** The spine of the scapula is a crest that runs from the medial border to the edge of the scapula, where it expands to form the **acromion** (ah-CRO-me-on) **process.** Another projection from the scapula is the **coracoid** (COR-uh-coyd) **process,** which projects anteriorly from the scapula.

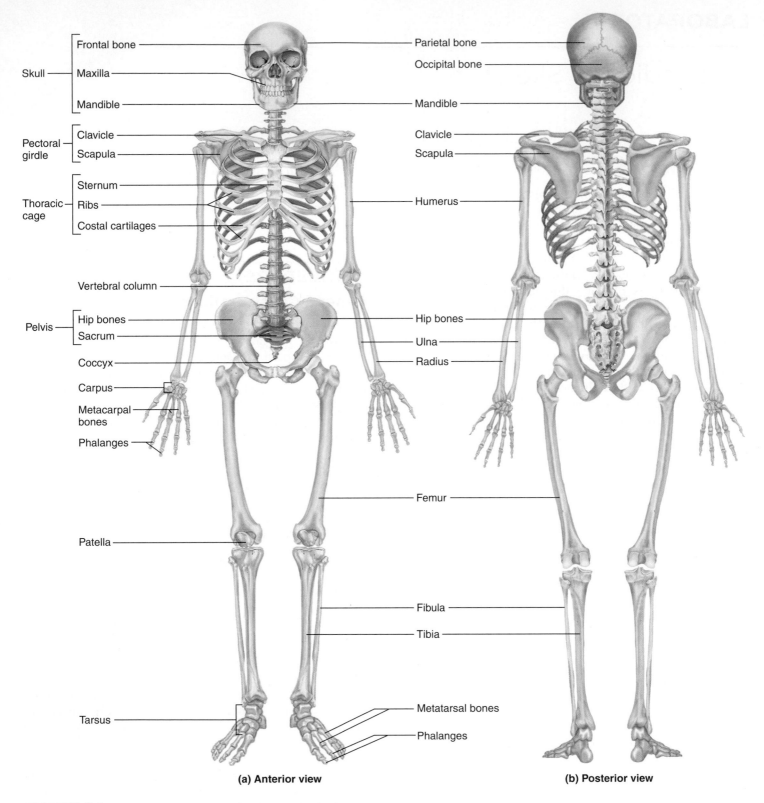

(a) Anterior view

(b) Posterior view

FIGURE 9.1 Bones of the Appendicular Skeleton, Shaded in Green

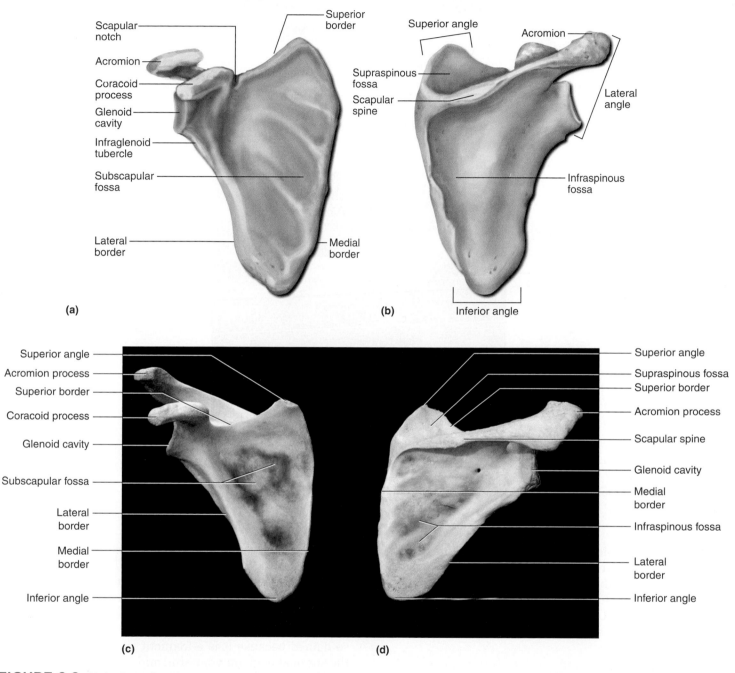

FIGURE 9.2 Right Scapula Diagram (a) anterior view; (b) posterior view. Photograph (c) anterior view; (d) posterior view.

The scapula also has two sharp angles known as the **inferior angle** and **superior angle.** The humerus moves in the shallow depression of the scapula known as the **glenoid cavity,** or glenoid fossa. Just below the glenoid cavity is a bump known as the **infraglenoid tubercle,** an attachment point for the triceps brachii muscle. The features of the scapula are important because numerous muscles attach to them. You should also be able to distinguish a right scapula from a left one. The spine of the scapula is pos-

terior, the inferior angle is less than 90 degrees, and the glenoid cavity is lateral for articulation with the humerus. Locate the features of the scapula on figure 9.2.

Clavicle

The **clavicle** is commonly known as the collarbone. It is a small bone that serves as a strut between the scapula and the sternum. The blunt, vertically truncated end of

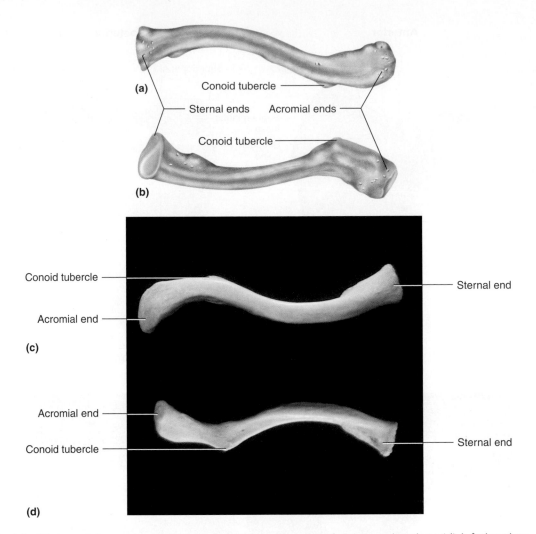

(a)
Conoid tubercle

Sternal ends   Acromial ends

Conoid tubercle

(b)

Conoid tubercle

Acromial end

(c)

Sternal end

Acromial end

Conoid tubercle

Sternal end

(d)

FIGURE 9.3 Clavicle Diagram (a) superior view; (b) inferior view. Photograph (c) superior view; (d) inferior view.

the clavicle is the **sternal end** and the horizontally flattened end is the **acromial end.** Note the **conoid** (CON-oyd) **tubercle** on the inferior surface of the clavicle. Examine the clavicles in the lab and compare them to figure 9.3. The arm transmits force to the clavicle, sandwiched between the scapula and the sternum.

Upper Extremity

Humerus

The **humerus** is a long bone that articulates with the scapula at its proximal end and with the radius and ulna at its distal end. The **head** of the humerus is a hemispheric structure that ends in a rim known as the **anatomical neck.** The anatomical neck is the site of the **epiphyseal line.** The humerus has two large proximal processes, the **greater tubercle,** the largest and most lateral process, and the **lesser tubercle,** which is medial and smaller. These processes are attachment points for muscles originating from the scapula. Between the tubercles is the **intertubercular sulcus (groove),** an elongated depression through which

one of the biceps brachii tendons passes. The intertubercular groove is anterior, and the head of the humerus is medial. These two features can orient you to the right or left humerus. Examine figure 9.4 and locate these features. Below the tubercles of the humerus is a constricted region known as the **surgical neck.** The surgical neck is so named because it is a frequent site of fracture. Locate the surgical neck on your arm, and in the following space name a commonly known muscle that occurs nearby.

? Muscle near surgical neck: _____ 1

A roughened area on the lateral surface of the humerus is the **deltoid tuberosity,** named for the attachment of the deltoid muscle. The majority of the length of the humerus is called the **shaft,** and a small hole penetrating the shaft is the **nutrient foramen.** Nutrient foramina are holes that occur in many bones and conduct blood vessels into these bones. At the distal end of the humerus are the regions of articulation with the bones of the forearm. On the lateral

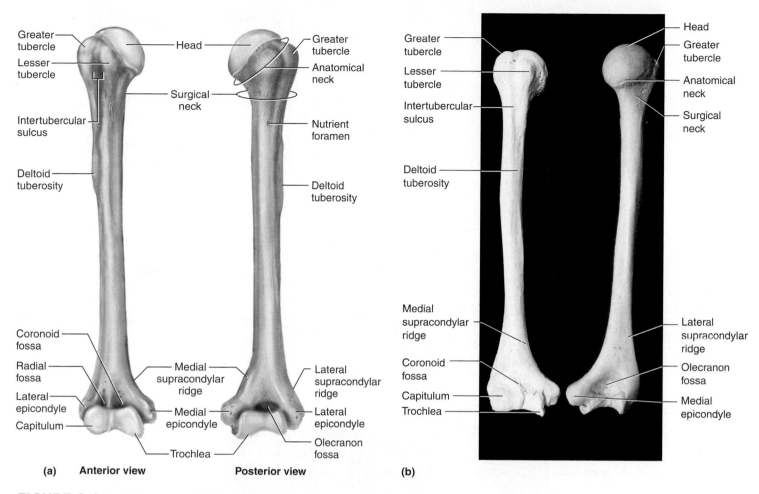

FIGURE 9.4 **Right Humerus** (a) Diagram; (b) photograph.

side is a round hemisphere known as the **capitulum** (ca-PIT-you-lum). This is where the head of the radius fits onto the humerus. On the medial side is an hourglass-shaped structure called the **trochlea** (TROW-klee-uh). The trochlea is where the ulna articulates with the humerus. The capitulum and trochlea represent the **condyles** (KON-dials) of the humerus. To the sides of these condyles are the **epicondyles** (EP-ee-KON-dials). The medial epicondyle is the bump that you can palpate (feel) medial to the elbow. The ridge that forms a wing of bone, proximal to the epicondyle, is the **medial supracondylar ridge.** The lateral epicondyle also has a **lateral supracondylar ridge.** The epicondyles and supracondylar ridges are points of attachments for muscles that manipulate the forearm and hand. At the distal end are two depressions in the humerus. The one on the anterior surface is the **coronoid fossa** and the one on the posterior surface is the **olecranon** (uh-LEC-ruh-non) **fossa.** Locate these structures on specimens in the lab and on figure 9.4.

Bones of the Forearm

Two bones make up the skeletal portion of the forearm. These are the **radius,** lateral (on the thumb side), and the

ulna, a medial bone. The radius has a proximal **head** that looks like a wheel. Distal to this is the **radial tuberosity,** where the biceps brachii muscle inserts, and at the most distal end is the **styloid process** of the radius. On the distal end of the radius is a medial depression called the **ulnar notch.** This is where the distal part of the ulna joins with the radius. Examine figure 9.5 for these structures and compare them to the bones in the lab.

The ulna has a U-shaped depression (some students remember this as "U for ulna") on the proximal portion, the **trochlear** (TROK-lee-ur) **notch.** The most proximal portion of the ulna is the **olecranon process,** commonly known as the elbow. The process distal to the trochlear notch is the **coronoid process,** which ensures a relatively tight fit with the humerus and, along with the **tuberosity of the ulna,** provides an area for muscle attachment. The ulna has a *distal* **head,** unlike the radius (the head of the radius is proximal), and a **styloid** (STY-loyd) **process** near the head. Locate these structures in figure 9.5. On the proximal part of the ulna, where the head of the radius fits into the ulna, is a depression known as the **radial notch.** If you know that the trochlear notch is anterior and the radial notch is lateral, you can determine if you are looking at a left or right ulna. In the following space,

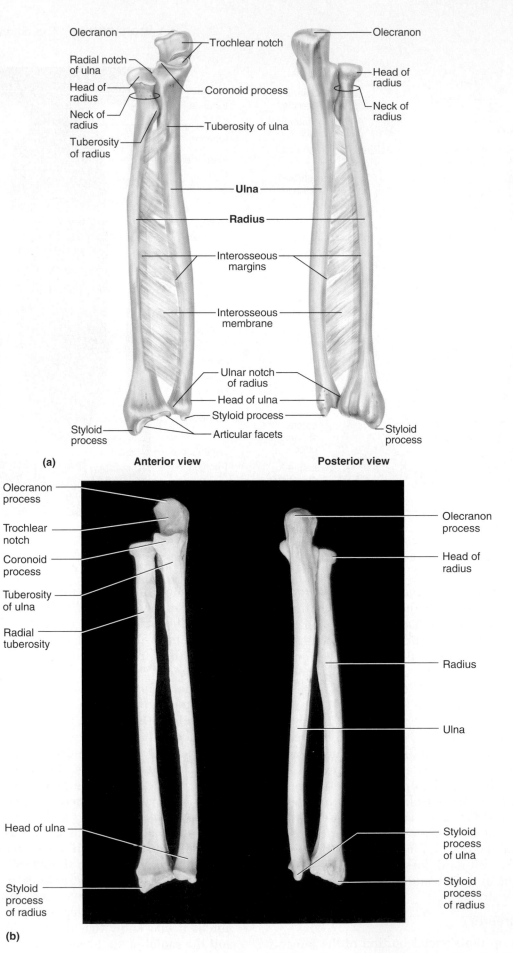

Anterior view

Posterior view

(a)

Olecranon — Trochlear notch
Radial notch of ulna
Head of radius — Coronoid process
Neck of radius — Tuberosity of ulna
Tuberosity of radius

Olecranon
Head of radius
Neck of radius

Ulna

Radius

Interosseous margins

Interosseous membrane

Ulnar notch of radius
Head of ulna
Styloid process
Styloid process — Articular facets
Styloid process

Olecranon process
Trochlear notch
Coronoid process
Tuberosity of ulna
Radial tuberosity

Olecranon process
Head of radius

Radius

Ulna

Head of ulna

Styloid process of ulna

Styloid process of radius

Styloid process of radius

(b)

FIGURE 9.5 **Right Radius and Ulna** (a) Diagram; (b) photograph.

list what side of the body the bones of the forearm you are studying come from.

? Side of the body: _____ 2

Bones of the Hand

Each hand consists of three groups of bones, the **carpals**, the **metacarpals**, and the **phalanges** (singular, *phalanx*).

The human hand is composed of eight carpal bones (see figure 9.6). These bones can be roughly aligned in two rows. The proximal row from lateral to medial includes the **scaphoid** (navicular), **lunate, triquetrum** (tri-KWĒ-trum) (triangular), and the **pisiform** (PIE-sih-form). The distal row includes the **trapezium** (greater multangular), the **trapezoid** (lesser multangular), the **capitate,** and the **hamate.** On the hamate is a hooklike projection called the **hamulus** of the **hamate.** The carpal bones were named for their shape, with the scaphoid

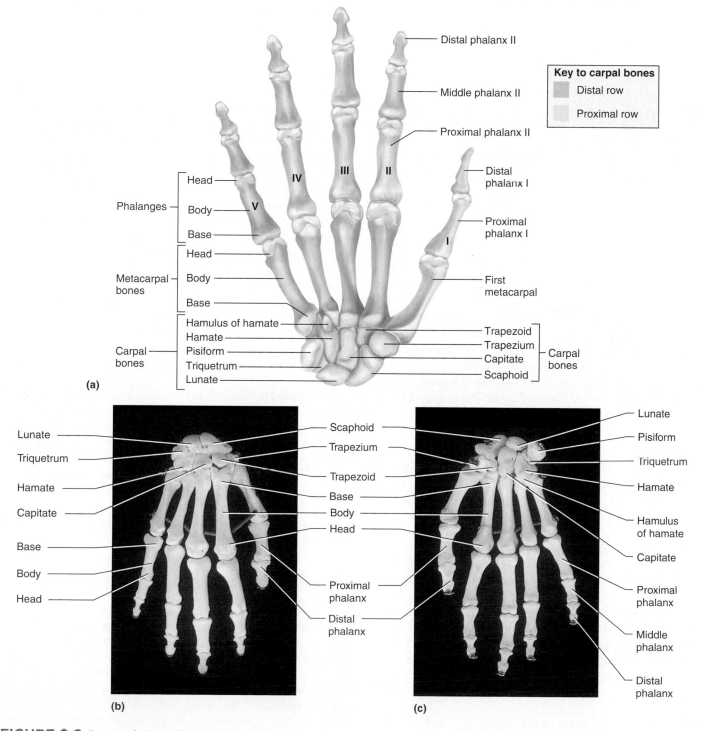

FIGURE 9.6 Bones of the Right Hand Diagram (a) anterior view. Photograph (b) posterior view; (c) anterior view.

resembling a boat (*scaphos* = boat), the lunate named for the crescent moon, triquetrum (a triangle), pisiform being pea-shaped, the trapezium (rhymes with *thumb* and named for a shape), the trapezoid (named for a shape), the capitate being "headlike," and the hamate (*hamus* = hook). Another way to put the bones in proper order is to use a mnemonic device that uses the first letter of each carpal bone (*S* for scaphoid, *L* for lunate, etc.) in a sentence: "Say Loudly To Pam, Time To Come Home." The carpal bones serve as regions of attachment for forearm muscles and for intrinsic muscles of the hand.

The **metacarpals** are the bones of the palm. Each metacarpal consists of a proximal **base**, a **body**, and a distal **head**. The metacarpals are labeled by Roman numerals I to V, with I being proximal to the thumb and V being proximal to the little finger. Metacarpals II to V have little movement, but metacarpal I allows for significant movement of the thumb. Locate metacarpal I on the skeletal material and on your own hand and note the substantial range of movement. This type of movement is covered in greater detail in Laboratory Exercise 12. Examine the carpals and metacarpals in the lab and in figure 9.6.

Distal to the metacarpals are the **phalanges.** There are 14 bones that make up this group. The thumb is called the pollex. Each finger of the hand has three phalanges, a **proximal, middle,** and **distal** phalanx, except for the thumb. The thumb has only a proximal and distal phalanx. You should be able to identify each bone by noting the digit it comes from and whether it is proximal, middle, or distal. The thumb is the first phalanx and the little finger is the fifth phalanx. Therefore, the tip of the little finger is the distal phalanx V and the base of the middle finger is the proximal phalanx III. Like the metacarpals, each phalanx has a **base, body,** and **head** and, also like the metacarpals, phalanges are classified as long bones. Locate the various bones of the fingers in material in the lab and in figure 9.6.

Pelvic Girdle

The pelvic girdle consists of the hip bones (os coxae), or **innominate, bones.** Each hip bone results from the fusion of three bones, the **ilium** (ILL-ee-um), the **ischium** (ISS-kee-um), and the **pubis** (PYOU-bis). The hip bones are joined in the front by the pubic symphysis, a fibrocartilage pad. Below the pubic symphysis is the **subpubic angle** (pubic arch). This angle is greater than 90 degrees in females and less than 90 degrees in males. The pelvic girdle consists of the two hip bones, but the **pelvis** consists of the two hip bones and the sacrum (part of the vertebral column). The large hole at the inferior part of the hip bone is the **obturator** (OB-tur-aye-tur) **foramen.** This space is covered by the **obturator membrane.**

Ilium

The most superior hip bone is the ilium. If you feel the top part of your hip you are feeling the **iliac crest,** a long,

crescent-shaped ridge of the ilium. The lateral side is the outer surface of the ilium or the **iliac blade.** The two processes that jut from the ilium in the front are the **anterior superior iliac spine** and the **anterior inferior iliac spine.** These are attachment points for some of the thigh flexor muscles. The posterior ilium has the **posterior superior iliac spine** and the **posterior inferior iliac spine.** Below the posterior inferior iliac spine is a large depression known as the **greater sciatic notch.** The outer surface of the ilium is an attachment point for the gluteal muscles. The interior, shallow depression of the ilium is known as the **iliac fossa.** Locate these areas in figure 9.7.

Ischium and Pubis

The ischium is inferior to the ilium and is the part of the hip bone on which you sit. The **ischial spine** is a sharp projection on the ischium, and the **ischial tuberosity** is an attachment site for the hamstring muscles. Anterior and superior to the ischial tuberosity is a cuplike depression known as the **acetabulum** (a region in which all three hip bones are joined). The acetabulum is the socket into which the femur fits. The ischium connects to the pubis by an elongated portion of bone known as the **ischial ramus.** This ramus connects to the **inferior pubic ramus,** which is posterior to the pubic symphysis.

If the hip bones are articulated, you can easily see the upper basin of the pelvis, known as the **false pelvis.** The false pelvis is medial to the iliac fossa. There is a rim of bone that separates the upper basin from a deeper, smaller basin known as the **true pelvis.** This rim is known as the **pelvic brim** in an intact pelvis or **arcuate line** on an isolated hip bone. The true pelvis is medial to the obturator foramen.

Examine the disarticulated hip bone in lab. You can determine if you have a left or right bone and the sex of the individual. In a female pelvis the greater sciatic notch has a fairly broad angle, which approximates the arc inscribed by your outstretched thumb and index finger. In a male the greater sciatic notch is less than 90 degrees and approximates the angle if you were to separate the index finger and the middle finger. As you examine the bones in the lab look at figure 9.7 and locate the various terms listed on the illustration.

Lower Extremity
Femur

The **femur** is the longest and heaviest bone in the body, and it is the only bone of the thigh. It articulates proximally with the hip bone and inferiorly with the **tibia.** The femur has a proximal, medial spherical **head,** which inserts into the **acetabulum** of the hip bone, and an elongated, constricted **neck.** On the head of the femur is a depression known as the fovea capitis. It is a region where a ligament attaches from the femur to the acetabulum. Below the

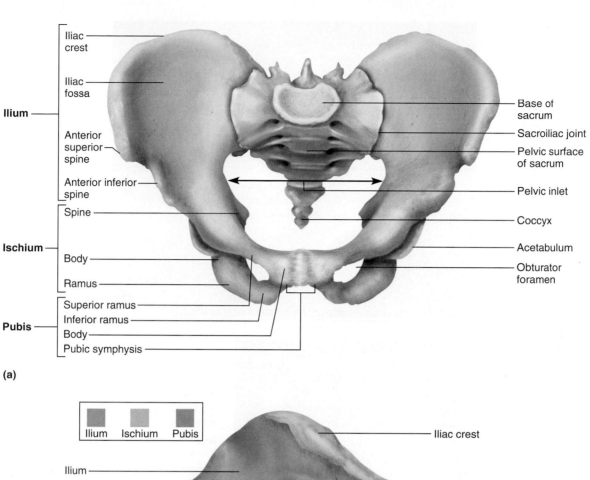

Ilium
- Iliac crest
- Iliac fossa
- Anterior superior spine
- Anterior inferior spine

Ischium
- Spine
- Body
- Ramus

Pubis
- Superior ramus
- Inferior ramus
- Body
- Pubic symphysis

- Base of sacrum
- Sacroiliac joint
- Pelvic surface of sacrum
- Pelvic inlet
- Coccyx
- Acetabulum
- Obturator foramen

(a)

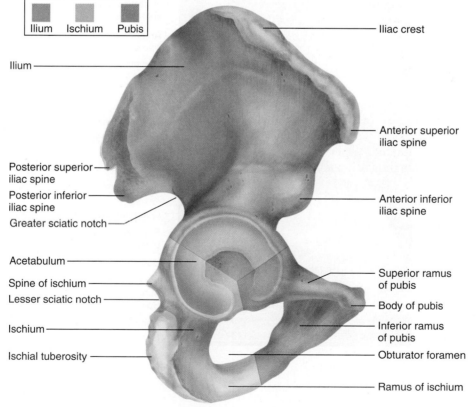

Ilium Ischium Pubis

- Ilium
- Posterior superior iliac spine
- Posterior inferior iliac spine
- Greater sciatic notch
- Acetabulum
- Spine of ischium
- Lesser sciatic notch
- Ischium
- Ischial tuberosity

- Iliac crest
- Anterior superior iliac spine
- Anterior inferior iliac spine
- Superior ramus of pubis
- Body of pubis
- Inferior ramus of pubis
- Obturator foramen
- Ramus of ischium

(b)

FIGURE 9.7 Bones of the Pelvic Girdle Diagram (a) anterior view; (b) lateral view, right hip. Photograph (c) anterior view; (d) lateral view, right hip.

Continued

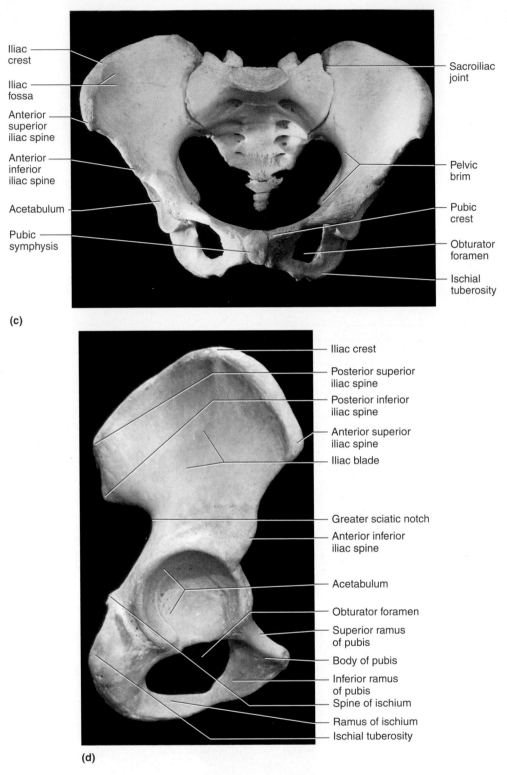

(c)

(d)

FIGURE 9.7 *Continued*

neck are two large processes called the **trochanters,** areas of muscle attachment. The larger, anterior process is the **greater trochanter** and the smaller, posterior process is the **lesser trochanter.** Locate these structures in figure 9.8. You should also find the **intertrochanteric crest,** a posterior ridge between the two trochanters and the anterior

intertrochanteric line. In the following space, determine which is more broad, the intertrochanteric crest or the intertrochanteric line.

Broadest surface feature: _____

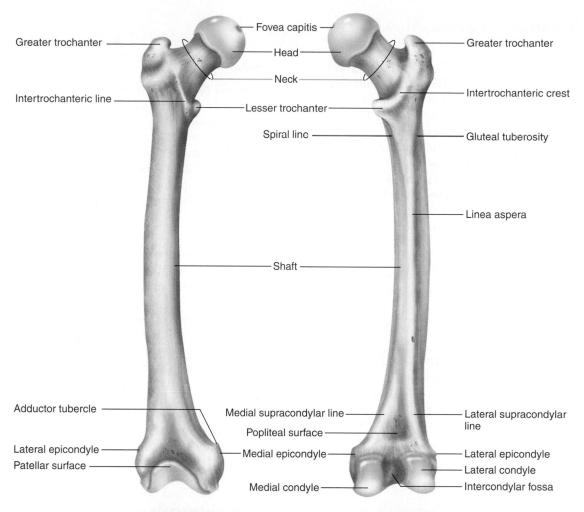

Greater trochanter — — Fovea capitis —
— Head — — Greater trochanter
Intertrochanteric line — — Neck —
— Intertrochanteric crest
— Lesser trochanter —
Spiral linc — — Gluteal tuberosity
— Linea aspera

Shaft

Adductor tubercle — Medial supracondylar line — — Lateral supracondylar line
Popliteal surface —
Lateral epicondyle — — Lateral epicondyle
Patellar surface — Medial epicondyle — — Lateral condyle
Medial condyle — — Intercondylar fossa

(a)   **Anterior view**   **Posterior view**

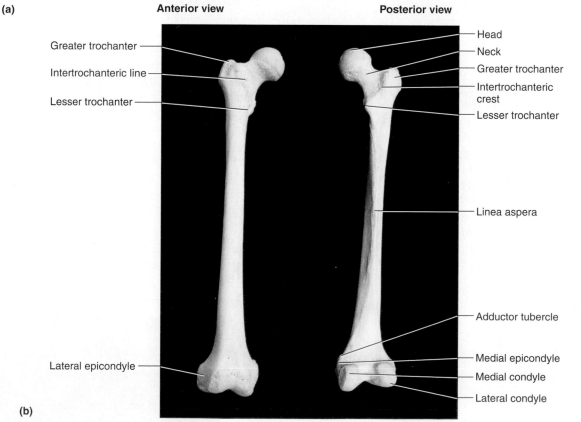

Greater trochanter — — Head
Intertrochanteric line — — Neck
Lesser trochanter — — Greater trochanter
— Intertrochanteric crest
— Lesser trochanter

— Linea aspera

— Adductor tubercle

Lateral epicondyle — — Medial epicondyle
— Medial condyle
— Lateral condyle

(b)

FIGURE 9.8 Right Femur (a) Diagram; (b) photograph.

The **shaft** of the femur is curved and bows anteriorly with the posterior shaft marked by the **linea aspera** (meaning rough line). At the proximal portion of the linea aspera is a roughened area called the **gluteal tuberosity,** an attachment site for the gluteus maximus muscle. The distal portion of the femur consists of **medial** and **lateral condyles.** These two smooth processes come into contact with the condyles of the tibia. To the sides of the condyles are bulges known as the **epicondyles.** Locate the **lateral epicondyle** and the **medial epicondyle** on the femur. Also note the location of the **adductor tubercle,** a small, triangular process proximal to the medial epicondyle. This is one of the attachment points for one of the adductor muscles. The head of the femur is medial, and the linea aspera is posterior. These features help distinguish whether you have a left or right femur.

Patella

The **patella** is a bone formed in the tendon that runs from the quadriceps muscles on the anterior thigh to the tibia. The patella is a specific type of bone known as a **sesamoid bone.** Sesamoid bones develop in tendons. The patella ossifies between ages three and six. When the leg is flexed, as when kneeling, the patella protects the ligaments of the knee joint. Examine the patella in the lab and compare it to figure 9.9.

Bones of the Leg

Tibia

The largest bone of the leg is the **tibia,** the weight-bearing bone of the leg inferior to the femur. It is the medial bone of the leg. The **tibial condyles** are separated by the **intercondylar eminence,** and they articulate with the condyles of the femur. The tibia is roughly triangular in cross section. The ridge on the anterior tibia is known as the **anterior crest.** This is the crest that is bruised when you run into something, such as a coffee table in the middle of the night. There is a proximal process on the anterior surface of the tibia known as the **tibial tuberosity.** This is the attachment point for the patellar ligament. At the distal end of the tibia is an extension of bone known as the **medial malleolus** (MAH-lee-OH-lus). This bump is one part of the ankle joint as it articulates with the talus of the foot. Locate the structures of the tibia in the lab and in figure 9.10.

Fibula

The **fibula** is smaller than the tibia (you can remember that it is smaller than the tibia because you "tell a little fib"). The fibula is lateral to the tibia. The fibula has a proximal **head** and a distal process called the **lateral malleolus.** The lateral malleolus is the other part of the leg that forms a joint with the talus. The fibula is not a weight-bearing bone but is an attachment point for muscles. Examine the bones in the lab and compare the fibula to figure 9.10.

Foot

There are seven **tarsal bones** of the foot, including the **talus,** which is the bone of articulation with the leg. Directly below the talus is the **calcaneus** (cal-CAY-nee-us), commonly known as the heel bone. The bone anterior to both the talus and the calcaneus is the **navicular.** The foot has three **cuneiform** (cue-NEE-ih-form) bones, known as the **medial,** or first, **cuneiform;** the **intermediate,** or second, **cuneiform;** and the **lateral,** or third, **cuneiform.** Locate these on specimens in the lab and in figure 9.11. The lateral cuneiform bone is *not* the most lateral bone in the foot but is the most lateral of the cuneiform bones. The bone lateral to the third cuneiform is the **cuboid.** One way to remember the tarsal bones is with the mnemonic "children that never march in line cry." The beginning letter of each word is the same as a tarsal bone (calcaneus, talus, navicular, medial, intermediate, and lateral cuneiform and cuboid).

The **metatarsals** are similar to the metacarpals in that the first metatarsal is under the largest digit, the big toe, and the fifth metatarsal is under the smallest digit. The pattern of the phalanges of the foot is the same as in the hand, with toes 2 to 5 having a **proximal, middle,** and **distal phalanx** each and the big toe, or **hallux,** having a proximal and distal phalanx. Examine the material in the lab and compare the metatarsals and phalanges to figure 9.11.

When you have finished locating the major features of the bones of the appendicular skeleton, test yourself and your lab partner by pointing to various structures and seeing how accurate you are at identifying the structures.

STUDY HINTS

Use your time in lab to *find* the material at hand. Once you are at home, draw or trace material from your manual and name all of the parts.

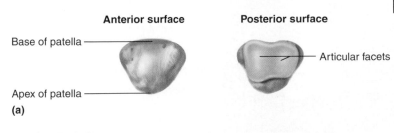

Anterior surface   **Posterior surface**

Base of patella

Articular facets

Apex of patella

(a)

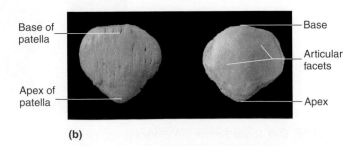

Base of patella

Base

Articular facets

Apex of patella

Apex

(b)

FIGURE 9.9 **Patella, Anterior and Posterior Views** (a) Diagram; (b) photograph.

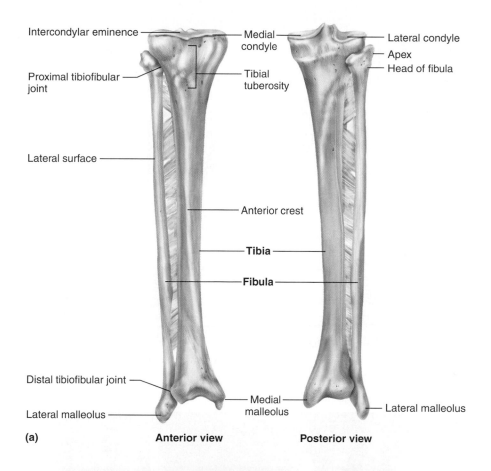

Intercondylar eminence

Proximal tibiofibular joint

Lateral surface

Anterior crest

Tibia

Fibula

Distal tibiofibular joint

Lateral malleolus

Medial condyle

Tibial tuberosity

Medial malleolus

Lateral condyle

Apex

Head of fibula

Lateral malleolus

(a)   **Anterior view**   **Posterior view**

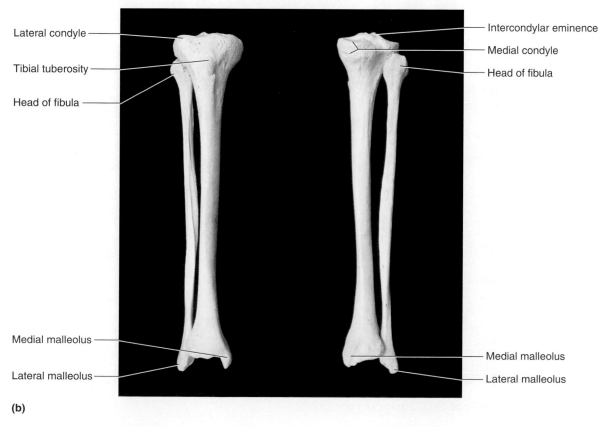

Lateral condyle

Tibial tuberosity

Head of fibula

Medial malleolus

Lateral malleolus

Intercondylar eminence

Medial condyle

Head of fibula

Medial malleolus

Lateral malleolus

(b)

FIGURE 9.10 Right Tibia and Fibula (a) Diagram; (b) photograph.

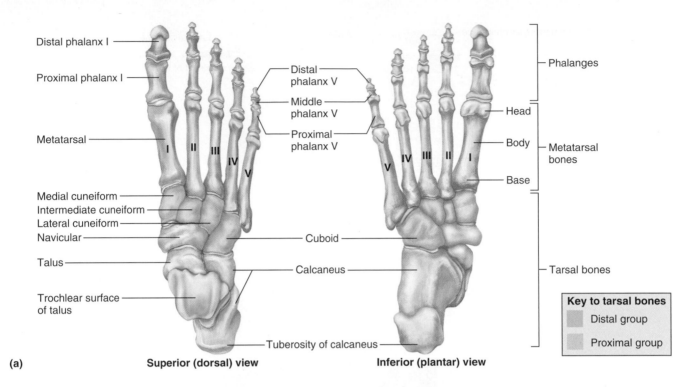

(a)   **Superior (dorsal) view**   **Inferior (plantar) view**

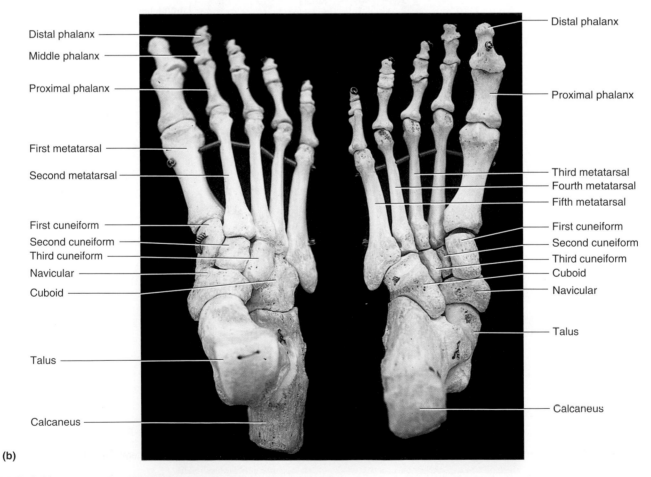

(b)

FIGURE 9.11 Bones of the Right Foot (a) Diagram; (b) photograph.

REVIEW SECTION

Appendicular Skeleton

Name _____ *Date* _____

Lab Section _____ *Time* _____

Review Questions

1. Name the anterior depression on the scapula.

2. The humerus fits into what specific part of the scapula?

3. What specific part of the clavicle attaches to the scapula?

4. Frequently, the clavicle is broken when the arms are extended to brace a fall. Explain how hitting the ground with your hands can fracture the clavicle.

5. The epiphyseal line on the humerus has what other name?

6. The condyles of the humerus have specific names. What are they?

7. Name the ridge of bony tissue proximal to the lateral condyle of the humerus.

8. What is the name of the lateral bone of the forearm?

9. Name the depression on the ulna into which the humerus inserts.

10. Name the bony process that extends distally from the head of the ulna.

11. How does the head of the ulna differ in position from the head of the radius?

12. Each metacarpal bone consists of three major regions. What are they?

13. Name the carpal bone at the base of the thumb.

14. The _____ joins the two hip bones at the anterior junction of these bones.

15. What curved area occurs directly on the hip bone below the posterior inferior iliac spine?

16. What is the cuplike depression of the hip bone into which the head of the femur fits?

17. Explain the difference between the false pelvis and the true pelvis.

18. Name the roughened line that runs along the length of the posterior femur.

19. Name the sesamoid bone that forms in the quadriceps tendon.

20. What is the name of the process on the distal portion of the tibia?

21. What is the function of the fibula?

22. What is another name for the heel bone?

23. What is the name of the bone of the foot that joins with the tibia and fibula?

24. Match the bone with the region it comes from.

    a. hallux   hip bone

    b. ilium   hand

    c. clavicle   upper extremity

    d. scaphoid   pectoral girdle

    e. radius   foot

25. How many bones are in the ankle versus the number of bones in the wrist?

26. Label the parts of the scapula in the following illustration. Use the terms provided.

    medial border   acromion process

    scapular spine   infraspinous fossa

    supraspinous fossa   inferior angle

    coracoid process   lateral border

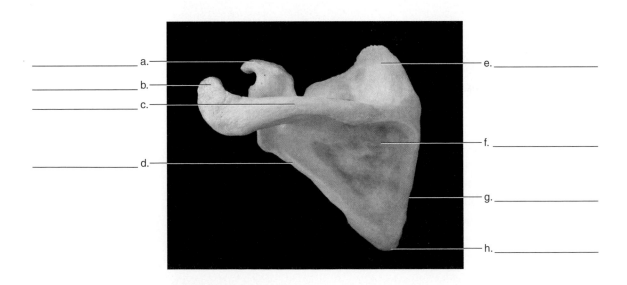

27. Label the parts of the ulna as illustrated, and determine if the bone is from the left or right side of the body.

olecranon process coronoid process
styloid process   trochlear notch
head   radial notch

28. Name the bone at the tip of the middle finger.

29. Label the following illustration of the foot.

calcaneus   medial cuneiform
cuboid   intermediate cuneiform
talus   lateral cuneiform
navicular

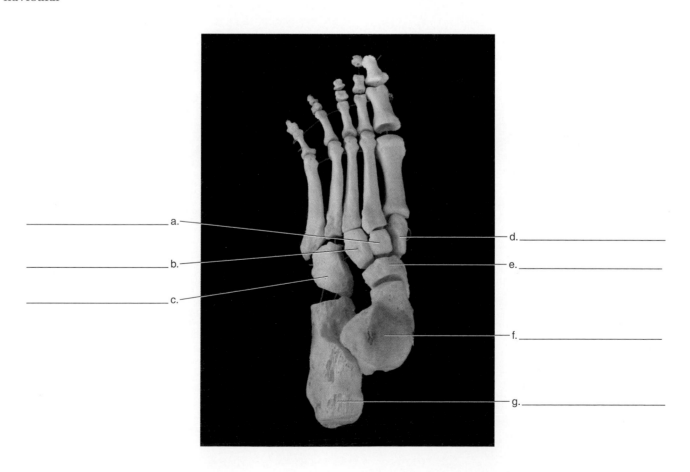

30. Label the following illustration of the hand.

lunate   trapezoid

hamate   capitate

pisiform

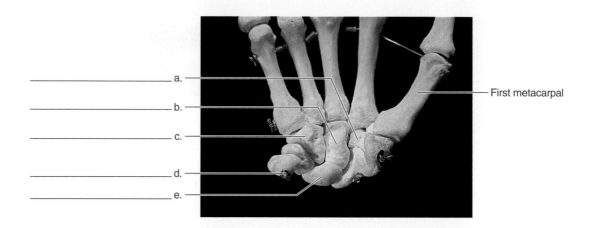

a.

b.

c.

First metacarpal

d.

e.

? Chapter Summary Data

Use this section to record your results from questions within the exercise.

1. _____   2. _____

LABORATORY

Axial Skeleton, Vertebrae, Ribs, Sternum, Hyoid

INTRODUCTION

The axial skeleton consists of 80 bones, including the skull, hyoid bone, vertebrae, sternum, and ribs. In this exercise you examine all parts of the axial skeleton except for the skull. Details of the skull are covered separately in Laboratory Exercise 11. There are five regions of the vertebral column. There are cervical (SERVE-ik-ul), thoracic (thor-AH-sic), and lumbar vertebrae and vertebrae that make up the sacrum and the coccyx. There are 33 individual vertebrae, and they form the vertebral column. Some fuse into larger structures, such as the sacrum. The thoracic cage consists of the 12 pairs of ribs and the sternum, which protect the lungs and heart yet provide flexibility during breathing. The hyoid bone is a small, floating bone between the floor of the mouth and the upper anterior neck. These topics are discussed further in the Saladin text in chapter 8, "The Skeletal System."

OBJECTIVES

At the end of this exercise you should be able to

1. locate and name all the bones of the vertebral column on an articulated skeleton;
2. name the significant surface features of individual vertebrae;
3. describe the differences among cervical, thoracic, and lumbar vertebrae and name and distinguish between the atlas and the axis;
4. locate the major features of the hyoid bone;
5. name the markings on the rib and distinguish a left rib from a right rib;
6. point out the three major regions of the sternum.

MATERIALS

Articulated skeleton or plastic casts of a skeleton

Disarticulated skeleton or plastic casts of bones

Charts of the skeletal system

Plastic straws cut on a bias or pipe cleaners for pointer tips

Foam pads of various sizes (to protect bone from hard countertops)

PROCEDURE

Locate the bones of the axial skeleton in the lab and on figure 10.1. In this exercise you study the bones of the axial skeleton by examining both the isolated bones and the articulated skeleton. Hold each bone up to the skeleton to see how it is positioned in relation to the other bones of the body. When examining bones in the lab do not use your pen or pencil to locate a structure. These leave marks on the bones. Use a cut plastic straw or a pipe cleaner to point out structures. Your instructor may want you to place real bone material on foam pads to cushion the bone from the tabletop.

Vertebral Column

The **vertebral column** is significantly different in humans than in other mammals in that we are the only habitually bipedal mammal. In humans the vertebrae increase in size from the cervical to the lumbar vertebrae. This is due to the increase in weight on the lower vertebrae. Between the vertebrae are fibrocartilaginous pads known as **intervertebral discs.** Obtain a vertebra

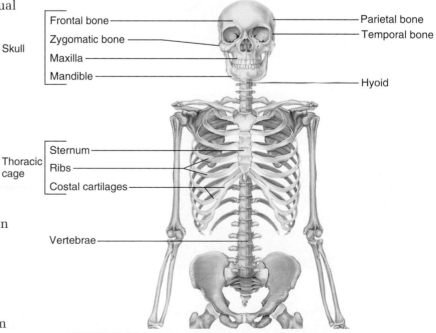

FIGURE 10.1 Major Bones of the Axial Skeleton

and examine it for the typical features (see figure 10.2). Place the vertebra in front of you so you can see through the large hole known as the **vertebral foramen.** Note that the large **body** of the vertebra supports the weight of the vertebral column and is in contact with the intervertebral discs. Note also the vertebral foramen, a hole where the spinal cord is located, and the **vertebral** (or **neural) arch,** which consists of two **pedicles** (PED-ik-uls) and two **laminae** (LAM-in-aye). The pedicles are

the parts of the arch that extend from the body of the vertebra to the two lateral projections called the **transverse processes.** Each lamina (LAM-in-uh) is a broad, flat structure between the transverse process and the dorsal **spinous process.**

If you rotate the vertebra as illustrated in figure 10.2, you should be able to see the vertebral body in lateral view, with the **superior articular process** and the **superior articular facet** visible. The superior articular facet is a smooth face that articulates with the vertebra above. By comparison, there is also an **inferior articular process** and an **inferior articular facet** that articulates with the vertebra below. If you put two adjacent vertebrae together you can see the **intervertebral foramina,** holes where spinal nerves are located. These features are seen in figure 10.3.

After examining a representative vertebra, look at the vertebrae from each region of the spinal column and note the specific features of each region. Compare these to an articulated vertebral column (see figure 10.4).

Spinal Curvatures

The spine has four curvatures, which alternate from superior to inferior. The **cervical curvature** is convex (bowed forward) when seen from anatomical position. The **thoracic curvature** is concave, while the **lumbar curvature** is convex and the **sacral (pelvic) curvature** is concave. These allow for balance in an upright posture. Locate the curvatures in figure 10.5. In addition to the spinal curvatures described, there are abnormal spinal curvatures. **Scoliosis** is a lateral curvature. **Kyphosis** (KY-foh-sis) is an exaggerated thoracic curvature, and **lordosis** is an exaggerated lumbar curvature (see figure 10.5).

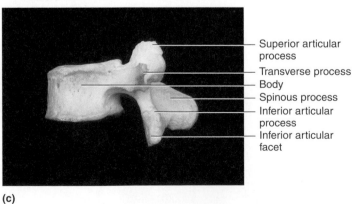

(a)

(b)

(c)

FIGURE 10.2 Features of a Typical Vertebra (a) Diagram, superior view. Photograph (b) superior view; (c) lateral view.

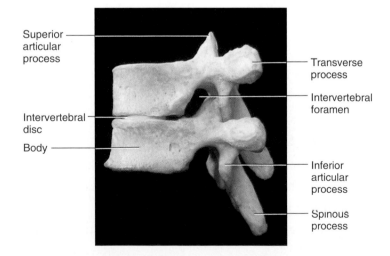

FIGURE 10.3 Articulated Vertebrae, Lateral View

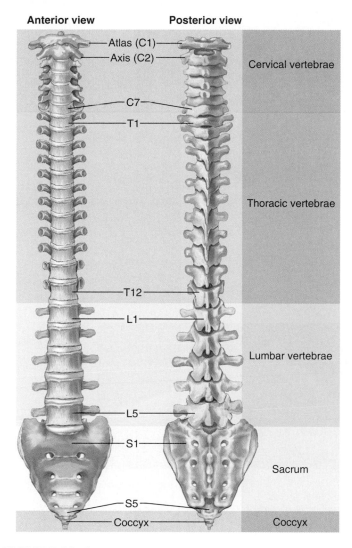

FIGURE 10.4 Overview of the Vertebral Column

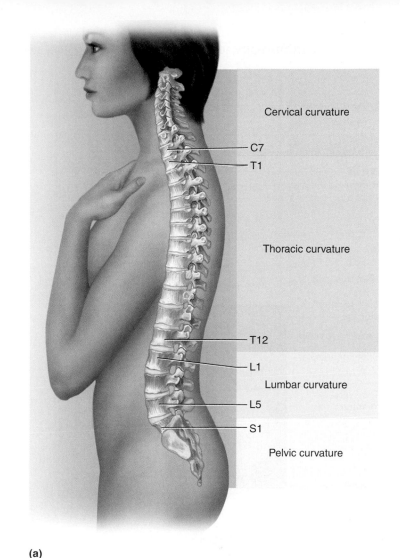

(a)

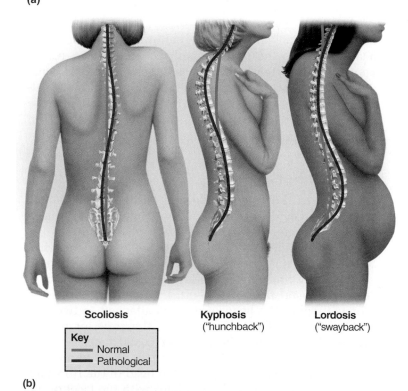

Scoliosis

Kyphosis ("hunchback")

Lordosis ("swayback")

Key
— Normal
— Pathological

(b)

FIGURE 10.5 Spinal Curvatures (a) Lateral view; (b) abnormal spinal curvatures.

Cervical Vertebrae

There are seven **cervical vertebrae** (C1–C7), and these can be distinguished from all other vertebrae in that each vertebra has three foramina. The **vertebral foramen** is the largest opening, and the two **transverse foramina** are specific to the cervical vertebrae. The transverse foramina house the vertebral arteries and vertebral veins. Some of the cervical vertebrae (C2–C6) have **bifid** (BY-fid) **spinous processes** (meaning split in two), and the bodies of the cervical vertebrae are less massive than those below. Examine the isolated cervical vertebrae in the lab and compare them to figure 10.6.

The first cervical vertebra (C1) is known as the **atlas** and is the only cervical vertebra without a body. The atlas carries the weight of the head and was named after the Greek mythological figure Atlas, a giant who carried the weight of the world on his shoulders. The atlas joins with the head and allows you to nod your head to indicate "yes." The second cervical vertebra (C2) is the **axis,** and it has a unique process called the **dens** or **odontoid process,** that runs superiorly through the atlas. The odontoid process allows

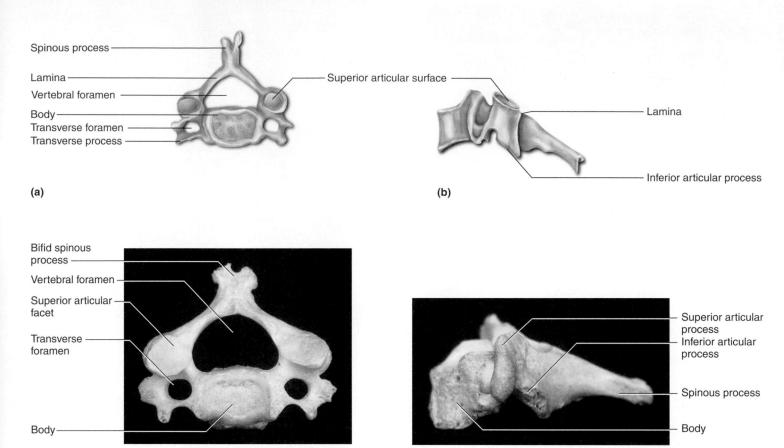

FIGURE 10.6 **Cervical Vertebra** Diagram (a) superior view; (b) lateral view. Photograph (c) superior view; (d) lateral view.

the atlas to rotate on the axis and allows you to rotate your head to indicate "no." The seventh cervical vertebra (C7) is known as the **vertebra prominens.** It has a **spinous process** that projects sharply in a posterior direction and can be palpated (felt) as a significant bump at the posterior base of the neck. Look at an atlas, an axis, and a vertebra prominens in the lab and compare them to figure 10.7.

Thoracic Vertebrae

There are 12 **thoracic vertebrae.** The thoracic vertebrae are distinguished from all other vertebrae by their markings on the lateral posterior surface of the body of the vertebrae, which are attachment points for the ribs. Sometimes a rib attaches to just one vertebra, in which case the marking on the vertebral body is known as a **complete costal facet** (FASS-et). In some sections of the thoracic region the head of a rib is attached to two vertebrae. The point of attachment at the superior part of the vertebra is called the **superior costal facet,** and the attachment of the rib to the inferior part of the vertebra is called the **inferior costal facet.** The head of the rib spans both of the vertebrae and articulates with the facet of the superior and inferior vertebrae. Vertebrae T1–T10 have

transverse costal facets on the terminal portions of the transverse processes. The thoracic vertebrae also have longer spinous processes than the cervical vertebrae, and the spinous processes of the thoracic vertebrae tend to angle in a more inferior direction. Examine figure 10.8 and note the characteristics of the vertebrae as seen in the lab.

Lumbar Vertebrae

There are five **lumbar vertebrae.** The lumbar vertebrae are distinguished from the other vertebrae by having neither transverse foramina nor rib facets. The spinous processes of the lumbar vertebrae tend to be more horizontal than those of the thoracic vertebrae, and the bodies of the lumbar vertebrae are larger, since they carry more weight than the thoracic vertebrae, yet the twelfth thoracic and the first lumbar vertebrae are remarkably similar. The last thoracic vertebra has rib facets on it, and the first lumbar does not. Look at the lumbar vertebrae in the lab and compare them to figure 10.9.

Sacrum

The **sacrum** (SAY-krum, SACK-rum) is a large, wedge-shaped bone composed of five fused vertebrae. The lines

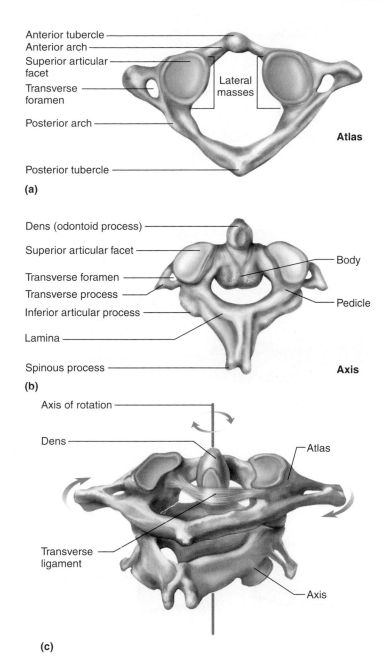

(a)

Anterior tubercle
Anterior arch
Superior articular facet
Transverse foramen
Posterior arch
Posterior tubercle
Lateral masses
Atlas

(b)

Dens (odontoid process)
Superior articular facet
Transverse foramen
Transverse process
Inferior articular process
Lamina
Spinous process
Body
Pedicle
Axis

(c)

Axis of rotation
Dens
Transverse ligament
Atlas
Axis

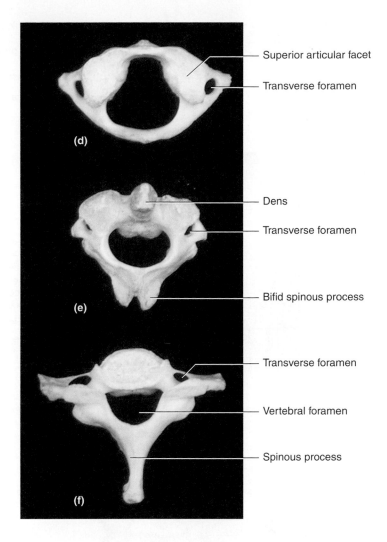

(d)
Superior articular facet
Transverse foramen

(e)
Dens
Transverse foramen
Bifid spinous process

(f)
Transverse foramen
Vertebral foramen
Spinous process

FIGURE 10.7 **Atlas, Axis, and Vertebra Prominens** (a) Atlas, superior view; (b) axis, supero-posterior view; (c) atlas and axis joined. Photographs of three vertebrae, superior view (d) atlas; (e) axis; (f) vertebra prominens.

of fusion are called **transverse lines,** and these may be seen on both the anterior and the posterior sides. The sacrum is shaped like a shallow, triangular bowl. If you place the sacrum in front of you as you would a cereal bowl, the shallow depression is the **anterior surface.** The two rows of holes you see are the **anterior sacral foramina.** The holes on the back are known as the **posterior sacral foramina.** Compare a bone in the lab to figure 10.10.

The **sacral promontory** is a rim on the anterior superior part of the sacrum, and the **alae** (AIL-ee) (singular *ala*) are two expanded regions of the sacrum lateral to the promontory. The roughened areas lateral to the alae are the **auricular** (aw-RIC-you-lur), **surfaces** of the sacrum; each

joins with the ilium to form the **sacroiliac** (SACK-ro-ILL-ee-ac) **joint.** As additional force is applied to the sacrum, it wedges into the ilium. If you examine the posterior surface of the sacrum you will notice the posterior sacral foramina, the **median** and **lateral sacral crests,** and the superior and inferior openings of the **sacral canal.** The **sacral hiatus** is the inferior opening of the sacrum. As with the other vertebrae, the **superior articular process** and the **superior articular facets** join with the next most superior vertebra (the fifth lumbar vertebra). These features can be seen in figure 10.10. The sacrum in humans is wedge-shaped; this allows the upper body to carry more weight than if the sacrum were box-shaped, as it is in many quadrupeds. Hold a block of wood or a box between your hands, as

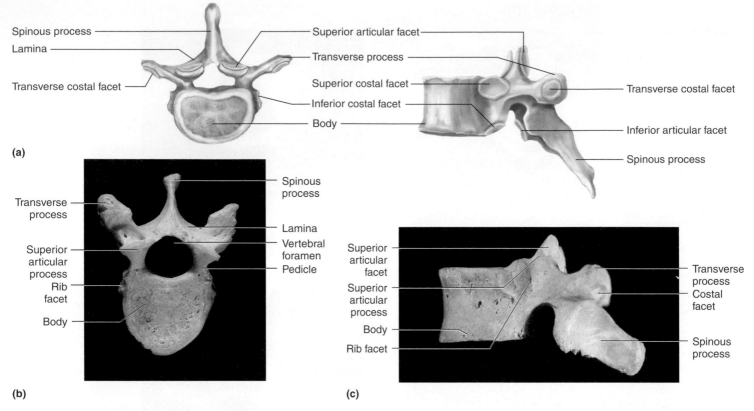

FIGURE 10.8 Thoracic Vertebra Diagram (a) superior and lateral views. Photograph (b) superior view; (c) lateral view.

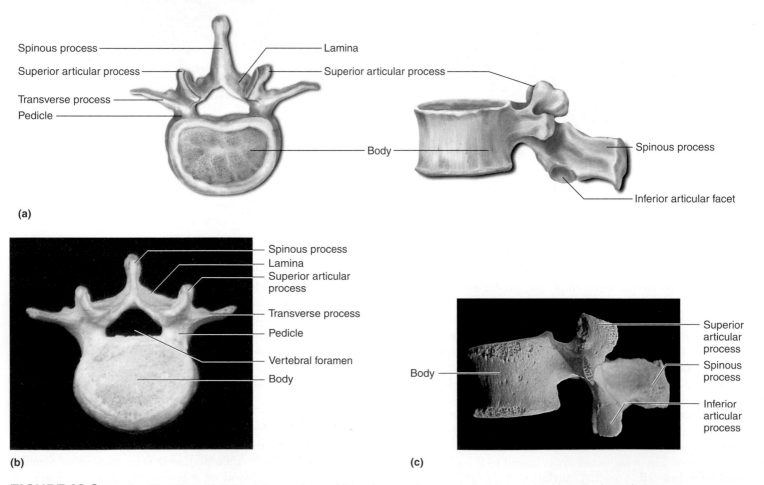

FIGURE 10.9 Lumbar Vertebra Diagram (a) superior and lateral views. Photograph (b) superior view; (c) lateral view.

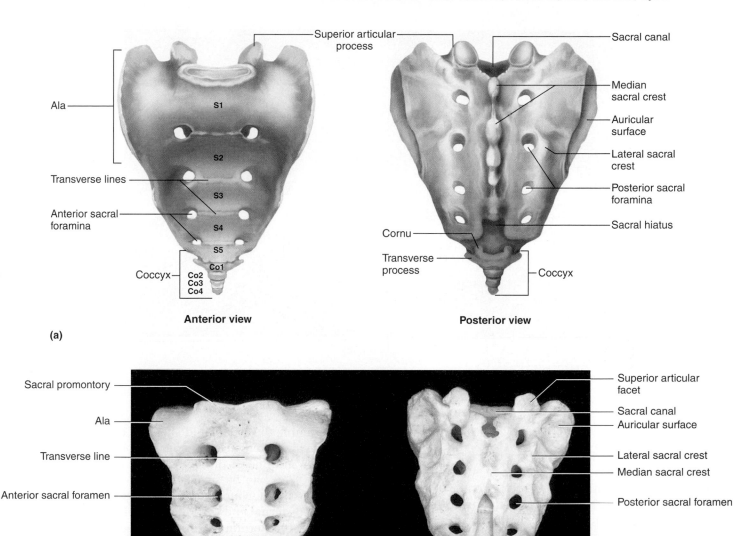

FIGURE 10.10 Sacrum and Coccyx (a) Diagram; (b) photograph.

illustrated in figure 10.11*a*. Have your lab partner push down on the box and see how much force is required for the box to slip through your hands. This would be similar to the forces that would act on a rectangular sacrum. Turn the box or block of wood so that it forms a wedge in your hands, as illustrated in figure 10.11*b*. Have your lab partner push down on the box. Is it easier or harder to dislodge the "sacrum" in this way?

Coccyx

The **coccyx** is the terminal portion of the vertebral column, and it consists of usually four fused vertebrae. The coccyx is fused with the sacrum in some individuals, especially females who have had several children. Examine the coccyx in the lab and compare it to figure 10.10.

Hyoid

The **hyoid** is a floating bone found at the junction of the floor of the mouth and the neck. The hyoid may be anchored by muscles from the anterior, posterior, or inferior directions; it aids tongue movement and swallowing. Locate the central **body** of the hyoid, the **greater cornua** (COR-new-uh) (singular, *cornu* [COR-new]), and the **lesser cornua** on the material in the lab and in figure 10.12.

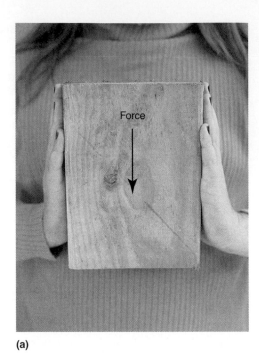

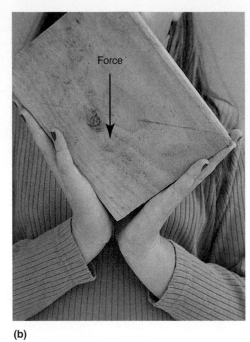

(a)　　　　　　　　　　　　　　　　(b)

FIGURE 10.11 **Forces Acting on the Sacrum** (a) Rectangular sacrum; (b) wedge-shaped sacrum.

Ribs

There are 12 pairs of **ribs** in the human, and these, along with the sternum, make up the thoracic cage. Each rib has a **head** that articulates with the body of one or more vertebrae. On the head is a **facet,** the site of articulation with the vertebral body. A constricted region near the head is the **neck** of the rib. Locate this on figure 10.13. The process near the neck on ribs 1–10 is the **tubercle** of the rib. This tubercle articulates with the transverse process of the vertebra.

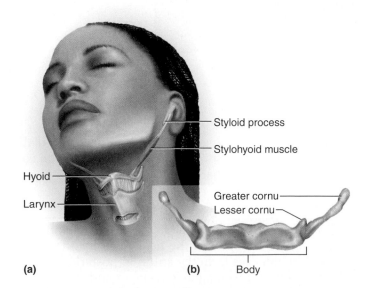

FIGURE 10.12 **Hyoid Bone, Anterior View** (a) In situ; (b) details of hyoid.

Examine the articulated skeleton in the lab and notice that the **shaft** of the ribs bends at about the same distance from the midline of the body as the inferior angle of the scapula. This can be seen on ribs 2–10 and is known as the **angle of the rib** or the **costal angle.** Some ribs have a truncated **sternal end** that attaches to costal cartilage prior to joining the sternum. The superior edge of the rib is more blunt than the inferior edge. The depression that runs along the inferior side of each rib is known as the **costal groove.** With the costal groove in an inferior position and the blunt sternal end toward the midline, determine whether you are looking at a left rib or a right rib.

The first seven pairs of ribs are **true ribs.** A true rib is one that attaches to the sternum by its own cartilage. These are known also as **vertebrosternal ribs,** as they attach to the vertebra posteriorly and to the sternum anteriorly. Ribs 8–12 are **false ribs,** because they do not attach to the sternum by their own cartilage. Ribs 8–10 attach to the sternum by way of the cartilage of rib 7. These ribs are also called **vertebrochondral ribs** for their posterior attachment to the vertebrae and their anterior attachment to the cartilage of rib 7. Ribs 11 and 12 do not attach to the sternum and are known as **floating ribs** (a specific type of false rib) or **vertebral ribs.** Examine the articulated skeleton in the lab and compare it to figure 10.14.

Sternum

The **sternum** is composed of three fused bones. The superior segment is the **manubrium** (ma-NOO-bree-um) and

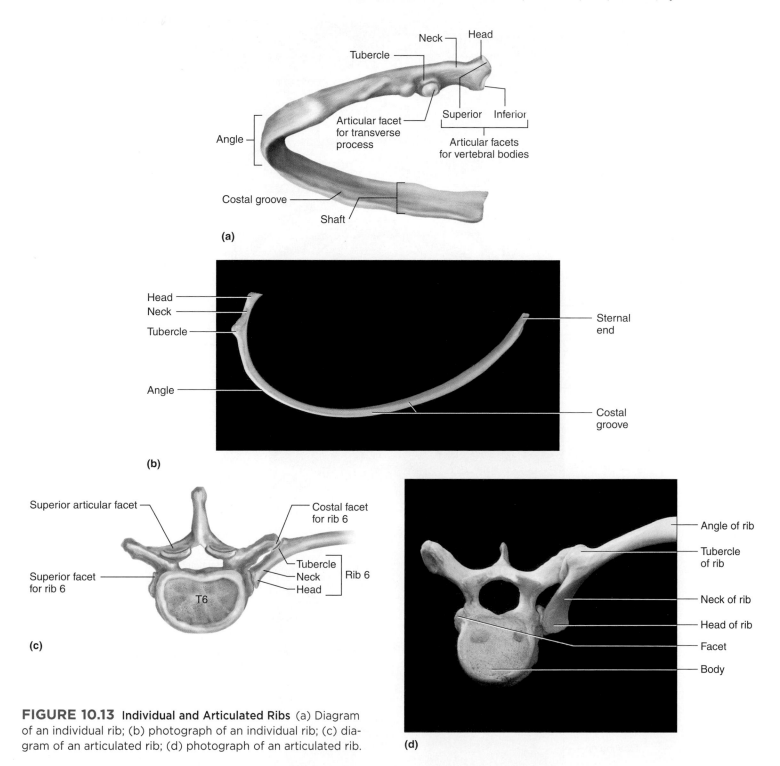

FIGURE 10.13 Individual and Articulated Ribs (a) Diagram of an individual rib; (b) photograph of an individual rib; (c) diagram of an articulated rib; (d) photograph of an articulated rib.

the depression at the top of the manubrium is the **median suprasternal notch** (jugular notch). The lateral indentations are the **clavicular notches.** The main portion of the sternum is the **body,** and between the body and the manubrium is the **sternal angle.** This is a landmark for finding the second rib when using a stethoscope to listen to heart sounds. Locate the **costal notches** on the body of the sternum. These are where the cartilages of the ribs attach. The

narrow, bladelike part that is the most inferior segment of the sternum is the **xiphoid** (ZYE-foyd) **process.** Care must be taken when performing cardiopulmonary resuscitation (CPR) so that pressure is applied to the body of the sternum and not to the xiphoid process. If the force is applied to the xiphoid process it could be fractured and driven into the liver. Examine the structures of the sternum on lab specimens and in figure 10.14.

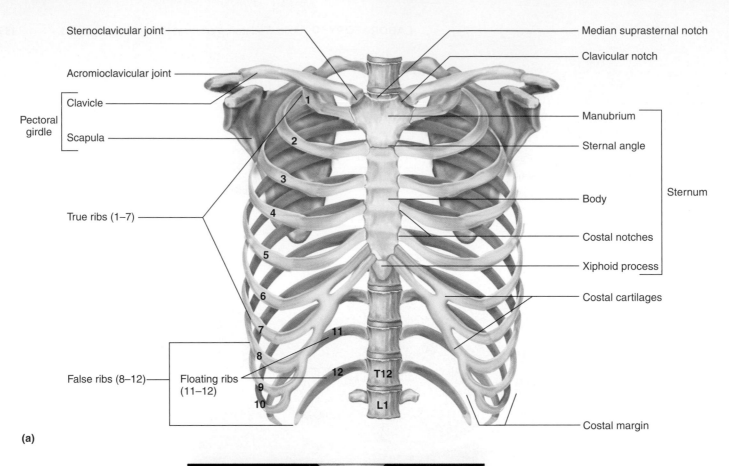

(a)

Sternoclavicular joint

Acromioclavicular joint

Pectoral girdle
- Clavicle
- Scapula

True ribs (1–7)

False ribs (8–12)

Floating ribs (11–12)

Median suprasternal notch

Clavicular notch

Manubrium

Sternal angle

Body — Sternum

Costal notches

Xiphoid process

Costal cartilages

Costal margin

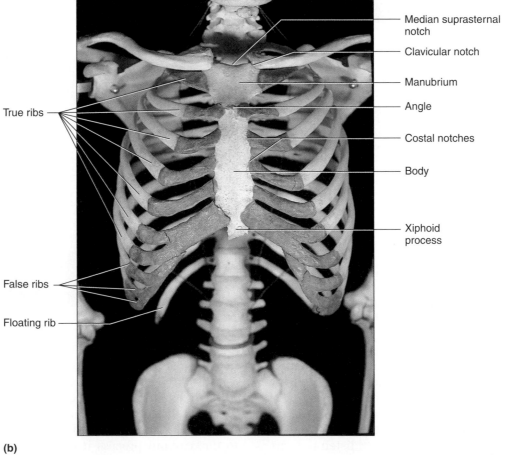

(b)

True ribs

False ribs

Floating rib

Median suprasternal notch

Clavicular notch

Manubrium

Angle

Costal notches

Body

Xiphoid process

FIGURE 10.14 Thoracic Cage (a) Diagram; (b) photograph.

REVIEW SECTION

Axial Skeleton, Vertebrae, Ribs, Sternum, Hyoid

Name ———————————————————— *Date* ——————

Lab Section ———————————————————— *Time* ——————

Review Questions

1. Label the following illustration using the terms provided.

body	pedicle	transverse process
lamina	spinous process	vertebral arch

a. ——————————

b. ——————————

c. ——————————

d. ——————————

e. ——————————

f. ——————————

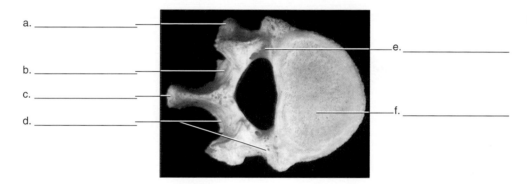

2. What part of a rib articulates with the transverse process of a vertebra?

3. The most inferior portion of the sternum is the

  a. body.

  b. manubrium.

  c. angle.

  d. xiphoid.

4. Distinguish between the posterior sacral foramina and the sacral canal.

5. What two features do cervical vertebrae have that no other vertebrae have?

6. Determine from what part of the spinal column the vertebra in the photograph comes. How can you tell?

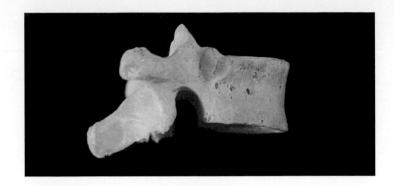

7. What is the name of the superior portion of the sternum?

8. What structures make up the vertebral arch?

9. The superior articular process of a vertebra articulates with what specific structure?

10. How many lumbar vertebrae are in the human body?

11. Name the second cervical vertebra.

12. Which spinal curvature is the most superior?

13. Does the hyoid bone have any solid bony attachments?

14. A rib that attaches to the sternum by the cartilage of another rib has what name?

15. Is the angle of the rib on the anterior or posterior side of the body?

LABORATORY

Axial Skeleton—Skull

INTRODUCTION

The skull consists of two major regions, the **cranium** and the **face.** The skull is the most complex region of the skeletal system and not only houses the brain but contains a significant number of the sense organs as well. The cranium consists mostly of flat bones, while the face is composed of irregular bones. In this exercise you learn the details of the skull and the anatomy of the fetal skull. These topics are discussed in the Saladin text in chapter 8, "The Skeletal System."

OBJECTIVES

At the end of this exercise you should be able to

1. list all the bones of the cranium and the major features of those bones;
2. list all the bones of the face and the major features of those bones;
3. name all the bones that occur singly or in pairs in the skull;
4. locate the major foramina of the skull;
5. name the major sutures of the skull;
6. list all the fontanels of the fetal skull.

MATERIALS

Disarticulated skull, if available

Articulated skulls with the calvaria cut

Foam pads to cushion skulls from the desktop

Fetal skulls

Pipe cleaners or plastic straws for pointer tips

PROCEDURE

General Considerations

Familiarize yourself with the bones of the skull in the cranium and the face. The skull bones in the cranium are listed with a number after the name of the bone indicating whether the bone occurs *singly* or as a *pair.* Take a skull back to your lab table and locate the bones listed in figure 11.1.

Bones of the Cranium

frontal (1)	parietal (2)
occipital (1)	temporal (2)
sphenoid (1)	ethmoid (1)

You can remember the bones of the cranium with the mnemonic "of pets." Each letter represents a cranial bone ('*o*' for *occipital,* '*f*' for *frontal,* etc.).

The bones of the face are listed next. Locate these bones on the skull in the lab, as shown in figures 11.1, 11.2, and 11.3.

Bones of the Face

maxilla (2)	nasal (2)
mandible (1)	palatine (2)
vomer (1)	zygomatic (2)
lacrimal (2)	inferior nasal concha (plural, *conchae*) (2)

The facial bones can be remembered by this mnemonic device: "Manny makes naturally insane zigzags pleasing Vera Lynn." The first names (Manny and Vera) represent the single bones of the face (mandible and vomer) and the other bones are paired.

As you locate the cranial and facial bones on a skull in the lab, make sure you do not poke pencils, pens, or fingers into the delicate regions of the eyes or nasal cavity. You may break the delicate structures in these regions. Be very careful handling skulls. Once you have found the major bones of the skull use the following descriptions and illustrations to find the structures of the skull. The skull is described from a series of views, and you should locate the structures listed for those views.

Lateral View

Locate the cranial bones from a lateral view. These are the **frontal, parietal, occipital, temporal, sphenoid,** and **ethmoid** bones. Find the coronal and lambdoid sutures, as well as the **squamous suture,** which separates the **temporal bone** from the **parietal bone.** You may be able to see

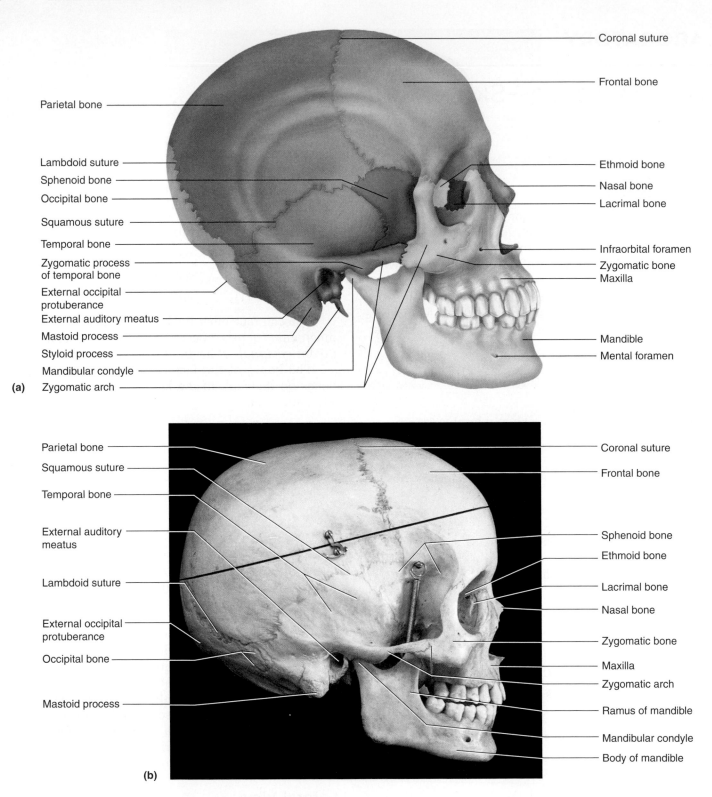

(a)

(b)

FIGURE 11.1 **Skull, Lateral View** (a) Diagram; (b) photograph.

a bump at the back of the occipital bone from this angle. This is known as the **external occipital protuberance.** The hole on the side of the head where the ears attach is the **external auditory meatus.** The large process posterior and inferior to this opening is the **mastoid process.** A long, thin spine medial to the mastoid process is the **styloid process,** which has muscles that connect it to the hyoid bone, tongue, and larynx. The sphenoid bone can also be seen from this view just anterior to the temporal bone. A bony process is found lateral to the sphenoid bone. This

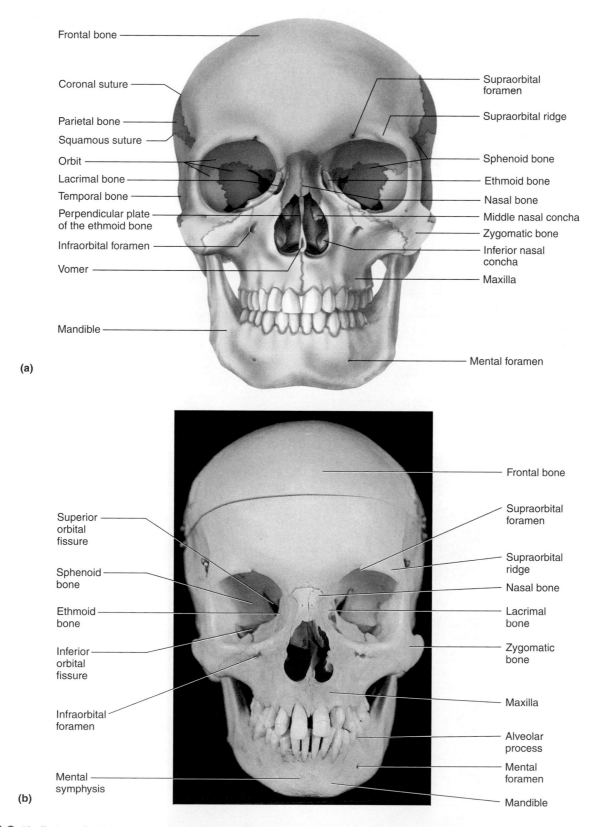

Frontal bone

Coronal suture

Parietal bone

Squamous suture

Orbit

Lacrimal bone

Temporal bone

Perpendicular plate of the ethmoid bone

Infraorbital foramen

Vomer

Mandible

Supraorbital foramen

Supraorbital ridge

Sphenoid bone

Ethmoid bone

Nasal bone

Middle nasal concha

Zygomatic bone

Inferior nasal concha

Maxilla

Mental foramen

(a)

Superior orbital fissure

Sphenoid bone

Ethmoid bone

Inferior orbital fissure

Infraorbital foramen

Mental symphysis

Frontal bone

Supraorbital foramen

Supraorbital ridge

Nasal bone

Lacrimal bone

Zygomatic bone

Maxilla

Alveolar process

Mental foramen

Mandible

(b)

FIGURE 11.2 Skull, Anterior View (a) Diagram; (b) photograph.

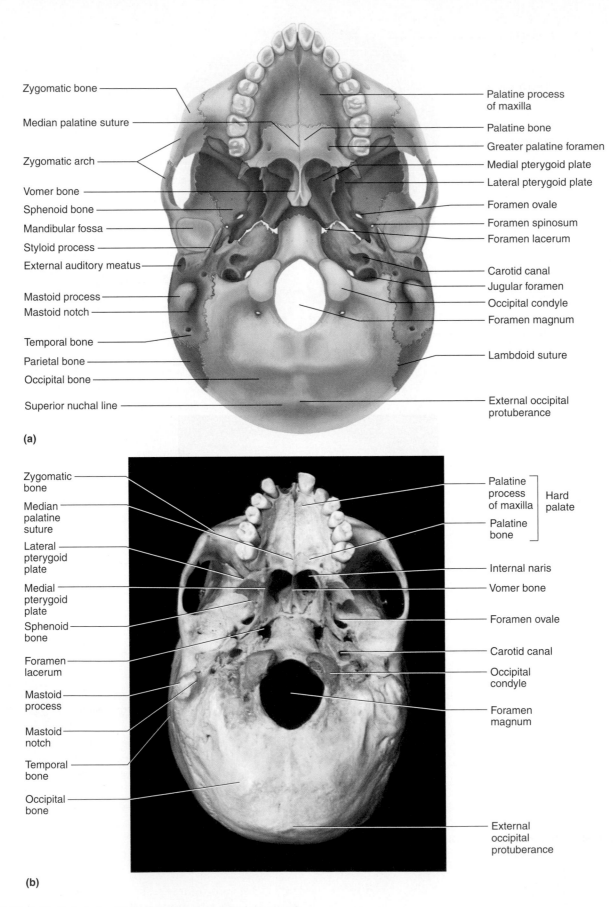

Zygomatic bone

Median palatine suture

Zygomatic arch

Vomer bone

Sphenoid bone

Mandibular fossa

Styloid process

External auditory meatus

Mastoid process

Mastoid notch

Temporal bone

Parietal bone

Occipital bone

Superior nuchal line

Palatine process of maxilla

Palatine bone

Greater palatine foramen

Medial pterygoid plate

Lateral pterygoid plate

Foramen ovale

Foramen spinosum

Foramen lacerum

Carotid canal

Jugular foramen

Occipital condyle

Foramen magnum

Lambdoid suture

External occipital protuberance

(a)

Zygomatic bone

Median palatine suture

Lateral pterygoid plate

Medial pterygoid plate

Sphenoid bone

Foramen lacerum

Mastoid process

Mastoid notch

Temporal bone

Occipital bone

Palatine process of maxilla

Palatine bone

Hard palate

Internal naris

Vomer bone

Foramen ovale

Carotid canal

Occipital condyle

Foramen magnum

External occipital protuberance

(b)

FIGURE 11.3 Skull, Inferior View (a) Diagram; (b) photograph.

is the **zygomatic arch,** which makes up the upper part of the cheek. The zygomatic arch is composed of part of the temporal bone and part of the **zygomatic bone,** commonly known as the cheekbone. It can be seen as forming the lateral wall of the orbit. On the inner wall of the orbit is the ethmoid bone, the lacrimal bone, the maxilla, and the nasal bone. Examine these structures in figure 11.1.

Anterior View

The large bone that makes up the forehead is the **frontal bone.** It is a single bone that makes up the superior portion of the **orbits** (the eye sockets) and has two ridges above the eyes called **supraorbital ridges** (the eyebrows). There is a hole in each of these ridges called the **supraorbital foramina,** which allow for nerves and arteries to reach the face. Most of what you can see inferior to the frontal bone are facial bones. The bony part of the nose is composed of the paired **nasal bones,** which join with the cartilage that forms the tip of the nose. Posterior to the nasal bones are thin strips of the upper **maxillae,** bones that hold the upper teeth; posterior to the maxillae are the thin **lacrimal bones,** which contain the **nasolacrimal duct,** a tube that drains tears into the nose. Locate these bones in figure 11.2. Posterior to the lacrimal bones is the **ethmoid bone.** It is very delicate and frequently broken on mishandled skulls. Posterior to the ethmoid is the **sphenoid bone,** which forms the posterior wall of the orbit and contains not only the **optic canal** (a passageway for the optic nerve) but also the **superior orbital fissure** and the **inferior orbital fissure.** Note the bones on the lateral side of the orbit. These are the **zygomatic bones.**

The major bone below the orbit is the maxilla, which also makes up the floor of the orbit. The two maxillae have sockets called **alveoli** (singular, *alveolus*), which contain the upper teeth. Extensions of bone between each pair of sockets are called **alveolar processes.** The **infraorbital foramen** is a small hole below the eye in the maxilla and is a passageway for nerves and blood vessels. The most inferior bone of the face is the **mandible,** commonly known as the jaw. The mandible also has alveoli and alveolar processes (see figure 11.2). The mandible begins as two bones in utero and fuses at the midline of the chin. This fusion of the **mental symphysis** joins the two mandible bones into one. On the lateral aspects of the mandible are the **mental foramina,** which conduct nerves and blood vessels to the tissue anterior to the jaw.

Inferior View

Place the skull in front of you with the mandible removed (see figure 11.3). The largest hole in the skull, the **foramen magnum,** should be close to you. The foramen magnum is in the occipital bone and is the boundary between the brain and the spinal cord. Adjacent to the foramen magnum are the **occipital condyles,** processes that articulate with the first cervical vertebra. A small bump at the posterior part of

the occipital bone is the external occipital protuberance, an attachment site for muscles. At the junction of the occipital bone and the temporal bone is the **jugular foramen,** a hole through which passes the internal jugular vein. If you carefully insert a pipe cleaner into the jugular foramen and turn the skull over, you will see that the jugular foramen leads to the posterior part of the skull.

You can see the mastoid process of the temporal bone and a depression just medial to the process known as the **mastoid notch.** Find the styloid processes in your specimen; they may be hard to locate because they frequently get broken in lab specimens. The styloid process is an attachment point for muscles that move the tongue, larynx, or hyoid. Medial to the styloid process is the **carotid canal,** through which passes the internal carotid artery to the brain. If you *carefully* insert a pipe cleaner into the carotid canal of a real skull you will notice that the canal bends at about a 90-degree angle; if you turn the skull over you will note that the opening occurs in the middle of the skull. You cannot do this with most plastic casts of skulls. At the junction of the temporal bone and the sphenoid bone is the **foramen lacerum,** which is next to the carotid canal. The temporal bone also has a **mandibular fossa,** the articulation site of the mandible. The zygomatic process of the temporal bone can also be seen from this view.

From the inferior view the sphenoid bone can be seen as a bone that runs from one side of the skull to the other. The part of the sphenoid seen from the lateral view is one of the **greater wings** of the sphenoid. Two pairs of flattened process can also be seen in the view, the **lateral pterygoid plate** and the **medial pterygoid plate** (*pterygoid* means winglike). These are attachments for muscles that extend from the sphenoid to the mandible. Just posterior to the pterygoid processes is the **foramen ovale,** which conducts one of the branches of the trigeminal nerve to the mandible.

In the midline of the skull and sometimes confused as part of the sphenoid is the **vomer.** This is a single bone of the face that forms part of the **nasal septum.** The holes on either side of the vomer are the **internal nares.** Connected to the vomer and forming part of the **hard palate** is the **palatine bone.** The palatine bones are **L**-shaped bones with a horizontal plate and a vertical plate. The horizontal plates normally join together at the **median palatine suture.** If this suture does not fuse completely at birth an individual has a cleft palate. The rest of the hard palate consists of horizontal shelves of the maxillae. These form the **palatine processes of the maxilla.**

The major openings of the skull are presented in table 11.1. Locate the openings and note the number of structures that pass through these holes.

Superior View

Locate the major suture lines of the skull from the superior view (see figure 11.4). The frontal bone is separated from the pair of parietal bones by the **coronal suture.**

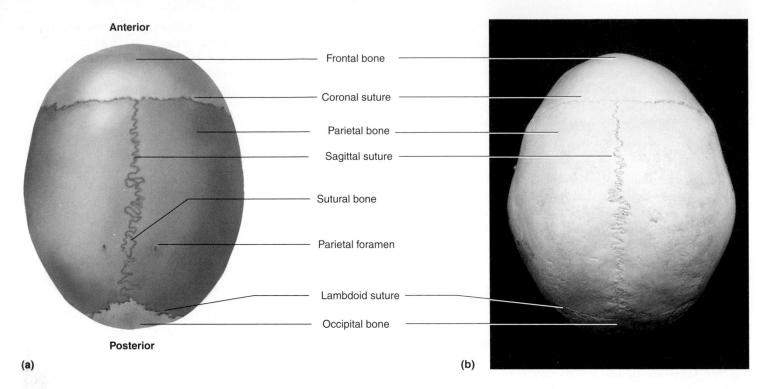

FIGURE 11.4 Skull, **Superior View** (a) Diagram; (b) photograph.

TABLE 11.1	Openings of the Skull
Opening	**Function or Structure in Opening**
Foramen magnum	Spinal cord vertebral arteries
Jugular foramen	Internal jugular vein, vagus, and other nerves
Carotid canal	Internal carotid artery
Stylomastoid canal	Facial nerve exits skull
Foramen lacerum	Closed by cartilage
Foramen ovale	Mandibular branch of trigeminal nerve
Foramen spinosum	Meningeal blood vessels
Foramen rotundum	Maxillary branch of trigeminal nerve
Optic canal	Optic nerve
Superior orbital fissure	Nerves to the eye and face
Inferior orbital fissure	Maxillary branch of trigeminal nerve
Mandibular foramen	Mandibular branch of trigeminal nerve
Mental foramen	Mental nerve and blood vessels
Supraorbital foramen	Supraorbital nerve and artery for face
Infraorbital foramen	Infraorbital nerve and artery for face
External acoustic meatus	Opening for sound transmission
Internal acoustic meatus	Vestibulocochlear nerve and facial nerve

The parietal bones are separated from one another by the **sagittal suture.** The parietal bones are separated from the occipital bone by the **lambdoid suture** (named after the Greek letter lambda), which is **Y**-shaped. There may be small bones between the occipital bone and the parietal bones (or between other skull bones), and these are known as **sutural,** or **Wormian, bones.**

Interior of the Cranium

With the top of the skull removed you can see that the cranial cavity is divided into three major regions. These are the **anterior cranial fossa,** a depression anterior to the lesser wings of the sphenoid; the **middle cranial fossae,** which lie between the lesser wings of the sphenoid and the petrous portion of the temporal bone; and the **posterior cranial fossa,** posterior to the petrous portion of the temporal bone. Examine figure 11.5 for a view of the interior of the cranium.

Beginning with the anterior cranial fossa you should find the centrally located **ethmoid bone.** A sharp ridge known as the **crista galli** projects from the main portion of this bone. The small, horizontal plate of bone with numerous holes lateral to the crista galli is the **cribriform plate.** The holes in this plate are called the **olfactory foramina.** The nerves passing through these holes transmit the sense of smell from the nose to the brain. The nerves synapse in the olfactory bulbs, which are superior to the cribriform plate. If the skull was cut close to the

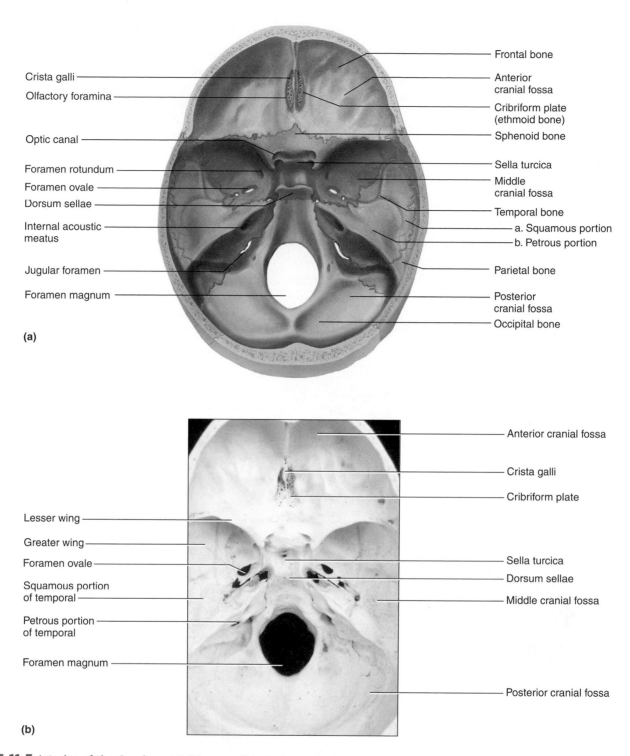

FIGURE 11.5 **Interior of the Cranium** (a) Diagram; (b) photograph.

orbit, you can see the **frontal sinus** as a hollow space in the anterior portion of the frontal bone.

The dividing line between the anterior and middle cranial fossae is the sphenoid bone. Locate the lesser wings of the sphenoid and the **sella turcica** just posterior to it. The sella turcica has a small depression in which the pituitary gland sits. The posterior, raised part of the sella turcica is known as the **dorsum sellae.** The greater wings

of the sphenoid are more inferior than the lesser wings, and each one contains the **foramen rotundum,** which takes a branch of the trigeminal nerve to the maxilla. The **foramen ovale** can be seen from this view as well.

Behind the sphenoid bone is the temporal bone, which has a flattened lateral section known as the **squamous portion** and a heavier mass of bone known as the **petrous portion.** The petrous portion divides the middle and posterior cranial fossae. The petrous portion also has a hole in the posterior surface, the **internal acoustic meatus.** This is a passageway for the nerves that come from the inner ear.

The posterior cranial fossa is located dorsal to the petrous portion of the temporal bone. It contains the foramen magnum and the jugular foramina. Most of this fossa is formed by the occipital bone.

Midsagittal Section of the Skull

Be extremely careful with the midsagittal section of the skull because many of the internal structures are fragile. Use figure 11.6 as a guide.

Locate the **nasal septum,** composed of the **vomer,** the **perpendicular plate** of the ethmoid bone, and the **nasal cartilage** (absent in skull preparations). If the

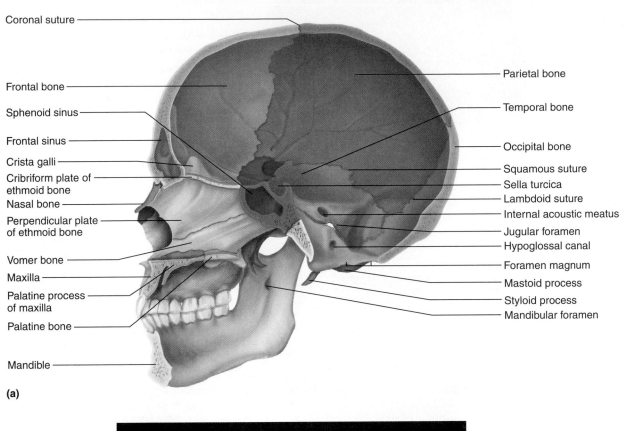

Coronal suture
Frontal bone
Sphenoid sinus
Frontal sinus
Crista galli
Cribriform plate of ethmoid bone
Nasal bone
Perpendicular plate of ethmoid bone
Vomer bone
Maxilla
Palatine process of maxilla
Palatine bone
Mandible

Parietal bone
Temporal bone
Occipital bone
Squamous suture
Sella turcica
Lambdoid suture
Internal acoustic meatus
Jugular foramen
Hypoglossal canal
Foramen magnum
Mastoid process
Styloid process
Mandibular foramen

(a)

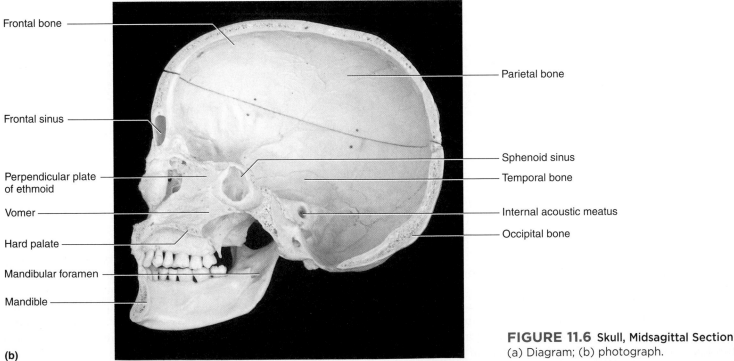

Frontal bone
Frontal sinus
Perpendicular plate of ethmoid
Vomer
Hard palate
Mandibular foramen
Mandible

Parietal bone
Sphenoid sinus
Temporal bone
Internal acoustic meatus
Occipital bone

(b)

FIGURE 11.6 Skull, Midsagittal Section
(a) Diagram; (b) photograph.

nasal septum is removed, you should be able to see the **superior nasal concha** and the **middle nasal concha** of the ethmoid bone. Below these is the **inferior nasal concha,** a separate, distinct bone. Look for the junction between the palatine bone and the palatine process of the maxilla. These two bony plates make up the hard palate. If the mandible is present, locate the **mandibular foramen** on the inner aspect of the mandible. It transmits branches of the trigeminal nerve and blood vessels to the mandible.

Sinuses

There are numerous sinuses and air cells in the skull. These sinuses provide shape to the skull while decreasing its weight and some add resonance to the voice. The **paranasal sinuses** occur around the region of the nose and are named for the bones in which they are found. They include the **frontal sinus,** the **maxillary sinus,** the **ethmoid sinus** (or air cells), and the **sphenoid sinus.** These sinuses may fill with fluid when a person has a cold and harbor bacteria in secondary infections. Locate the sinuses in skulls in the lab and compare them to figure 11.7.

Fontanels

The **fontanels** are the "soft spots" of an infant's skull. There are four types of fontanels, and some of them allow for the passage of the skull through the birth canal by enabling the bones of the cranium to slide over one another. After birth the fontanels allow for further expansion of the skull. The **anterior (frontal) fontanel** is an area between the frontal bone and the parietal bones. The **poste-**

rior (occipital) fontanel is between the occipital bone and the parietal bones. The **sphenoid (anterolateral) fontanels** are paired structures on each side of the skull and are located superior to the **sphenoid** bone, and the **mastoid (posterolateral) fontanels** are paired structures posterior to the temporal bone. Most fontanels fuse before 1 year of age, although the frontal fontanel may fuse as late as age 2. Locate these structures in figure 11.8 and on the material available in the lab.

Select Individual Bones of the Skull

Mandible

The mandible has a **condyloid process** with a terminal **mandibular condyle,** which articulates with the temporal bone; a **coronoid process,** which lies medial to the zygomatic arch; and the **mandibular notch,** a depression between the condyloid process and coronoid process of the mandible. The vertical section of the mandible is known as the **ramus of the mandible** (*ramus* = branch), and the horizontal portion of the mandible is the **body of the mandible.** The **angle** of the mandible is at the posterior of the mandible at the junction of the body and the ramus. On the inside of each ramus of the mandible is the **mandibular foramen,** a conduit for an artery, a vein, and a nerve. The parts of the mandible can also be identified in figures 11.1 and 11.9.

Ethmoid

The **ethmoid bone** is a cranial bone located in the middle of the skull. The **perpendicular plate** of the ethmoid can be seen in midsagittal view or from the anterior view through the external nares. The **orbital plate** is the part

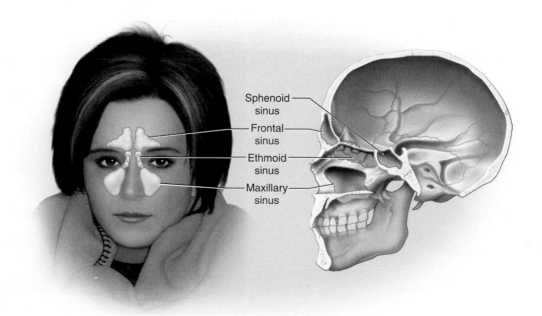

Sphenoid sinus
Frontal sinus
Ethmoid sinus
Maxillary sinus

FIGURE 11.7 Sinuses of the Skull

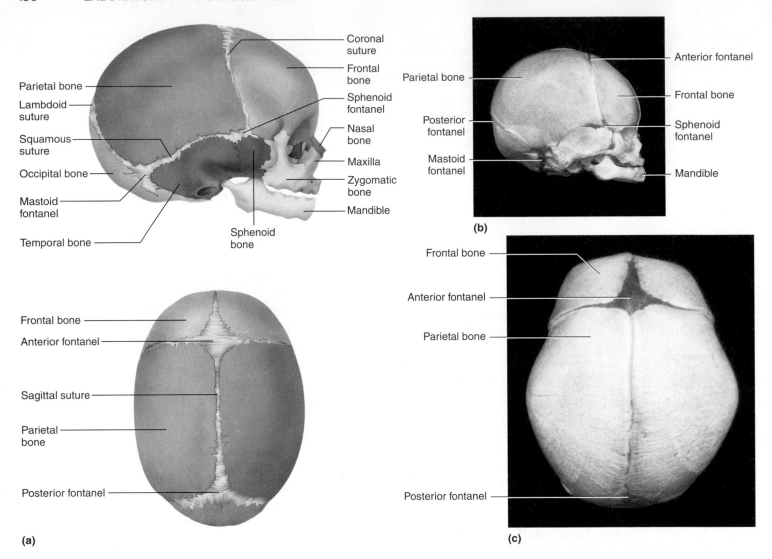

FIGURE 11.8 Fetal Skull and Fontanels (a) Diagram of lateral and superior views. Photograph (b) lateral view; (c) superior view.

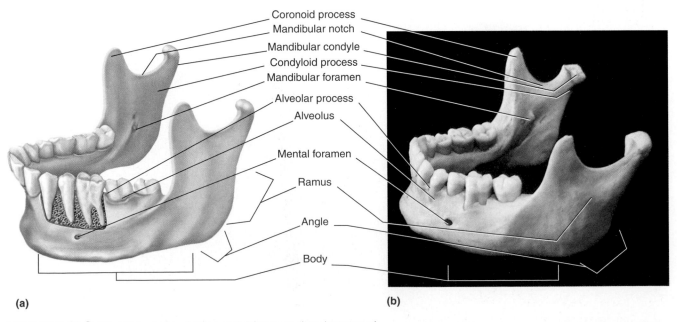

FIGURE 11.9 Mandible, Lateral View (a) Diagram; (b) photograph.

of the ethmoid that lines the medial wall of the orbit. The **middle nasal conchae** can be seen from the nasal cavity as well, but the **superior nasal conchae** are best seen by looking at an inferior view of the skull through the internal nares or at a midsagittal view with the nasal septum removed. Examine isolated ethmoid bones in the lab and find the **crista galli, cribriform plate,** and other structures, as shown in figure 11.10.

Sphenoid

The **sphenoid bone** is seen in figure 11.11. Examine an isolated sphenoid bone in the lab and locate the **greater wings,** the **lesser wings,** the **medial** and **lateral pterygoid plates,** the **sella turcica,** the **dorsum sellae,** and other features, as seen in figure 11.11. The sella turcica has a small depression in it called the **hypophyseal fossa** in which the pituitary gland sits.

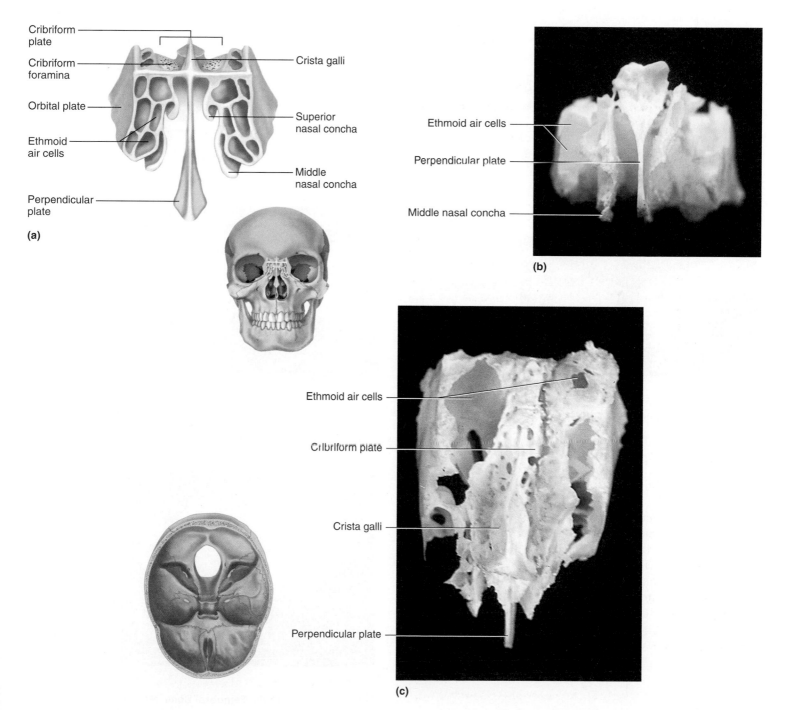

FIGURE 11.10 Ethmoid Bone Diagram (a) anterior view, Photograph (b) anterior view; (c) superior view.

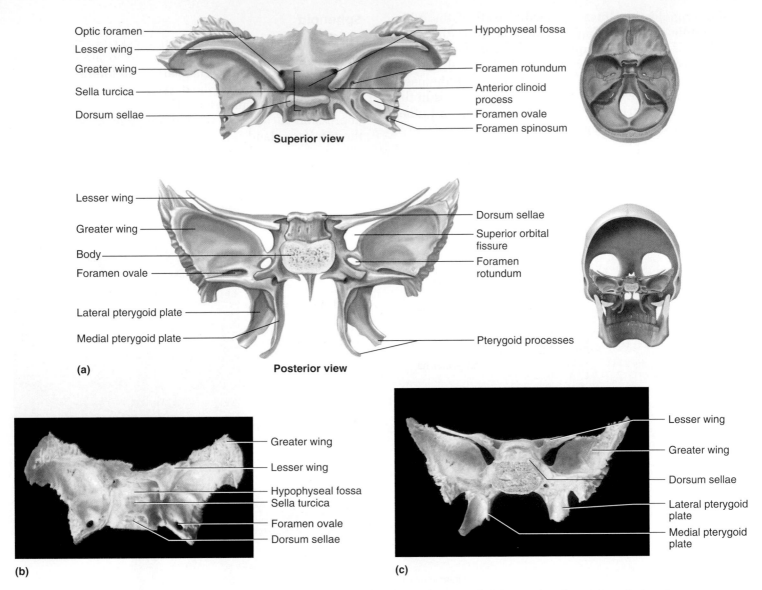

FIGURE 11.11 Sphenoid Bone Diagram (a) superior and posterior views. Photograph (b) superior view; (c) posterior view.

Temporal

The temporal bone is a paired cranial bone that has a squamous portion that is the lateral part of the bone and forms part of the cranial vault. There is also a medial part of the temporal called the petrous portion. The petrous portion contains the ear ossicles and the opening of the internal acoustic meatus, as seen in the medial view of the temporal bone in figure 11.12. Examine an isolated temporal bone and locate the zygomatic process, which articulates with the zygomatic bone, the mastoid process, which can be palpated (felt) as a bump posterior to the ear, and other features in figure 11.12.

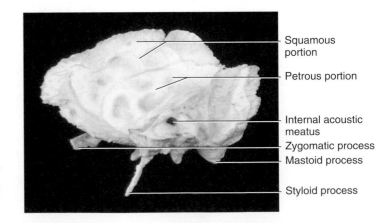

FIGURE 11.12 Right Temporal Bone, Medial View

REVIEW SECTION

Axial Skeleton—Skull

Name _____ *Date* _____

Lab Section _____ *Time* _____

Review Questions

1. The eyebrows are superficial to what bone?

2. What is the common name for the zygomatic bone?

3. What is the name of the bony process posterior to the earlobe?

4. The hard palate is made up of what bones?

5. What are the names of the major paranasal sinuses?

6. The mandible fits into what part of the temporal bone to form the jaw joint?

7. What bone is found just posterior to the ethmoid bone in the orbit?

8. The sella turcica is found in what bone?

9. What is the name of the bone that makes up most of the temple?

10. What are the names of the bones that surround the opening of the nose?

11. The upper teeth are held by what bones?

12. In what bone would you find the foramen magnum?

13. What is the name of the bone that makes up most of the posterior surface of the orbit?

14. What are the two bony structures that make up the nasal septum?

15. The mastoid process is located on which bone?

16. The sagittal suture separates the _____ from the _____.
    a. sphenoid, ethmoid
    b. left parietal, right parietal
    c. frontal, parietal
    d. parietals, occipital

17. Which bone does *not* occur in the orbit?
    a. maxilla
    b. zygomatic
    c. ethmoid
    d. sphenoid
    e. temporal

18. Which bone is *not* a paired bone of the skull?
    a. zygomatic
    b. temporal
    c. lacrimal
    d. vomer

19. Based on what you know about the maxillary sinus, why would a significant impact to the maxilla create a more difficult situation for healing than would the fracture of a long bone?

20. Label the following illustration using the terms provided.

vomer   occipital condyle   carotid canal
foramen magnum   internal nares   palatine
palatine process of maxilla   external occipital protuberance   mastoid process
zygomatic   jugular foramen   mastoid notch

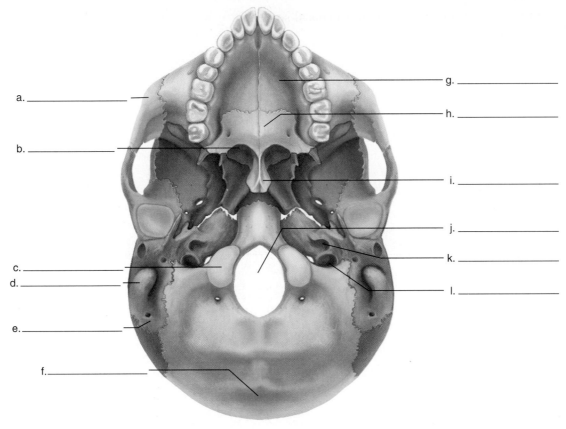

a. _____
b. _____
c. _____
d. _____
e. _____
f. _____
g. _____
h. _____
i. _____
j. _____
k. _____
l. _____

21. Label the following illustration using the terms provided.

ramus   mandibular notch   body
mental foramen   condyloid process   angle
coronoid process

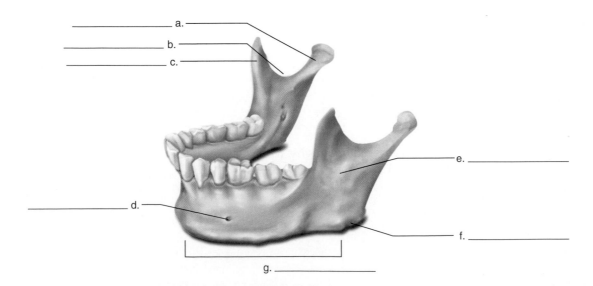

a. _____
b. _____
c. _____
d. _____
e. _____
f. _____
g. _____

LABORATORY

Articulations

INTRODUCTION

The study of the joints between bones is called **arthrology.** It is important not only as the point of interplay between the skeletal system and the muscular system but also due to the significant trauma or disease that can occur in joints. Joints themselves are known as **articulations.** Diseases such as arthritis affect millions of people worldwide. Artificial replacement of some joints, such as the hip and knee, is common if the joints cause pain or lack mobility.

In this exercise you study the structure and function of various types of joints in the body. Joints are classified according to their physical composition. As you study the joints in the lab, use the joints in your body as reference and mimic the action of a specific articulation. These topics are covered in the Saladin text in chapter 9, "Joints."

OBJECTIVES

At the end of this exercise you should be able to

1. distinguish among fibrous, cartilaginous, and synovial joints;

2. discuss the nature of a synovial joint;

3. locate the fibrous capsule, synovial membrane, synovial fluid, and articular cartilage in a dissected synovial joint;

4. list five types of synovial joints;

5. explain the structure of the knee, hip, jaw, ankle, and shoulder.

MATERIALS

Mammal joint with intact synovial capsule

Dissection tray with scalpel or razor blades, blunt probe, and protective gloves

Waste container

Model or chart of joints, including those of the shoulder, knee, hip, ankle, and jaw

Articulated skeleton

PROCEDURE

Types of Joints

The four major groups of joints classified according to composition are bony, fibrous, cartilaginous, and synovial joints. Bony joints occur when two bones fuse together to form one. Fibrous joints are typically composed of connective tissue fibers between two bones. They permit little movement. Cartilaginous joints consist of cartilage between two bones and are generally more movable than fibrous joints, although some cartilaginous joints may have no movement at all. Synovial joints have the most complex structure, including a joint capsule, an inner membrane, and synovial fluid, and they are the most movable of the joints (table 12.1).

Bony Joints

A **bony joint,** or **synostosis** (SIN-oss-TOE-sis), occurs when two separate bones fuse to become one. These joints are immovable joints. When a growth plate fuses in a long bone, a synostosis forms.

Fibrous Joints

Fibrous joints connect one bone to another with collagenous fibers. If the fibers are close together, the joint does not allow for much, if any, movement between the bones. One example of this type of joint is called a **suture,** and it occurs between adjacent bones in the cranium. In these joints the bones of the cranium are tightly bound together by dense, fibrous connective tissue. As a person approaches the age of 35 or so, some of the sutures of the skull begin to fuse from the region closest to the brain toward the superficial surface of the skull. This fusion leads to the complete union of two bones, or **synostosis.** The two frontal bones in the fetal skeleton fuse together as a synostosis and form a single frontal bone. Examine a skull in the lab and locate the sutures. Examine the different types of sutures in figure 12.1.

Another type of fibrous joint is a **gomphosis** (gom-FOE-sis), which is represented by teeth in the sockets of the maxilla and the mandible. This type of joint is like a peg in a socket. A gomphosis connects the bone of the jaw to the tooth by fibrous connective tissue called **periodontal ligaments.** Examine a jaw or complete skull in the lab and locate a gomphosis. Compare your specimen to figure 12.2.

TABLE 12.1	Classification of Joints
Bony Joints—Joints Composed of Two Fused Bones	**Examples**
Synostosis	Fusion of two frontal bones into one bone
Fibrous Joints—Joints Held Together by Collagenous Fibers	
Suture	Frontal and parietal bones
Gomphosis	Teeth and mandible
Syndesmosis	Distal tibia and distal fibula
Cartilaginous Joints—Joints Held Together by Cartilage	
Synchondrosis	Epiphyseal plate, humeral head and shaft
	Costal cartilages, ribs and sternum
Symphysis	Joint between two vertebral bodies
Synovial Joints—Joints Enclosed by Synovial Capsule	
Plane	Between carpal bones
Hinge	Humerus and ulna
Pivot	Atlas and axis
Condyloid	Radius and scaphoid
Saddle	Trapezium and first metacarpal
Ball-and-socket	Acetabulum and femur

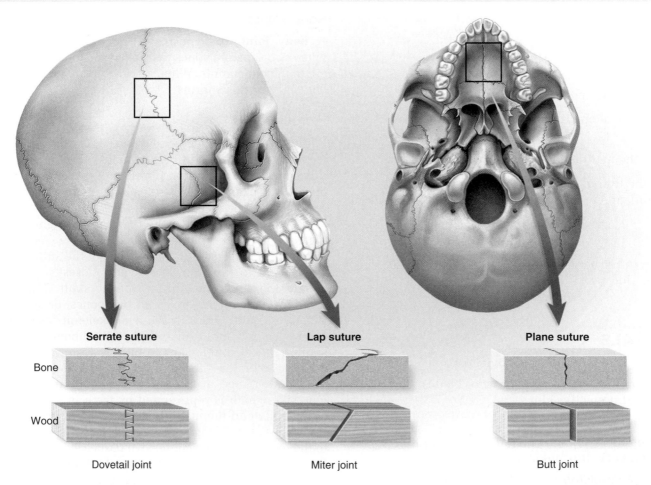

FIGURE 12.1 Sutures Examples and locations in the skull, along with cross sections and analogous wood joints.

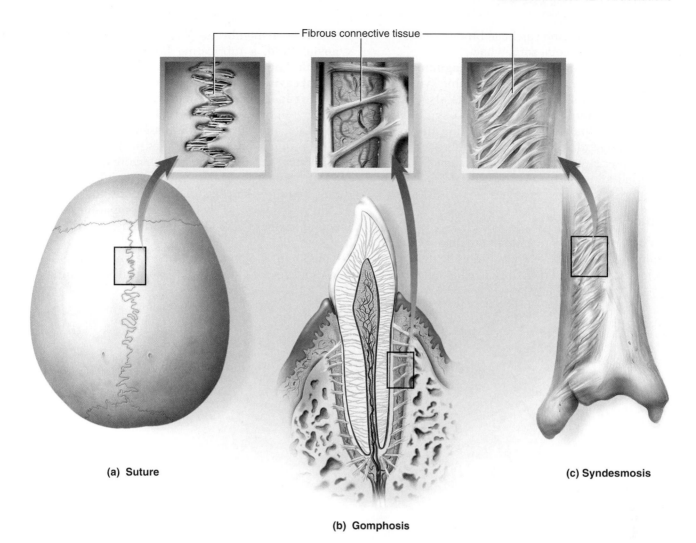

Fibrous connective tissue

(a) Suture

(b) Gomphosis

(c) Syndesmosis

FIGURE 12.2 **Fibrous Joints** (a) Suture; (b) gomphosis; (c) syndesmosis.

Another type of fibrous joint is called a **syndesmosis** (SIN-dez-MO-sis). The fibrous connective tissue is longer in these joints than in sutures or gomphoses. An example of a syndesmosis is the connection between the distal radius and ulna or the distal tibia and fibula. Examine an articulated skeleton in the lab and compare it to figure 12.2.

Cartilaginous Joints

If bones are held together by cartilage, the articulation is known as a **cartilaginous joint.** As in fibrous joints, if the cartilage is thin between bones, the joint is immovable. Cartilaginous joints are known as **synchondroses** (SIN-kon-DRO-sees) (singular, *synchondrosis*). If the cartilage is long, there is more movement in the joint. In one case, the cartilaginous joint is a phase in the development of the skeleton. This is an immovable joint called the **epiphyseal** (EP-ih-FIZ-ee-ul) **plate,** which is illustrated in figure 12.3. This joint eventually fuses to form a single bone.

Another example of a synchondrosis is the articulation between the true ribs and the sternum by the **costal**

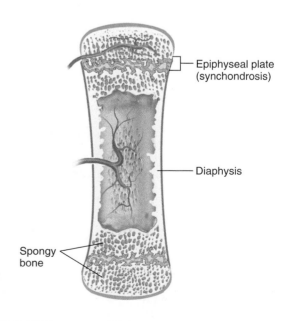

Epiphyseal plate
(synchondrosis)

Diaphysis

Spongy
bone

FIGURE 12.3 **Synchondrosis** Immovable epiphyseal plate.

cartilages. In this joint, hyaline cartilage binds the ribs to the sternum yet lets them move somewhat.

A **symphysis** (SIM-fih-sis) is a **fibrocartilaginous** pad between the pubic bones (for example, the pubic symphysis) or in the intervertebral discs. These joints also allow for some movement. The physical stress on a symphysis is greater than that on the costal cartilages, and the joint reflects this in its fibrocartilage composition. Fibrocartilage endures much greater stress than hyaline cartilage. Locate a symphysis in lab and compare it to figure 12.4, which compares symphyses to costal cartilages.

Synovial Joints

Joints that allow for extensive movement are called **synovial joints.** The outer part of the synovial joint is the **joint capsule,** which is made of an outer **fibrous capsule** and an inner **synovial membrane.** The synovial membrane secretes **synovial fluid,** a lubricating liquid that reduces friction inside the joint. The space inside the joint is called the **synovial cavity,** and each bone of the joint ends in a hyaline cartilage cap called the **articular cartilage.** There are other structures in synovial joints that are characteristic of specific joints, and these are discussed later in this exercise. The shape of the bones determines the type of movement at the articulation, and in general the more movable a joint is, the less stable it is. The stability of a joint depends on the number and types of ligaments, tendons, and muscles and on the way the bones fit together. Compare the features of the joint in figure 12.5 to models or charts in the lab.

Dissection of a Synovial Joint   To understand the structure of a synovial joint, it is beneficial to examine a fresh

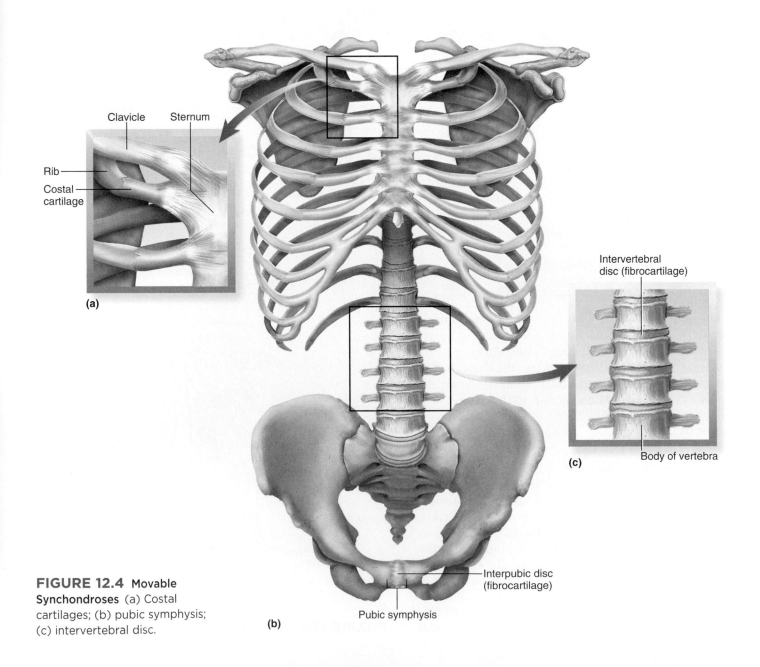

FIGURE 12.4 Movable Synchondroses (a) Costal cartilages; (b) pubic symphysis; (c) intervertebral disc.

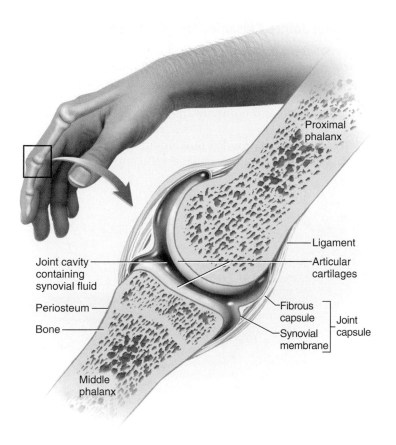

FIGURE 12.5 Synovial Joint Structure

or recently thawed mammal joint. Wear protective gloves as you dissect the joint, and wash your hands thoroughly with soap and water after the dissection. Place the joint in front of you on a dissecting tray and cut into the joint capsule with a scalpel or razor blade.

Note the tough, white material that surrounds the joint. This is the joint capsule, and it may be fused with **ligaments** that bind the bones of the joint together. Notice the synovial fluid, which is a slippery substance that provides a slick feel to the inside of the capsule. Once you cut into the joint, examine the articular cartilage found on the ends of the bones. *Carefully* cut into this cartilage with a scalpel or razor blade, and notice how the material chips away from the bone. When you have finished with the dissection, make sure to rinse off the dissection equipment, dispose of the joint in the appropriate animal waste container, and wash your hands.

Modified Synovial Structures Bursae and tendon sheaths are modified synovial structures. **Bursae** (singular, *bursa*) are small synovial sacs between tendons and bones or other structures. The bursae cushion the tendons as they pass over the other structures. **Tendon sheaths** are modified synovial structures, such as those encircling the tendons that pass through the palm of the hand. Tendon sheaths lubricate the tendons as they slide past one another. Examine figure 12.6 for these structures.

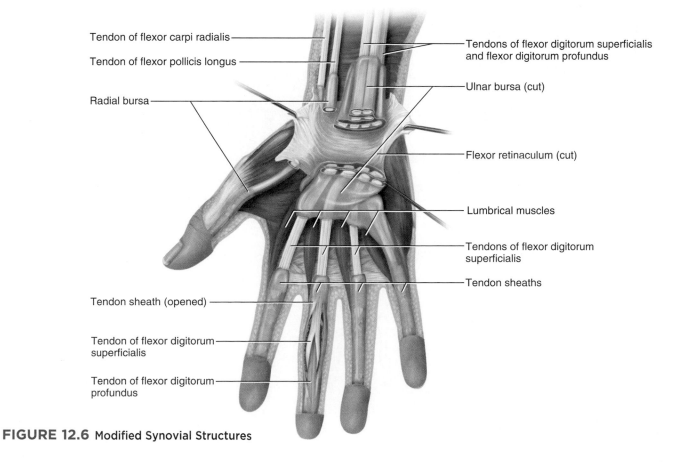

FIGURE 12.6 Modified Synovial Structures

Joints Classified by Movement

Articulations may be immovable, semimovable, or freely movable. Immovable joints are known as **synarthrotic joints.** The bones are tightly bound by fibers or hyaline cartilage. Semimovable joints are known as **amphiarthrotic joints,** and they may be fibrous or cartilaginous. Freely movable joints are known as **diarthrotic joints,** and they are always synovial joints. Specific types of synovial joints are discussed next.

Synovial Joints Classified by Movement

The synovial joints are classified according to the type of movement they allow between the articulating bones. The joints are listed here in the general order of least movable to most movable. Examine the joints on an articulated skeleton in lab and compare them to figure 12.7.

1. **Plane, joints** allow for movement between two flat surfaces, such as between the superior and

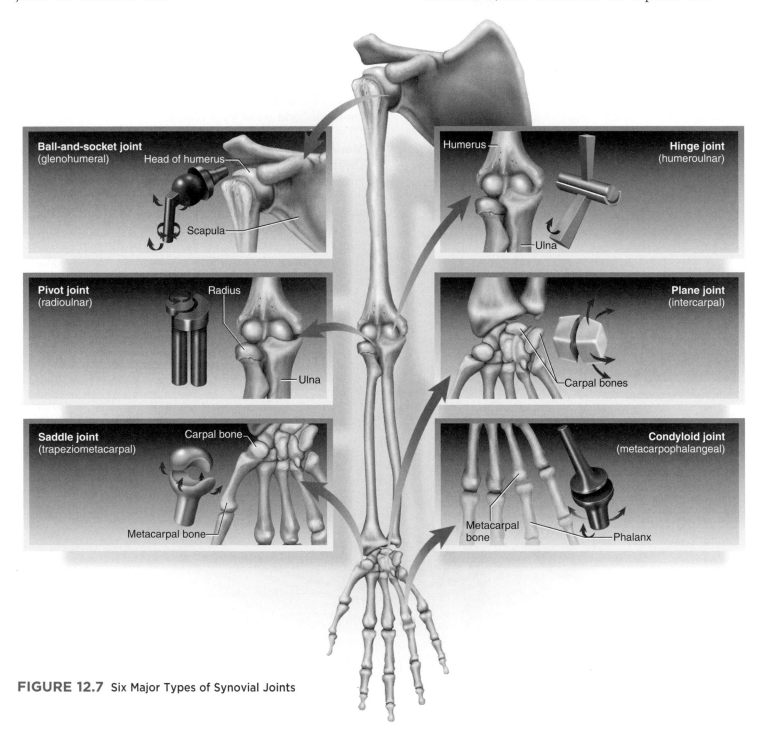

FIGURE 12.7 Six Major Types of Synovial Joints

inferior facets of adjacent vertebrae or intertarsal or intercarpal joints.

2. **Hinge joints** allow for angular movement, such as in the elbow, in the knee, or between the phalanges of the fingers. You can increase or decrease the angle of the two bones with this joint.

3. **Pivot joints** allow for rotational movement between two bones, such as in the movement of the atlas and the axis when moving the head to indicate "no." They also occur at the proximal radius and ulna.

4. **Condyloid** or **ellipsoid joints** allow significant movement in two planes, such as at the base of the phalanges. Condyloid joints consist of a convex surface paired with a concave surface. The junction between the radius and scaphoid bone is a good example of a condyloid joint. Others are found between the atlas and occipital bone or the metacarpals and phalanges of digits 2 through 5.

5. **Saddle joints** have two concave surfaces that articulate with one another. An example of a saddle joint is that between the trapezium and the first metacarpal of the thumb. This provides for more movement in the thumb than the condyloid joint of the wrist.

6. **Ball-and-socket joints** consist of a spherical head in a round concavity, such as in the shoulder and the hip. There is extensive movement in these joints, yet they are inherently less stable due to the freedom of movement that they afford.

Monaxial joints move only in one plane. Hinge, pivot, and plane joints are monaxial joints. Biaxial joints move in two planes. Condyloid and saddle joints are examples of biaxial joints. Multiaxial joints move in many planes. Ball-and-socket joints are multiaxial joints.

Specific Joints of the Body

Temporomandibular

The **temporomandibular joint (TMJ)** is the only diarthrotic joint of the skull. The **articular disc** is a pad of fibrocartilage that provides a cushion between the mandibular condyle and the temporal bone. This joint is both a hinge and a plane joint. Numerous ligaments strengthen this joint (see figure 12.8). Examine a skull with the mandible attached for the nature of the temporomandibular joint.

Shoulder Joint

The **shoulder joint** is known as the **glenohumeral joint** because the glenoid fossa articulates with the head of

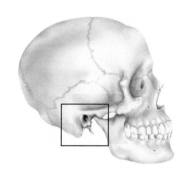

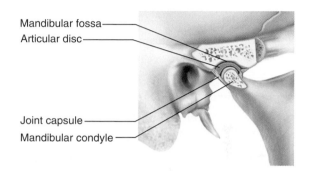

Mandibular fossa
Articular disc
Joint capsule
Mandibular condyle

FIGURE 12.8 Temporomandibular Joint

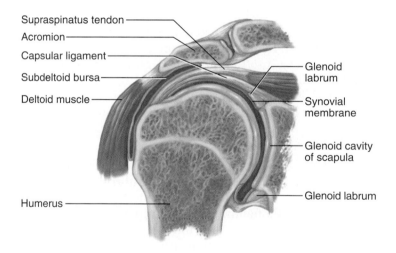

Supraspinatus tendon
Acromion
Capsular ligament
Subdeltoid bursa
Deltoid muscle
Glenoid labrum
Synovial membrane
Glenoid cavity of scapula
Glenoid labrum
Humerus

FIGURE 12.9 Glenohumeral Joint, Coronal Section

the humerus. The shallow glenoid fossa is deepened by the **glenoid labrum,** a cartilaginous ring that surrounds the cavity. Numerous bursae, ligaments, a tough joint capsule, and the **rotator cuff muscles** also stabilize the joint. Compare figure 12.9 to a model or an actual joint in the lab.

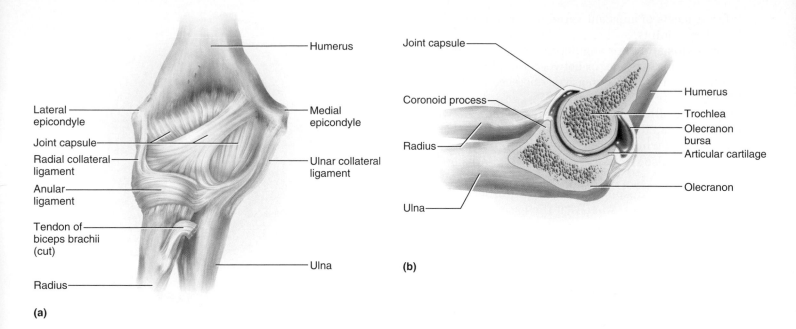

FIGURE 12.10 Elbow Joint (a) Anterior view; (b) sagittal section.

Elbow Joint

The **elbow (humeroulnar** and **humeroradial) joint** is a complex joint having both hinge and pivot characteristics. The ulna locks into the humerus tightly but the **anular ligament** that wraps around the radius is commonly torn from the radial head when the forearm is abruptly pulled (see figure 12.10).

Hip Joint

The **hip joint** is known as the **acetabulofemoral joint.** As with the shoulder joint, the **acetabular labrum** deepens the hip socket. Numerous ligaments, including the iliofemoral, pubofemoral, and ischiofemoral, bind the femur to the hip bone. The **round ligament** is a band of dense connective tissue that attaches the acetabulum to the **fovea capitis** of the femur. Compare the models, charts, or specimens in the lab to figure 12.11.

Knee Joint

The **tibiofemoral joint,** also known as the **knee joint,** is the largest, most complex joint of the body. It consists of several major ligaments, including the **tibial (medi-**

al) **collateral ligament, the fibular (lateral) collateral ligament, the anterior cruciate ligament,** and the **posterior cruciate ligament.** Another important structure is the **patellar tendon,** which runs from the quadriceps

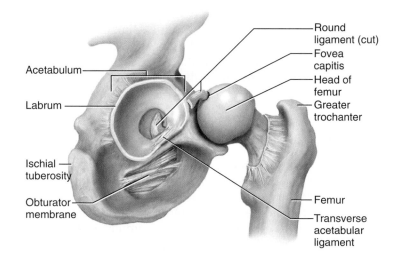

FIGURE 12.11 Acetabulofemoral Joint

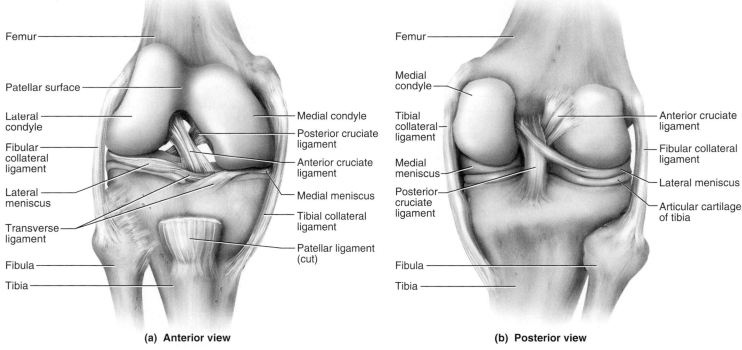

(a) Anterior view

Femur
Patellar surface
Lateral condyle
Fibular collateral ligament
Lateral meniscus
Transverse ligament
Fibula
Tibia
Medial condyle
Posterior cruciate ligament
Anterior cruciate ligament
Medial meniscus
Tibial collateral ligament
Patellar ligament (cut)

(b) Posterior view

Femur
Medial condyle
Tibial collateral ligament
Medial meniscus
Posterior cruciate ligament
Fibula
Tibia
Anterior cruciate ligament
Fibular collateral ligament
Lateral meniscus
Articular cartilage of tibia

(c) Sagittal section

Femur
Bursa under lateral head of gastrocnemius
Joint capsule
Articular cartilage
Meniscus
Tibia
Quadriceps femoris
Quadriceps femoris tendon
Suprapatellar bursa
Prepatellar bursa
Patella
Synovial membrane
Joint cavity
Infrapatellar fat pad
Superficial infrapatellar bursa
Patellar ligament
Deep infrapatellar bursa

FIGURE 12.12 Tibiofemoral Joint (a) Anterior; (b) posterior; (c) sagittal.

femoris muscle to the patella. The **patellar ligament** connects the patella and the tibial tuberosity. The **medial** and **lateral menisci** are wedge-shaped pads of fibrocartilage that provide a cushion between the femur and tibia. The knee is primarily a hinge joint with a little lateral movement allowed. Compare specimens in the lab to figure 12.12.

Ankle Joint

The **talocrural joint,** or ankle joint, is formed from the tibia, the fibula, and talus. The lateral malleolus of the distal fibula and the medial malleolus of the distal tibia are on each side of the talus, and all three bones share a joint capsule. The tibia and fibula are connected by an **anterior tibiofibular ligament** and a **posterior tibiofibular ligament.** The tibia is connected to the calcaneus, talus, and navicular by fibers of the **medial ligament,** and the fibula is connected to the calcaneus and talus by fibers of the **lateral ligament.** A sprained ankle is the stretching of these fibers; it frequently occurs when the foot is excessively inverted or everted. Examine figure

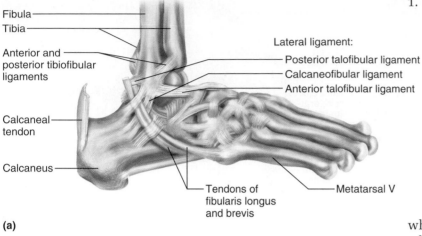

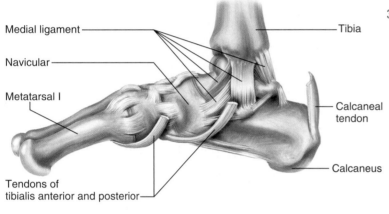

(a)

(b)

FIGURE 12.13 **Ankle Joint** (a) Lateral view; (b) medial view.

12.13 and models in lab, or a dissected ankle in a cadaver, and find the bones and major ligaments of the ankle.

Levers

Levers are mechanical devices that increase power or speed. The musculoskeletal system can be viewed as various lever systems that provide a mechanical advantage for the body. There are three parts to a lever: the fulcrum, the resistance arm, and the effort arm. The fulcrum is the point of movement or rotation of the lever. The resistance arm is the part of the lever that is to be moved, and the effort arm is that part to which an action is applied to move the lever (see figure 12.14). There are three classes of levers and these are explained next.

1. A **first-class lever** is one where the effort is on one side of the fulcrum and the resistance is on the other side. A classic example of this is a see-saw, and a common example in humans is the junction between the head and the atlas. Examine an articulated skeleton in lab. The atlas is the fulcrum, the chin is the resistance, and the neck muscles attaching to the back of the head are the effort (see figure 12.15a).

2. A **second-class lever** has the fulcrum at one end, the resistance in the middle, and the effort at the other end. A wheelbarrow is a good representative of a second-class lever. The wheel is the fulcrum, the load in the basin of the wheelbarrow is the resistance, and the effort is your arms pulling up on the handles of the wheelbarrow (see figure 12.15b).

3. A **third-class lever** has the fulcrum at one end, the effort in the middle, and the resistance at the other end. Opening up a trap door from the top is an example of a third-class lever. The hinge on the door represents the fulcrum, the handle in the middle of the door is the effort, and the weight of the door is the resistance. Most of the joints of the body involve third-class levers and a common example is flexing the forearm. The fulcrum is the elbow, the effort is the biceps brachii muscle, and the resistance is the weight of the forearm. Examine this joint of an articulated skeleton (see figure 12.15c).

Answer the following questions to see if you understand the concepts of levers.

1. When you are lying on your back with your thighs and legs straight and you elevate your lower limbs, what kind of lever is this? _____

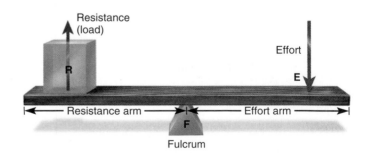

FIGURE 12.14 **Lever System**

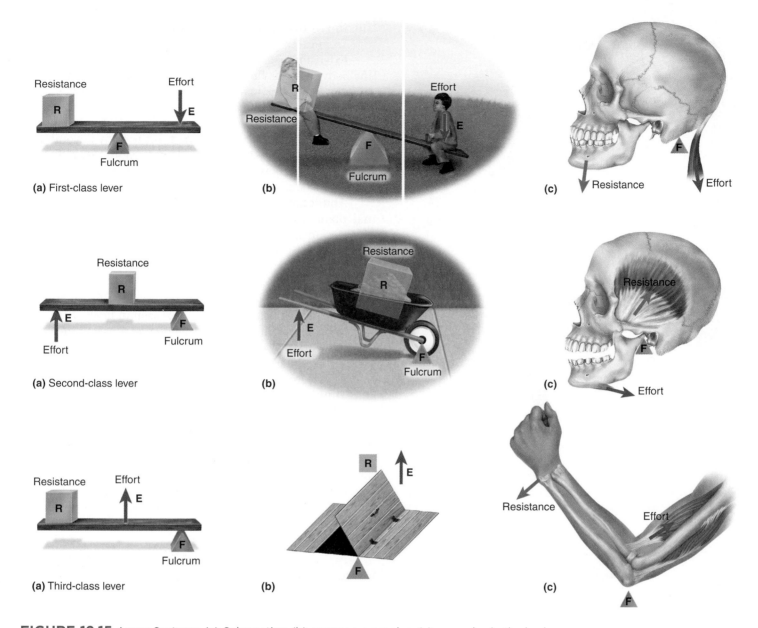

FIGURE 12.15 Lever Systems (a) Schematics; (b) common examples; (c) examples in the body.

2. The triceps brachii muscle on the back of your arm is attached to the tip of your elbow. When you use this muscle to straighten your arm, what kind of lever is this? _____

Actions at Joints

Many different kinds of movements occur at joints. These movements, controlled by muscles, are called actions. Actions can decrease a joint angle, increase a joint angle, and cause rotation at a joint, among other movements. The specific types of actions are as follows.

Flexion is a decrease in the joint angle from anatomical position. If you bend your elbow, you are flexing your forearm. Flexion of the thigh is in the anterior direction, yet flexion of the leg is in the *posterior* direction. Bending forward at the waist is flexion of the vertebral column. Looking at your toes is

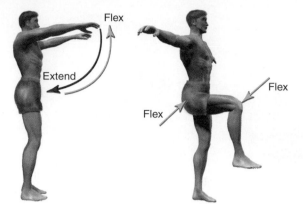

(a) Flexion and extension of the arm   **(b)** Flexion of the thigh and leg

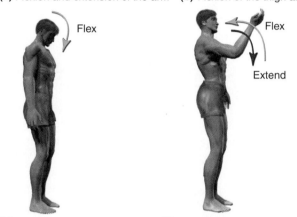

(c) Flexion of the head   **(d)** Flexion and extension of the forearm

FIGURE 12.16 **Flexion and Extension of Selected Joints**
(a) Flexion and extension of the arm; (b) flexion of the thigh and leg; (c) flexion of the head; (d) flexion and extension of the forearm.

flexion of the head. Examine figure 12.16 for examples of flexion.

Extension is a return to anatomical position of a part of the body that was flexed. If you are looking at your toes and lift your head back to anatomical position, you are extending your head. If you straighten your knee after it is bent (flexed), you are extending the leg. Examine figure 12.16 for examples of extension. Extension of the part of the body beyond anatomical position is known as **hyperextension.** When you are about to roll a bowling ball and your arm reaches the very back of the arc, you are hyperextending your arm.

Abduction is movement of the limbs in the coronal plane away from the body (*abduct* = to take away). Abduction is taking away a part of the body in a lateral direction. This can be seen in figure 12.17.

Adduction is the return of the part of the body to anatomical position after abduction. In doing "jumping jacks," you are abducting and adducting in series. Think of adduction as "adding" a limb back to the body, as seen in figure 12.17.

Rotation is the circular movement of a part of the body. **Lateral rotation** moves the anterior surface of the limb toward the lateral side of the body. **Medial rotation** turns the anterior surface of the limb toward the midline.

Supination is lateral rotation of the hand. The hands are supinated when the body is in anatomical position.

Pronation is medial rotation of the hands. When you turn your palms posteriorly from anatomical position, you are pronating your hands. Rotate various joints of the body and compare them to figure 12.18.

Circumduction is the movement of a muscle in a conical shape, with the point of the cone being proximal. Examine figure 12.18*c* for an example of circumduction.

Protraction is a horizontal movement in the anterior direction, as in jutting the chin forward.

Retraction is the reverse of protraction. The jaw that moves from anterior to posterior is retracted. These two actions are illustrated in figure 12.19.

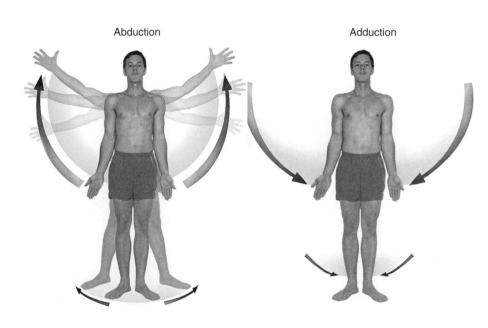

Abduction   Adduction

FIGURE 12.17 **Abduction and Adduction of the Arms and Thighs**

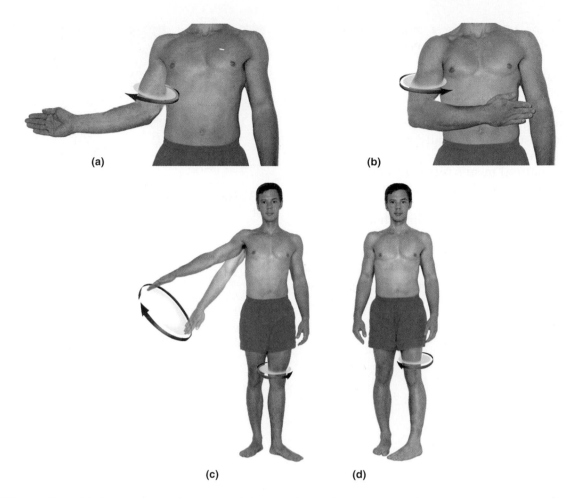

FIGURE 12.18 **Rotation of Joints** (a) Lateral rotation of the arm; (b) medial rotation of the arm; (c) circumduction of the arm and lateral rotation of the thigh; (d) medial rotation of the thigh.

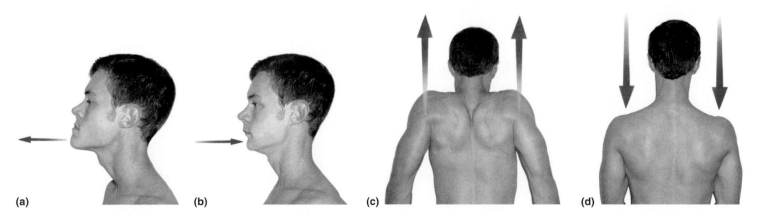

FIGURE 12.19 **Protraction, Retraction, Elevation, and Depression** (a) Protraction of the jaw; (b) retraction of the jaw; (c) elevation of the scapulae; (d) depression of the scapulae.

Elevation means to move in a superior direction. Elevation of the shoulders occurs when you shrug your shoulders.

Depression is the opposite of elevation; it is movement in the inferior direction. Elevation and depression are seen in figure 12.19.

Inversion (supination) refers to movement of the feet. Turning the soles of the feet medially so they face each other is known as inversion.

Eversion (pronation) means turning the soles of the feet laterally. Inversion and eversion can be seen in figure 12.20.

Fixing a muscle prevents motion in either direction. This is done with opposing muscles contracting simultaneously. The muscle that has the main force on a joint is called the **prime mover.** Muscles that assist with the prime mover are **synergists,** while those that oppose the muscle are **antagonists.**

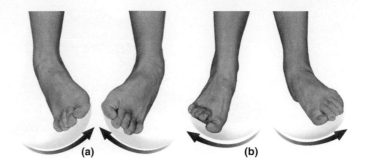

FIGURE 12.20 **Inversion and Eversion** (a) Inversion of the feet; (b) eversion of the feet.

REVIEW SECTION

Articulations

Name _____ *Date* _____

Lab Section _____ *Time* _____

Review Questions

1. Which one of the following joints has the greatest range of movement?

    a. gomphosis

    b. suture

    c. synchondrosis

    d. hinge

2. In which of these joints would you find a meniscus?

    a. cartilaginous

    b. fibrous

    c. synovial

3. Label the following illustration using the terms provided.

    fibrous capsule

    bone

    synovial membrane

    synovial cavity

    articular cartilage

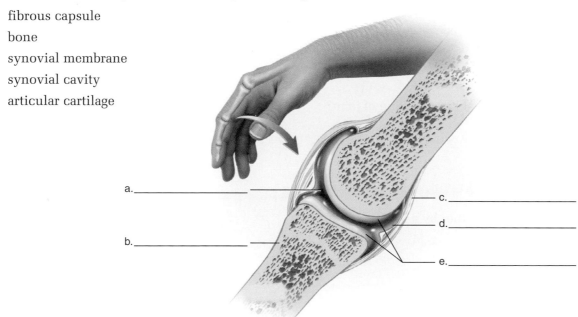

a._____

b._____

c._____

d._____

e._____

4. Match the joint in the left column with the type of joint in the right column.

    _____ acetabulofemoral   a. hinge

    _____ radiocarpal   b. ball-and-socket

    _____ tibiofemoral   c. plane

    _____ intertarsal   d. condyloid

5. Rank the following joints from least movable to most movable, with 1 being the least movable and 5 being the most movable.

plane   saddle   suture   syndesmosis   ball-and-socket

6. What is the function of the meniscus in the knee?

7. What is the function of the labrum in the glenohumeral joint?

8. Bones held together by cartilage are known as _____ joints.

9. A class of joint with great movement is known as a _____ .

10. The teeth are held in the jaw by what specific type of joint?

11. What is the name of a joint held together by a joint capsule?

12. The joint between the femur and the tibia is what specific type of joint?

13. Synovial fluid is secreted by what structure?

14. A skull suture is what type of joint, in terms of movement?

15. What type of joint is found at the wrist (between the radius and the scaphoid bones)?

16. The joint between the first metacarpal and the proximal phalanx is what type of joint?

17. When two bones fuse together into a single bone, this union is called a(n) _____.

18. What kind of joint is slightly movable and held together by fibrous connective tissue?

Notes

LABORATORY

Introduction to the Study of Muscles and Muscles of the Shoulder and Arm

INTRODUCTION

The next six exercises focus on skeletal muscles. There are over 600 skeletal muscles in the human body. The muscles are grouped according to their occurrence in particular areas of the body, and most muscles occur as pairs, with only a few occurring singly. In the following exercises you will learn about some of the major muscles of the body. Skeletal muscle is voluntary muscle in that it contracts when consciously stimulated by specific nerves. The muscular system functions in movement, maintenance of posture, generation of heat (shivering), and compression of the abdomen, among other functions. These topics are covered in more detail in the Saladin text in chapter 10, "The Muscular System."

The origin, insertion, action, and innervation are listed for about 100 muscles. Your instructor may wish to customize the list so that you learn specific muscles or specific things about each muscle.

Some students see the study of human musculature as a daunting task, while other students recall the muscle section of the course as their favorite part of the study of human anatomy and physiology. Studying muscles is more satisfactory if you set attainable goals. Begin your study of muscles early and not until the night before the lab exam! This is best accomplished by choosing a few muscles to study at a sitting and using flash cards as study aids. It is vital that you know the bones and bony markings before studying muscles. Look over Laboratory Exercises 9 through 11 if you need to review the skeletal markings. To help you visualize the attachment points, you may want to make a quick sketch of the bones and how the muscles attach to them.

A thorough study of muscles involves knowing not only the name of the muscle but also its origin, insertion, action, and the nerves that innervate it. The muscles are listed in this exercise by their name, followed by their origin. The **origin** is the attachment point of the muscle that does not move during muscular contraction. The muscles also are listed with their **insertion,** the attachment that moves during contraction. Each muscle has an action, which represents all the effects the muscle has on a part of the body (such as flexion of the arm). There are two schools of thought on action. One is that the action has an impact on a joint, such as flexion of the wrist. The other is that an action moves the region beyond a joint, such as flexion of the hand. This lab manual will mostly follow the latter. Finally, the **innervation** of a muscle is the specific nerve that controls it. Nerves are important in that, if the motor portion of a nerve is damaged, the muscle receiving the stimulus by that nerve cannot function.

Origins and Insertions

Examine an articulated skeleton as you study the muscles. Visualize the muscle as it attaches to the origin or the insertion. As you study the muscles locate them on your body and determine which part is the stable origin and which part is the moving insertion.

Actions

To fully understand muscles you must know their actions at joints and associate the actions with the body in anatomical position. When you are learning the action of a muscle, imagine a piece of string tied between the origin and the insertion of the muscle. Think of how the bones would move as you pulled the insertion closer to the origin. Mimic the action of the muscle as you learn it. When you perform the action of a particular muscle, you should be able to feel that muscle tighten. You should also review the actions at joints in Exercise 12.

In addition to moving a joint, muscles can also fix a joint. **Fixing** means to prevent motion in either direction. This is done with opposing muscles contracting simultaneously. The muscle that has the main force on a joint is called the **prime mover.** Muscles that assist the prime mover are **synergists,** while those that oppose the muscle are **antagonists.**

Muscle Nomenclature

Another aid to learning muscles is to understand how they are named. Muscles are named by a number of criteria:

Action: the *extensor* digitorum is a muscle that *extends* the fingers.

Origin: the *infraspinatus* muscle originates on the *infraspinous* fossa of the scapula.

Insertion: the flexor *hallucis* longus muscle inserts on the *hallux,* or big toe.

Fiber direction: the *transversus* abdominis muscle has fibers that run horizontally, or in a *transverse* direction.

Number of heads: the *triceps* brachii muscle has *three heads.*

Shape: the *deltoid* muscle is shaped like a *delta,* or triangle.

OBJECTIVES

At the end of this exercise you should be able to

1. locate the muscles of the shoulder and the arm on a torso model, chart, cadaver, or cat;
2. list the origin, insertion, and action of each muscle presented;
3. describe what nerve controls each muscle;
4. list what muscles function as synergists or antagonists to the prime mover;
5. name all the muscles that have an action on a joint, such as all the muscles that flex the arm;
6. apply your knowledge of cat musculature to human muscles.

MATERIALS

Human torso model

Human arm models

Human muscle charts

Articulated skeleton

Cadaver (if available)

Cat

Cat wetting solution

Materials for Cat Dissection

Dissection trays

Scalpel and two or three extra blades

Gloves (household latex gloves work well for repeated use)

Blunt (Mall) probe

String and tags

Pins

Forceps and sharp scissors

First aid kit in lab or prep area

Sharps container

Animal waste disposal container

PROCEDURE

Examination of Muscles

Look at the charts, models, and cadaver in the lab and locate the muscles presented next. The muscles of the shoulder and arm are described here if they have some action on the scapula or the humerus. You can find a complete listing of the muscles in table 13.1. Figure 13.1

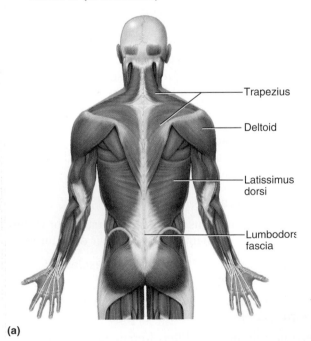

(a)

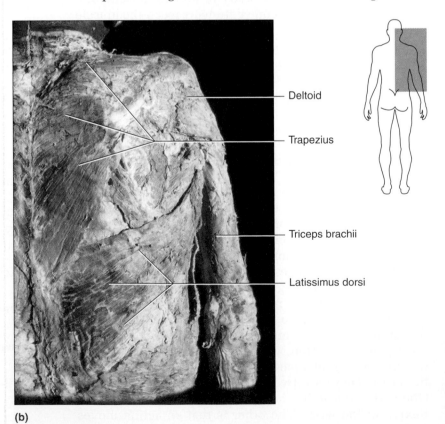

(b)

FIGURE 13.1 Superficial Back Muscles, Posterior View (a) Diagram; (b) photograph.

TABLE 13.1	Muscles of the Shoulder and Arm			
Name	**Origin**	**Insertion**	**Action**	**Innervation**
Superficial Muscles of the Shoulder				
Trapezius	Posterior occipital bone, ligamentum nuchae, C7–T12	Clavicle, acromion process, and spine of scapula	Extends and abducts head, rotates and adducts scapula, fixes scapula	Accessory nerve (XI), spinal nerves C2–4
Deltoid	Clavicle, acromion process, spine of scapula	Deltoid tuberosity of the humerus	Abducts arm; flexes, extends, medially, and laterally rotates arm	Axillary nerve
Pectoralis major	Clavicle, sternum, cartilages of ribs 1–7	Crest of greater tubercle of humerus	Flexes, adducts, and medially rotates arm	Medial and lateral pectoral nerves
Pectoralis minor	Ribs 3–5	Coracoid process of scapula	Depresses glenoid cavity, raises ribs 3–5	Medial pectoral nerve
Latissimus dorsi	T7–12, L1–5, S1–5, crest of ilium, ribs 10–12	Intertubercular groove of humerus	Extends, adducts, and medially rotates arm; draws shoulder inferiorly	Thoracodorsal nerve
Deep Muscles of the Shoulder				
Supraspinatus	Supraspinous fossa	Greater tubercle of humerus	Abducts arm, helps stabilize shoulder joint	Suprascapular nerve
Infraspinatus	Infraspinous fossa	Greater tubercle of humerus	Laterally rotates arm, stabilizes shoulder joint	Suprascapular nerve
Subscapularis	Subscapular fossa	Lesser tubercle of humerus	Medially rotates arm, stabilizes shoulder joint	Subscapular nerve
Teres minor	Lateral border of scapula	Greater tubercle of humerus	Laterally rotates and adducts arm, stabilizes shoulder joint	Axillary nerve
Teres major	Inferior angle of scapula	Crest of lesser tubercle of humerus	Extends, adducts, and medially rotates arm	Subscapular nerve
Muscles of the Arm				
Biceps brachii	Long head: superior margin of glenoid fossa Short head: coracoid process of scapula	Radial tuberosity	Flexes arm, flexes forearm, supinates hand	Musculocutaneous nerve
Triceps brachii	Infraglenoid tuberosity of scapula, lateral and posterior surface of humerus	Olecranon process, tuberosity of ulna	Extends and adducts arm, extends forearm	Radial nerve
Coracobrachialis	Coracoid process of scapula	Midmedial shaft of humerus	Flexes and adducts arm	Musculocutaneous nerve
Brachialis	Anterior, distal surface of humerus	Coronoid process of ulna	Flexes forearm	Musculocutaneous nerve
Brachioradialis	Lateral supracondylar ridge of humerus	Styloid process of radius	Flexes forearm	Radial nerve

T = thoracic vertebrae

C = cervical vertebrae

S = sacral vertebrae

L = lumbar vertebrae

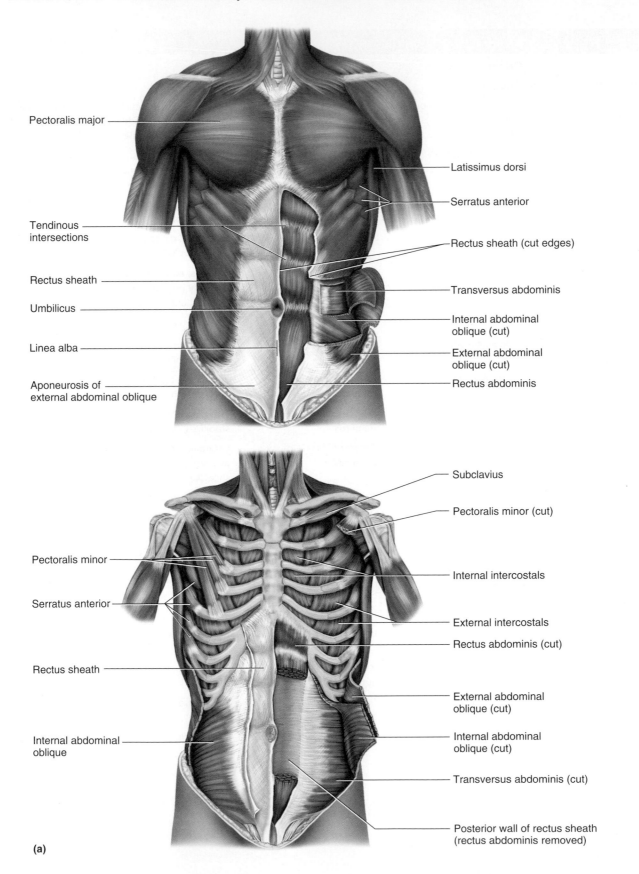

Pectoralis major

Latissimus dorsi

Serratus anterior

Tendinous intersections

Rectus sheath (cut edges)

Transversus abdominis

Rectus sheath

Internal abdominal oblique (cut)

Umbilicus

Linea alba

External abdominal oblique (cut)

Aponeurosis of external abdominal oblique

Rectus abdominis

Subclavius

Pectoralis minor (cut)

Pectoralis minor

Internal intercostals

Serratus anterior

External intercostals

Rectus abdominis (cut)

Rectus sheath

External abdominal oblique (cut)

Internal abdominal oblique

Internal abdominal oblique (cut)

Transversus abdominis (cut)

Posterior wall of rectus sheath (rectus abdominis removed)

(a)

FIGURE 13.2 Thoracic Muscles, Anterior View Diagram (a) superficial and deep muscles. Photograph (b) superficial muscles; (c) deep muscles.

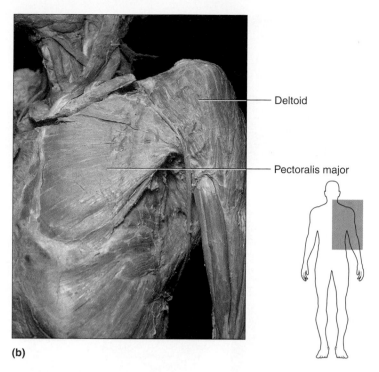

(b)

Deltoid

Pectoralis major

FIGURE 13.2 *Continued*

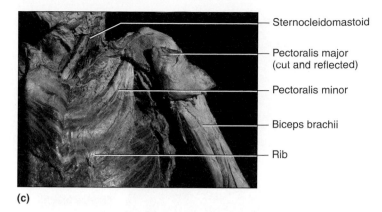

Sternocleidomastoid

Pectoralis major (cut and reflected)

Pectoralis minor

Biceps brachii

Rib

(c)

illustrates the superficial muscles of the back and shoulder. Locate the diamond-shaped **trapezius** (trah-PEE-zee-us) muscle, which has an origin in the midline of the vertebral column and head and inserts laterally. If the scapula is fixed, the head moves, yet if the vertebral column and head are fixed, the trapezius moves the scapula.

Another superficial muscle of the posterior surface is the **latissimus dorsi** (lah-TISS-ih-mus DOR-si) muscle, which arises from the vertebrae by way of a broad, flat lumbodorsal fascia. The latissimus dorsi originates on the back, but the insertion is on the anterior aspect of the humerus. It is a powerful extensor of the arm and has the common name of the swimmer's muscle (see figure 13.1).

The **deltoid** is located on top of the shoulder and is a major abductor of the arm. Locate the deltoid on material in the lab and compare it to figures 13.1 and 13.2. It is a fan-shaped muscle whose insertion partially covers the insertion of the **pectoralis** (PEC-tur-AL-is) **major** muscle seen in figure 13.2. The pectoralis major has fibers that run horizontally across the chest region, and it is a superficial muscle of the chest. Deep to the pectoralis major muscle is the **pectoralis minor** muscle, whose fibers run in a more vertical direction.

Deep to the trapezius and the deltoid are muscles that originate on the scapula proper. The **supraspinatus** (SOUP-rah-spin-AT-us) is named for its origin on the supraspinous fossa, and it is a synergist to the deltoid. The **infraspinatus** is a muscle originating on the infraspinous fossa and laterally rotates the arm. The **subscapularis** (sub-SCAP-you-LAR-is) is named for its origin on the subscapular fossa, and it medially rotates the arm.

The subscapularis is located on the *anterior* surface of the scapula, between the scapula and the ribs, and thus cannot be seen in a posterior view. Locate these muscles in the lab and compare them to figure 13.3.

The scapula gives rise to two other muscles, the teres minor and the teres major. The **teres** (TARE-eez) **minor** appears like a slip of the infraspinatus, while the **teres major** crosses on the medial side of the humerus parallel to the latissimus dorsi (see figure 13.3).

The **musculotendinous** (MUS-que-lo-TEND-in-us) **(rotator) cuff** muscles stabilize the shoulder joint. Muscles composing the cuff are the **supraspinatus, infraspinatus, subscapularis,** and **teres minor.** You can remember their names because the humerus "SITS" in the musculotendinous cuff. A rotator cuff injury is one where there is damage to any of these muscles.

The **biceps brachii** (BI-ceps BRAY-key-eye) muscle is a two-headed muscle of the arm. It has an action on the arm, yet the biceps brachii neither originates nor inserts on the humerus. The biceps brachii has a long tendon that runs between the intertubercular groove of the humerus and a short tendon that originates on the coracoid process. The **triceps** (TRY-ceps) **brachii** is the only major muscle on the posterior surface of the humerus. The triceps brachii is an antagonist to the biceps brachii.

Along the short head of the biceps brachii is a small slip of muscle known as the **coracobrachialis** (core-AK-oh-BRAY-key-AL-us). This muscle is a short, diagonal muscle of the arm. Underneath the biceps brachii on the anterior surface of the arm is the **brachialis** (BRAY-key-AL-us) muscle. The brachialis only crosses the elbow joint and thus flexes the forearm. The **brachioradialis** (BRAY-key-oh-RAY-dee-AL-us) is a distal muscle of the arm and is the most lateral muscle of the forearm. Examine figure 13.4 for these muscles and compare them to descriptions found in table 13.1.

Cat Dissection

The objective in using a cat for the exercise on musculature is to provide a study specimen for dissection and application to the human system. Differences occur

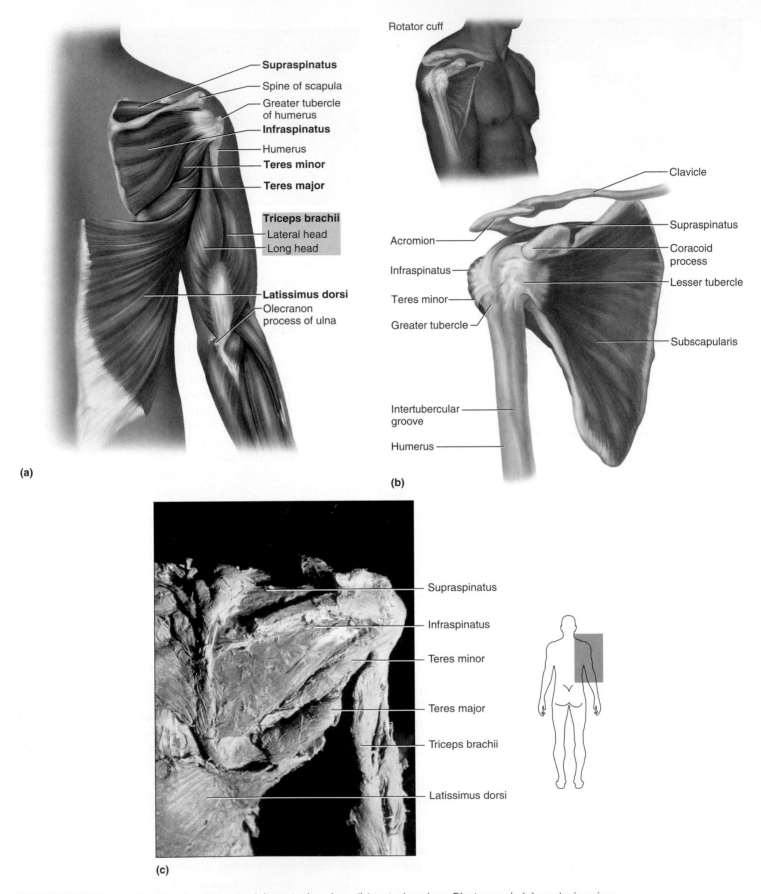

FIGURE 13.3 **Scapular Muscles** Diagram (a) posterior view; (b) anterior view. Photograph (c) posterior view.

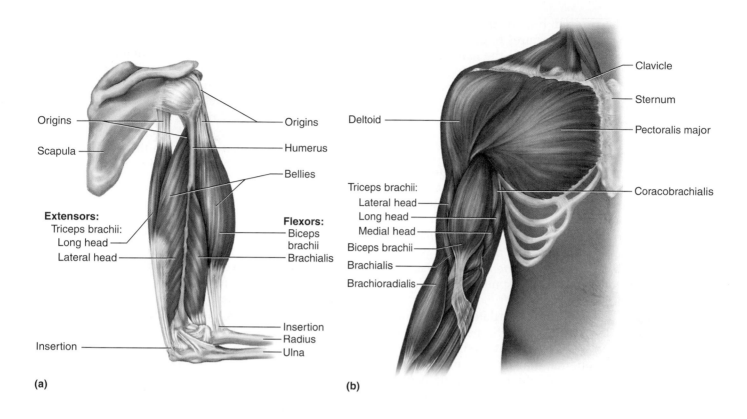

(a)

(b)

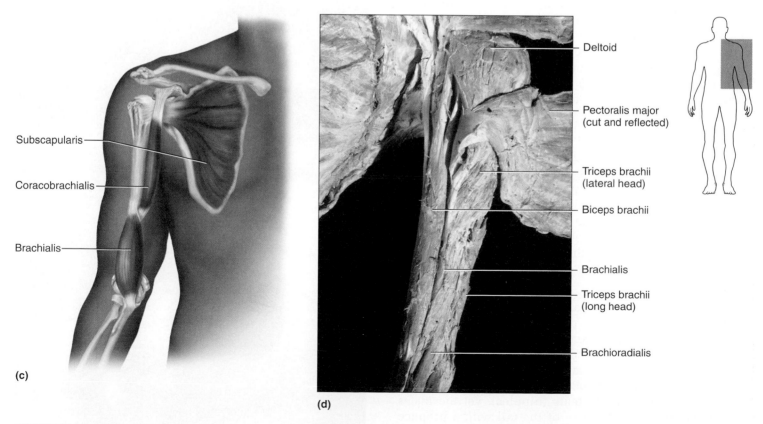

(c)

(d)

FIGURE 13.4 Muscles of the Arm Diagram (a) right lateral view; (b) superficial muscles, right anterior view; (c) deep muscles, right anterior view. Photograph (d) left anterior view.

between cat musculature and human musculature, but you should focus on similar structures to gain an appreciation of human musculature. Read all the material in the exercise prior to beginning the dissection. Dissection is a skill in which you try to separate the overlying structures from those underneath while keeping intact as much material as you can.

Dissection Concerns

Wear an apron or an old overshirt and safety glasses as you dissect. Dissection instruments are sharp, and you must be careful when using scalpels. Do not cut *down* into the specimen but rather lift structures gently and try to make incisions so that the scalpel blade cuts *laterally*. Cut *away* from yourself and your lab partners. *If you do cut yourself notify the instructor immediately!* Wash the cut with antimicrobial soap and seek medical advice to reduce the chance of infection.

The laboratory should be well ventilated, and if your eyes burn and you develop a headache get some fresh air for a moment. If you dissect without having your face directly over the specimen, that may help. Some students have allergic reactions to formaldehyde. This usually consists of a feeling of restriction of breath. If this happens notify your instructor.

At the end of the exercise wash your dissection equipment with soap and water, taking extra precaution with the scalpel blade. Place all used blades or sharp material in the **sharps container** in lab. Remove all the excess animal material on the dissection trays and put it in the appropriate animal waste container. *Do not dump excess animal material in the lab sinks!* Wash the dissection trays and place them in the appropriate area to dry.

Cat Care

Keep your cat in a plastic bag. It is recommended that you retain the skin of the cat as a protective wrapping when you are finished with the day's dissection or that you take a small towel (or an old T-shirt) and wrap the specimen in the cloth, soaking it in a cat wetting solution before you place it back in the bag. Usually, one lab period is required to skin the cat and prepare the specimen for further study.

External Features

Take a dissection tray and a cat specimen to your table. Remove the cat from the plastic bag and place it on the tray. You may have excess fluid in the plastic bag, and this should be disposed of properly as directed by your instructor. Place the cat on its back and determine whether you have a male specimen or a female specimen. Both sexes have multiple **teats** (nipples), yet the males have a **scrotum** near the base of the tail with a **prepuce** (penile foreskin) ventral to the scrotum. Females have a **urogenital opening** anterior to the **anus** without a scrotum and prepuce. Once you have identified the sex of your

specimen, compare yours to others in the class so that you can identify the sexes externally. If you have difficulty determining the sex, ask your instructor for help.

Examine the cat and notice the **vibrissae** (vib-RISS-ee), or whiskers, in the facial region. Other variations from humans are the presence of **claws** and **friction pads** on the extremities and a **tail.** Review the planes of the body as illustrated in Laboratory Exercise 2 before you begin the dissection. The terms used in that exercise will be of great importance in the dissection procedures.

Removal of the Skin

Begin your dissection by lifting the skin in the pectoral region with a forceps and making a small cut with a scalpel or sharp scissors in the midline. Work a blunt probe gently into the cut so that you free the skin from the underlying fascia and muscle somewhat. Be careful—the muscles are close to the skin. Insert your scalpel, blade side up, and make small incisions in the skin, cutting away from the underlying muscle as illustrated in figure 13.5.

Make a cut that runs up the midline of the sternal region until you reach the neck. Likewise, cut posteriorly until you reach an area craniad to the genital region. Leave the genitals intact and carefully make an incision that runs perpendicular to your first cut. Likewise, make

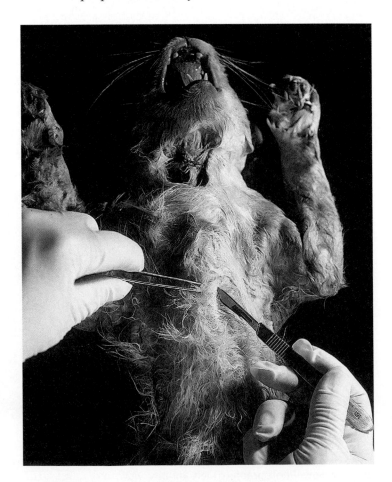

FIGURE 13.5 Lateral Cutting with a Scalpel

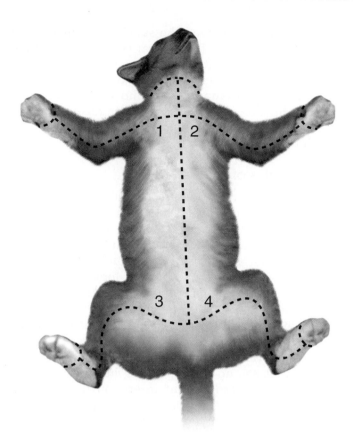

FIGURE 13.6 Removal of the Skin of the Cat After cutting down the midline, make incisions into the limbs (follow the numbers) and gently remove the skin from underlying structures.

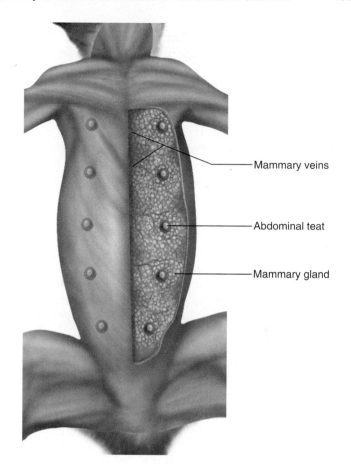

FIGURE 13.7 Mammary Glands of the Cat

Mammary veins

Abdominal teat

Mammary gland

a perpendicular incision along the upper thoracic region (see figure 13.6). Continue your caudal incision along the medial aspect of the thigh. Watch out for superficial blood vessels in this area (the great saphenous vein) and stop when you reach the knee. Cut the skin around the knee and begin removing it as a layer from the dorsal side of the cat. Cut the skin from the base of the tail and remove the skin from the back.

Once you reach the shoulders, turn the cat back to the ventral side and remove the skin on the medial side of the arm until you reach the elbow. Stop at this level and remove the skin from the lateral aspect of the arm, working back toward the shoulder. Be careful not to damage the superficial veins on the lateral side of the forearm. Continue your removal of the skin from the ventral body region. If your cat is a female, locate the mammary glands, elongated, beige, lobular tissue on each side of the midline on the ventral side (see figure 13.7).

Make an incision on the ventral side of the neck along the midline. Be careful—there are numerous blood vessels along the neck. Do *not* cut through these blood vessels. Continue up into the face and look for beige lumps of glandular tissue in the region of the mandible. These are the salivary glands. Cut the skin carefully from the face and remove it from the head by cutting around the ears (see figure 13.8).

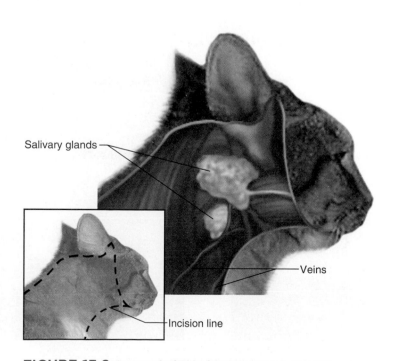

Salivary glands

Veins

Incision line

FIGURE 13.8 Removal of Skin from the Head of a Cat

You should be able to remove all of the skin from the cat at this time. Examine the undersurface of the skin and note the superficial muscles that cause the skin to move. These are the **cutaneous** (que-TAY-nee-us) **maximus** and the **platysma** (plah-TISS-mah).

Remove as much fat overlying the muscles as you can. Subcutaneous fat is variable from cat to cat, and your specimen may have little or may have significant amounts. The muscle also is covered by fascia, a connective tissue wrapping. Remove the fascia from the muscle so that the fiber direction is apparent.

Individual Muscles of the Cat

Begin your **dissection** of the muscles by understanding that the term *dissect* means to separate. When you isolate one muscle from another, locate the tendons of that muscle. The **tendon** is the attachment point of the muscle to a bone. Broad, flat tendons are known as **aponeuroses** (AH-poh-nyeu-ROH-seez).

Once you locate the muscle, tug gently on it to locate its **origin** and **insertion.** If you pull too hard, you may rip the muscle. The outer wrapping of the muscle is known as the **fascia,** and it should be removed to find the fiber direction of the muscle. The main part of the muscle is known as the **belly.** You may need to cut a muscle from time to time to locate deeper muscles. This is done by **transecting** the muscle, which is to cut the muscle in half perpendicular to the fiber direction. After transecting the muscle you may want to **reflect** it, or pull it toward its attachment site. When separating two muscles you may find a cottonlike material between the muscles. This is loose connective tissue that forms part of the fascia.

Once you have removed the skin from your cat and removed the superficial fascia, you should identify the major muscles of the cat. Look for the large **latissimus dorsi** muscle of the back and the **external abdominal oblique** muscle. You should also find the **deltoids** and **triceps brachii** muscles of the shoulder region and the **gluteus** and **biceps femoris** muscles of the hip and thigh region. Compare your cat to figure 13.9.

Pectoral Muscles of the Cat

There are four major muscles seen in a superficial view of the pectoral region. From anterior to posterior these are the pectoantebrachialis, pectoralis major, pectoralis minor, and xiphihumeralis. The **pectoantebrachialis** has no corresponding muscle in the human. It originates on the sternum and inserts on the forearm. Transect and reflect the pectoantebrachialis to see the **pectoralis major.** The pectoralis major originates on the sternum and inserts on the upper humerus. The **pectoralis minor** is a large muscle in cats and it inserts on the upper humerus. The **xiphihumeralis** (ZI-fee-HEU-mur-AL-is) is another cat muscle that has no corresponding human muscle; it originates on the sternum and inserts on the proximal humerus along with the pectoralis major and minor. Locate these muscles on your cat and in figure 13.10.

Muscles of the Back

In humans there is a singular trapezius and deltoid muscle on each side of the body. In cats the trapezius consists of three muscles, as does the deltoid. Examine the cat from the dorsal side and locate the **clavotrapezius,** the **acromiotrapezius,** and the **spinotrapezius.** All

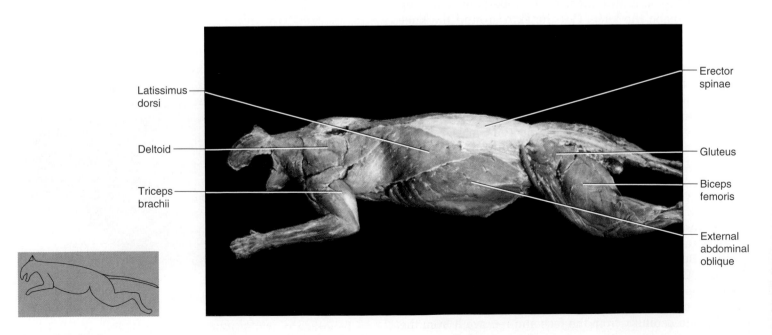

FIGURE 13.9 Major Muscles of the Cat

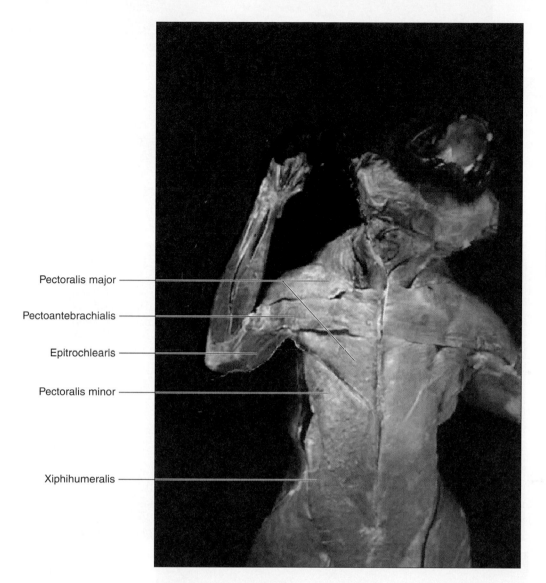

Pectoralis major

Pectoantebrachialis

Epitrochlearis

Pectoralis minor

Xiphihumeralis

FIGURE 13.10 Muscles of the Pectoral Region of the Cat

of these muscles originate on the vertebral column, with the clavotrapezius also originating on the occipital bone. The clavotrapezius inserts on the clavicle, the acromio-trapezius on the acromion process of the scapula, and the spinotrapezius on the spine of the scapula. Compare these muscles to figure 13.11.

The deltoid muscles are also named for their bony attachments. The **clavodeltoid** (clavobrachialis) originates on the clavicle, the **acromiodeltoid** on the acromion process, and the **spinodeltoid** on the spine of the scapula. Insertions of these muscles are on the arm or forelimb. Locate these muscles on the cat and compare them to figure 13.11.

Two other muscles of the region are the **latissimus dorsi** and the **levator scapulae ventralis.** These two muscles are similar to those in the human. Locate these muscles on the cat and compare them to figure 13.11.

The deep muscles of the scapula can be seen by reflecting the overlying muscles. The **supraspinatus,**

infraspinatus, subscapularis, teres major, and **teres minor** are roughly equivalent to the same muscles in the human. Locate the supraspinatus, infraspinatus, and teres major and find them in figure 13.12.

Forelimb Muscles

The muscles that have an action on the forelimb of the cat typically either flex the forelimb or extend the forelimb as their primary action. The **epitrochlearis** is a muscle that does not have a corresponding muscle in humans. The epitrochlearis is on the medial side of the humerus and extends the forelimb. The **biceps brachii** is also a medial muscle, and it flexes the forelimb. The **triceps brachii, anconeus, brachioradialis,** and **brachialis** are lateral or posterior muscles. The triceps brachii and the anconeus extend the forearm, while the brachialis flexes the forearm. Find the lateral muscles, using figures 13.10, 13.13, and 13.14 as a guide.

FIGURE 13.11 Superficial Muscles of the Shoulder of the Cat

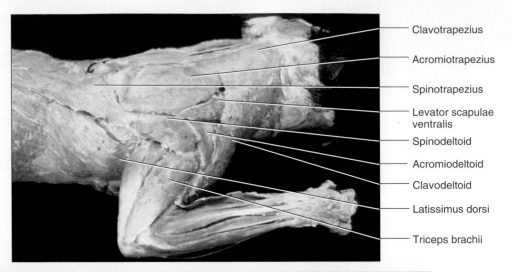

- Clavotrapezius
- Acromiotrapezius
- Spinotrapezius
- Levator scapulae ventralis
- Spinodeltoid
- Acromiodeltoid
- Clavodeltoid
- Latissimus dorsi
- Triceps brachii

FIGURE 13.12 Deep Muscles of the Scapula of the Cat

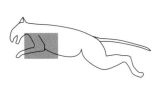

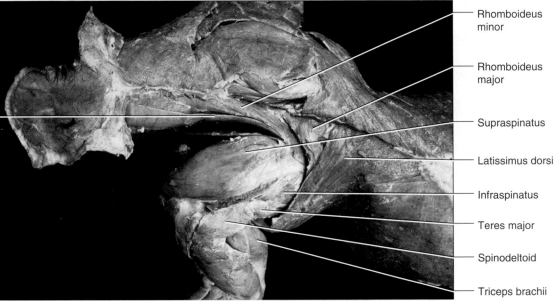

- Rhomboideus capitis
- Rhomboideus minor
- Rhomboideus major
- Supraspinatus
- Latissimus dorsi
- Infraspinatus
- Teres major
- Spinodeltoid
- Triceps brachii

FIGURE 13.13 Muscles of the Proximal Forelimb of the Cat, Ventral View

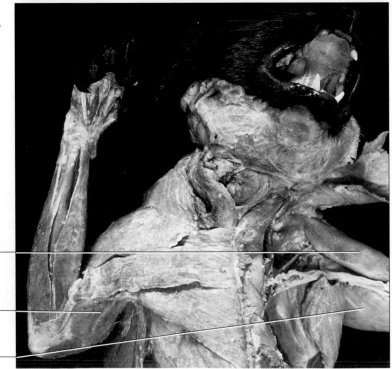

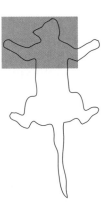

- Biceps brachii
- Epitrochlearis
- Triceps brachii

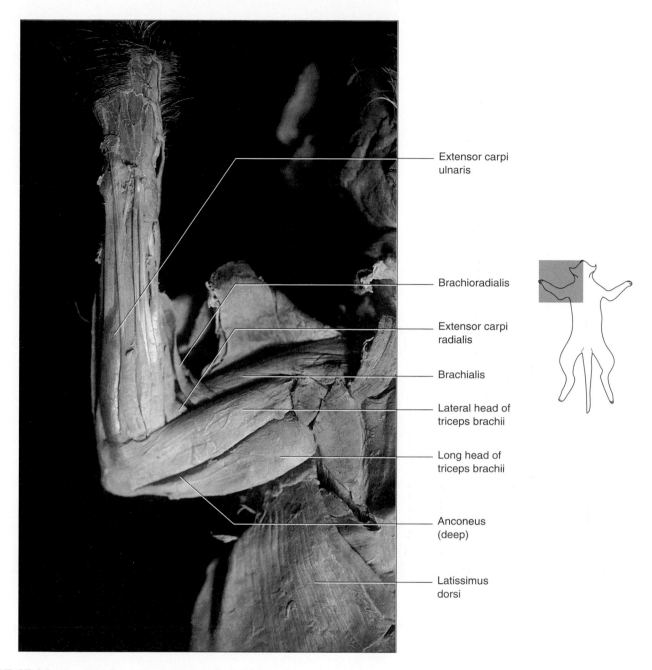

Extensor carpi ulnaris

Brachioradialis

Extensor carpi radialis

Brachialis

Lateral head of triceps brachii

Long head of triceps brachii

Anconeus (deep)

Latissimus dorsi

FIGURE 13.14 Muscles of the Proximal Forelimb of the Cat, Dorsal View

Notes

REVIEW SECTION

Introduction to the Study of Muscles and Muscles of the Shoulder and Arm

Name _____ *Date* _____

Lab Section _____ *Time* _____

Review Questions

1. In human muscles:

   a. What is the action of the deltoid muscle?

   b. Name the origin of the supraspinatus muscle.

   c. What is the insertion of the trapezius muscle?

   d. Does the biceps brachii muscle originate or insert on the humerus?

   e. What is the insertion of the pectoralis minor?

2. In cat muscles:

   a. What is the action of the epitrochlearis?

   b. Circle the muscle that does *not* correspond to a human muscle: biceps brachii, brachialis, xiphihumeralis, latissimus dorsi.

   c. How does the deltoid of the cat differ from the deltoid of the human?

3. Match each term on the left with a description on the right.

1.	dissect	a. what a muscle does
2.	flexion	b. to cut a muscle in half
3.	reflect	c. to stabilize a joint
4.	transect	d. to separate muscles
5.	action	e. to decrease a joint angle
6.	fixing	f. to pull back a muscle

4. What muscles are antagonists of the triceps brachii?

5. What is the origin of the trapezius?

6. The pectoralis major has what action?

7. What is the origin of the pectoralis minor?

8. The latissimus dorsi muscle has what action?

9. Name the insertion of the infraspinatus.

10. What is the insertion of the subscapularis?

11. What is the action of the triceps brachii?

12. What is the origin of the brachialis?

13. Name all the muscles that flex the arm.

14. As you hit a nail with a hammer, what arm muscle are you using?

15. As you look up at the ceiling, what back muscle are you using?

LABORATORY

Muscles of the Forearm and Hand

INTRODUCTION

The study of the muscles of the forearm and hand is important, particularly in rehabilitating limbs after surgery to relieve carpal tunnel syndrome. You should learn the origins and insertions as you learn the muscles of the forearm, because many of these muscles look similar. By knowing the origins and insertions of a muscle you will not easily mistake it for another muscle. These muscles are covered in the Saladin text in chapter 10, "The Muscular System." As in the previous exercise, you can compare the musculature of the human to that of the cat.

OBJECTIVES

At the end of this exercise you should be able to

1. locate the muscles of the forearm and hand on a model, chart, cadaver, or cat;
2. list the origin, insertion, and action of each muscle presented;
3. describe what nerve controls each muscle;
4. name all the muscles that have an action on a joint, such as all the muscles that flex the hand;
5. compare and contrast cat musculature and human musculature.

MATERIALS

Human torso model

Human arm models

Human muscle charts

Articulated skeleton

Cadaver (if available)

Cat

Materials for Cat Dissection

Dissection trays and equipment

Gloves

String

PROCEDURE

Review the muscle nomenclature and actions of muscles as outlined in Laboratory Exercises 12 and 13. Examine the following muscles on a model, chart, or cadaver and locate the origins and insertions of each. Refer to the following descriptions as you examine the muscles. The specific details of the muscles are provided in table 14.1. Review the bones of the hand as well as the bones of the arm and forearm in Laboratory Exercise 9 to precisely locate the bony origins and insertions.

The muscles of the forearm, due to their similar appearance, provide greater challenges in distinguishing one from another than do the muscles of the shoulder and arm. As you study these muscles, it is important that you determine the origin and insertion, as this will help you determine if you are looking at the right muscle.

The muscles of the forearm and hand are generally named for their action (*pronate* the hand or *flex* the digits) or for their insertion (*carpi* for inserting on carpals or metacarpals, *digitorum* for fingers, and *pollicis* for thumb). A few are named for the shape of the muscle (*teres* for round or *quadratus* for square).

Many of the anterior forearm muscles are innervated by the median nerve. As you examine these muscles look for the median nerve.

Much of the muscle mass for moving the fingers is located on the forearm. In this way the fingers move as if on puppet strings. If all the muscle mass for the hand were located on the hand proper, the hand would look like a softball. By having most of the muscle mass in the forearm, an efficiency of form allows for a powerful grip yet precise movements of the fingers.

Muscles That Supinate and Pronate the Hand

The first group of muscles in this study are those muscles that insert on the radius. The **supinator** (SOUP-in-AY-tor) is a muscle that originates on the arm and forearm; it supinates the hand. It wraps around the radius and is the deepest, proximal muscle of the forearm. Examine this muscle in the lab and compare it to figure 14.1.

The **pronator teres** (PRO-nay-tor TARE-eez) is a round muscle that pronates the hand. It originates on the medial

TABLE 14.1	Muscles of the Forearm and Hand			
Name	**Origin**	**Insertion**	**Action**	**Innervation**
Supinator	Lateral epicondyle of humerus, anterior ulna	Proximal radius	Supinates hand	Radial nerve
Pronator quadratus	Distal part of anterior ulna	Distal radius	Pronates hand	Median nerve
Pronator teres	Medial epicondyle of humerus, coronoid process of ulna	Lateral, middle shaft of radius	Pronates hand, flexes forearm	Median nerve
Palmaris longus	Medial epicondyle of humerus	Palmar aponeurosis	Flexes hand	Median nerve
Flexor carpi radialis	Medial epicondyle of humerus	Second and third metacarpals	Flexes and abducts hand	Median nerve
Flexor carpi ulnaris	Medial epicondyle of humerus, olecranon process and dorsal border of ulna	Pisiform, hamate, fifth metacarpal	Flexes and adducts hand	Ulnar nerve
Flexor digitorum superficialis	Medial epicondyle of humerus, proximal ulna, proximal radius	Middle phalanges of second through fifth digits	Flexes proximal and middle phalanges, flexes hand	Median nerve
Flexor digitorum profundus	Anterior, proximal surface of ulna; interosseus membrane	Distal phalanges of second through fifth digits	Flexes phalanges, flexes hand	Median and ulnar nerves
Flexor pollicis longus	Anterior portion of radius and interosseus membrane	Distal phalanx of pollex (thumb)	Flexes thumb	Median nerve
Flexor pollicis brevis	Trapezium	Proximal phalanx of first digit	Flexes thumb	Median and ulnar nerves
Abductor pollicis brevis	Scaphoid, trapezium	Proximal phalanx of first digit	Abducts thumb	Median nerve
Opponens pollicis	Trapezium	First metacarpal	Opposes thumb	Median nerve
Abductor digiti minimi	Pisiform	Proximal phalanx of fifth digit	Abducts fifth digit	Ulnar nerve
Flexor digiti minimi brevis	Hamulus of hamate	Proximal phalanx of fifth digit	Flexes fifth digit	Ulnar nerve
Opponens digiti minimi	Hamulus of hamate	Fifth metacarpal	Opposes fifth digit	Ulnar nerve
Extensor carpi radialis longus	Lateral supracondylar ridge of humerus	Second metacarpal	Extends and abducts hand	Radial nerve
Extensor carpi radialis brevis	Lateral epicondyle of humerus	Third metacarpal	Extends and abducts hand	Radial nerve
Extensor carpi ulnaris	Lateral epicondyle of humerus, proximal ulna	Fifth metacarpal	Extends and adducts hand	Radial nerve
Extensor digitorum	Lateral epicondyle of humerus	Middle and distal phalanges of second through fifth digits	Extends phalanges, extends hand	Radial nerve
Abductor pollicis longus	Posterior radius and ulna, interosseus membrane	First metacarpal	Abducts thumb	Radial nerve
Extensor pollicis longus and brevis	Posterior radius and ulna, interosseus membrane	Proximal and distal phalanges of pollex (thumb)	Extends thumb	Radial nerve

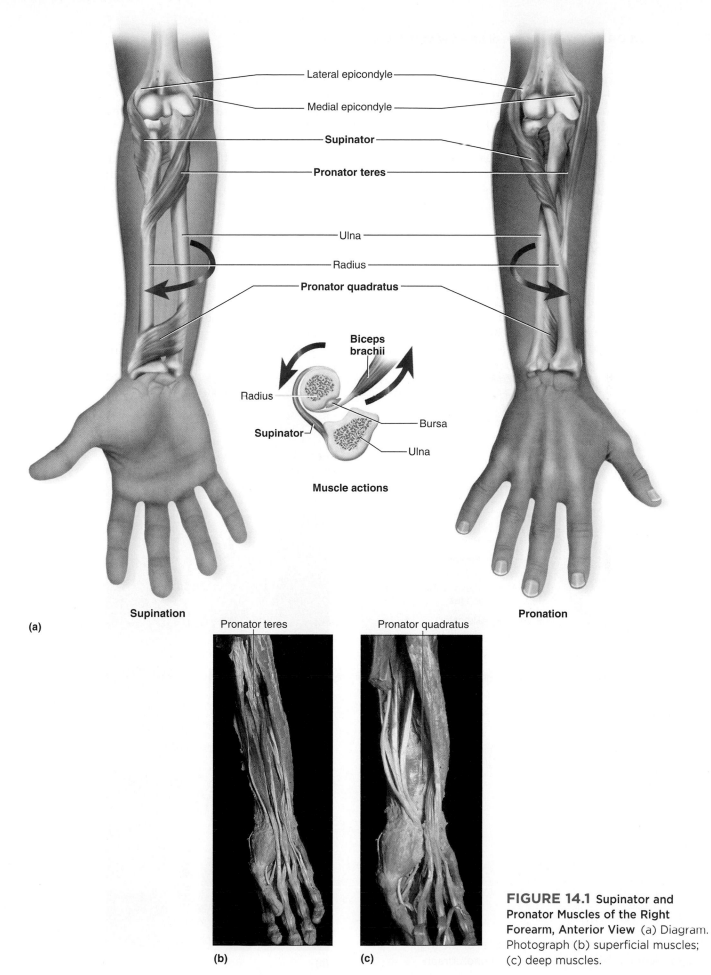

Lateral epicondyle

Medial epicondyle

Supinator

Pronator teres

Ulna

Radius

Pronator quadratus

Biceps brachii

Radius

Bursa

Supinator

Ulna

Muscle actions

Supination

Pronation

(a)

Pronator teres

Pronator quadratus

(b)

(c)

FIGURE 14.1 Supinator and Pronator Muscles of the Right Forearm, Anterior View (a) Diagram. Photograph (b) superficial muscles; (c) deep muscles.

side of the arm and forearm and inserts on the lateral side of the radius. It is a bit different from the other superficial forearm muscles in that the pronator teres runs at an oblique angle on the forearm, while the other muscles run parallel along the length of the forearm.

The **pronator quadratus** (quah-DRAY-tus) is a square muscle deep to the other forearm muscles on the distal part of the radius and the ulna. It pronates the hand. Compare the two pronator muscles in the lab to figure 14.1.

Flexor Muscles

The next muscles to be studied are grouped by their insertion on the hand. The superficial **palmaris** (pahl-MARE-us) **longus** muscle is absent in about 10% of the population. It is centrally located in the middle, anterior forearm and inserts into a broad, flat tendon known as the **palmar aponeurosis.** This aponeurosis has no

bony attachment but rather attaches to the fascia of the underlying muscles.

The **flexor carpi** (CAR-pie) **radialis** muscle inserts on the metacarpals on the radial side of the hand. The pulse of the radial artery is taken at the wrist just lateral to the tendon of the flexor carpi radialis muscle. Most of the flexor muscles of the hand originate from the medial epicondyle of the humerus, as seen with the flexor carpi radialis. The tendon of this muscle runs underneath a connective tissue band known as the **flexor retinaculum** (RET-in-AK-you-lum), which anchors the tendons to the wrist and prevents the tendons from pulling away from the wrist when the hand is flexed.

The **flexor carpi ulnaris** muscle also has an origin on the medial epicondyle of the humerus and inserts on the carpals and a metacarpal of the ulnar side of the hand. The flexor carpi ulnaris is a medial muscle of the forearm. Examine these muscles in the lab and in figure 14.2.

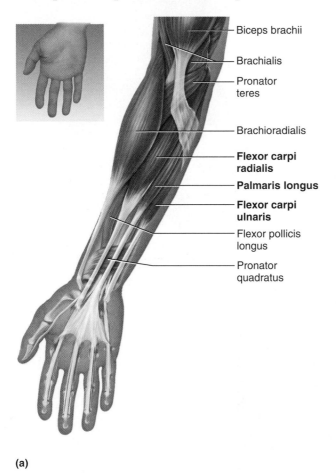

(a)

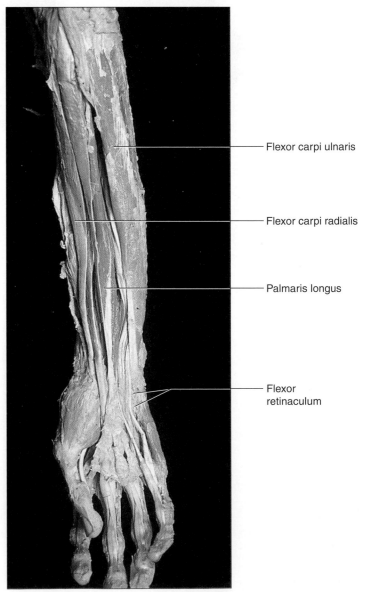

(b)

FIGURE 14.2 Superficial Flexor Muscles of the Right Forearm, Anterior View (a) Diagram; (b) photograph.

The **flexor digitorum superficialis** (SOUP-er-FISH-ee-AL-is) is a superficial flexor muscle of the digits. It is not the most superficial muscle of the forearm but is actually under the palmaris longus, the flexor carpi radialis, and the flexor carpi ulnaris. The flexor digitorum superficialis is superficial to the deep digit flexor, discussed next. As with the other flexor muscles, the flexor digitorum superficialis has an origin on the medial epicondyle of the humerus. The distal tendons form a V on the middle phalanges of the second through fifth digits.

The **flexor digitorum profundus** (pro-FUND-us) is a deep muscle of the digits. It is an exception to the other hand flexors in that it does *not* originate on the medial epicondyle of the humerus but on the ulna and the membrane between the radius and the ulna **(interosseus membrane).** It is a deep muscle of the forearm that runs underneath the flexor digitorum superficialis. At the insertion point the tendons of the flexor digitorum profundus run through the split tendons of the flexor digitorum superficialis and extend to the distal phalanges of the second through fifth digits.

The **flexor pollicis** (PAUL-eh-sis) **longus** originates on the radius and the interosseus membrane and runs along the lateral forearm to insert on the pad-side of the thumb. Flexion of the thumb is the curling of the thumb as if you were ready to flip a coin. Examine these muscles and compare them to figure 14.3.

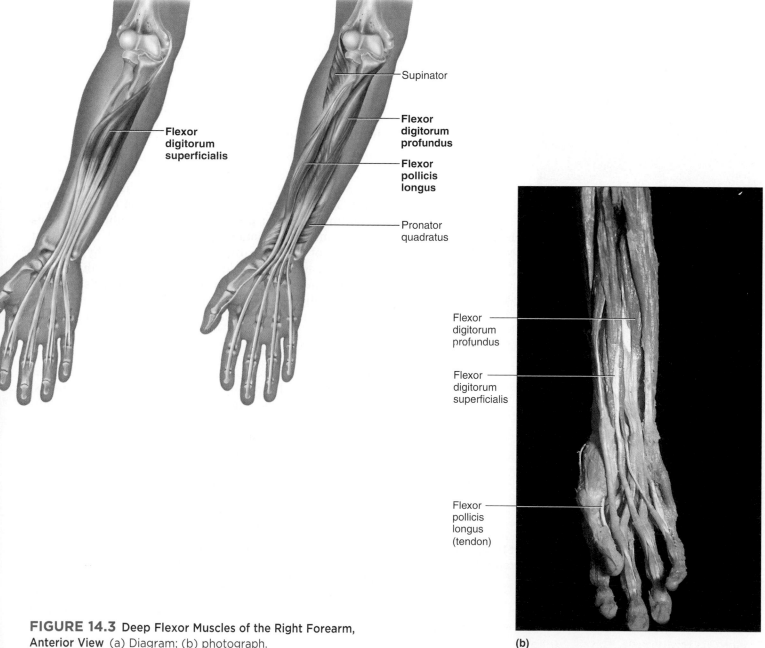

FIGURE 14.3 Deep Flexor Muscles of the Right Forearm, **Anterior View** (a) Diagram; (b) photograph.

(b)

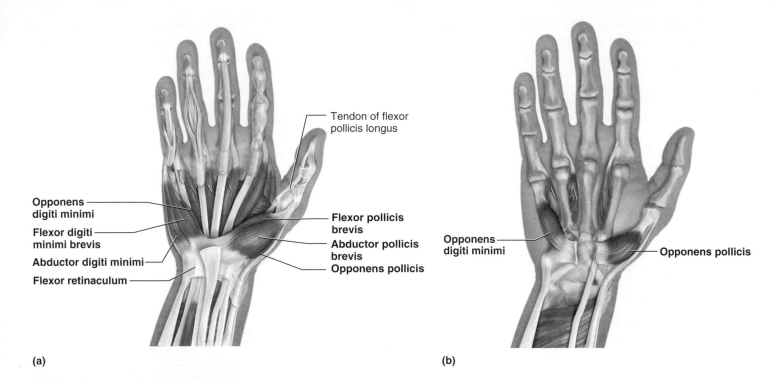

(a)

(b)

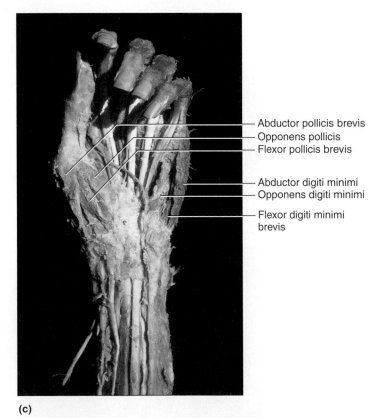

(c)

FIGURE 14.4 Intrinsic Muscles of the Right Hand Diagram
(a) superficial muscles; (b) deep muscles. Photograph
(c) superficial muscles.

Intrinsic Muscles of the Hand

Many muscles originate on the hand and have an action on the hand. Some of these arise on the pad of muscles at the base of the thumb known as the **thenar eminence.** These muscles are named for their action on the thumb, including the **flexor pollicis brevis,** the **abductor pollicis brevis,** and the **opponens pollicis.** The pad of tissue at the base of the fifth digit is known as the **hypothenar eminence,** and the muscles that arise there are the **abductor digiti minimi,** the **flexor digiti minimi brevis,** and the **opponens digiti minimi.** Find these muscles on models or cadavers in lab and compare them to figure 14.4.

Extensor Muscles

As a general rule, most of the extensors originate on the lateral epicondyle or lateral supracondylar ridge of the humerus. The extensor muscle tendons are held to the posterior surface of the wrist by a connective tissue band known as the **extensor retinaculum** (RET-in-AK-u-lum). The **extensor carpi radialis longus** muscle originates on the humerus and inserts on the second metacarpal (on the radial side) of the hand. The **extensor carpi radialis brevis** (BREV-is) is deep to the longus, and it inserts on the dorsum of the third metacarpal. Both of these muscles extend and abduct the hand. Examine these muscles in figure 14.5.

The **extensor carpi ulnaris** originates on the lateral epicondyle and inserts on the fifth metacarpal (on the ulnar side) of the hand. As it contracts the hand is extended and adducted. The **extensor digitorum** is a singular muscle on the back of the hand (remember there are two flexor digitorum muscles). As an extensor muscle it originates on the lateral epicondyle of the humerus. It inserts

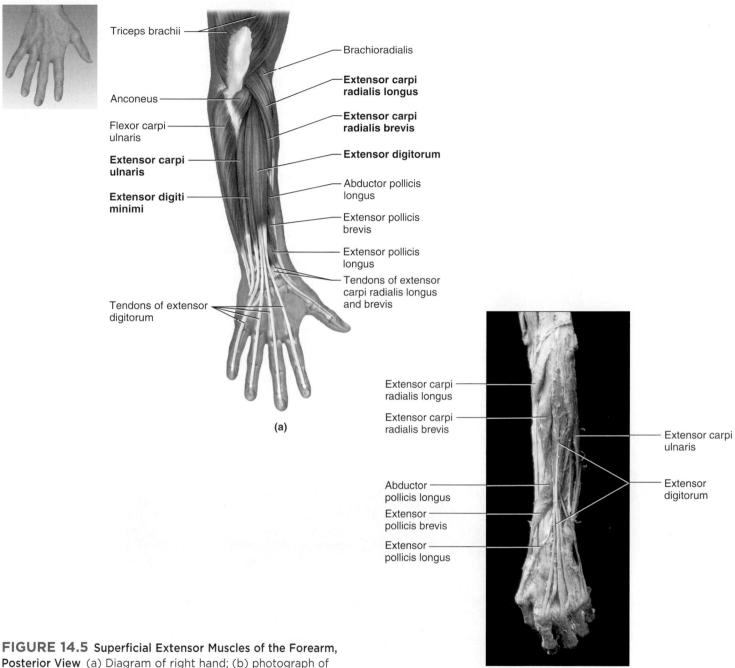

Triceps brachii

Brachioradialis

Extensor carpi radialis longus

Anconeus

Extensor carpi radialis brevis

Flexor carpi ulnaris

Extensor digitorum

Extensor carpi ulnaris

Abductor pollicis longus

Extensor digiti minimi

Extensor pollicis brevis

Extensor pollicis longus

Tendons of extensor carpi radialis longus and brevis

Tendons of extensor digitorum

(a)

Extensor carpi radialis longus

Extensor carpi radialis brevis

Extensor carpi ulnaris

Abductor pollicis longus

Extensor digitorum

Extensor pollicis brevis

Extensor pollicis longus

FIGURE 14.5 Superficial Extensor Muscles of the Forearm, Posterior View (a) Diagram of right hand; (b) photograph of left hand.

(b)

on the middle and distal phalanges of the second through fifth digits. The tendons of this muscle can be seen on the dorsal surface of the hand and in figure 14.5.

The **abductor pollicis longus** muscle inserts on the metacarpal of the thumb. By pulling your thumb ventrally away from the index finger you are abducting the thumb. There are two extensor pollicis muscles, the **extensor pollicis longus** and the **extensor pollicis brevis.** Extension of the thumb is done when you flip a coin. At the end of the flip the thumb is extended. The tendons of the extensor pollicis muscles and the abductor pollicis muscles form a depression in the form of a triangle at the base of the thumb. This depression is known as the *anatomical snuff box*. These muscles can be seen in figures 14.5 and 14.6.

Cat Dissection

If you have not already removed the skin from the forelimb of the cat, you should do so now. This can be accomplished by making a longitudinal incision along the length of the forelimb. Be careful not to cut the tendons, blood vessels, or nerves. Pull the skin off as if you were removing a pair of knee socks. As you get to the tips of the digits cut the skin from them. Pay particular attention and keep the tendons

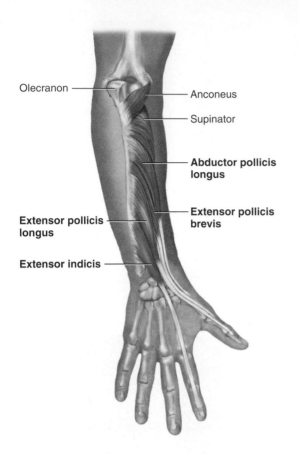

Olecranon

Anconeus

Supinator

Abductor pollicis longus

Extensor pollicis longus

Extensor pollicis brevis

Extensor indicis

FIGURE 14.6 Deep Extensor Muscles of the Right Thumb, Posterior View Diagram.

intact. It is a good procedure to dissect only one side of the cat at a time. If you make an error on one side, you will have the other side to dissect. As you cut through the pad on the ventral side of the paw of the cat, make sure that you do not cut through the tendons of the flexor digitorum muscles.

As you make the dissection in the cat, you can leave many of the muscles of the forelimb intact. Deeper muscles can be seen by moving the more superficial muscles to the side.

Superficial Muscles on the Medial Aspect of the Forelimb

Most of the muscles of the forelimb of the cat run parallel to the radius and ulna. An exception to this is the **pronator teres,** a small slip of muscle that runs obliquely down the forelimb. It pronates the wrist. The **palmaris longus** is a broad, flat muscle, superficially located on the forelimb with insertions into the digits. In humans the palmaris longus terminates at the palmar aponeurosis.

The **flexor carpi radialis** is a thin muscle inserting on the second and third metacarpals. It is named for its action, its insertion, and its location. The **flexor carpi ulnaris** has an origin on the humerus and ulna and an insertion on medial metacarpals and carpals. The **flexor digitorum superficialis** (flexor digitorum sublimis) occurs in the forelimb as a middle-level muscle, underneath the palmaris longus. It originates on fascia of other forearm muscles (usually, muscles originate on bone). Examine the superficial muscles of the medial forelimb and compare them to figure 14.7.

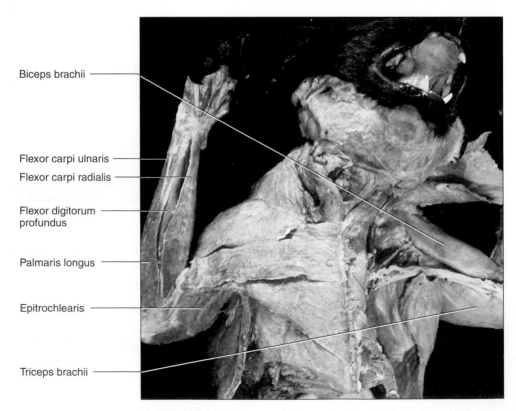

Biceps brachii

Flexor carpi ulnaris

Flexor carpi radialis

Flexor digitorum profundus

Palmaris longus

Epitrochlearis

Triceps brachii

FIGURE 14.7 Superficial Muscles of the Right Medial Forelimb of the Cat

Deep Muscles on the Medial Aspect of the Forelimb

The **supinator** is a deep muscle that runs diagonally from the lateral epicondyle of the humerus to the proximal radius. It is the deepest of the proximal forelimb muscles. The **flexor digitorum profundus** is an extensive muscle that inserts on the first through fifth digits. It replaces the flexor pollicis longus for the thumb flexion, as this muscle is absent in cats. The **pronator quadratus** is a square muscle located between the radius and ulna deep to the flexor digitorum profundus. Find the deep muscles in the cat and compare them to figure 14.8.

Muscles on the Lateral Aspect of the Forelimb

The lateral muscles of the forelimb are the extensor group. The **extensor carpi radialis longus** muscle is deep to the brachioradialis and inserts on the second metacarpal. The **extensor carpi radialis brevis** is underneath the extensor carpi radialis longus and inserts on the third metacarpal.

Locate the **extensor carpi ulnaris,** next to the extensor digitorum lateralis, inserting on the fifth metacarpal. The **extensor digitorum communis** is on the lateral aspect of the forelimb. It is a broad muscle inserting by tendons on the second through fifth digits. Locate these muscles on the cat and in figure 14.9.

The **extensor digitorum lateralis** is specific to the cat and inserts with the tendons of the digitorum communis to the digits. The **extensor pollicis brevis** is a well-developed muscle in the cat, while the abductor pollicis longus is absent in cats.

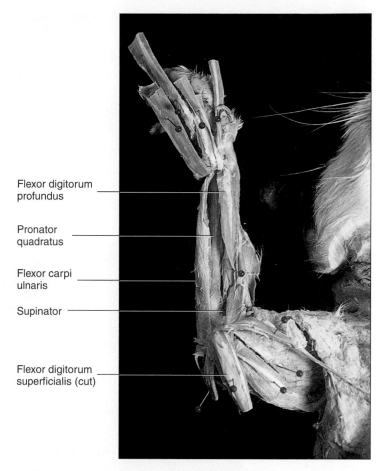

Flexor digitorum profundus

Pronator quadratus

Flexor carpi ulnaris

Supinator

Flexor digitorum superficialis (cut)

FIGURE 14.8 Deep Muscles of the Right Medial Forelimb of the Cat

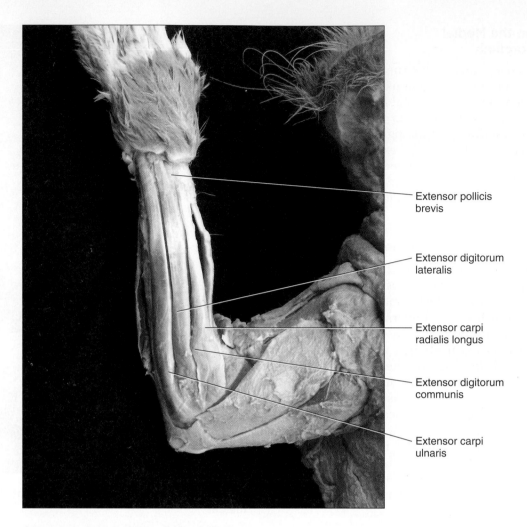

Extensor pollicis
brevis

Extensor digitorum
lateralis

Extensor carpi
radialis longus

Extensor digitorum
communis

Extensor carpi
ulnaris

FIGURE 14.9 Lateral Muscles of the Left Forelimb of the Cat

REVIEW SECTION

Muscles of the Forearm and Hand

Name _____ *Date* _____

Lab Section _____ *Time* _____

Review Questions

1. What is the origin of the flexor carpi ulnaris in humans?

2. What is an antagonist to the supinator muscle?

3. Where does the flexor digitorum superficialis of the human insert?

4. What is the insertion of the extensor carpi ulnaris muscle?

5. Which muscle is more developed in cats, the flexor digitorum superficialis or the flexor digitorum profundus?

6. What is an antagonist to the extensor pollicis longus and brevis muscles?

7. What is the action of the flexor carpi radialis muscles?

8. Name all the muscles that extend the hand.

9. What muscles flex the hand?

10. What muscles extend the thumb?

11. Where are the extensor carpi muscles found, on the anterior or posterior side of the forearm?

12. Label the muscles in the following illustration.

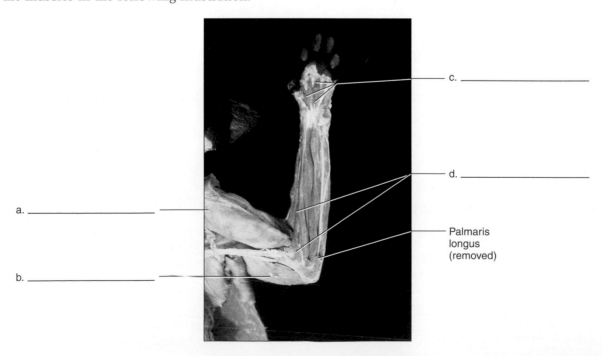

c. _____

d. _____

a. _____ — _____

Palmaris
longus
(removed)

b. _____ — _____

LABORATORY

Muscles of the Hip and Thigh

INTRODUCTION

The muscles of the hip and thigh are primarily muscles of locomotion. They can be divided into muscles that flex or extend the thigh, adduct or abduct the thigh, and flex or extend the leg. Lateral and medial rotations of the thigh are other actions that occur in muscles of this group. These muscles are discussed in the Saladin text in chapter 10, "The Muscular System."

The size and shape of the muscles of the hip and thigh in humans are different from other mammals in that humans are bipedal. This two-legged walking habit is much different from a quadruped's locomotion with respect to balance and forward movement. When a quadruped walks, typically three legs remain on the ground. When a biped walks, the instability of being on one leg makes the support limb that much more important. As you study these muscles think about what happens to the center of balance when you move a leg to walk or run.

OBJECTIVES

At the end of this exercise you should be able to

1. locate the muscles of the hip and thigh on a model, chart, cadaver, or cat;
2. list the origin, insertion, and action of each muscle presented;
3. describe what nerve controls each muscle;
4. list what muscles function as synergists or antagonists to the prime mover;
5. name all the muscles that move a joint, such as all of the muscles that flex the thigh.

MATERIALS

Human torso model

Human leg models

Human muscle charts

Articulated skeleton

Cadaver

Cat

Materials for Cat Dissection

Dissection trays

Scalpel or razor blades

Gloves

Blunt probe

Pins

PROCEDURE

Review the muscle nomenclature and the actions as outlined in Laboratory Exercises 12 and 13. Examine muscle models or charts in the lab and locate the muscles described in the following section and in table 15.1. Correlate the shape of the muscle with the name as you study the models or charts. You may want to look at an articulated skeleton as you review the origins and insertions of the muscles so that you can better see the muscle attachment points. The descriptions in the text portion of this exercise can help you understand the nature of the muscle, while table 15.1 gives you the particular information about the muscle.

Once you have learned the origin and the insertion, you should be able to understand the action of the muscle. This is done, in part, by imagining how the bony attachments of the origin and insertion would come together if the muscle pulled them closer to one another.

Muscles of the Hip and Thigh

The **iliopsoas** (ILL-ee-oh-SO-az) muscle is a major flexor of the thigh. It originates inside the abdominopelvic cavity and crosses over the pelvic brim to insert on the posterior aspect of the femur. The iliopsoas is two muscles, the **iliacus** (ILL-ee-AK-us) and **psoas major.** A companion muscle is the **psoas minor.** In this exercise we treat these muscles as one. Examine the material in the lab and compare it to figures 15.1 and 15.2.

The **sartorius** (sar-TOR-ee-us) is a slender muscle that runs from a superior, lateral origin and inserts in an inferior, medial location. When a tailor hems a pair of pants, he or she sits cross-legged. This motion mimics the

TABLE 15.1	Muscles of the Hip and Thigh			
Name	**Origin**	**Insertion**	**Action**	**Innervation**
Iliopsoas	T12, L1–5, sacrum, iliac crest, iliac fossa	Lesser trochanter of femur	Flexes thigh and lumbar vertebrae	Femoral nerve, spinal nerves L1–3
Sartorius	Anterior superior iliac spine	Proximal, medial tibia	Flexes and laterally rotates thigh, flexes leg	Femoral nerve
Gracilis	Symphysis pubis	Proximal, medial tibia	Adducts thigh, flexes leg	Obturator nerve
Pectineus	Pubis	Femur inferior to lesser trochanter	Adducts and flexes thigh	Femoral nerve
Adductor brevis	Pubis	Linea aspera at proximal portion of femur	Adducts, flexes, and laterally rotates thigh	Obturator nerve
Adductor longus	Pubis	Linea aspera at middle portion of femur	Adducts, flexes, and laterally rotates thigh	Obturator nerve
Adductor magnus	Pubis, ischium	Gluteal tuberosity, linea aspera, adductor tubercle of distal femur	Adducts, flexes, extends, and laterally rotates thigh	Obturator nerve
Quadriceps Femoris				
Rectus femoris	Anterior inferior iliac spine, margin of acetabulum	Tibial tuberosity by patellar tendon	Flexes thigh, extends leg	Femoral nerve
Vastus lateralis	Greater trochanter of femur, linea aspera of femur	Tibial tuberosity by patellar tendon	Extends leg	Femoral nerve
Vastus intermedius	Proximal, anterior femur	Tibial tuberosity by patellar tendon	Extends leg	Femoral nerve
Vastus medialis	Linea aspera, medial side	Tibial tuberosity by patellar tendon	Extends leg	Femoral nerve
Tensor fasciae latae	Iliac crest, anterior iliac surface	Iliotibial band of fasciae latae	Flexes, abducts, and medially rotates thigh	Superior gluteal nerve
Gluteus maximus	Outer iliac blade, iliac crest, sacrum, coccyx	Gluteal tuberosity of femur, iliotibial band of fasciae latae	Extends and laterally rotates thigh, braces knee	Inferior gluteal nerve
Gluteus medius	Outer iliac blade	Greater trochanter of femur	Abducts and medially rotates thigh	Superior gluteal nerve
Gluteus minimus	Outer iliac blade	Greater trochanter of femur	Abducts and medially rotates thigh	Superior gluteal nerve
Hamstrings				
Biceps femoris	Ischial tuberosity, linea aspera	Head of fibula, lateral condyle of tibia	Extends thigh, flexes leg	Sciatic nerve
Semitendinosus	Ischial tuberosity	Proximal, medial tibia	Extends thigh, flexes leg	Sciatic nerve
Semimembranosus	Ischial tuberosity	Medial condyle of tibia	Extends thigh, flexes leg	Sciatic nerve

action of the sartorius, named for the Latin word *sartor*, or tailor. The sartorius is the most superficial muscle on the anterior thigh. The most superficial muscle on the medial thigh is the **gracilis** (grah-SIL-us). The gracilis is a thin muscle that adducts the thigh. Locate these muscles in the lab and in figure 15.1.

The muscles in the medial aspect of the thigh are adductor muscles. The gracilis is an example of an adductor muscle, as is the **pectineus** (peck-TIN-ee-us) and the three **adductor muscles** proper. The gracilis is medial, the adductors are more lateral, and the pectineus is more lateral still. This sequence spells the word *GAP*. The pectineus is a small, quadrangular muscle, the

most proximal of the group. It is superficial to the adductor brevis muscle and proximal to the adductor longus muscle. Examine the material in the lab and compare the pectineus to figure 15.2.

The **adductor brevis** is the short adductor muscle deep in the medial aspect of the thigh. It can be seen if the pectineus is reflected. The adductor brevis originates on the pubis and inserts on the proximal third of the thigh. The **adductor longus** also has a pubic origin but its insertion is on the middle third of the thigh. The largest of the adductor muscles is the **adductor magnus.** It originates along the length of the inferior hip bone (pubis and ischium). The adductor magnus inserts on the femur

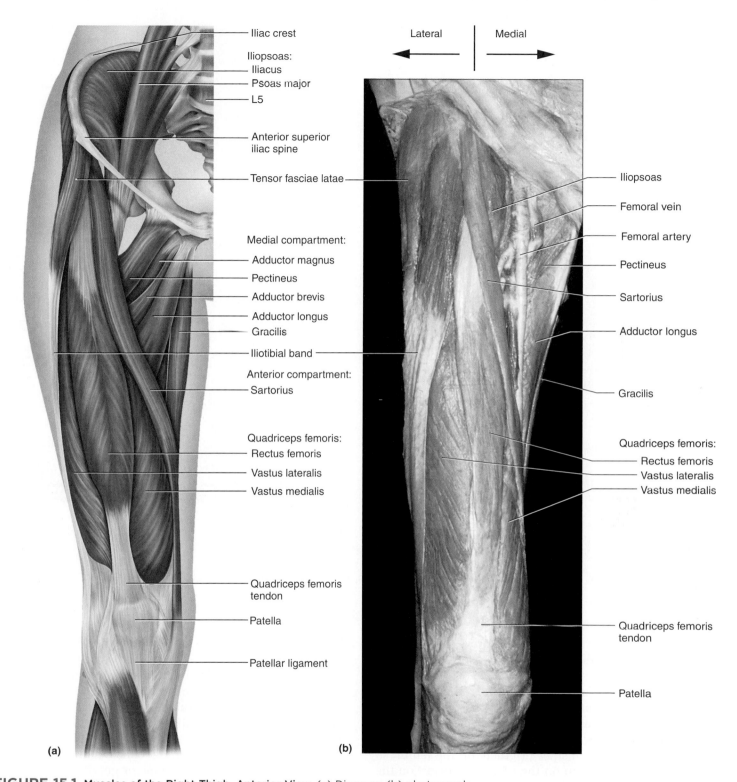

Lateral | Medial

(a) **(b)**

FIGURE 15.1 Muscles of the Right Thigh, Anterior View (a) Diagram; (b) photograph.

from the gluteal tuberosity, along the length of the linea aspera to the adductor tubercle. Examine these muscles in figure 15.2.

The muscles of the anterior aspect of the thigh belong to the quadriceps group. These muscles all extend the leg, while only one of the group flexes the thigh. The

quadriceps muscles all have a common tendon of insertion, called the patellar tendon. This tendon inserts on the tibial tuberosity. The most superficial muscle of the group is the **rectus femoris** (FEM-or-us) (see figures 15.1 and 15.3). It is a muscle that originates directly inferior to the sartorius. The rectus femoris runs straight down the

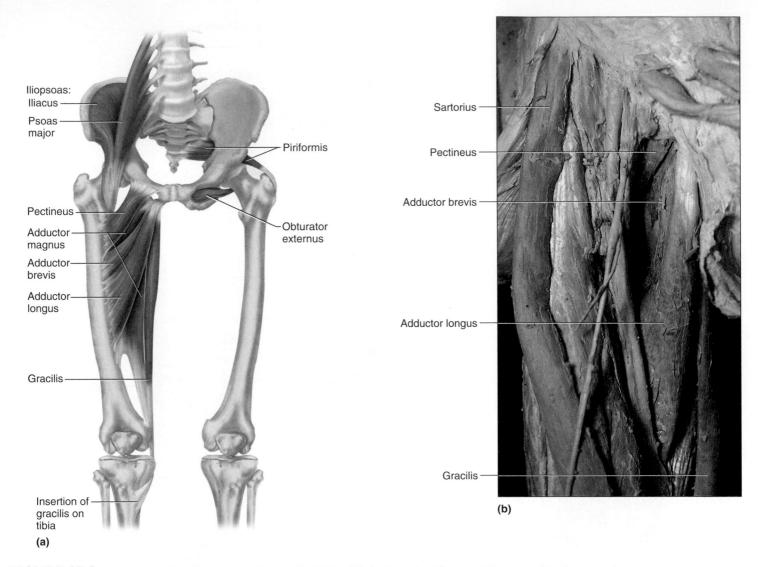

Iliopsoas:
Iliacus
Psoas major
Piriformis
Pectineus
Obturator externus
Adductor magnus
Adductor brevis
Adductor longus
Gracilis
Insertion of gracilis on tibia
(a)

Sartorius
Pectineus
Adductor brevis
Adductor longus
Gracilis
(b)

FIGURE 15.2 **Iliopsoas and Adductor Muscles of the Right Thigh, Anterior View** (a) Diagram; (b) photograph.

femur (*rectus* = straight). Because it crosses the hip joint, the rectus femoris is the only quadriceps muscle to flex the thigh.

The three other quadriceps muscles are the vastus muscles. The **vastus lateralis** is so named due to its lateral position. The **vastus intermedius** is deep to the rectus femoris, and the **vastus medialis** is the most medial of the vastus muscles. These muscles originate on the femur, so they do not have an action on the hip. They insert on the tibia and extend the leg. Locate these muscles in the lab and compare them to figure 15.3.

The **tensor fasciae latae** (TEN-sur FASH-ee-ee-LAY-tee) is a lateral muscle of the thigh that attaches to a thick band of connective tissue that runs down the lateral aspect of the thigh like a stripe on a pair of tuxedo pants. This is known as the **iliotibial tract,** a thickened portion of a broader sheet of connective tissue known as the fasciae latae. The fasciae latae covers the lateral aspect of the

thigh. The tensor fasciae latae abducts the thigh. Examine the material in the lab and compare it to figure 15.4.

The gluteus muscles in humans are different than in other mammals because we are bipedal. The largest of the gluteal muscles is the **gluteus maximus.** It extends the thigh in standing from a sitting position or in actions such as climbing up stairs. The **gluteus medius** and the **gluteus minimus** both aid in keeping balance during walking in that they maintain the center of gravity. The gluteus medius is superficial to the gluteus minimus, but they both originate on the outer iliac blade. When you move one leg to walk, gravity pulls the body in the direction of the lifted leg. The gluteus medius and minimus on the opposite side of the body contract to maintain upright posture. Compare these muscles to figure 15.4.

The **hamstring muscles** are so named because these muscles from a pig traditionally were tied together by their tendons (the hamstrings) and hung in a smokehouse

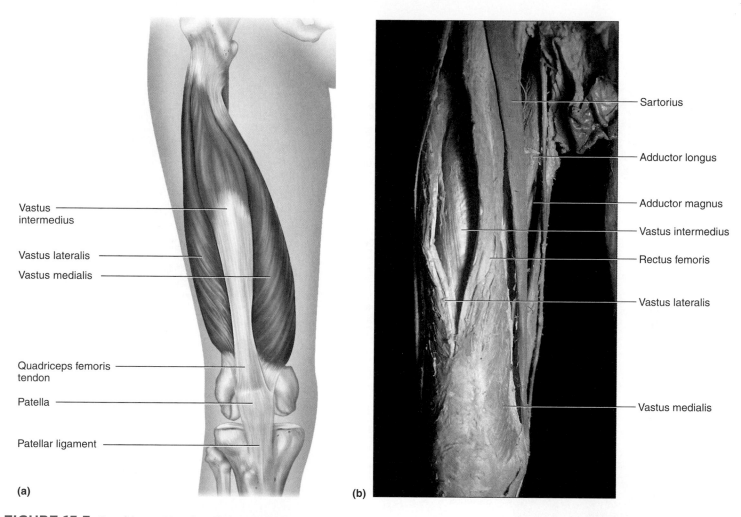

FIGURE 15.3 **Quadriceps Muscles of the Right Thigh, Anterior View** (a) Diagram (rectus femoris removed); (b) photograph.

to cure. Three muscles make up the hamstrings: the biceps femoris, the semitendinosus, and the semimembranosus. All three of these muscles cross the hip joint, are extensors of the thigh, and flex the leg.

The **biceps femoris** is the "two-headed muscle of the femur." One head originates with the other hamstring muscles, on the ischial tuberosity, while the second head originates on the femur. The biceps femoris is a muscle that inserts laterally on the proximal leg. The **semitendinosus** also originates on the ischial tuberosity, and it can be distinguished by the long, pencil-like distal tendon. As opposed to the biceps femoris, the semitendinosus inserts medially. The **semimembranosus** also originates on the ischial tuberosity and has a broad, flat, membranous tendon on the proximal part of the muscle. The semimembranosus inserts medially on the leg along with the semitendinosus (remember the two "semis" go together and the membranosus is medial). Examine these muscles in figure 15.5.

Cat Musculature

The cat has many muscles of the hip and thigh that are good models for studying human muscles. Because cats are quadrupeds, there is a different size or placement of some of the thigh muscles. In the dissection be careful that you *leave the blood vessels and nerves intact.* You will be looking at these structures in later exercises.

If you have not already done so, remove the skin from the lower limb of the cat. Pay particular attention to the **great saphenous vein,** which runs under the skin and should be preserved. Be careful cutting the skin from the distal portions of the leg so that you do not cut through the tendons of the foot. Remove any excess fat and fascia from the muscles as you dissect the material. If you need to cut a muscle, bisect it perpendicular to the fiber direction so that half of it is attached to the origin side and half of it is attached to the insertion side.

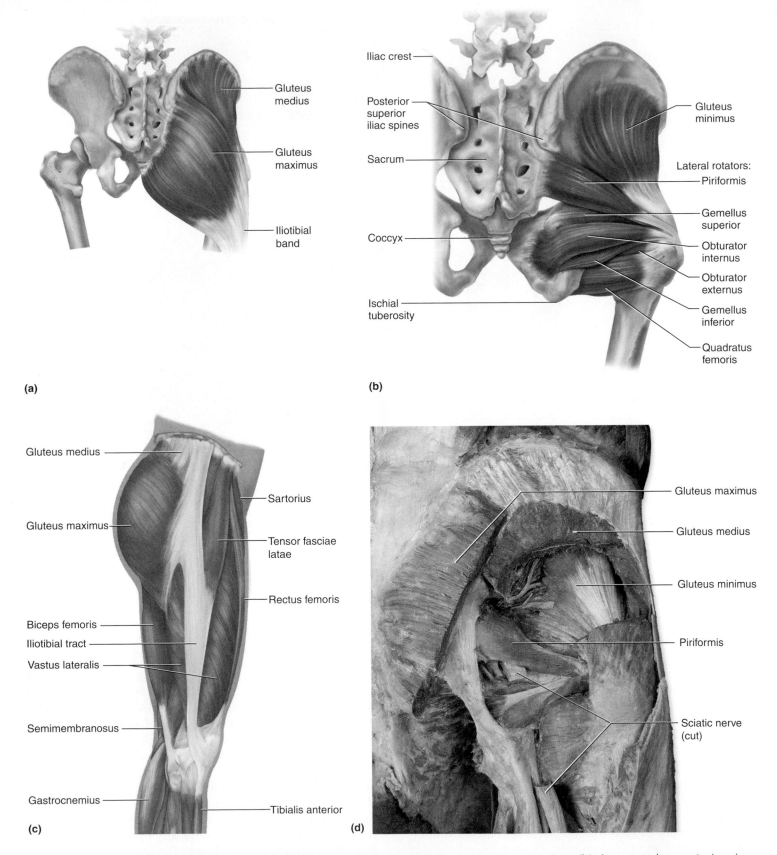

(a)

Gluteus medius

Gluteus maximus

Iliotibial band

(b)

Iliac crest

Posterior superior iliac spines

Sacrum

Coccyx

Ischial tuberosity

Gluteus minimus

Lateral rotators:
Piriformis

Gemellus superior

Obturator internus

Obturator externus

Gemellus inferior

Quadratus femoris

(c)

Gluteus medius

Gluteus maximus

Biceps femoris

Iliotibial tract

Vastus lateralis

Semimembranosus

Gastrocnemius

Sartorius

Tensor fasciae latae

Rectus femoris

Tibialis anterior

(d)

Gluteus maximus

Gluteus medius

Gluteus minimus

Piriformis

Sciatic nerve (cut)

FIGURE 15.4 Muscles of the Right Hip and Thigh Diagram (a) superficial muscles, posterior view; (b) deep muscles, posterior view; (c) lateral muscles. Photograph (d) lateral and posterior muscles.

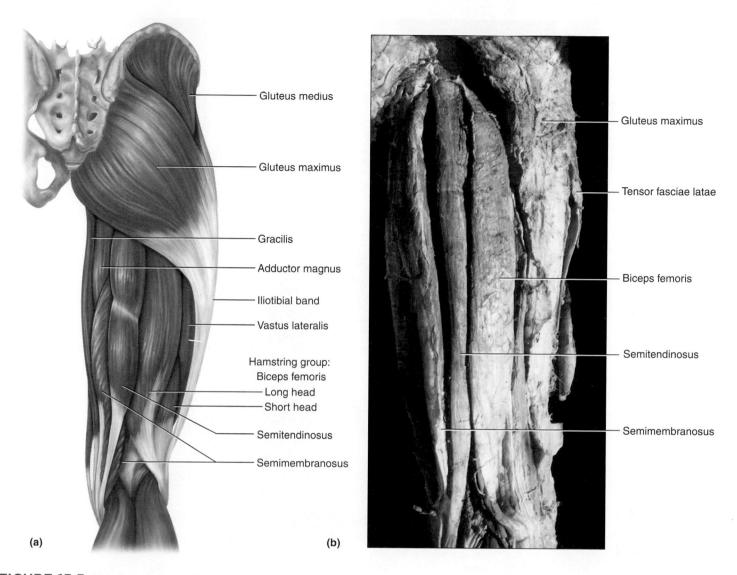

Gluteus medius

Gluteus maximus

Gracilis

Adductor magnus

Iliotibial band

Vastus lateralis

Hamstring group:
Biceps femoris
Long head
Short head
Semitendinosus

Semimembranosus

Gluteus maximus

Tensor fasciae latae

Biceps femoris

Semitendinosus

Semimembranosus

(a) (b)

FIGURE 15.5 Muscles of the Right Thigh, Posterior View (a) Diagram; (b) photograph.

Medial Muscles of the Thigh of the Cat

Two major muscles on the medial aspect of the thigh in the cat are the **sartorius** and the **gracilis** muscles. Locate these muscles in figure 15.6 and note that they are much broader in the cat than in the human.

Cut through the sartorius and gracilis and reflect the ends of these muscles. You should be able to see the deeper muscles of the thigh, including the **vastus medialis,** the **adductor femoris** (a large muscle specific to the cat), and the **semimembranosus.** Locate the external portion of the **iliopsoas** muscle, on the medial aspect of the thigh. The distal portion of the **semitendinosus** can also be seen from this view. Examine figure 15.6 for the deep muscles of the medial thigh and locate the **rectus femoris, vastus medialis,** and **vastus lateralis.** Move the rectus femoris to see the **vastus intermedius** muscle. The **adductor longus** is a thin

muscle anterior to the **adductor femoris,** and these can be seen on the medial aspect of the thigh.

Lateral Muscles of the Thigh of the Cat

On the lateral aspect of the thigh are numerous muscles, including the **biceps femoris,** the **tensor fasciae latae;** the **gluteus muscles,** and a muscle specific to the cat, the **caudofemoralis.** The biceps femoris is the largest and most lateral muscle of the thigh. Bisect the biceps femoris and reflect the muscle. The **gluteus medius** is larger than the **gluteus maximus** in cats due to the lengthening of the pelvic girdle. The **semimembranosus** is a large muscle in cats (much larger than the semitendinosus), and it can be seen from the lateral side of the cat. The adductor magnus and adductor brevis are not found in the cat. Examine these muscles in figure 15.7.

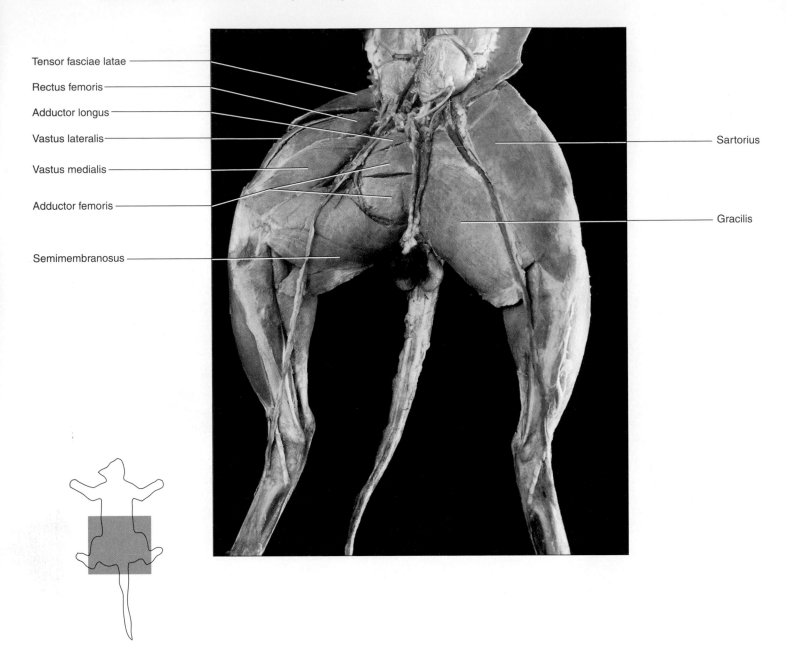

Tensor fasciae latae

Rectus femoris

Adductor longus

Vastus lateralis

Vastus medialis

Adductor femoris

Semimembranosus

Sartorius

Gracilis

FIGURE 15.6 **Thigh Muscles of the Cat** Superficial muscles (left side of cat); deep muscles (right side of cat).

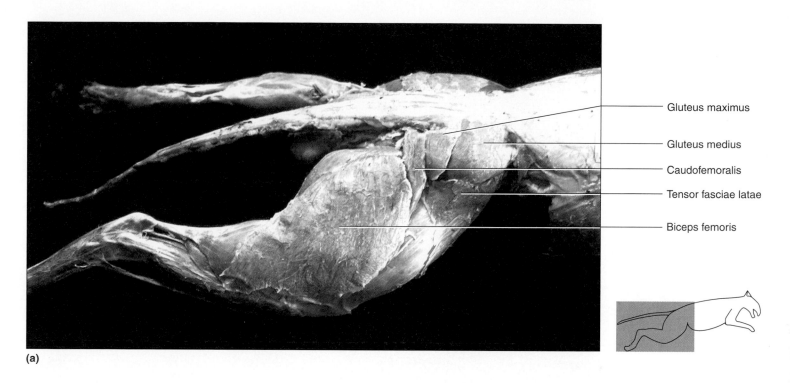

(a)

- Gluteus maximus
- Gluteus medius
- Caudofemoralis
- Tensor fasciae latae
- Biceps femoris

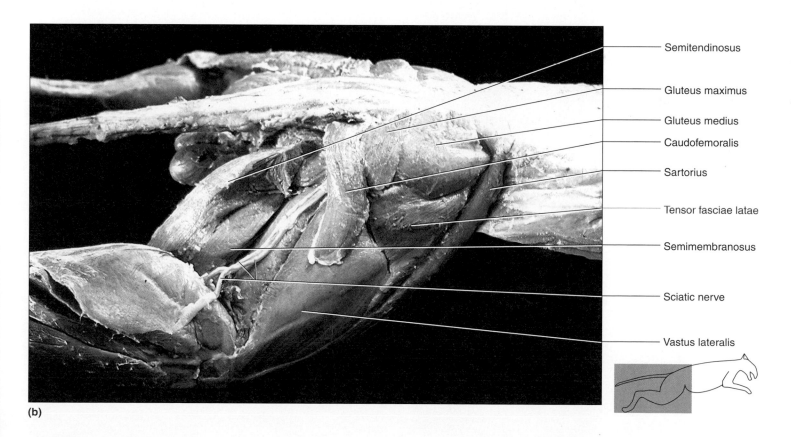

(b)

- Semitendinosus
- Gluteus maximus
- Gluteus medius
- Caudofemoralis
- Sartorius
- Tensor fasciae latae
- Semimembranosus
- Sciatic nerve
- Vastus lateralis

FIGURE 15.7 **Lateral Thigh Muscles of the Cat** (a) Superficial muscles of the right side; (b) deep muscles of the right side.

Notes

REVIEW SECTION

Muscles of the Hip and Thigh

Name _____ Date _____

Lab Section _____ Time _____

Review Questions

1. If you were to ride a horse, what muscles would you use to keep your seat out of the saddle as you ride?

2. How do the gluteus medius and gluteus minimus prevent you from toppling over as you walk?

3. What is a muscle that is an antagonist to the biceps femoris muscle?

4. What are two muscles that are synergists with the biceps femoris muscle?

5. What is the insertion of all the muscles of the quadriceps group?

6. How does the action of the rectus femoris differ from those of the other quadriceps muscles?

7. How many adductor muscles are there?

8. List two muscles in this exercise responsible for thigh flexion.

9. Where do the hamstring muscles originate as a group?

10. What is the action of the vastus lateralis?

11. Which muscle group is found on the anterior part of the thigh?

12. Is abduction of the thigh movement away from or toward the midline?

13. What muscle flexes the lumbar vertebrae as part of its action?

14. Label the muscles in the following illustration. Use the terms provided.

vastus lateralis   sartorius

gracilis   vastus medialis

rectus femoris   tensor fasciae latae

adductor femoris   semimembranosus

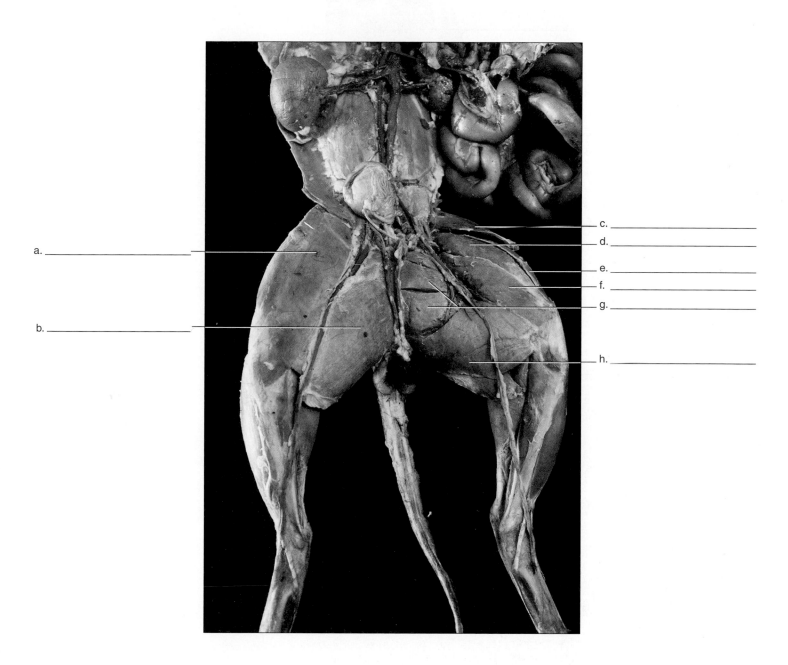

a. _____

b. _____

c. _____

d. _____

e. _____

f. _____

g. _____

h. _____

15. Label the muscles in the following illustration.

a. _____

b. _____

c. _____

d. _____

e. _____

f. _____

g. _____

h. _____

i. _____

j. _____

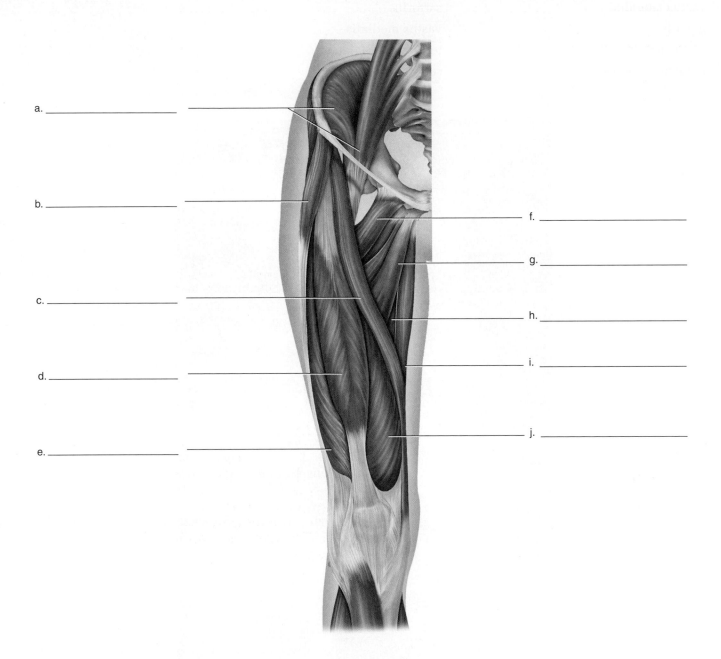

LABORATORY

Muscles of the Leg and Foot

INTRODUCTION

The muscles of the leg and foot are primarily muscles of locomotion. They originate typically from the femur, tibia, or fibula and insert on the bones of the foot. Actions on the leg are either flexion or extension, whereas actions on the foot are **dorsiflexion** (decreasing the angle between the shin and dorsum or top of the foot) or **plantar flexion** (the action done as you stand on your toes to reach something high). Review the other actions of the leg and foot in Exercise 12. The muscles in this exercise are covered in the Saladin text in chapter 10, "The Muscular System." Some of the muscles of the feet are analogous to those of the hand, such as the extensor digitorum muscles.

OBJECTIVES

At the end of this exercise you should be able to

1. locate the muscles of the leg and foot on a torso model, chart, cadaver, or cat;
2. list the origin, insertion, and action of each muscle presented;
3. describe what nerve controls each muscle;
4. list what muscles function as synergists or antagonists to the prime mover;
5. name all the muscles that have an action on a joint, such as all the muscles that flex the leg.

MATERIALS

Human torso model

Human leg models

Human muscle charts

Articulated skeleton

Cadaver

Cat

Materials for Cat Dissection

Dissection trays

Scalpel or razor blades

Gloves

Blunt probe

Pins

PROCEDURE

Examine the models and charts in the lab as you study the following descriptions. If you have a cadaver, pay particular attention to the origins and the tendons of insertion of the muscles of the leg and foot. Table 16.1 provides the details of the muscles in this exercise, while the text descriptions point out major features of the muscles. When you are studying the origins and insertions of the muscles, it helps to have an articulated skeleton so that you can locate the bony markings as you study the muscle attachments. The actions of the muscles will be easier when you have identified the origins and the insertions.

Posterior Muscles

The **gastrocnemius** is a thin calf muscle that is the most superficial of the posterior leg group. The gastrocnemius crosses the knee joint and flexes the leg. It inserts on the calcaneus by way of the calcaneal tendon and plantar flexes the foot. The **calcaneal tendon** is also known as the Achilles tendon. Deep to the gastrocnemius is the **soleus** muscle. Unlike the gastrocnemius, the soleus does not originate on the femur; therefore, it has no action on the leg. It inserts on the calcaneus, sharing the calcaneal tendon with the gastrocnemius, and it plantar flexes the foot. The **plantaris** is a weak plantar flexor of the foot. The **popliteus** is a small muscle that crosses the knee joint. It unlocks the knee joint. These muscles can be seen in figure 16.1.

The **tibialis posterior** is a large muscle deep to the soleus that plantar flexes and inverts the foot. The tendon of the muscle runs along the medial aspect of the ankle. Near the tibialis posterior is the **flexor digitorum longus,** which is a muscle that inserts on the distal phalanges of all of the digits of the foot except for the hallux. The flexor digitorum longus flexes (curls) the toes in addition to plantar flexing and inverting the foot. The **flexor hallucis longus** flexes the hallux (big toe) and aids the flexor digitorum longus in inverting the foot. Locate these muscles in the lab and compare them to figure 16.1.

Anterior Muscles

The **tibialis anterior** is located just lateral to the crest of the tibia on the anterior side of the leg. The tendon of the tibialis anterior crosses to the medial side of the foot and inserts on the first metatarsal and first cuneiform. As the tibialis

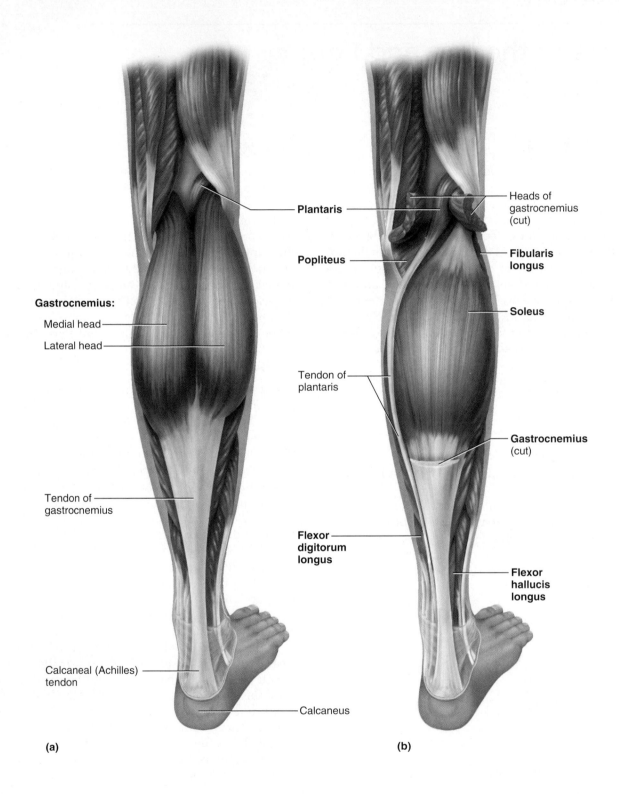

Plantaris

Popliteus

Gastrocnemius:

Medial head

Lateral head

Tendon of
gastrocnemius

Calcaneal (Achilles)
tendon

Heads of
gastrocnemius
(cut)

Fibularis
longus

Soleus

Tendon of
plantaris

Gastrocnemius
(cut)

Flexor
digitorum
longus

Flexor
hallucis
longus

Calcaneus

(a)

(b)

FIGURE 16.1 **Muscles of the Right Leg and Foot, Posterior View** (a) Superficial muscles of the leg; (b) middle-level muscles of the leg; (c) deep muscles of the leg; (d) photograph.

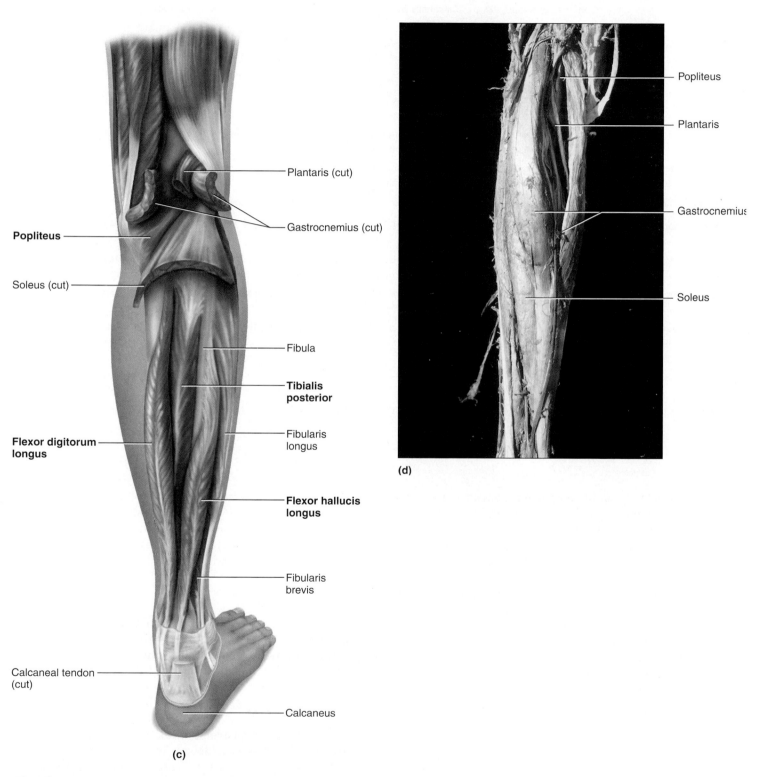

Popliteus

Plantaris (cut)

Gastrocnemius (cut)

Soleus (cut)

Fibula

Tibialis posterior

Flexor digitorum longus

Fibularis longus

Flexor hallucis longus

Fibularis brevis

Calcaneal tendon (cut)

Calcaneus

Popliteus

Plantaris

Gastrocnemius

Soleus

(c)

(d)

FIGURE 16.1 *Continued*

TABLE 16.1	Muscles of the Leg and Foot			
Name	**Origin**	**Insertion**	**Action**	**Innervation**
Gastrocnemius	Condyles of femur	Calcaneus by calcaneal tendon	Flexes leg, plantar flexes foot	Tibial nerve
Soleus	Posterior, proximal tibia and fibula	Calcaneus by calcaneal tendon	Plantar flexes foot	Tibial nerve
Plantaris	Distal femur	Calcaneus	Flexes leg, plantar flexes foot	Tibial nerve
Popliteus	Lateral condyle of femur	Proximal tibia	Flexes leg	Tibial nerve
Tibialis posterior	Proximal portion of tibia and fibula	Second through fourth metatarsals and some tarsals	Plantar flexes and inverts foot	Tibial nerve
Flexor digitorum longus	Posterior tibia	Distal phalanges of second through fifth digits	Flexes toes, plantar flexes and inverts foot	Tibial nerve
Flexor hallucis longus	Midshaft of fibula	Distal phalanx of hallux	Flexes hallux, inverts foot	Tibial nerve
Tibialis anterior	Lateral condyle and proximal tibia	First metatarsal and first cuneiform	Dorsiflexes and inverts foot	Deep fibular nerve
Extensor digitorum longus	Lateral condyle of tibia, shaft of fibula	Middle and distal phalanges of second through fifth digits	Extends toes, dorsiflexes foot	Deep fibular nerve
Extensor hallucis longus	Anterior shaft of fibula	Distal phalanx of hallux	Extends hallux, dorsiflexes foot	Deep fibular nerve
Fibularis longus	Head and shaft of fibula, lateral condyle of tibia	First metatarsal, first cuneiform	Plantar flexes and everts foot	Superficial fibular nerve
Fibularis brevis	Shaft of fibula	Fifth metatarsal	Plantar flexes and everts foot	Superficial fibular nerve
Fibularis tertius	Anterior, distal fibula	Dorsum of fifth metatarsal	Dorsiflexes and everts foot	Deep fibular nerve

anterior contracts, it dorsiflexes the foot. Because the insertion of this muscle is medial, it also inverts the foot.

The **extensor digitorum longus** muscle is lateral to the tibialis anterior, and the tendons of this muscle splay out and insert on the middle and distal phalanges of all the digits of the foot except the hallux. The **extensor hallucis longus** extends the hallux and is a synergist to the extensor digitorum longus in dorsiflexion of the foot. Examine these muscles in the lab and compare them to figure 16.2. These are major muscles of the foot. There are also smaller muscles such as the extensor digitorum brevis (figure 16.2) and the extensor hallucis brevis.

The fibularis (peroneus) muscles are named for originating on and running the length of the fibula. The **fibularis longus** has a tendon that travels from the lateral side of the foot underneath to the medial side, crossing under the arch of the foot. The **fibularis brevis** parallels the fibularis longus except that the tendon inserts on the lateral side of the foot at the fifth metatarsal. Both of these muscles have tendons that hook posterior to the lateral

malleolus, and they both plantar flex and evert the foot. The **fibularis tertius** does not arch behind the lateral malleolus, and it inserts on the dorsum of the fifth metatarsal. When it contracts it dorsiflexes and everts the foot. Examine these muscles in figure 16.3.

Cat Dissection

If you have not done so already, remove the skin from the lower portion of the hind limb of the cat. Be careful cutting the skin from the distal portions of the leg so that you do not cut through the tendons of the foot.

Posterior Muscles

Examine the large **gastrocnemius** on the posterior aspect of the leg. The **soleus** is deeper to the gastrocnemius and inserts communally with the gastrocnemius on the calcaneus. In cats the soleus has only one point of origin, the fibula, while in humans there are two bones of origin.

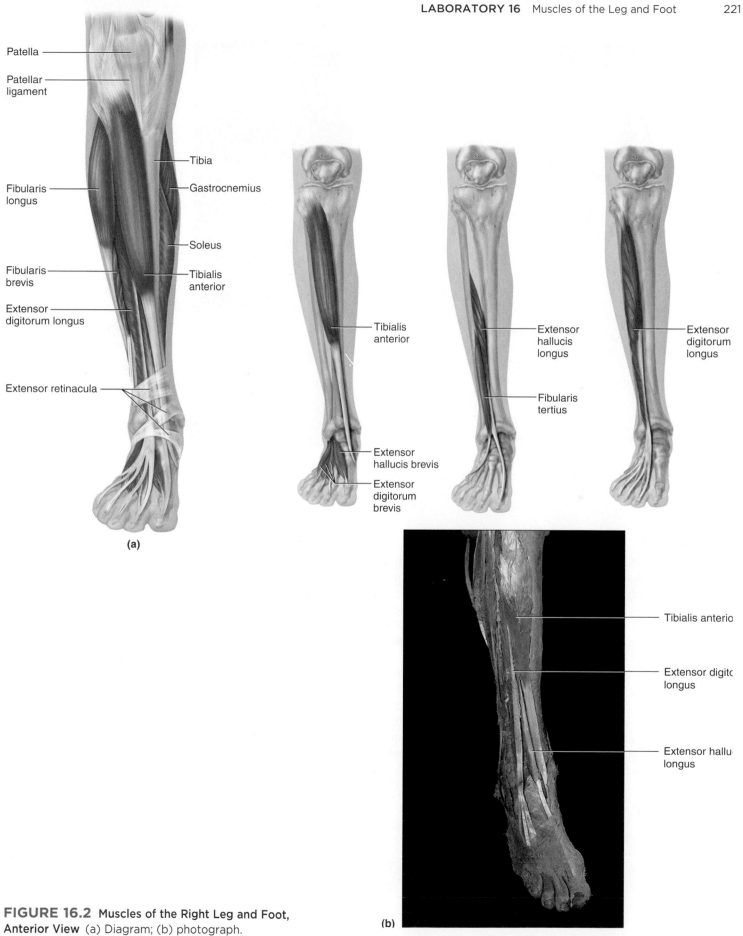

FIGURE 16.2 Muscles of the Right Leg and Foot, Anterior View (a) Diagram; (b) photograph.

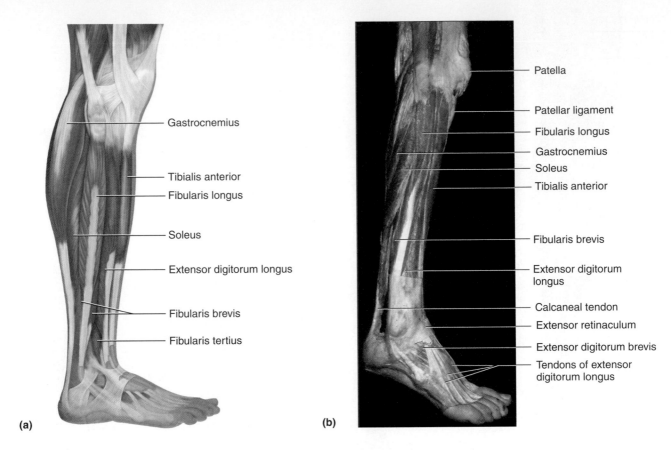

FIGURE 16.3 Muscles of the Right Leg and Foot, Lateral View (a) Diagram; (b) photograph.

Examine figure 16.4 with your dissection. Cut through the calcaneal tendon and lift the gastrocnemius and soleus so that you can study the underlying muscles.

The **popliteus** is a small, triangular muscle that crosses the knee joint. Do not damage the nerves and blood vessels that pass over the popliteus because you will study them in the subsequent exercises.

The **flexor digitorum longus** is a muscle that runs along the medial side of the hind limb and flexes the digits of the cat. It joins with the **flexor hallucis longus,** which inserts on all the digits of the hind limb. The **tibialis posterior** is a narrow muscle that inserts on the tarsal bones of the foot. Locate these muscles in the cat and compare them to figure 16.5.

Anterior Muscles

The **tibialis anterior** is a large muscle of the lower limb and inserts on the dorsum of the foot. The **extensor digitorum longus** originates on the femur in cats and inserts on the distal phalanges in all of the digits in the cat. The extensor hallucis longus is not found in cats. These muscles can be seen in figure 16.6. Examine these muscles and isolate them in your dissection.

The **fibularis longus** extends along the length of the fibula with the **fibularis brevis.** The fibularis longus inserts at the base of the metatarsals, while the fibularis brevis inserts on the fifth metatarsal. The **fibularis tertius** inserts into the tendon of the extensor digitorum muscle. These muscles can be seen in figure 16.7.

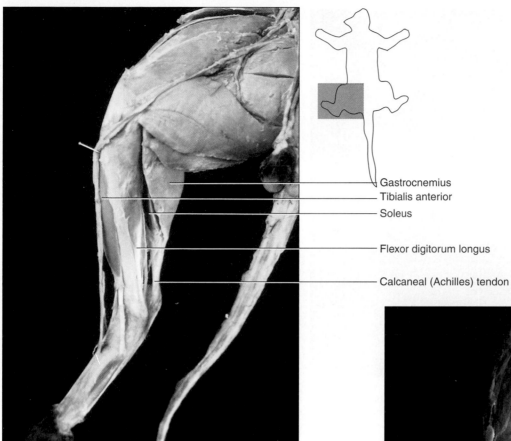

FIGURE 16.4 Superficial Muscles of the Right Leg of the Cat, Medial View

Gastrocnemius
Tibialis anterior
Soleus

Flexor digitorum longus

Calcaneal (Achilles) tendon

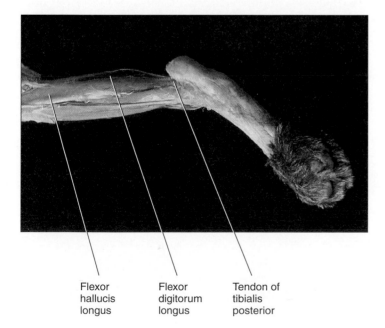

Flexor hallucis longus

Flexor digitorum longus

Tendon of tibialis posterior

FIGURE 16.5 Deep Muscles of the Left Leg of the Cat, Lateral View

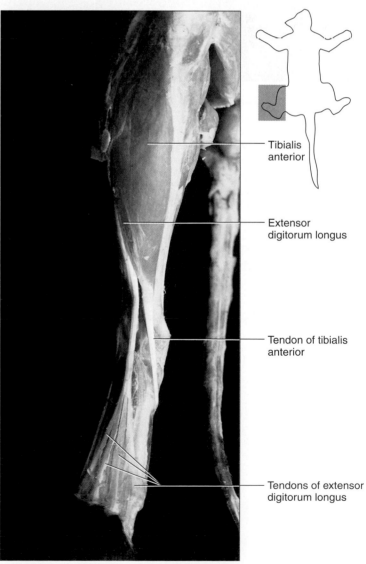

Tibialis anterior

Extensor digitorum longus

Tendon of tibialis anterior

Tendons of extensor digitorum longus

FIGURE 16.6 Muscles of the Right Leg of the Cat, Anterior View

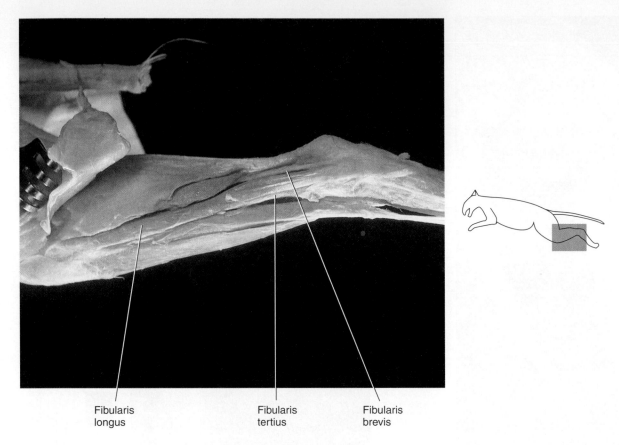

Fibularis
longus

Fibularis
tertius

Fibularis
brevis

FIGURE 16.7 Muscles of the Left Leg of the Cat, Lateral View

REVIEW SECTION

Muscles of the Leg and Foot

Name _____ *Date* _____

Lab Section _____ *Time* _____

Review Questions

1. What is the origin of the gastrocnemius?

2. What is the insertion of the tibialis anterior in humans?

3. How does the action of the fibularis longus in humans differ from that of the fibularis tertius?

4. What is the action of the extensor hallucis longus?

5. The calf is made of what two major muscles?

6. Label the illustration using the terms provided.

    extensor digitorum longus

    soleus

    tibialis anterior

    fibularis longus

    extensor hallucis longus

7. Plantar flexion occurs by what muscles?

8. What muscle extends the toes?

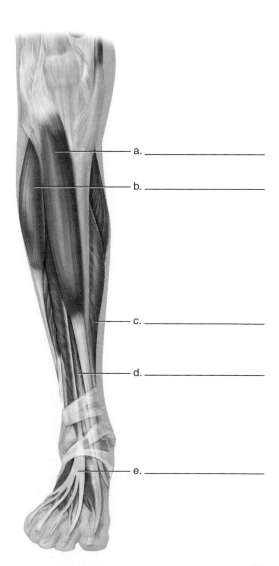

a. _____

b. _____

c. _____

d. _____

e. _____

9. Name a muscle in this exercise that dorsiflexes the foot.

10. Plantar flexion and eversion of the foot occurs by what muscle?

11. What is the insertion of the soleus?

12. What muscles do you use when you rise up on your toes?

13. What is the insertion of the fibularis tertius muscle?

14. What is the insertion of the flexor digitorum longus?

15. Label the muscles in the following illustration.

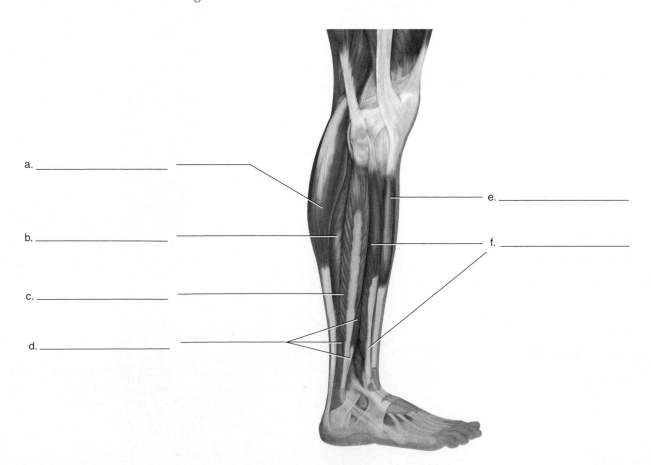

a. _____

b. _____

c. _____

d. _____

e. _____

f. _____

LABORATORY

Muscles of the Head and Neck

INTRODUCTION

The major muscles of the head and neck are numerous and can be grouped into a few functional classifications. Some are chewing muscles or are used to manipulate the food in the mouth. Other muscles function in facial expression or in the closing of the eyes or mouth. Still others move the head or neck or move the hyoid or the larynx in speech or swallowing. These muscles are discussed in the Saladin text in chapter 10, "The Muscular System."

OBJECTIVES

At the end of this exercise you should be able to

1. locate the muscles of the head and neck on a torso model, chart, cadaver, or cat;
2. list the origin, insertion, and action of each muscle presented;
3. describe what nerve controls each muscle;
4. name all the muscles that have an action on a joint, such as all the muscles that flex the head;
5. reproduce the actions of select muscles on your head or neck.

MATERIALS

Human torso model or head and neck model

Human muscle charts

Articulated skeleton

Cadaver

Cat

Materials for Cat Dissection

Dissection trays

Scalpel or razor blades

Gloves

Blunt probe

Pins

PROCEDURE

Review the muscle nomenclature and the actions as outlined in Laboratory Exercises 12 and 13. Examine a model or chart of the human musculature and cadaver (if available) as you read the following descriptions. The details of the muscles are listed in table 17.1. Examine a skull or an articulated skeleton and review the bony markings as you study the origins and insertions of the muscles in this exercise. Once you know the origins or insertions, the actions should be more comprehensible.

Muscles of the Neck

The **levator scapulae** (le-VAY-tur SCAP-u-lay) is named for what it does, elevates the scapula. The levator scapulae originates on the lateral side of the neck and inserts on the upper scapula. If the neck is fixed, the levator scapulae raises the scapula, as in shrugging. If the scapula is fixed, it rotates or abducts the neck.

The **scalenes** (skah-LEENS) muscles are found on the lateral side of the neck. They are bounded by the sternocleidomastoid in the front and the levator scapulae in the back. They rotate the neck or elevate the ribs. Examine figure 17.1 for these muscles.

The **sternocleidomastoid** (STER-no-KLY-do MAS-toyd) muscle rotates the head in a unique way. Place your hand on the right sternocleidomastoid and turn your head to the right. Notice how the muscle does not contract. Now turn your head to the left, and you can feel the muscle contract. The right sternocleidomastoid turns the head to the left and the left sternocleidomastoid turns the head to the right.

The **sternohyoid, sternothyroid,** and **omohyoid** are named for their origins and insertions. In the case of the sternohyoid and sternothyroid, the sternum anchors the stable part of the muscle (the origin), and the hyoid bone and thyroid cartilage of the larynx move when the respective muscles contract. In the case of the omohyoid, the term *omo* means shoulder, and the scapula anchors the stable part of the muscle. The hyoid is depressed when the omohyoid contracts. Examine figure 17.2 for muscles of the anterior neck.

The **platysma** (plah-TIS-mah) is a broad, thin muscle that has a soft origin (on the fascia of the pectoral and deltoid muscles). It inserts on the mandible and skin of

TABLE 17.1	Muscles of the Neck and Head			
Name	**Origin**	**Insertion**	**Action**	**Innervation**
Neck Muscles				
Levator scapulae	C1–4	Upper vertebral border of scapula	Elevates scapula, abducts and rotates neck	Spinal nerves C3–5 and dorsal scapular nerve
Scalenes (anterior, middle, and posterior)	Transverse process of cervical vertebrae	Ribs 1 and 2	Flexes and rotates neck, elevates ribs 1 and 2	Spinal nerves C4–8
Sternocleidomastoid	Sternum, clavicle	Mastoid process of temporal	Abducts, rotates, and flexes head	Spinal nerves C2–4, accessory nerve (XI)
Sternohyoid	Manubrium of sternum	Hyoid	Depresses hyoid	Spinal nerves C1–3
Sternothyroid	Manubrium of sternum	Thyroid cartilage of larynx	Depresses thyroid cartilage	Spinal nerves C1–3
Omohyoid	Superior surface of scapula	Hyoid	Depresses hyoid	Spinal nerves C1–3
Platysma	Fascia covering pectoralis major and deltoid	Mandible and skin of lower region of face	Depresses lower lip, opens jaw	Facial nerve (VII)
Digastric	Inferior, distal margin of mandible (anterior belly), mastoid notch of temporal (posterior belly)	Hyoid	Elevates, protracts, and retracts hyoid; opens mandible;	Trigeminal (V) and facial (VII) nerves
Mylohyoid	Inner, inferior margin of mandible	Body of hyoid and median raphe	Elevates hyoid and tongue	Trigeminal nerve (V)
Head Muscles				
Frontalis	Galea aponeurotica	Skin superior to orbit	Raises eyebrows, draws scalp anteriorly	Facial nerve (VII)
Occipitalis	Occipital and temporal bone	Galea aponeurotica	Draws scalp posteriorly	Facial nerve (VII)
Temporalis	Temporal fossa	Coronoid process and ramus of mandible	Closes mandible	Trigeminal nerve (V)
Masseter	Zygomatic arch	Angle and ramus of mandible	Closes mandible	Trigeminal nerve (V)
Pterygoids (medial and lateral)	Pterygoid processes of sphenoid bone	Medial ramus of mandible	Medial gliding of mandible (for chewing)	Trigeminal nerve (V)
Orbicularis oculi	Frontal and maxilla on medial margin of orbit	Skin of eyelid	Closes eyelid	Facial nerve (VII)
Orbicularis oris	Fascia of facial muscles near mouth	Skin of lips	Closes lips	Facial nerve (VII)
Corrugator supercilii	Medial portion of frontal bone	Skin of eyebrows	Adducts eyebrows (pulls them medially)	Facial nerve (VII)
Risorius	Fascia of masseter	Skin at angle of mouth	Abducts corner of mouth (draws edge of mouth lateral)	Facial nerve (VII)
Mentalis	Anterior mandible	Skin of chin below lower lip	Protrudes lower lip	Facial nerve (VII)
Buccinator	Maxilla and mandible near molar teeth	Orbicularis oris	Compresses cheek	Facial nerve (VII)
Zygomaticus (major and minor)	Zygomatic bone	Muscle and skin at angle of mouth	Elevates corners of mouth (in smiling and laughing)	Facial nerve (VII)
Depressor labii inferioris	Lateral mandible	Muscles and skin of lower lip	Depresses lower lip	Facial nerve (VII)
Levator labii superioris	Maxilla and zygomatic bones	Muscle and skin of lips	Elevates upper lip, flares nostril	Facial nerve (VII)

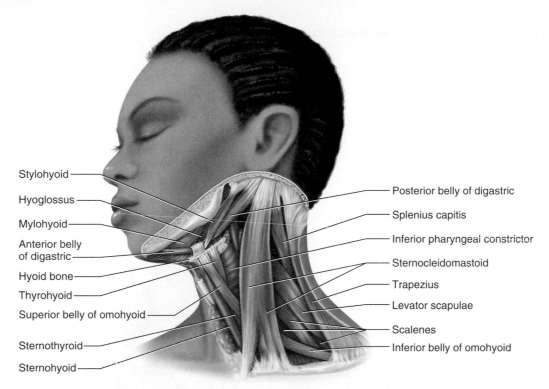

Stylohyoid
Hyoglossus
Mylohyoid
Anterior belly of digastric
Hyoid bone
Thyrohyoid
Superior belly of omohyoid
Sternothyroid
Sternohyoid

Posterior belly of digastric
Splenius capitis
Inferior pharyngeal constrictor
Sternocleidomastoid
Trapezius
Levator scapulae
Scalenes
Inferior belly of omohyoid

FIGURE 17.1 Muscles of the Neck, Lateral View

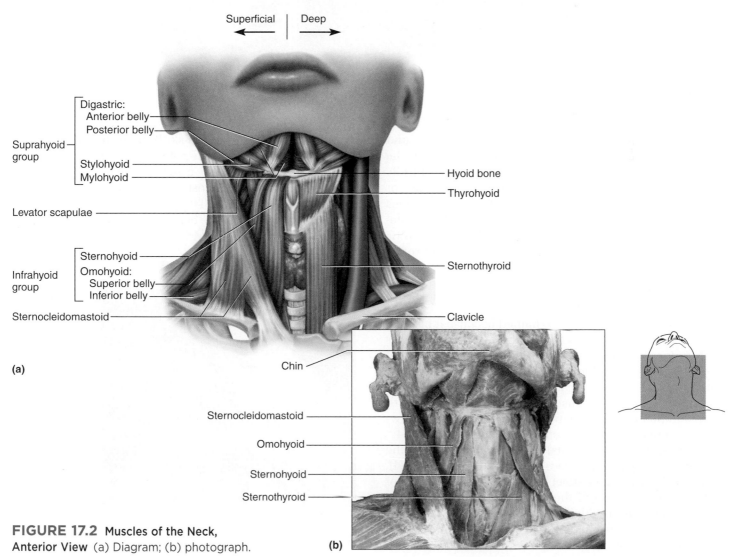

Superficial | Deep

Digastric:
  Anterior belly
  Posterior belly
Suprahyoid group
  Stylohyoid
  Mylohyoid

Levator scapulae

Infrahyoid group
  Sternohyoid
  Omohyoid:
    Superior belly
    Inferior belly
Sternocleidomastoid

Chin

Hyoid bone
Thyrohyoid

Sternothyroid

Clavicle

(a)

Sternocleidomastoid
Omohyoid
Sternohyoid
Sternothyroid

(b)

FIGURE 17.2 Muscles of the Neck,
Anterior View (a) Diagram; (b) photograph.

229

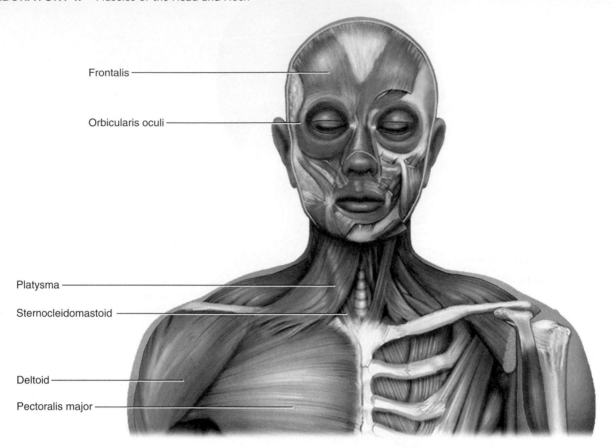

FIGURE 17.3 Platysma, Anterior View

the lips and can be seen if you elevate your chin and subsequently pout. The thin wings that stick out on the side of your neck are the edges of the platysma muscle. Examine the platysma in figure 17.3 and in the models or on the cadaver in the lab. There are numerous muscles of the neck involved in moving the head. The splenius capitis (see figure 17.1) and the semispinalis capitis (see figure 18.3) are both deep to the trapezius and rotate and extend the head.

Muscles of the Head

The **digastric** (di-GAS-trik) is so named because it has two bellies. Few muscles open the mandible. The digastric is one that does. In addition, the digastric has significant action on the hyoid, which is important in tongue movement for speech and swallowing. Deep to the digastric is the **mylohyoid,** a broad muscle of the floor of the mouth that aids in pushing the tongue superiorly when swallowing. These muscles can be seen in figures 17.2 and 17.4.

The **frontalis** (fron-TAL-is) muscle originates on the **galea aponeurotica** (GALE-ee-uh AP-oh-nu-ROT-ih-KAH), a broad, flat, tendinous sheet on the superior aspect of the skull. The insertion of the frontalis is on the eyebrow region. The **occipitalis** (AUK-sip-ih-TAL-us) is a functional continuation of the frontalis muscle in that it attaches to the posterior part of the galea aponeurotica. If the galea is

fixed to the posterior of the skull, then the frontalis raises the eyebrows. If the galea is not so anchored, then the scalp is brought forward, as in frowning. The occipitalis pulls the galea aponeurotica posteriorly, and the scalp attached to the galea goes with it. These muscles are sometimes discussed as one muscle, the **occipitofrontalis.** Examine these muscles in figure 17.5.

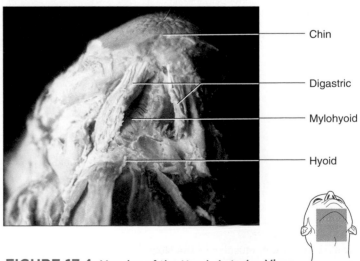

FIGURE 17.4 Muscles of the Head, Anterior View

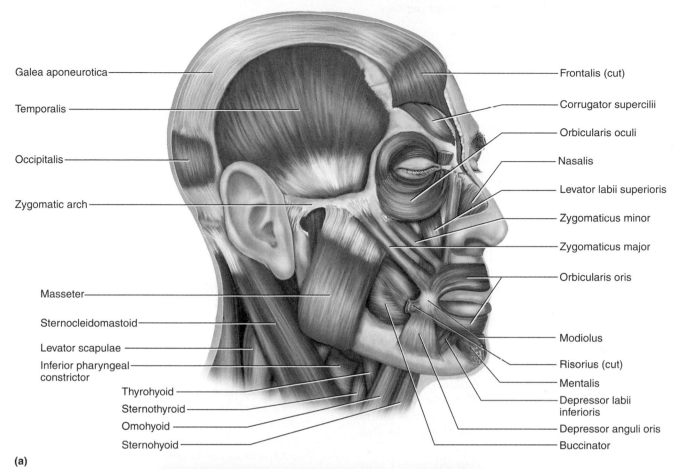

Galea aponeurotica

Temporalis

Occipitalis

Zygomatic arch

Masseter

Sternocleidomastoid

Levator scapulae

Inferior pharyngeal constrictor

Thyrohyoid

Sternothyroid

Omohyoid

Sternohyoid

Frontalis (cut)

Corrugator supercilii

Orbicularis oculi

Nasalis

Levator labii superioris

Zygomaticus minor

Zygomaticus major

Orbicularis oris

Modiolus

Risorius (cut)

Mentalis

Depressor labii inferioris

Depressor anguli oris

Buccinator

(a)

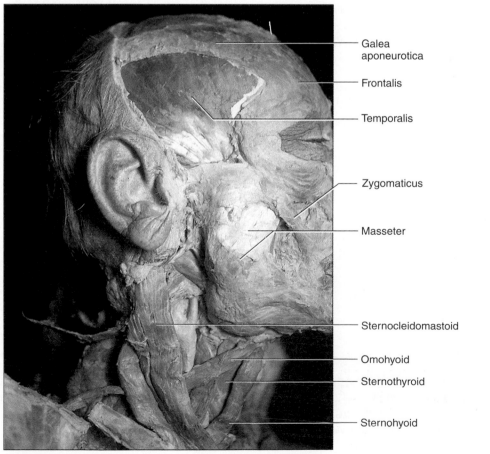

Galea aponeurotica

Frontalis

Temporalis

Zygomaticus

Masseter

Sternocleidomastoid

Omohyoid

Sternothyroid

Sternohyoid

FIGURE 17.5 Muscles of the Head, Lateral View (a) Diagram; (b) photograph.

(b)

The **masseter** (MASS-ih-tur) is a large muscle of the head. It acts to powerfully close the jaws. If you place your fingers on the ramus of the mandible and clench your teeth, you can feel the masseter tighten.

The **temporalis** is a powerful muscle that closes the jaw and is a synergist to the masseter. It is a chewing muscle, or a muscle of *mastication.* The temporal fossa is so named because it is a depression medial to the zygomatic arch. The temporalis muscle is deep to the zygomatic arch and inserts on the coronoid process and on the superior and medial surface of the ramus of the mandible.

The **pterygoid** (TARE-ih-goyd) muscles are deep muscles that originate on the sphenoid bone and insert laterally on the mandible. They pull the jaw horizontally, which helps in rotatory chewing, characteristic of a person chewing gum. These muscles can be seen in figures 17.5 and 17.6.

The **orbicularis oculi** and the **orbicularis oris** are sphincter muscles that close the eyes and the mouth, respectively. Sphincter muscles act as the strings of a drawstring purse. The orbicularis oculi has a medial, bony origin and an insertion on the eyelid. The muscles ring the eye and close the eyelids. The orbicularis oris originates on fascia and facial muscles near the mouth and inserts on the skin of the lips, thus closing the lips. Examine the facial muscles in figure 17.7.

The **corrugator supercilii** (SOUP-ur-SIL-ee-ee) muscle has a medial point of origin between the eyebrows and inserts laterally. It furrows the eyebrows.

The **zygomaticus** (ZY-goh-MAT-ih-cus) **major** and the **zygomaticus minor** elevate the corners of the mouth by pulling them superiorly and laterally, as in smiling or laughing. They are named for their origin on the zygomatic bone. The **risorius** (rise-OH-ree-us) is known as the laughing muscle because it pulls the lips laterally. It does not have a bony point of origin but attaches to the fascia of the masseter (see figure 17.7).

The **mentalis** originates on the chin (anterior mandible) and is another of the pouting muscles. The **buccinator** (BUK-sin-ay-tur) muscle of the cheek runs in a horizontal direction. It puckers the cheeks, as in trumpet playing, and pushes food toward the molars in chewing.

The **depressor labii** (LABE-ee-ee) **inferioris** pulls the lower lip inferiorly, as when pouting. It is named for its action (depressing the lower lip), while the **levator labii superioris** muscle raises the skin of the upper lip and expands the nostrils, as in the expression of showing extreme disgust. These muscles are seen in figure 17.7.

Activities

1. When you turn your head to the left, which sternocleidomastoid muscle contracts?
2. Tilt your head so that your chin is elevated and pout your lower lip. What muscle forms a thin membrane along the anterolateral neck?
3. Purse your lips and feel the buccinator muscle as it contracts in the cheek.
4. Clench your teeth and palpate the temporalis muscle and the masseter.

Examine figure 17.8 for surface views of facial muscles. Fill in which muscles are represented in the photographs.

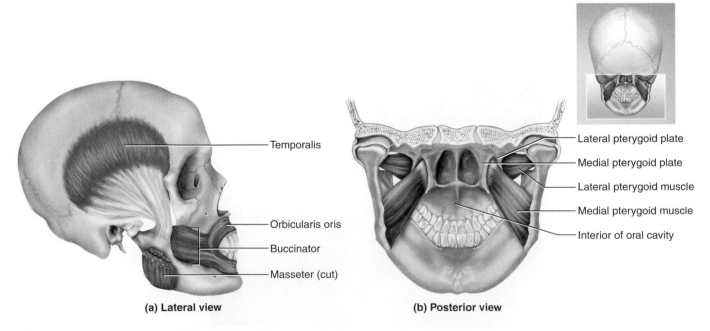

(a) Lateral view

— Temporalis

— Orbicularis oris

— Buccinator

— Masseter (cut)

(b) Posterior view

— Lateral pterygoid plate

— Medial pterygoid plate

— Lateral pterygoid muscle

— Medial pterygoid muscle

— Interior of oral cavity

FIGURE 17.6 Muscles of the Head, Lateral and Posterior Views

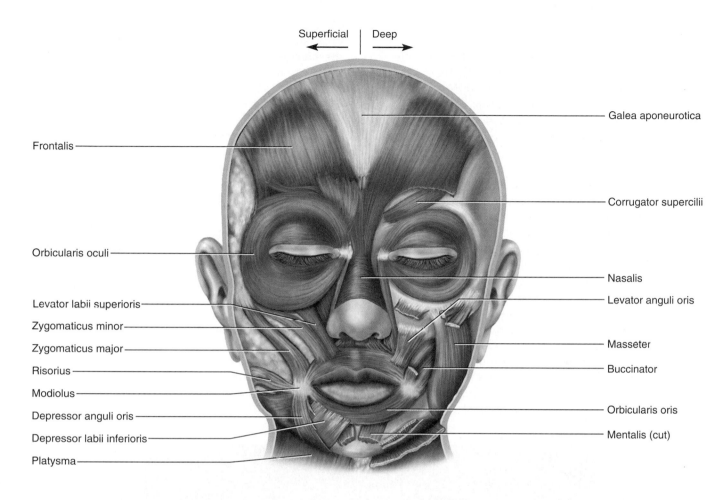

Superficial | Deep

Frontalis

Orbicularis oculi

Levator labii superioris
Zygomaticus minor
Zygomaticus major
Risorius
Modiolus
Depressor anguli oris
Depressor labii inferioris
Platysma

Galea aponeurotica

Corrugator supercilii

Nasalis
Levator anguli oris

Masseter
Buccinator

Orbicularis oris
Mentalis (cut)

FIGURE 17.7 Muscles of the Face, Anterior View

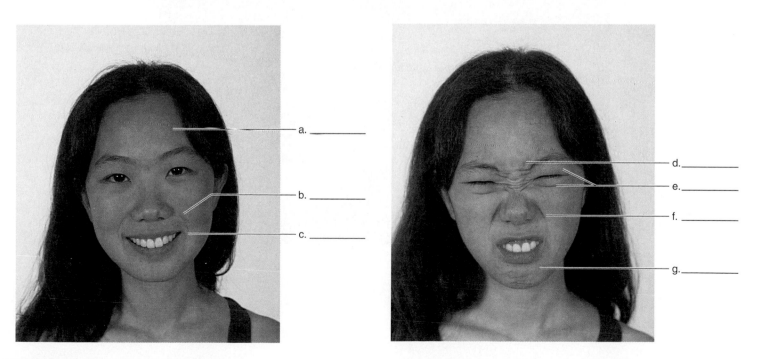

a. _____

b. _____

c. _____

d. _____

e. _____

f. _____

g. _____

FIGURE 17.8 Muscles of Facial Expression

Cat Dissection

The head and neck dissection of the cat is beneficial, but you must take care to not cut through the digestive structures, such as the salivary glands, and the circulatory structures, such as the veins and arteries. You will study these structures in later exercises. A dorsal neck muscle of the cat is the **levator scapulae,** found deep to the trapezius. You will need to cut the clavotrapezius to see the levator scapulae. In cats there is an additional muscle called the levator scapulae ventralis, which inserts on the scapular spine. Examine figure 17.9 for the levator scapulae muscles. The **scalenes** can be dissected by reflecting the pectoralis minor. Notice how the scalenes are composed of separate slips of muscle that run from the ribs to the neck. In humans the muscle is more lateral than in cats. Compare your specimen to figure 17.9.

The remainder of the muscles you will study in this exercise are seen from a ventral aspect. The **platysma** in the cat was removed during the skinning process, and it will not be seen unless you kept the skin with the cat.

In the cat the sternocleidomastoid consists of two muscles, the **sternomastoid** and the **cleidomastoid.** The sternomastoid extends from the sternum to the mastoid process of the skull, and the cleidomastoid runs from the clavicle to the mastoid process. Underneath the sternomastoid is the most medial muscle of the neck group, the **sternohyoid.** The **sternothyroid** is deeper and more lateral than the sternohyoid. These muscles can be seen in figure 17.10.

The **digastric** muscle runs parallel to the lower edge of the mandible and underneath the submandibular gland. Dissect only one side of the head, leaving the structures on the other side intact for study of the digestive system. Deep to the digastric is the **mylohyoid,** a broad muscle. Notice how the muscle fibers run transverse to the direction of the digastric. Compare your dissection to figure 17.10.

The **temporalis** is located more dorsally than the masseter and can be dissected by removing the skin and fascia anterior to the ear. The **masseter** is a large, well-developed muscle in the cat that originates on the zygomatic arch and inserts on the lateral surface of the mandible. Examine these muscles in figure 17.11.

The **pterygoids** are usually not dissected because you have to cut through the ramus of the mandible to examine them. The muscles of facial expression are generally not studied in the cat. These muscles are small and are frequently removed with the skin. You should study these muscles on human models or a cadaver (if available).

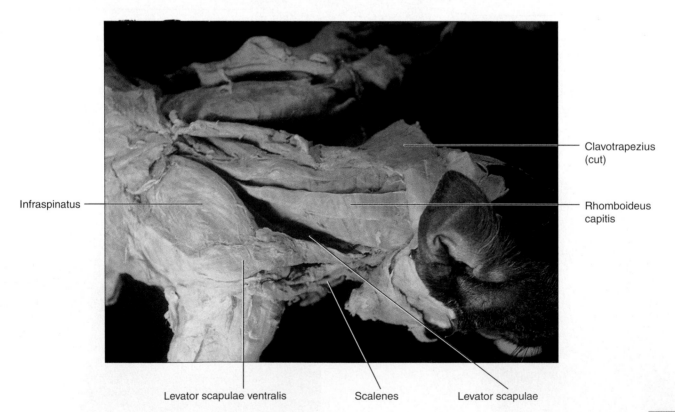

FIGURE 17.9 Muscles of the Neck of the Cat, Dorsolateral View

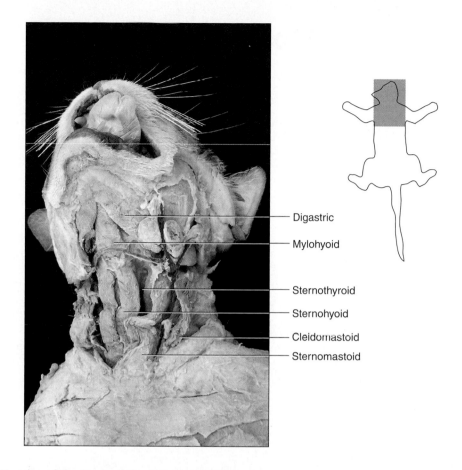

— Digastric

— Mylohyoid

— Sternothyroid

— Sternohyoid

— Cleidomastoid

— Sternomastoid

FIGURE 17.10 Muscles of the Neck of the Cat, Ventral View

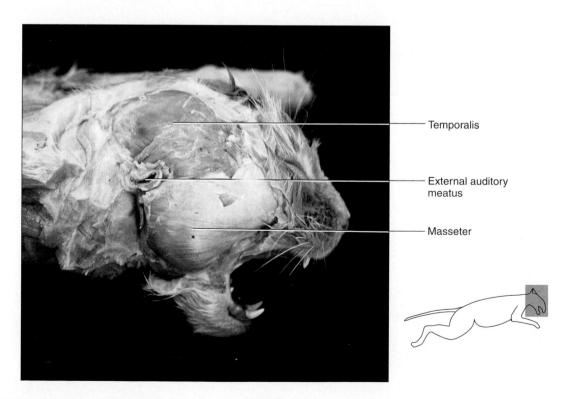

— Temporalis

— External auditory meatus

— Masseter

FIGURE 17.11 Muscles of the Head of the Cat, Lateral View

Notes

REVIEW SECTION

Muscles of the Head and Neck

Name _____ *Date* _____

Lab Section _____ *Time* _____

Review Questions

1. What is the origin of the masseter muscle?

2. What muscle is a synergist of the masseter muscle?

3. Where is the origin of the levator scapulae muscle?

4. What kind of muscle is the orbicularis oculi or orbicularis oris muscle in terms of function?

5. Label the following illustration.

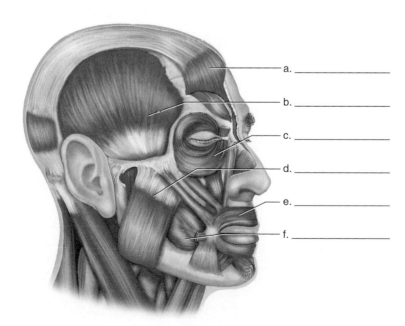

a. _____

b. _____

c. _____

d. _____

e. _____

f. _____

6. What muscle originates on the temporal fossa?

7. Name two muscles that close the jaw.

8. Where does the sternocleidomastoid muscle insert?

9. What muscle closes the lips?

10. Where does the orbicularis oculi insert?

11. What is the insertion of the temporalis?

12. Name a muscle that closes the eye.

13. What is the action of the sternocleidomastoid?

LABORATORY

Muscles of the Trunk

INTRODUCTION

The muscles of the trunk can be grouped into a few functional areas. These are the abdominal muscles, which tighten the abdomen; the respiratory muscles, which assist in breathing; the postural muscles of the back; and the muscles that act on the scapula or on the head and neck. These muscles are discussed in the Saladin text in chapter 10, "The Muscular System."

OBJECTIVES

At the end of this exercise you should be able to

1. locate the muscles of the trunk on a torso model, chart, cadaver, or cat;
2. list the origin, insertion, and action of each muscle;
3. describe what nerve controls each muscle;
4. list what muscles function as synergists or antagonists to the prime mover;
5. name all the muscles that have a particular action on the torso, such as all the muscles that compress the abdomen.

MATERIALS

Human torso model

Human muscle charts

Articulated skeleton

Cadaver

Cat

Materials for Cat Dissection

Dissection trays

Scalpel or razor blades

Gloves

Blunt probe

Pins

PROCEDURE

Review the muscle nomenclature and the actions as outlined in Laboratory Exercises 12 and 13. Examine a torso model or chart in the lab and locate the muscles described next and in table 18.1. Correlate the shape or fiber direction of the muscle with the name and begin to visualize the muscles as you study the models or charts. You may want to look at an articulated skeleton as you study the origins and insertions of the muscles so that you can better see the muscle attachment points. The descriptions in the text portion of this exercise can help you understand the nature of the muscle, while table 18.1 gives you specific information about the muscle.

Once you have learned the origin and the insertion, you should be able to understand the action of the muscle. This is done, in part, by imagining how the bony attachments of the origin and insertion would come together if the muscle pulled them closer to one another.

Anterior Muscles

The muscles of the abdomen compress the viscera, which aid in breathing and food regurgitation. These muscles are also involved in the **valsalva maneuver,** which consists of taking a deep breath and contracting these muscles during defecation, urination, and childbirth. The **external abdominal oblique** is a broad, superficial muscle of the abdomen with fibers that run from a superior direction to an inferior, medial direction. The **internal abdominal oblique** is deep to the external abdominal oblique and has fiber directions that run perpendicular to the external abdominal oblique. Locate the abdominal muscles on models in the lab and compare them to figure 18.1.

The deepest of the abdominal muscles is the **transversus abdominis,** which has fibers running in a horizontal direction. The **rectus abdominis** (*rectus* means straight) muscle runs vertically up the abdomen. The rectus abdominis has small connective tissue bands called **tendinous intersections** located horizontally across the muscle, dividing it into small segments. If the abdominal fat is minimal and the muscles are well developed, the "washboard stomach," or "six-pack," is apparent due to the muscle fibers increasing in girth while the tendinous intersections remain undeveloped.

The **intercostal** muscles and the diaphragm are respiratory muscles. Normally, the **diaphragm** is responsible for about 60% of the resting breath volume, while the external intercostals contribute to the remaining volume. The diaphragm is a domed muscle that has a peripheral origin. The insertion of the diaphragm is central at the

TABLE 18.1	Muscles of the Trunk			
Name	**Origin**	**Insertion**	**Action**	**Innervation**
Muscles of Thorax, Abdomen, and Pelvis				
External abdominal oblique	Ribs 5–12	Linea alba, iliac crest, pubis	Compresses abdominal wall, laterally rotates trunk	Intercostal nerves T7–12
Internal abdominal oblique	Inguinal ligament, iliac crest	Linea alba, ribs 10–12	Compresses abdominal wall, laterally rotates trunk	Intercostal nerves T7–12, spinal nerve L1
Transversus abdominis	Inguinal ligament, iliac crest, ribs 7–12	Linea alba, crest of pubis	Compresses abdominal wall, laterally rotates trunk	Intercostal nerves T7–12, spinal nerve L1
Rectus abdominis	Crest of pubis, symphysis pubis	Cartilages of ribs 5–7, xiphoid process	Flexes vertebral column, compresses abdominal wall	Intercostal nerves T6–12
Serratus anterior	Ribs 1–8	Vertebral border and inferior angle of scapula	Abducts scapula (moves scapula away from spinal column)	Long thoracic nerve
External intercostals	Inferior border of a rib	Superior border of rib below	Elevates ribs (increases volume in thorax)	Intercostal nerves
Internal intercostals	Inferior border of a rib	Superior border of rib below	Depresses ribs (decreases volume in thorax)	Intercostal nerves
Diaphragm	Xiphoid process, lower ribs, upper lumbar vertebrae	Central tendon	Inspiration (contraction), expiration (relaxation)	Phrenic nerve
Deep Muscles of the Back				
Erector spinae				
Iliocostalis *Longissimus* *Spinalis*	Vertebral column, ilium, ribs	Ribs, vertebral column, occipital and temporal bones	Extends and rotates vertebral column and head	Numerous spinal nerves
Multifidus	Iliac crest, vertebral column	Vertebral column above origins	Extends and rotates vertebral column	Numerous spinal nerves
Quadratus lumborum	Posterior iliac crest	T12 and L1–4, rib 12	Extends and abducts vertebral column	T12, L1–4
Rhomboideus major	Spines of T2–5	Lower one-third of vertebral border of scapula	Adducts scapula (draws scapulae together)	Dorsal scapular nerve
Rhomboideus minor	Ligamentum nuchae, spines C7–T1	Vertebral border of scapula at scapular spine	Adducts scapula (draws scapulae together)	Dorsal scapular nerve
Splenius	Ligamentum nuchae, C7–T6	C2–4, occipital and temporal bone	Extends and rotates head	Middle and lower cervical nerves
Semispinalis	C7–T12	Occipital bone, T1–4	Extends head and vertebral column, rotates vertebral column	Cervical and thoracic spinal nerves

base of the mediastinum. If you think of the diaphragm as a trampoline, the outer springs represent the origin, while the center (where you jump) represents the insertion. Examine the models in the lab and compare them to figures 18.1 and 18.2.

The **serratus anterior** is a broad, fan-shaped muscle that has slips of muscle originating on the upper ribs. These slips of muscles unite and insert on the medial border of the scapula and the inferior angle of the scapula. As the muscle contracts it pulls the scapula toward the front of the ribs.

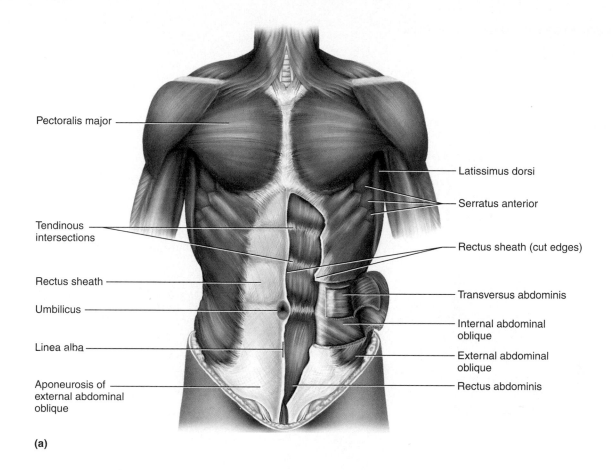

(a)

Pectoralis major

Latissimus dorsi

Serratus anterior

Tendinous intersections

Rectus sheath (cut edges)

Rectus sheath

Transversus abdominis

Umbilicus

Internal abdominal oblique

Linea alba

External abdominal oblique

Aponeurosis of external abdominal oblique

Rectus abdominis

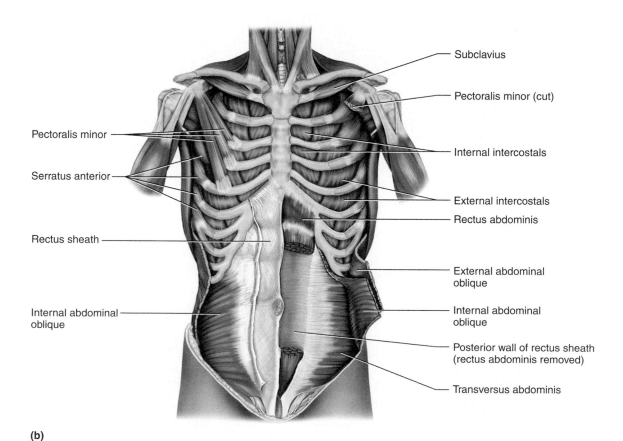

(b)

Subclavius

Pectoralis minor (cut)

Pectoralis minor

Serratus anterior

Internal intercostals

Rectus sheath

External intercostals

Rectus abdominis

Internal abdominal oblique

External abdominal oblique

Internal abdominal oblique

Posterior wall of rectus sheath (rectus abdominis removed)

Transversus abdominis

FIGURE 18.1 Muscles of the Abdomen, Anterior View Diagram (a) superficial muscles; (b) deep muscles; (c) photograph.

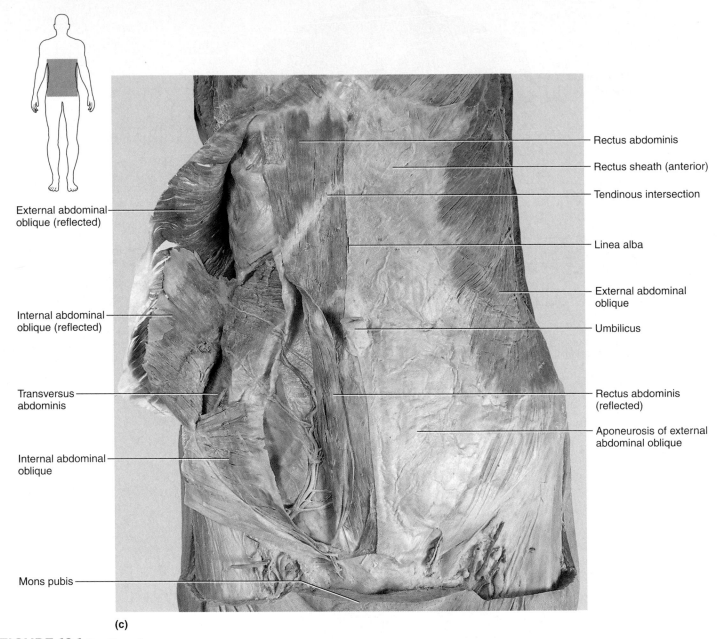

External abdominal oblique (reflected)

Internal abdominal oblique (reflected)

Transversus abdominis

Internal abdominal oblique

Mons pubis

Rectus abdominis

Rectus sheath (anterior)

Tendinous intersection

Linea alba

External abdominal oblique

Umbilicus

Rectus abdominis (reflected)

Aponeurosis of external abdominal oblique

(c)

FIGURE 18.1 *Continued*

The intercostal muscles do contribute to the breathing volume at rest, but they also contribute to a greater increase in the movement of the thorax during times of exercise. There is some debate as to the functions of the intercostals. Some evidence suggests that both the intercostals are involved in inhalation. Other evidence suggests that the **external intercostal** is involved in inhalation, while the **internal intercostal** is involved in exhalation. In this exercise we treat the external intercostals as being responsible for inhalation, while the internal intercostals are involved in exhalation. Locate the intercostal muscles and compare them to figures 18.1b and 18.3.

Posterior Muscles

The **erector spinae** make up most of the postural muscles of the back. The erector spinae are actually many muscles that occur between individual vertebrae or between the

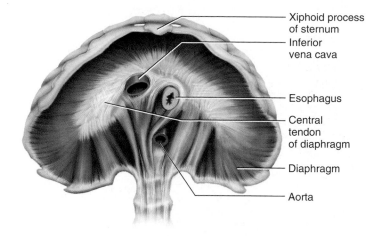

- Xiphoid process of sternum
- Inferior vena cava
- Esophagus
- Central tendon of diaphragm
- Diaphragm
- Aorta

FIGURE 18.2 Muscles of the Thorax Diagram, inferior view.

vertebrae and the ribs. These muscles are grouped conveniently into long strap muscles known collectively as the erector spinae. There are three major groups of erector spinae muscles, the **spinalis,** the **longissimus,** and the **iliocostalis.** The **multifidus** is a related muscle and, when grouped with the erector spinae, makes up the *s.l.i.m.* muscles as you move from medial to lateral and then inferior. Another muscle that extends the vertebral column is the **quadratus lumborum,** a square muscle that runs

from the iliac crest to the lower vertebrae and twelfth rib. These muscles can be seen in figure 18.4.

The **rhomboideus muscles** occur deep to the trapezius. Reflection of the trapezius is necessary to see the rhomboideus muscles. These muscles originate on the vertebral column and insert on the medial border of the scapula. As they contract they pull the scapulae together, adducting the scapulae. Other deep muscles of the back are the **splenius** and the **semispinalis** muscles. These extend and rotate the head and vertebral column. Compare the material in lab to figure 18.5.

Cat Dissection

Anterior Muscles

Place the cat on its back and examine the abdominal muscles. The abdominal muscles in the cat are similar to those in the human in that the **external abdominal oblique** is a broad, superficial muscle on the ventral abdomen. Carefully cut through the external abdominal oblique to reveal the **internal abdominal oblique,** as illustrated in figure 18.6. Deep to this is the **transversus abdominis,** and it can be seen by carefully dissecting the internal abdominal oblique. If you cut too deeply you will enter the abdominal cavity, so be careful in this part of the dissection. The **rectus abdominis** is a muscle that runs from the pubic region to the sternum.

Move to the thoracic region and examine the muscle of the lateral thorax dorsal to the xiphihumeralis. This is the **serratus anterior** (serratus ventralis in the cat),

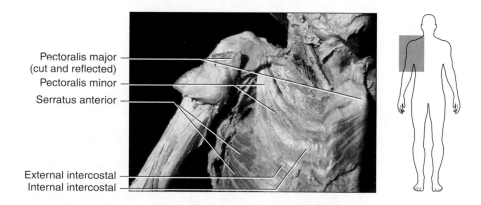

- Pectoralis major (cut and reflected)
- Pectoralis minor
- Serratus anterior
- External intercostal
- Internal intercostal

FIGURE 18.3 Muscles of the Thorax Serratus anterior and intercostal muscles, anterior view.

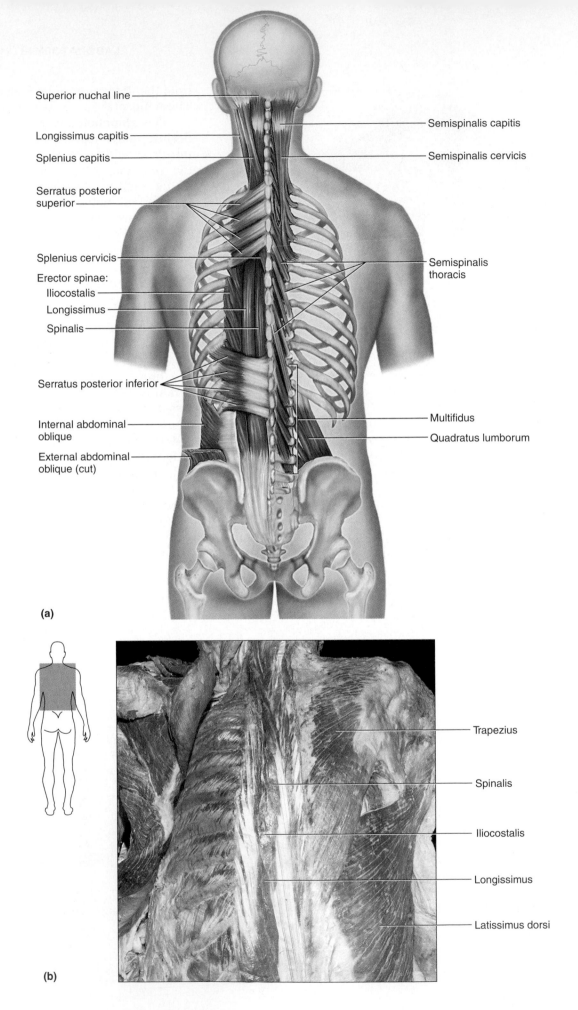

Superior nuchal line

Longissimus capitis

Splenius capitis

Serratus posterior superior

Splenius cervicis

Erector spinae:
  Iliocostalis
  Longissimus
  Spinalis

Serratus posterior inferior

Internal abdominal oblique

External abdominal oblique (cut)

Semispinalis capitis

Semispinalis cervicis

Semispinalis thoracis

Multifidus

Quadratus lumborum

(a)

Trapezius

Spinalis

Iliocostalis

Longissimus

Latissimus dorsi

(b)

FIGURE 18.4 Deep **Muscles of the Back, Posterior View** (a) Diagram; (b) photograph.

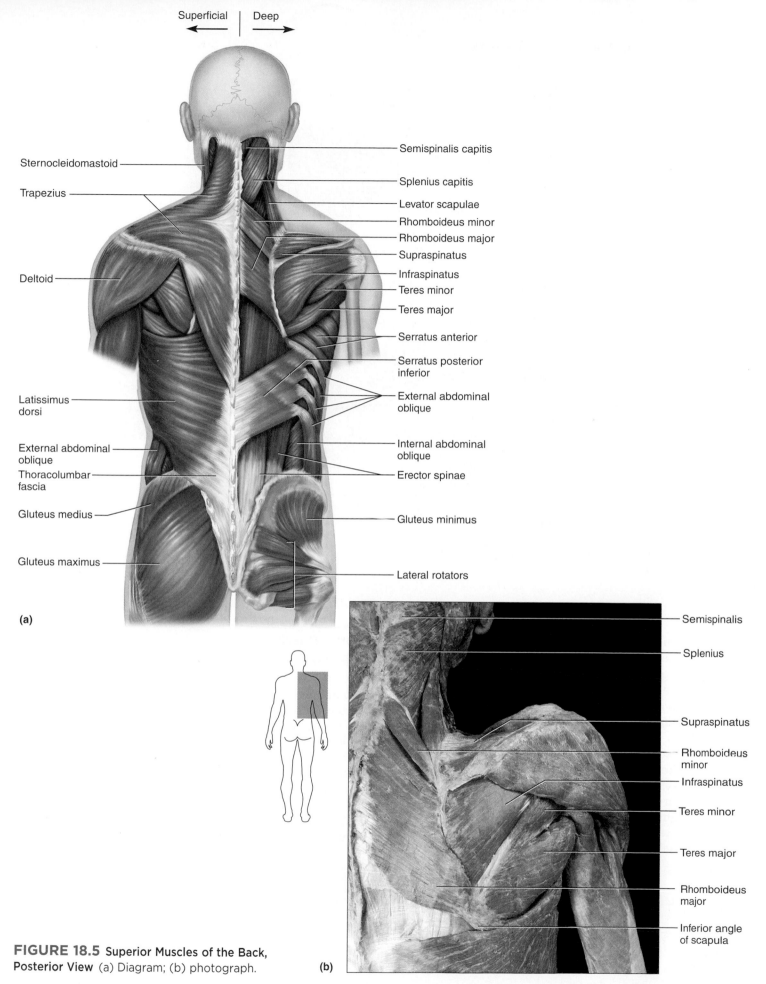

Superficial ← | → Deep

Sternocleidomastoid

Trapezius

Deltoid

Latissimus dorsi

External abdominal oblique

Thoracolumbar fascia

Gluteus medius

Gluteus maximus

Semispinalis capitis

Splenius capitis

Levator scapulae

Rhomboideus minor

Rhomboideus major

Supraspinatus

Infraspinatus

Teres minor

Teres major

Serratus anterior

Serratus posterior inferior

External abdominal oblique

Internal abdominal oblique

Erector spinae

Gluteus minimus

Lateral rotators

(a)

Semispinalis

Splenius

Supraspinatus

Rhomboideus minor

Infraspinatus

Teres minor

Teres major

Rhomboideus major

Inferior angle of scapula

(b)

FIGURE 18.5 Superior Muscles of the Back, Posterior View (a) Diagram; (b) photograph.

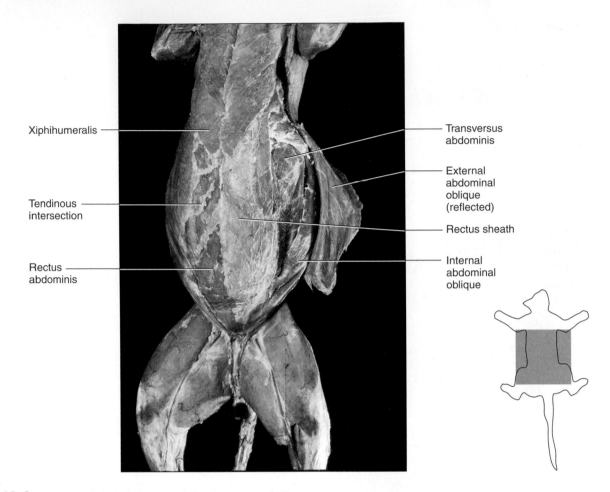

Xiphihumeralis

Tendinous intersection

Rectus abdominis

Transversus abdominis

External abdominal oblique (reflected)

Rectus sheath

Internal abdominal oblique

FIGURE 18.6 Muscles of the Abdomen of the Cat, Lateral View

and you should see the scalloped edges of the muscle. Carefully separate this muscle from the others and trace its insertion to the scapula. To see the intercostal muscles you will have to bisect the superficial chest muscles. If you have not done so already, cut through the middle of the belly of the pectoral muscles, exposing the ribs of the cat. Carefully remove the outer layer of fascia from the muscle between the ribs and locate the **external intercostal** muscle. You should be able to cut part of this muscle away and expose the **internal intercostal** muscle. The fibers run perpendicular to one another. Do not look for the diaphragm at this time. You can see it in Laboratory Exercise 39, when you examine the lungs. Examine these thoracic muscles in figure 18.7.

Dorsal Muscles

Place the cat so that you can examine the dorsal surface. You will need to dissect the trapezius carefully to see the **rhomboideus** muscles. These muscles originate on the vertebral column and insert on the scapula. If you examine the muscles of the neck and head, you should see the **splenius** muscle. These are shown in figure 18.8.

You will need to bisect the latissimus dorsi and the posterior portion of the external abdominal oblique to see the **erector spinae** muscles. The relative position of the cat erector spinae can be seen in figure 18.9. Compare this figure to your dissection. Locate the **iliocostalis, longissimus, spinalis,** and **multifidus** in the cat.

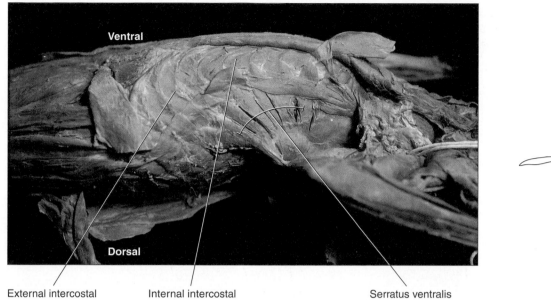

External intercostal   Internal intercostal   Serratus ventralis

FIGURE 18.7 Muscles of the Thorax of the Cat, Lateral View

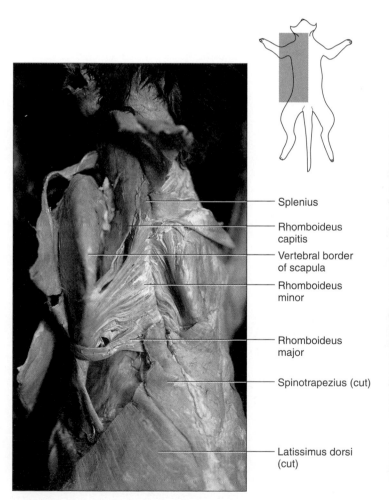

- Splenius
- Rhomboideus capitis
- Vertebral border of scapula
- Rhomboideus minor
- Rhomboideus major
- Spinotrapezius (cut)
- Latissimus dorsi (cut)

FIGURE 18.8 Anterior, Deep Muscles of the Back of the Cat, Dorsal View

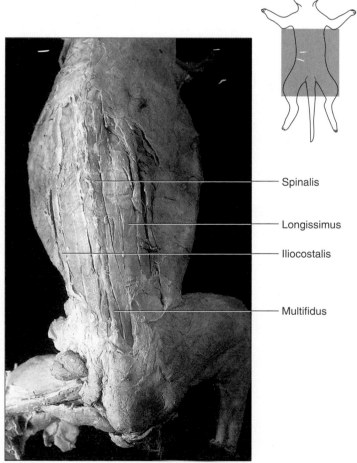

- Spinalis
- Longissimus
- Iliocostalis
- Multifidus

FIGURE 18.9 Posterior, Deep Muscles of the Back of the Cat, Dorsal View

Notes

REVIEW SECTION

Muscles of the Trunk

Name _____ *Date* _____

Lab Section _____ *Time* _____

Review Questions

1. Compression of the abdominal wall occurs by what four muscles?

2. Which is the deepest anterior abdominal muscle?

3. The tendinous intersections are found in what muscle?

4. How does the action of the rectus abdominis differ from that of the other abdominal muscles?

5. What is the action of the serratus anterior muscle?

6. What is the physical relationship of the intercostal muscles to each other?

7. How does the serratus anterior function as an antagonist to the rhomboideus muscles?

8. What is the action of the intercostal muscles?

9. What muscle inserts on the central tendon?

10. Name *five* muscles that extend the vertebral column.

11. What is the action of the quadratus lumborum?

12. Reviewing *all* the muscles you have studied so far, which ones allow you to look up at the ceiling when you are standing?

13. Which is the more superior muscle, the rhomboideus major or the rhomboideus minor?

14. What is the origin of the rhomboideus major muscle?

15. Label the muscles in the following illustration using the terms provided.

rectus abdominis   transversus abdominis
external abdominal oblique   serratus anterior
internal abdominal oblique   external intercostal
internal intercostal

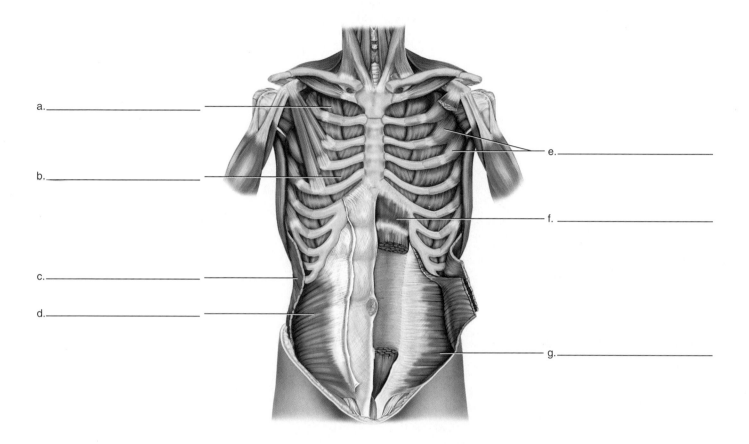

a._____

b._____

c._____

d._____

e._____

f. _____

g._____

16. In the cat, how do the abdominal muscles compare to those in the human in terms of relative position?

Notes

LABORATORY

Muscle Physiology

INTRODUCTION

Skeletal muscles contract due to stimulation by nerves controlling them. In this exercise you explore the nature of skeletal muscle contraction as initiated by external electrical stimulation and apply this information to the functions of the skeletal muscle in your body. These topics are covered in the Saladin text in chapter 11, "Muscular Tissue."

Two events are fundamental to an understanding of muscle physiology. One is an electrical event that occurs in muscle membranes, and the other is the physical contraction of the muscle. The normal contraction of skeletal muscle occurs when an electrochemical **nerve impulse** (action potential) travels down an axon and reaches the **synapse** between the nerve and the muscle. This synapse between the corresponding neuron and muscle fiber is known as the **neuromuscular junction. Acetylcholine (ACh)** is released by the neuron and stimulates an action potential that travels along the length of the muscle fiber.

When the action potential reaches the **transverse (T) tubules** of the muscle, **calcium** is released from the sarcoplasmic reticulum and the **actin** and **myosin filaments (myofilaments)** join together, producing a power stroke in the muscle cell (see figure 19.1). As a muscle fiber is **depolarized** the muscle fiber contracts. Refer to your text for a more complete description of these events.

Entire muscles in the body exhibit a **graded response,** where muscles gradually increase from slight to more forceful contractions. This is due to the presence of multiple fibers in a contracting muscle, and the overall contractile strength of the entire muscle is determined by the number of muscle fibers in that muscle.

When a muscle is stimulated with a single, brief electrical impulse, the muscle undergoes a contraction known as a **twitch.** In a twitch the muscle has three phases: a latent phase, a contraction phase, and a relaxation phase. After the initial **stimulus** the muscle undergoes a **latent phase** (see figure 19.2). This is the time when the action potential travels across the muscle cell membrane and calcium ions are released from the sarcoplasmic reticulum. After this short time the muscle goes through the **contraction phase** as the myofilaments slide across one another and the muscle shortens. Finally, there is a **relaxation phase,** characterized by the muscle returning to a resting state.

The impulse that causes a muscle to contract must exceed a **threshold** value before any contraction can occur.

A subthreshold stimulus will not elicit a response in the muscle. A stimulus above threshold level that occurs too soon after a preliminary stimulus also does not cause the muscle to contract. The time when a stimulus, delivered just after a previous stimulus, produces no contraction is known as the **refractory period.** If a stimulus is applied shortly after the refractory period, a muscle contracts and the contraction strength is more pronounced. This effect is called **temporal summation** or **wave summation** (see figure 19.3). If you continue to stimulate a muscle there is a brief increase in the strength of the individual contractions for the first several contractions. This is known as **treppe** (trep-aye).

If rapid, repeated stimuli are sent to a muscle, then the muscle produces a series of contractions called **incomplete tetanus.** If the **frequency** (number of pulses per second) of the stimuli increases, the contractions fuse in a smooth contraction of the muscle known as **complete tetanus.**

Tetanus, as used in this exercise, is seen as muscles are stimulated electrically, and the word is used differently than the disease tetanus, described in your lecture text. Examine figure 19.4 for recordings of incomplete tetanus and complete tetanus.

In this exercise you examine the muscular contraction in a simulated frog or a real frog and what occurs as voltage is increased or the frequency is changed. Due to the decline in some frog populations (such as leopard frogs in North America) simulated activities may be preferable to the use of live frogs. Bullfrogs are not only plentiful but a pest species in many areas and, as of today, may be used without significant impact to their population levels.

OBJECTIVES

At the end of this exercise you should be able to

1. demonstrate the procedure to determine threshold stimulus;
2. differentiate between tetanus and treppe;
3. describe incomplete tetanus and complete tetanus;
4. demonstrate maximum recruitment with lab equipment and a frog;
5. compare the lab experiments on frogs to muscular contractions in humans.

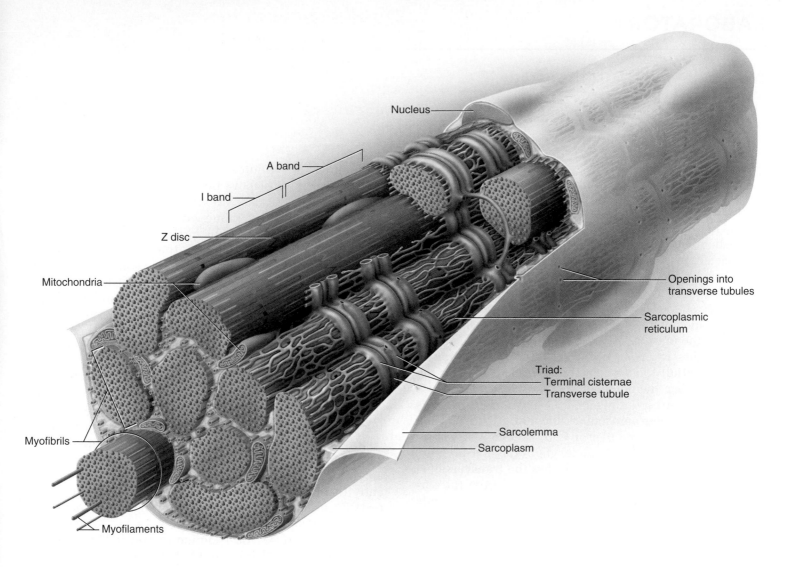

FIGURE 19.1 Overview of a Muscle Cell

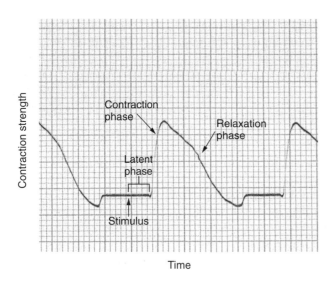

FIGURE 19.2 Three Phases of a Muscle Twitch

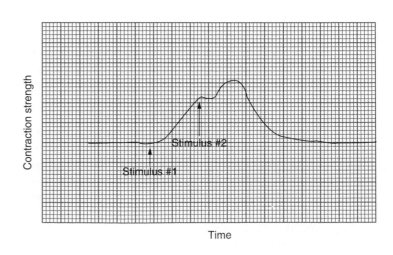

FIGURE 19.3 Wave Summation

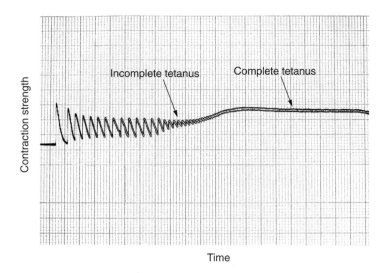

FIGURE 19.4 Incomplete and Complete Tetanus

MATERIALS

Virtual Experimental Lab

Ph.I.L.S. CD

Compatible computer

Frog (one per experiment)

Frog Ringer's solution in dropper bottles (150 mL per experiment)

Thread

Duograph, physiograph, computer

Myograph transducer

Stimulator and cables

Scissors

Clean, live animal dissection pan

Sharp pithing probes

Glass hooks

Scalpel

Pins

PROCEDURE

Simulated Frog Muscle Experiment— Physiology Interactive Lab Simulations

Load the Physiology Interactive Lab Simulations (Ph.I.L.S.) disc. You will need Macromedia® Flash 6 Player to run the program.

Stimulus-Dependent Force Generation

Under *Skeletal Muscle Functions,* select

1. Stimulus-Dependent Force Generation

Most of the folder tabs are inactive in this screen. The folder for "main menu" takes you back to the original screen where you have to start over.

Read the text and press the "*continue*" tab under **THE AIM.** You need to answer questions posed in the tutorial so that you understand the background information. The questions precede the reading material so that you can select an answer; if you select the wrong answer, you can select another response. When you find the correct response it will show up in the text on the screen. Once you answer all the questions for the section you will see a screen that simulates a muscle, electrodes to stimulate the muscle, and a recording input that records the "pull" of the muscle after it has been stimulated.

There is a yellow screen that leads to the wet lab preparation. The **WET LAB** tab is now active. Read the wet lab text and watch the movies that show how the frog is normally prepared. You can use the tabs in the movie viewer window to stop, rewind, or restart the movie.

After you have watched the wet lab for the frog preparation, return to the stimulator apparatus. Notice how the simulated frog muscle is attached to the clamp as the real frog was in the movie. Hook up the apparatus by dragging the appropriate electrical connections to the apparatus. The red and blue wires deliver electrical impulses to the frog muscle and the black wire records how much force the muscle exerts.

When you adjust the voltage in the control panel to 1 and click the "start" arrow you will see a graph showing the tension and the time. Move the **arrow cursor** to the top of the curve and click it there. Two numbers show up; the amp indicates the force of the contraction and the time indicates the number of seconds elapsed. If you click on the start button more than once you will notice a slight variation in the force of the contraction.

Hit the start button again and move the cursor to the top of the curve. To the right of the start button is the **journal entry** button. Click on this and notice how the data are entered for 1 volt. Go back to the voltage and reduce the voltage to zero. Hit the start button again, move the cursor to the line where the force is generated, click on that spot, and enter the data. Do this repeatedly until you fill in the entire journal from zero to 1.6 volts. Notice that at low voltages the force line stays flat. This is because muscles have a threshold of contraction below which no contraction occurs. Record the threshold voltage (when you first see a contraction) for the muscle in the following space. Enter the data in the following space and in the Chapter Summary Data section at the back of this exercise.

? Threshold voltage: _____ 1

What happens to the muscle contraction strength as you increase the voltage? Record your response in the following space.

Although it was not demonstrated in this exercise, there is a point at which the increase in voltage does not lead to an increase in contraction strength. This is known as **maximum recruitment,** and it is the voltage where all the muscle fibers are contracting.

Phases of Muscle Contraction

If you move the cursor over to the vertical black bar to the left of the graph you will see that the time will read zero. Notice how there is a "latent phase" from when the stimulus was delivered to when the muscle started to contract. The **lag,** or **latent, phase** is the time after the muscle is stimulated when calcium diffuses from the transverse tubules to the myofibrils. Click on the graph at the junction where you see the muscle contract and enter the value in the following space and at the end of the exercise.

? Latent phase: _____ 2

The left side of the curve is steeper than the right side of the curve. The left side of the curve is known as the **contraction phase.** How long does this phase take? It is measured by subtracting the time at the end of the latent phase from the time measured at the top of the curve. Enter the time in the following space and at the end of the exercise.

? Contraction phase: _____ 3

Move the cursor to the part of the graph where the curve reaches the baseline again and record the **relaxation phase.** This phase is determined by subtracting the time at the peak of the curve from the time obtained at the end of the contraction. Enter the time of the relaxation phase in the following space and at the end of the exercise.

? Relaxation phase: _____ 4

The Length-Tension Relationship

Under *Skeletal Muscle Functions,* select

2. The Length-Tension Relationship

As you read the AIM section you can click on the links to graphics. The animations are good at illustrating muscle contraction. Select "View Animation" and when the window opens there is a "play" arrow to start the animation.

Once you finish reading and viewing the graphics and animations you will see the experimental apparatus. The **WET LAB** folder is the same as the previous lab, so if you looked at it during the last simulation you do not need to repeat it here. You can click on the yellow box to remove it from the screen.

Follow the directions to set up the data acquisition unit, and set the voltage to maximum (1.6 volts). Click the start button, place the cursor at the top of the curve, and record the data in the journal. At the upper end of the frog apparatus there is a wheel with up and down arrows. You may need to drag the journal box to another part of the computer screen to see it. Move the arrow up so that you record the increase in length in 0.5 mm increments. This takes about two clicks of the upper arrow.

What occurs to the strength of contraction as you stretch the muscle from 26 mm to 30 mm?

Where do you predict the greatest overlap of actin and myosin myofilaments? Why?

3. Principles of Summation and Tetanus

Most muscles do not contract by a single twitch. Neurons typically send many impulses that repeatedly stimulate the muscle producing a smooth contraction. As in the previous exercises you will need to read the material presented and answer the questions before doing the experiment. Once you get to the experimental stage, turn on the power button to the data acquisition unit, drag the appropriate cables to their inputs, and set the voltage to 2. Move the cursor to the lower panel labeled "interval" and decrease the interval until the waves begin to merge. This is known as summation and you can get a visual demonstration of what this looks like by clicking on the highlighted term "summation."

What value did you receive for summation?

? Summation: _____ 5

As you decrease the interval between stimuli, notice how the peaks begin to merge. Instead of producing two single twitches the muscle produces a longer contraction. This condition is known as tetanus, and incomplete tetanus is measured when the second peak is larger than the first peak. Record your value for incomplete tetanus in the following space and at the end of the exercise.

? Incomplete tetanus: _____ 6

When the individual peaks merge into a smooth line, complete tetanus has occurred. Record this frequency in the following space.

? Complete tetanus: _____ 7

Real Frog Experiment

You may do this experiment in small groups or your instructor may elect to do a demonstration for the class. If you are doing this experiment as a group, *read the entire exercise first* and then follow the directions.

Frog Preparation

If the frog has not been pithed, follow the directions under number 1. If the frog has been pithed, begin at number 2.

1. The most humane way to conduct frog muscle experiments is to pith the frog quickly by inserting a sharp probe into the braincase.
   a. To do this, firmly grasp the frog and bend the head under your index finger (see figure 19.5).
   b. Insert the probe into the braincase and twirl it around in a conical manner, destroying the brain.

This is known as a **single pith.** The frog will be killed at this point yet still have reflexes in the lower limbs.
   c. To stop the lower limb reflexes, insert the sharp probe into the vertebral canal and run it toward the caudal end of the frog. This is known as a **double pith.** Be careful not to thrust the probe into your hand as you try to locate the vertebral canal. The correct positioning is illustrated in figure 19.5.
2. Once the frog is pithed, carefully snip its skin above the hip joint and peel it back to the foot (see figure 19.6).
   a. Separate the posterior muscles of the thigh, and locate the **sciatic nerve,** which appears as a thin, white glossy thread that runs along the lateral aspect of the femur. Always keep the sciatic nerve moist during this experiment. You may have to use a scalpel and tease the muscles away from the sciatic nerve.
   b. Using a glass hook, carefully lift the sciatic nerve away from the thigh muscles, keeping the nerve moist with frog Ringer's solution (see figure 19.7).
   c. **Ligate** the nerve by tying a thread around the proximal end of the nerve, near the sacrum, and cut the nerve above the ligature.

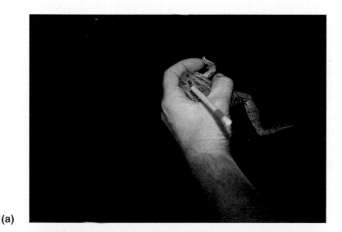

(a)

(b)

FIGURE 19.5 Pithing a Frog (a) Single pith; (b) double pith.

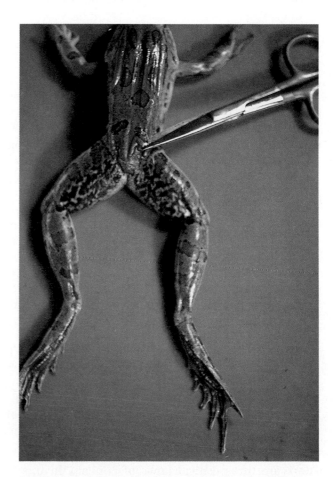

FIGURE 19.6 Preparation of the Frog, Skin Removal

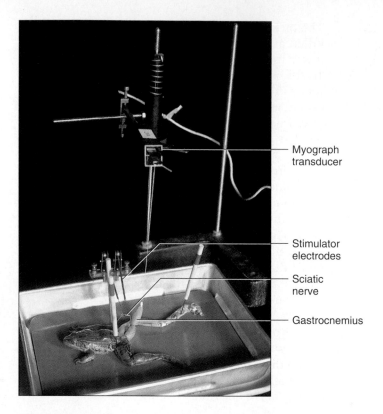

Myograph transducer

Stimulator electrodes

Sciatic nerve

Gastrocnemius

FIGURE 19.7 Frog Hookup to Recording Apparatus

d. Separate the thigh muscles from the femur and remove them, leaving the femur exposed. Be careful not to cut or damage the sciatic nerve as you do this.

e. Locate the **gastrocnemius muscle** of the frog and tie the tendon of the muscle with thread.

f. Cut the calcaneal tendon distal to the ligature and lift the gastrocnemius away from the other muscles and from the tibiofibula (a fused bone in frogs).

g. Cut the muscles and the tibiofibula just distal to the knee so that you have the femur, the sciatic nerve, the gastrocnemius, and the knee joint intact (see figure 19.7).

h. Anchor the femur to a board or femur clamp, or mount it on a tray and lay the sciatic nerve on the gastrocnemius muscle.

i. Keep the muscle and the nerve moist during the entire experiment. Do not tug on the nerve but gently lay it on the gastrocnemius muscle.

j. Attach the thread tied to the calcaneal tendon to the end of a myograph transducer leaf (see figure 19.7). This should be connected to a recording device, such as a physiograph, duograph, kymograph, or physiology computer. If you lightly tap on the leaf of the muscle transducer you should see a response in your recording apparatus.

There are three critical areas of concern in this lab. These are the preparation of the muscle, the stimulation of the muscle, and the recording of the response. Failure in any of these three areas will cause poor results or no results. The first of these areas is the preparation of the muscle, which has already been outlined. The second and third areas are discussed next.

Stimulator Setup

The preparation of the stimulator first involves determining how many stimuli you deliver in a particular time. Most stimulators can deliver repeating stimuli or a single stimulus (pulse). These are measured as the **frequency** of the stimulus, and a good frequency to start out with is two pulses per second. Another important factor is the **duration** of the stimulus. This is how long the stimulus is delivered to the sciatic nerve. Durations of 2 to 10 milliseconds usually produce good results.

Determination of Threshold Stimulus

1. You can determine the threshold voltage by keeping the voltage at zero and examining the muscle while the frequency and duration are set as described.

2. Lay the sciatic nerve on the stimulator electrodes and slowly increase the voltage until you see the contraction of the muscle. If you have reached 5 to 8 volts and you still have no response, turn the voltage down, shut the stimulator off, and recheck your connections and settings. Once you see the muscle contract, then the lowest voltage that produces a response is known as the **threshold stimulus.**

3. Record this value in the space provided and at the end of the exercise.

? Threshold stimulus: _____ 8

Multiple Motor Unit (MMU) Summation

A nerve fiber and all the muscle fibers it attaches to are called a **motor unit.** The strength of contraction in muscle increases as more motor units are stimulated. The smoothness of muscle contractions is brought about by activating different motor units at different times so that there is asynchronous firing of muscle fibers in a specific muscle. In this experiment you examine the effect of increasing voltage and its impact on increasing the number of motor units stimulated in a muscle. There is a point where an increase in voltage will not stimulate any more motor units. This point is known as the **maximum recruitment.**

Recording Apparatus Setup

Your lab may be equipped with one or more different physiological recorders. Follow your instructor's directions to set up the apparatus if you are to do the experiment in groups or pay close attention if your instructor demonstrates the experiment. Pay particular attention to the settings of the apparatus.

1. To demonstrate this, set the duration of the pulse to 2 milliseconds, keep the frequency at one or two pulses per second, and set the stimulus on repeat.
2. Begin the recording and increase the voltage until you see a response (threshold) in the muscle. Continue to increase the voltage slowly, and you should see the contraction force increase, as indicated by an increase in the height of the tracing.
3. As you continue to increase the voltage slowly, the contraction tracings will not get any higher. The minimum voltage it takes to produce the maximum height is known as the maximum recruitment voltage.
4. Record the **maximum recruitment voltage** in the space provided and at the end of the exercise.

❓ Maximum recruitment voltage: _____ 9

Treppe

1. Set the voltage reading at the maximum recruitment voltage and the frequency at one pulse per second.
2. As you stimulate the muscle, there is a stepwise increase in the peaks of the contractions as the muscle contracts over time. This increase in strength of contraction is known as **treppe.** This is thought to be due to the increased availability of calcium in the muscle fibers. As an increased frequency of stimulation occurs, not all the calcium is returned to the sarcoplasmic reticulum and more is available in the cytoplasm. This increase in calcium allows for more binding of myosin heads and stronger contractions. Another factor may be that, as the friction in the muscle cell increases the temperatures, the enzymes are more active.

Phases of Muscle Contraction

1. If you increase the chart speed to 50 mm per second you can obtain tracings of the muscle where the latent phase, contraction phase, and relaxation phase can be seen. If you can record the time when you stimulate the muscle, then you can calculate the latent phase of muscle contraction.
2. Make a tracing of the contraction and compare it to figure 19.8.

If the chart speed is 50 mm per second in your tracing, what is the length of the latent phase? This assumes that you have a mark indicating the time of stimulation.

❓ Latent phase: _____ 10

How long is the contraction phase? Record the length of time.

❓ Contraction phase: _____ 11

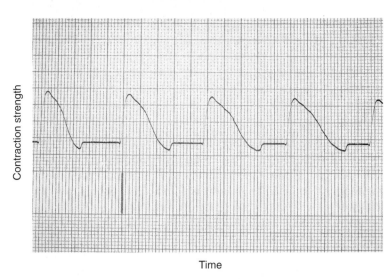

FIGURE 19.8 Phases of Muscle Contractions in a Single Twitch Contraction

How long is the relaxation phase? Record the length of time.

❓ Relaxation phase: _____ 12

Human Muscle Physiology

In this part of the exercise you study the effects of muscle recruitment and fatigue. These experiments are best done with the use of a device that can measure electrical activity, such as a physiograph or computer, and a hardware unit, such as the Biopac MP30 electrode lead set, disposable electrodes, and the EMG1 student module.

Muscle Recruitment

Attach the appropriate recording electrodes to the forearm (a positive, negative, and ground) and connect the electrodes to the physiograph or hardware unit. You will first need to establish a baseline recording by adjusting the sensitivity of the equipment to register the maximum grip strength that you can produce. To do this you should grip your hand tightly while recording the activity and make sure that you have adequate displacement displayed. Ask your instructor for specific directions for the equipment in your lab.

Once you have produced a reasonable recording for baseline data, begin the recording by slightly gripping your hand. Pause for a second and produce a slightly stronger grip. Pause once more for a second and close your hand as tightly as possible. Stop the recording.

Examine the height of the displacement of the physiograph recording or the electronic recording produced. As more and more muscle fibers are recruited, the increase in the electrical activity produces a stronger signal.

What do you predict to be the difference in your favored arm (for example, the right arm in a right-handed person) versus your weak arm?

Produce a recording of your *maximum* grip strength in your other arm and compare the two recordings.

Muscle Fatigue

You can examine the effects of muscle fatigue by using the same setup but, instead of using grip strength, examine the nature of muscles when they contract maximally for long periods. In this exercise you should hold a 5- or 10- pound weight for a moment to establish your baseline recording. Once you have produced an adequate recording begin the experiment as follows.

Hold a weight (5-, 10-, or 15- lb) with your forearm held at a 90-degree angle to your body. Begin the recording and wait for the muscle to begin to fatigue. This is seen when the forearm begins to drop somewhat. Stronger individuals should hold heavier weights. When the weight can no longer be held and the forearm begins to fall significantly stop the recording.

How does the strength of the signal compare at the beginning of the recording to the end of the recording?

Does the stability of the recording stay the same or is there variation in the electrical activity?

REVIEW SECTION

Muscle Physiology

Name _____   Date _____

Lab Section _____   Time _____

Review Questions

1. Define subthreshold stimulus.

2. Describe complete tetanus.

3. What is maximum recruitment?

4. In the following illustration, place an "A" on the latent period, a "B" on the contraction phase, and a "C" on the relaxation phase.

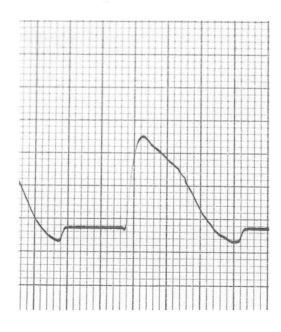

5. What was the threshold stimulus you obtained in lab?

6. What was the voltage at which you first got maximum recruitment?

7. Explain a muscle spasm in terms of recruitment of muscle fibers.

8. A skeletal muscle is stimulated to contract by what structure?

9. What chemical crosses the synapse, causing a muscle to contract?

10. Where is calcium released to cause muscle contraction?

11. What are the two types of filaments found in muscle cells that cause muscle contraction?

12. Name the first phase after a stimulus in a muscle contraction.

13. What happens to the strength of contraction during wave summation?

14 As an athlete "warms up" before exercising, the muscles increase in temperature. What effect does this have on various phases of muscle contraction?

? Chapter Summary Data

Use this section to record your results from questions within the exercise.

1. _____	7. _____
2. _____	8. _____
3. _____	9. _____
4. _____	10. _____
5. _____	11. _____
6. _____	12. _____

LABORATORY

Introduction to the Nervous System

INTRODUCTION

The function of the nervous system is communication between the various regions of the body, coordination of body functions (as in digestion or walking), orientation to the environment, and assimilation of information. The functional unit of the nervous system is the **neuron.** It is the cell that carries out the activity of nervous tissue. Neurons are located in the nerves of the body, ganglia, the spinal cord, and the brain. **Neuroglia** (glial cells) are the supporting cells of the nervous tissue. They aid the neurons in increasing the speed of neuron transmission, providing nutrients to the neurons, and protecting the neurons. These topics are further discussed in the Saladin text in chapter 12, "Nervous Tissue." In this exercise you learn about the basic structure of the nervous system and the cells that are part of the nervous system.

OBJECTIVES

At the end of this exercise you should be able to

1. describe the three parts of the neuron;
2. list the main divisions of the nervous system;
3. group the organs of the nervous system into the main divisions;
4. describe the functions of the various neuroglia.

MATERIALS

Charts or models of the nervous system

Charts or models of neurons

Microscopes

Prepared Slides

Spinal cord smear

Longitudinal section of nerve

Neuroglia

Cerebrum

Cerebellum

PROCEDURE

Divisions

The nervous system can be divided into two general divisions. The **central nervous system (CNS)** consists of the brain and spinal cord. The other division is the **peripheral nervous system (PNS),** which consists of cranial nerves, spinal nerves, ganglia, and somatic nerves (for example, the sciatic nerve). The peripheral nervous system is subdivided into sensory divisions (those that transmit sensations to the CNS) and motor divisions (those that carry information away from the CNS). Each one of these divisions has a somatic division (involving the muscles and/or skin, joints, and bones) and a visceral division (usually involving internal organs). You can read about these in more detail in your lecture text.

Examine the models or charts in the lab for the divisions of the nervous system. Compare the material in the lab to figure 20.1.

Neuron

The **neuron** is a remarkable cell. The nerves in your thigh and leg are composed of neuron fibers, and the neurons that pick up sensation in your toes continue as *single cells* up the leg and thigh to **synapse** (join) with other neurons in the lower back. When you look at prepared slides of neurons in the microscope in this exercise, remember that some of these neurons are of great length.

Neurons consist of three main parts, the **axon;** the **dendrite;** and the **nerve cell body,** or **soma.** Examine figure 20.2 and models or charts in the lab for the structure of neurons. Impulses that reach neurons stimulate dendrites or the nerve cell body. Dendrites are so named because they have branching structures that resemble a tree (*dendros* = tree). Nerve cell bodies consist of the **neuroplasm** (cytoplasm of the neuron), **Nissl bodies** (rough endoplasmic reticulum of the neuron), and the **nucleus.** The triangular region of the nerve cell body devoid of Nissl bodies is the **axon hillock,** and it leads to the axon that exits the nerve cell body.

Axons are long processes of neurons that transmit action potentials and are frequently wrapped in myelin sheaths. These sheaths are discussed later in the exercise.

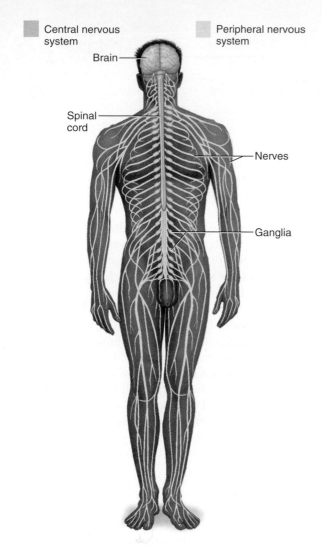

Central nervous system

Peripheral nervous system

Brain

Spinal cord

Nerves

Ganglia

FIGURE 20.1 Divisions of the Nervous System

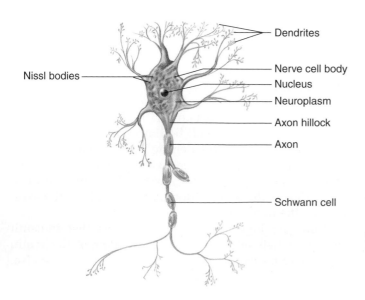

Dendrites

Nerve cell body

Nucleus

Neuroplasm

Axon hillock

Axon

Schwann cell

Nissl bodies

FIGURE 20.2 Parts of the Neuron

Functions of Neurons

There are three types of neurons based on function. **Sensory (afferent) neurons** conduct impulses *to* the central nervous system. They convey information from the external or internal body environment to the spinal cord and/or the brain. **Motor (efferent) neurons** conduct impulses *away from* the central nervous system to organs or glands that carry out an activity. **Interneurons,** or **association neurons,** occur *between* sensory and motor neurons. They transmit information to the brain for processing.

Neuron Shapes

Neurons can be classified according to shape. A **multipolar neuron** consists of several dendritic processes, a single nerve cell body, and a single axon. The majority of the neurons of the body (such as those in the brain and spinal cord) are multipolar neurons. Compare material in the lab to figure 20.3. **Bipolar neurons** are so named because the nerve cell body has two poles. Dendrites receive information and conduct it to one pole of the nerve cell body. At the other pole an axon leaves the nerve cell body and transmits the impulse away from the cell body. Bipolar neurons are found in nerves conducting the senses of smell and vision.

Unipolar, or **pseudounipolar, neurons** have a cell body with a single process attached to it. The somatic sensory nerves that take information to the spinal cord are unipolar neurons. In most neurons the dendrites take information to the cell body and axons take information away from the cell body. In unipolar neurons there are dendrites that receive information and axons, then send that information either to the cell body or directly to the axon that takes information to the spinal cord.

Histology of the Neuron

Examine a prepared slide of a spinal cord smear under the microscope and locate the purple, star-shaped structures under low power. These are the nerve cell bodies of multipolar neurons. Switch to high power and locate the Nissl bodies, the nucleus, and the axon hillock. If you find the axon hillock you should be able to see the axon leading away from the hillock. The other processes attached to the nerve cell body are dendrites. The small nuclei scattered throughout the smear belong to glial cells in the spinal cord. Compare your slide to figure 20.4.

Synapses

Neurons transmit information **electrochemically** along the length of the axon to the **synaptic knob.** Most neurons are not physically attached to one another but communicate by chemical signals that flow through the junction of the neurons. The junction is called the **synapse,** and the chemicals, such as acetylcholine, that move across the synaptic cleft are called **neurotransmitters.** There are a

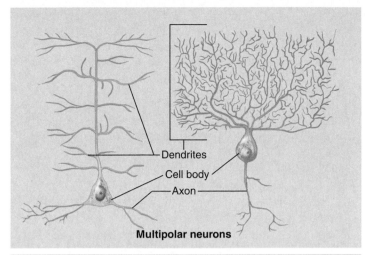

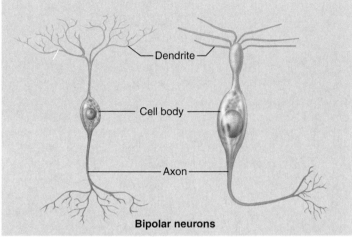

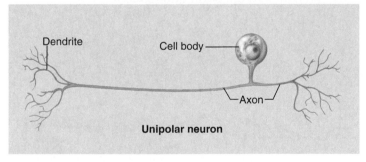

FIGURE 20.3 Neuron Shapes

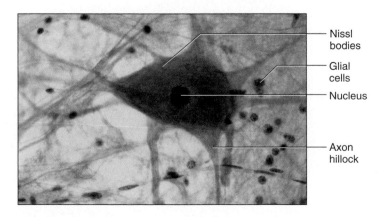

FIGURE 20.4 Spinal Cord Smear

few instances where an **electrical synapse** occurs, such as at gap junctions. In electrical synapses the cells are physically connected and stimulation of one cell causes excitation in the adjacent cell. These occurrences are less common and will not be covered in this exercise. Figure 20.5 illustrates a synapse.

Neuroglia

Numerous cells aid the functioning of the neuron. These cells are called the **neuroglia,** or **glial cells.**

PNS Neuroglia

The common glial cell of the peripheral nervous system is the **Schwann cell,** or **neurolemmocyte** (see figure 20.2). These glial cells wrap around the axon, leaving small gaps between successive cells called the **nodes of Ranvier.** Schwann cells wrap around the axon, much as a thin strip of paper can be wrapped around a pencil. The neurolemmocyte consists of a significant amount of a lipoprotein material called **myelin,** and the series of neurolemmocytes produces a **myelin sheath.**

The neurolemmocyte nodes allow for the nerve transmission to jump from node to node, thus increasing the transmission speed of the neuron. This type of jumping transmission is called **saltatory conduction.**

Histology of the Neurolemmocyte

Examine a prepared slide of a longitudinal section of nerve under high power. You should be able to see the axon fibers as long, dark threads in the microscope. The clear areas on each side of the axon fibers are the myelin sheaths. If you scan the slide closely, you should be able to see the junction of two neurolemmocytes. The gap

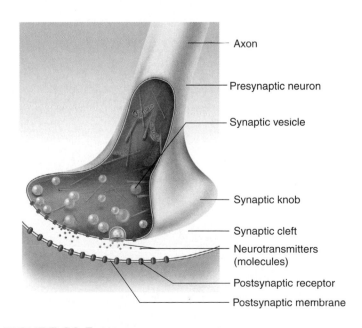

FIGURE 20.5 Synapse

between them is the node of Ranvier. Compare your slide to figure 20.6.

CNS Neuroglia

Other types of neuroglia include myelinating fibers in the CNS. Neurolemmocytes occur in the PNS, and **oligodendrocytes** are their counterpart in the CNS. Myelinated nerve fibers appear white, so this type of nervous tissue is called **white matter. Unmyelinated** fibers form the portion of the nervous tissue known as **gray matter.** Unlike the neurolemmocytes, oligodendrocytes frequently wrap around several neurons, and the white matter of the spinal cord and brain is due to the lipoprotein of the oligodendrocytes.

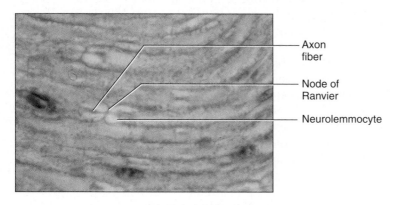

Axon
fiber

Node of
Ranvier

Neurolemmocyte

FIGURE 20.6 Nerve, Longitudinal Section (400×)

The most common glial cells in the CNS are **astrocytes** and they have numerous functions. They stimulate the capillaries in the brain to form the blood-brain barrier, which limits the entrance of microbes into the brain tissue. They also feed the neurons, release nerve growth factors, play a role in synaptic transmission, and form scar tissue when neurons are damaged.

Microglia are small, phagocytic glial cells. They are the brain's resident immunocompetent cells. When they are in their resting phase (see figure 20.7), they are actively sampling the brain for tissue damage or infection. In their reactive phase they resemble macrophages as they engulf microbes or damaged tissue. Examine figure 20.7 for examples of microglia and other glial cells.

The last of the glial cells covered in this exercise are the **ependymal cells.** These cells line the ventricles of the brain and secrete cerebrospinal fluid (CSF).

Specialized Neurons

The cerebral cortex of the brain contains **pyramidal cells** that have extensive dendritic branches. Examine a prepared slide of the cerebrum and locate the pyramidal cells. Compare them to figure 20.8.

In the cerebellum **Purkinje cells** are common. They occur in the gray matter of the cerebellum and have branching dendrites. Examine a prepared slide of Purkinje cells and compare them to figure 20.8.

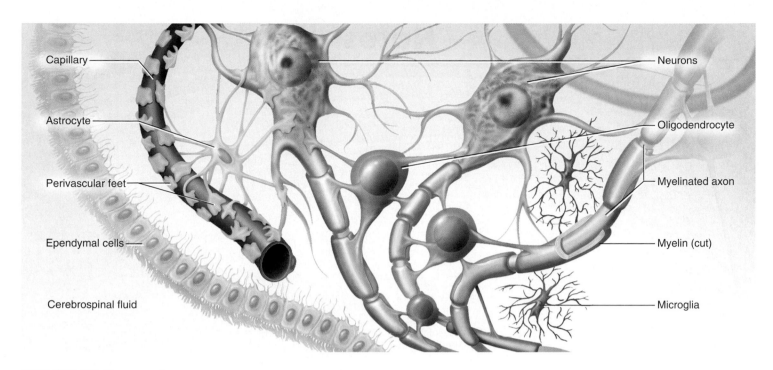

Capillary

Astrocyte

Perivascular feet

Ependymal cells

Cerebrospinal fluid

Neurons

Oligodendrocyte

Myelinated axon

Myelin (cut)

Microglia

FIGURE 20.7 Neuroglia

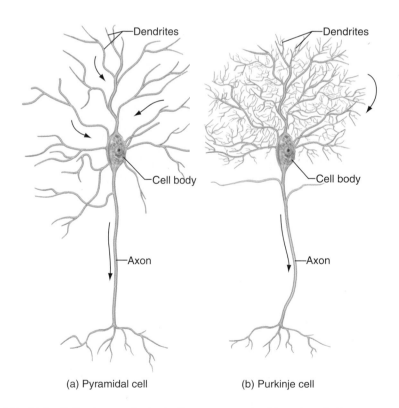

Dendrites

Dendrites

Cell body

Cell body

Axon

Axon

(a) Pyramidal cell

(b) Purkinje cell

FIGURE 20.8 Pyramidal and Purkinje Cells Arrows indicate the movement of impulses in the neurons.

Notes

REVIEW SECTION

Introduction to the Nervous System

Name _____ Date _____

Lab Section _____ Time _____

Review Questions

1. What type of neuron (multipolar, bipolar, or unipolar) is represented by the drawing? Label the parts with the terms provided.

   axon   dendrite
   nerve cell body   Nissl body

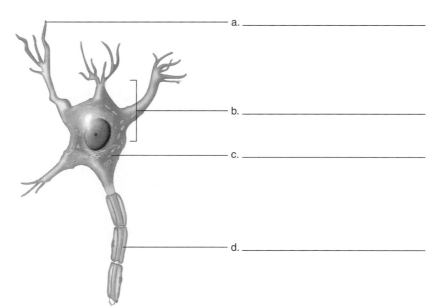

a. _____

b. _____

c. _____

d. _____

Neuron type _____

2. What type of neuron (multipolar, bipolar, or unipolar) is represented by the drawing? Label the parts with the terms provided.

   axon   dendrite
   nerve cell body   nucleus

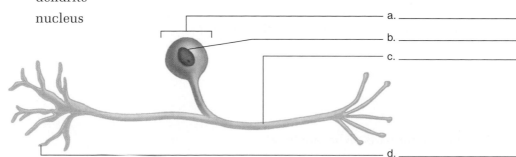

a. _____

b. _____

c. _____

d. _____

Neuron type _____

3. Describe the function of

   a. an astrocyte:

   b. an ependymal cell:

   c. an oligodendrocyte:

4. The brain belongs to what division of the nervous system?

5. A spinal nerve belongs to what division of the nervous system?

6. To what major division of the nervous system does the spinal cord belong?

7. What does CNS stand for?

8. What kind of cell performs the main function of the nervous system?

9. The nucleus is found in what specific part of the neuron?

10. A neuron has three main parts. What are they?

11. What is another name for an efferent neuron?

12. If a neuron has a soma with a dendrite on one side and an axon on the other, what kind of neuron is it?

13. Two adjacent neurons that communicate with one another are separated by a space. What is this space called?

14. How are neuroglia different from neurons in terms of function?

15. In which one of the nervous system divisions are neurolemmocytes found?

16. Myelin is made of what kind of material?

Notes

LABORATORY

The Brain and Cranial Nerves

INTRODUCTION

Two specific traits distinguish humans from other animals. One is our upright posture, and the other is the extensive development of our brain. In this exercise you examine the anatomy of the human brain and the cranial nerves associated with it. You also compare the human brain to a sheep brain and note similarities and differences. Not only is there a difference in the size of the brains between humans and sheep but also a difference in the relative size of various structures of the brain and the position of some of the anatomy of the brain. You may want to review the planes of sectioning in Exercise 2.

The brain, part of the central nervous system, is located in the cranial cavity of the skull and weighs approximately 1.4 kilograms (3 pounds). The brain is derived from three embryonic regions, each of which further develops into more specific areas. The three embryonic regions of the brain are the **prosencephalon,** or **forebrain;** the **mesencephalon,** or **midbrain;** and the **rhombencephalon,** or **hindbrain.** Cranial nerves, though belonging to the peripheral nervous system, are studied along with the brain, as they are closely associated with it. Knowledge of the anatomy of the brain is important in locating the cranial nerves. These topics are covered in the Saladin text in chapter 14, "The Brain and Cranial Nerves."

OBJECTIVES

At the end of this exercise you should be able to

1. name the three meninges of the brain and their location relative to one another;
2. locate the three major regions of the brain;
3. name the main structures in each of the three regions of the brain;
4. describe the function of the major structures in the brain;
5. trace the path of cerebrospinal fluid through the brain;
6. list the major blood vessels that take blood to or from the brain.

MATERIALS

Models and charts of the human brain

Preserved human brains (if available)

Cast of the ventricles of the brain

Chart, section, or illustration of the brain in coronal and transverse sections

Sheep brains

Dissection trays

Scalpels or razor blades

Gloves

Blunt probes

PROCEDURE

Meninges

There are three membranes, called **meninges,** that surround the brain. The outermost of these is the **dura mater,** a tough, dense connective tissue membrane that encircles the brain and has a series of shelves that extend into the brain. The next deeper layer is the **arachnoid mater,** a thin membrane. Between the dura mater and the arachnoid membrane is the subdural space. Deep to the arachnoid is the **subarachnoid space,** which contains **cerebrospinal fluid (CSF).** The deepest layer is the **pia mater,** a membrane directly on the outer surface of the brain. Locate the meninges in preserved brains in the lab (if available) and compare them to figure 21.1.

Overview of the Brain

The major regions of the brain can be subdivided into smaller areas. The forebrain is the largest region of the brain and consists primarily of the cerebrum and the diencephalon (see table 21.1 and figure 21.2). The midbrain is the smallest region of the brain and is located between the forebrain and the hindbrain. The hindbrain is composed of the pons, the medulla oblongata (me-DULL-ah OB-long-GAH-ta) and the cerebellum. The hindbrain is the most inferior portion of the brain and connects to the spinal cord at the level of the foramen magnum.

Blood Supply to the Brain

The blood vessels that supply nutrients and oxygen to the brain do not penetrate the brain tissue but form a meshwork around the brain and into the open spaces in the interior. The main arteries that provide blood to the

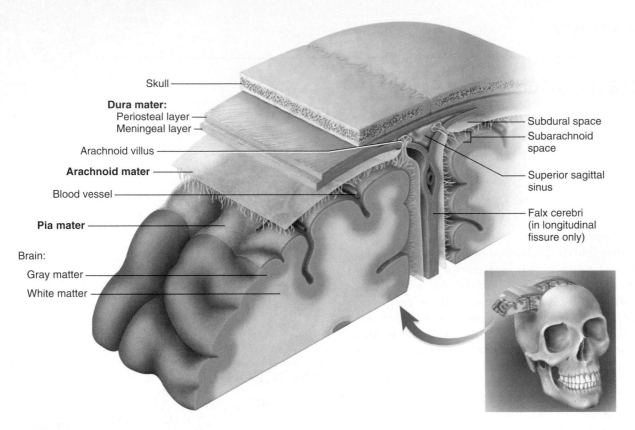

FIGURE 21.1 Meninges of the Brain

TABLE 21.1	Regions of the Brain
Forebrain	*Midbrain*
Telencephalon	Peduncles
Cerebrum (cerebral hemispheres)	Tectum
Cerebral cortex (gray matter)	Corpora quadrigemina
Basal nuclei (gray matter)	Superior colliculus
Corpus callosum	Inferior colliculus
Diencephalon	*Hindbrain*
Pineal body	Metencephalon
Thalamus	Pons
Hypothalamus	Cerebellum
Pituitary gland	Myelencephalon
Mammillary bodies	Medulla oblongata

brain are the **vertebral arteries** and the **internal carotid arteries.** Locate the vertebral arteries as illustrated in figure 21.3. These arteries pass through the transverse foramina of the cervical vertebrae and join to form the **basilar artery** at the base of the brain before branching into the **arterial circle** (circle of Willis) that forms a loop around the pituitary gland. The basilar artery gives

rise to the **cerebellar arteries,** which take blood to the cerebellum. A pair of vessels joins the arterial circle at the anterior end, and these are the internal carotid arteries. From the arterial circle numerous **cerebral arteries** take blood superiorly to the surface of the cerebrum and into the interior of the brain. These lead to capillaries where an exchange of blood gases, nutrients, and waste material takes place between the cardiovascular system and the specialized brain cells. Note the various arteries in figure 21.3 at the base of the brain.

The drainage of the brain occurs as veins take blood from the brain and pass through the subarachnoid membrane to the **venous sinuses** in the subdural spaces. The drainage of blood from the brain flows into the **internal jugular veins** on the return trip to the heart. Examine figure 21.4 for the venous drainage from the brain.

Ventricles of the Brain

The hollow neural tube that develops in the first trimester of pregnancy becomes the ventricles of the brain in the adult. The two ventricles that occupy the center of each cerebral hemisphere are known as the **lateral ventricles.** These ventricles receive **cerebrospinal fluid** from tufts of blood capillaries called **choroid plexuses.** In preserved brains the choroid plexuses are small, brown areas at the superior portions of the ventricles. Fluid from the lateral ventricles flows through the **interventricular foramina**

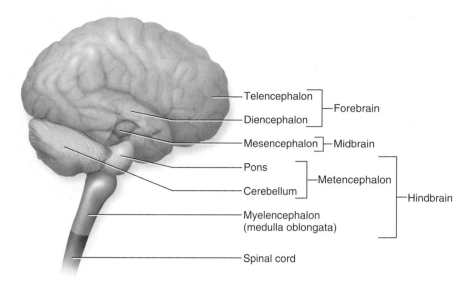

FIGURE 21.2 Major Regions of the Brain

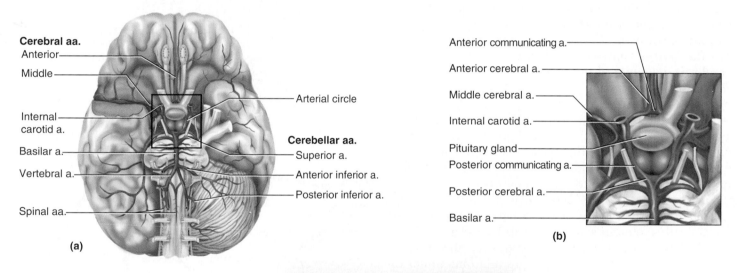

(a)

(b)

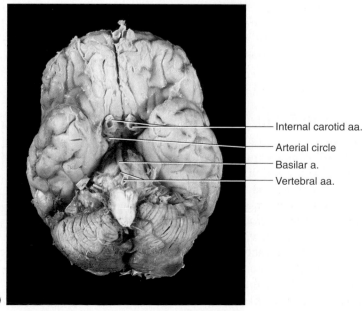

FIGURE 21.3 Brain with Arteries, Inferior View (a) Overview of arterial supply to brain; (b) close-up of arterial circle; (c) photograph.

(c)

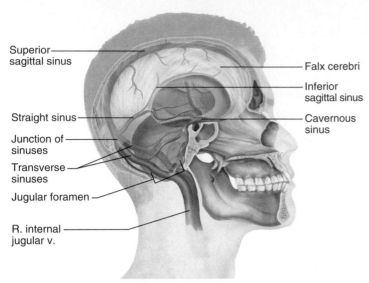

FIGURE 21.4 Major Venous Drainage of the Brain

(foramina of Monro) and into the **third ventricle.** The third ventricle occurs in the thalamus and receives CSF from choroid plexuses in that area. The third ventricle drains into the **fourth ventricle** by way of the **mesencephalic (cerebral) aqueduct.** If this duct becomes occluded, then CSF accumulates in the lateral and third ventricles. This increases the size of the ventricles, producing a condition known as **hydrocephaly.** Normally, the mesencephalic aqueduct is open and passes through the region of the midbrain. Posterior to the mesencephalic aqueduct is the fourth ventricle, which occupies a space anterior to the cerebellum. The fourth ventricle also has a choroid plexus that secretes CSF. Cerebrospinal fluid flows from the fourth ventricle to the central canal of the spinal cord, into the subarachnoid space of the spinal cord and brain. The CSF cushions and provides buoyancy to the brain. CSF flows through the arachnoid granulations under the dura mater of the skull to the venous sinuses, and the fluid returns to the rest of the cardiovascular system by the **internal jugular veins.** There are approximately 150 mL of CSF in the central nervous system, and it takes about 6 hours to circulate through the system. Locate the ventricles of the brain in figure 21.5.

Surface View of the Brain

Examine a model or chart of the brain and locate the **forebrain** and the **hindbrain.** In the forebrain you should examine the large **cerebrum,** which can be seen with folds and ridges known as **convolutions.** The ridges of the convolutions are called **gyri** (singular, *gyrus*), and the depressions are either **sulci** (singular, *sulcus*) or **fissures.** Fissures are deeper than sulci. The lateral view of the brain allows you to see the major lobes of each cerebral hemisphere. These lobes are named for the bones of the skull under which they lie. They are the **frontal, parietal, occipital,** and **temporal lobes.** The **lateral fissure** (sulcus) separates

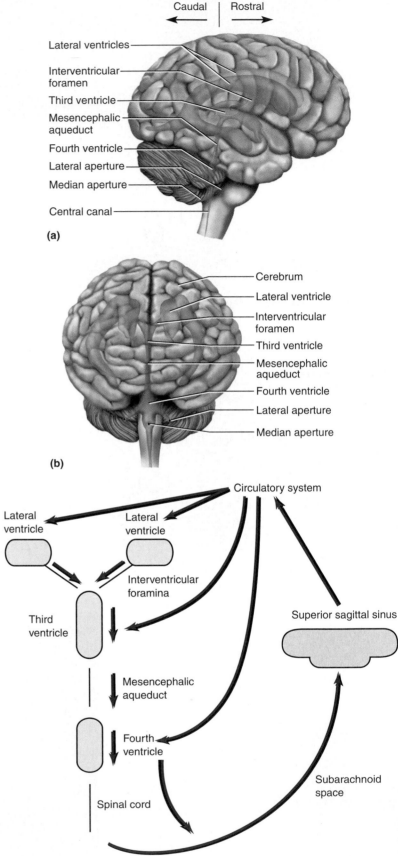

FIGURE 21.5 Ventricles of the Brain (a) Lateral view; (b) anterior view; (c) schematic presentation of the flow of cerebrospinal fluid.

the temporal lobe from the frontal and parietal lobes of the brain. You can locate these features in figure 21.6.

Frontal Lobe

The **frontal lobe** is responsible for many of the higher functions associated with being human. The frontal lobe is involved in intellect, abstract reasoning, creativity, social awareness, and language. An important area responsible for controlling the formation of speech is called the **Broca area,** or the **motor speech area.** This region is located on the left side of the brain in most people, whether they are right-handed or left-handed. This area controls the muscles involved in speech. It is usually located in the left frontal lobe. Locate the frontal lobe in figure 21.6. The posterior border of the frontal lobe is defined by the **central sulcus.**

To find the central sulcus, look for two convolutions that run from the superior portion of the cerebrum to the lateral fissure, more or less continuously. The gyrus anterior to the central sulcus is part of the frontal lobe and is known as the **precentral gyrus,** or the **primary motor cortex.** By constructing an image of the body on the brain, this

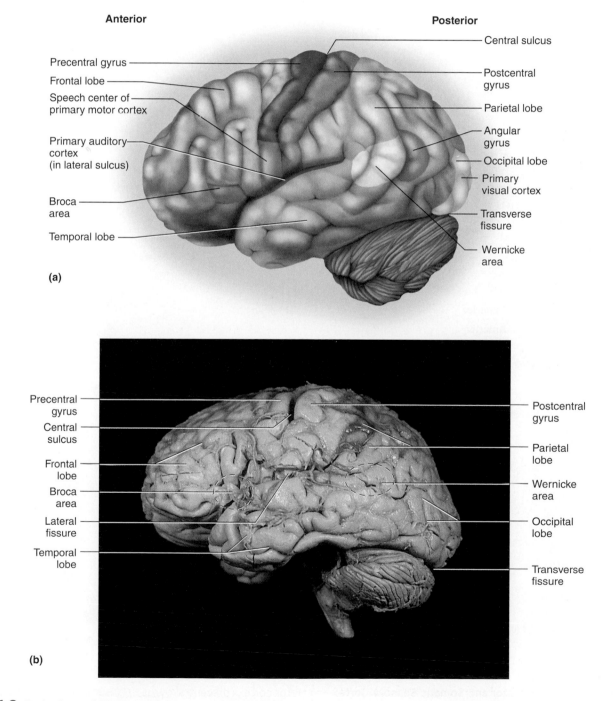

FIGURE 21.6 Brain, Lateral View (a) Diagram; (b) photograph.

figure produces the image of a person known as a homunculus. A **motor homunculus** is seen in figure 21.7a. How much of the precentral gyrus is dedicated to the face?

How much of the gyrus is dedicated to the hands?

How much is dedicated to the trunk?

Parietal Lobe

The gyrus posterior to the central sulcus is known as the **postcentral gyrus,** or the **primary somatic sensory cortex.** This is part of the parietal lobe and is involved in receiving somatic sensory information from the body. This area has also been mapped, and you can see a **sensory homunculus** represented in figure 21.7b. The primary sensory

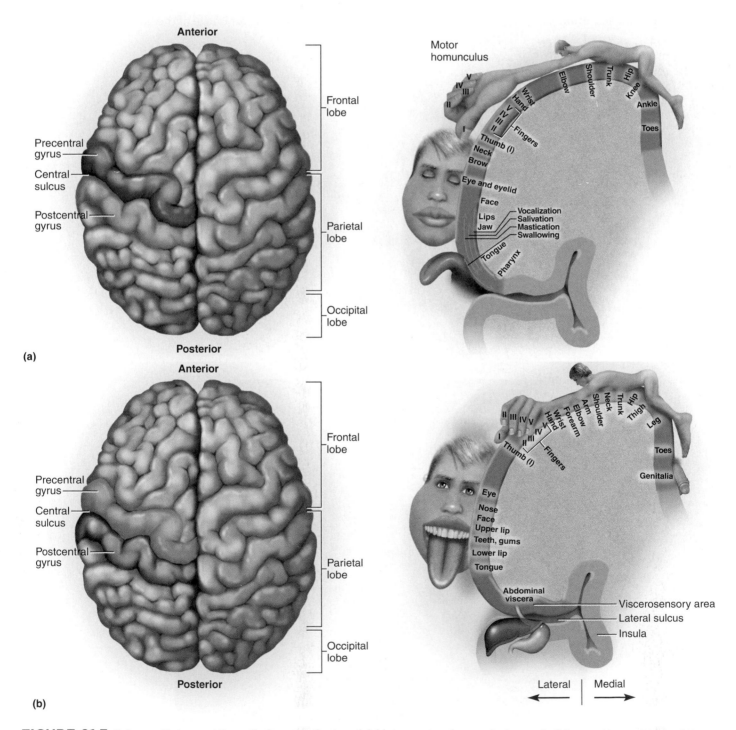

FIGURE 21.7 Primary Motor and Somatic Sensory Cortex (a) Motor cortex (precentral gyrus); (b) somatic sensory cortex (postcentral gyrus).

cortex receives information, yet the material is integrated just posterior to the sensory cortex in the **association areas.** The primary sensory cortex pinpoints the part of the body affected, and the association area interprets the type of sensation (pain, heat, cold, etc.). Locate the structures of the parietal lobe in figure 21.6.

The **Wernicke area** is involved in the formation of language, such as the recognition of written and spoken language and in the forming of coherent sentences. The Wernicke area extends into the temporal lobe as well.

Occipital Lobe

Posterior to the parietal lobe is the **occipital lobe** of the brain. The posterior occipital lobe is considered the primary visual area of the brain and damage to this lobe can lead to blindness. Shape, color, and distance of an object are perceived here. Recollection of past visual images occurs here as well. As you are reading these words your occipital lobe is receiving the information and transferring it to other regions, which convert the words to thought. Between the occipital lobe and the cerebellum is a **transverse fissure** that separates these two regions of the brain. Locate the occipital lobe and the cerebellum in figure 21.6.

Temporal Lobe

The **temporal lobe** is separated from the frontal and parietal lobes by the **lateral fissure** (sulcus). The temporal lobe contains an area known as the **primary auditory cortex,** which interprets hearing impulses sent from the inner ear. This auditory cortex distinguishes the nature of the sound (music, noise, speech) and the location, distance, pitch, and rhythm. The primary auditory cortex translates words into thought. The temporal lobe also has centers for the sense of smell **(olfactory centers)** and taste **(gustatory centers).** Deep to the temporal lobe is a small mass of cortical material called the insula.

Cerebral Hemispheres

Rotate the brain so that you are looking at it from a superior view, and find the **longitudinal fissure** that separates the cerebrum into the left and right cerebral hemispheres. The **left cerebral hemisphere** in most people is involved in language and reasoning. For example, the Broca area is typically on the left side of the brain. The **right cerebral hemisphere** of the brain is involved in space and pattern perceptions, artistic awareness, imagination, and music comprehension. This specialization where one hemisphere of the cerebrum is involved in a particular task is known as **cerebral asymmetry,** or **hemispheric dominance.** Examine the surface features of the brain in figure 21.8.

Inferior Aspect of the Brain
Forebrain

Examine the frontal and temporal lobes of the cerebrum. You may be able to see the **pituitary gland** if it has not been removed. The **optic chiasma** (KYE-as-muh) (*chiasma* = cross) is anterior to the pituitary and transmits visual impulses from the eyes to the brain. Two small

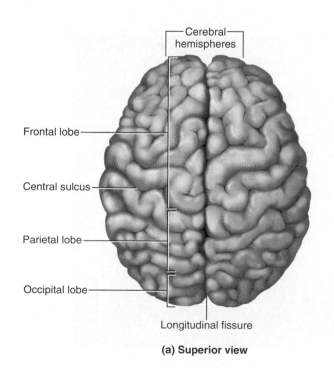

(a) Superior view

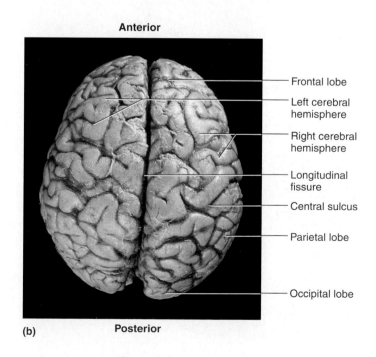

(b)

FIGURE 21.8 Brain, Superior View (a) Diagram; (b) photograph.

processes posterior to the pituitary are the **mammillary bodies,** which relay information from the limbic system to the fornix.

Hindbrain

From an inferior view of the brain locate the **pons, medulla oblongata,** and **cerebellum.** The cerebellum has much finer folds of neural tissue called **folia.** The medulla oblongata is located inferior to the cerebellum. The medulla oblongata has centers for respiratory rate control, control of blood pressure, and other vital centers. Some information from the right side of the body crosses over to the left brain in the medulla oblongata. Information coming from the left side of the body crosses over to the right side of the brain. The area where motor tracts cross over in the medulla is known as the **decussation of the pyramids.** The medulla oblongata terminates at the foramen magnum and the cervical region of the spinal cord continues inferior to the foramen magnum. The enlarged portion of the brain anterior to the medulla is the pons, which is a relay center for information. Examine these structures of the brain in figure 21.9.

Midsagittal Section of the Brain

Forebrain

Examine a midsagittal section of a brain, as illustrated in figure 21.10. This section is seen by cutting the brain through the longitudinal fissure. Locate the C-shaped **corpus callosum,** which connects the two cerebral hemispheres. The posterior portion of the corpus callosum is known as the **splenium,** and the anterior portion is the **genu.** Just inferior to the corpus callosum is the **septum pellucidum,** which separates the lateral ventricles of the brain. If the septum pellucidum is missing, you will be able to look into the lateral ventricle without obstruction. Locate these structures in figure 21.10.

Examine the material in the lab and locate the **diencephalon,** which consists of, in part, the thalamus and the hypothalamus. The **thalamus** forms the lateral wall around the third ventricle and is a relay center that receives almost all the sensory information from the body and projects it to the cerebral cortex.

Below the thalamus is the **hypothalamus,** which has numerous autonomic centers. The hypothalamus, in part, directs the ANS and is involved with the **pituitary gland** (in the hypothalamopituitary axis) in many endocrine functions. Centers for thirst, water balance, pleasure, rage, sexual desire, hunger, sleep patterns, temperature, and aggression are located in the hypothalamus.

You should also be able to locate the **mammillary bodies** on the inferior portion of the diencephalon and the **optic chiasma** just anterior to it. The **pineal gland** is located posterior to the thalamus and is an endocrine gland that secretes melatonin, a hormone that regulates daily rhythms. Both the pineal gland and the pituitary gland are covered more in depth in Laboratory Exercise 28.

Midbrain

The midbrain is a small area posterior to the diencephalon. This small area consists of the **cerebral peduncles,** which occupy an area anterior to the pons and on the

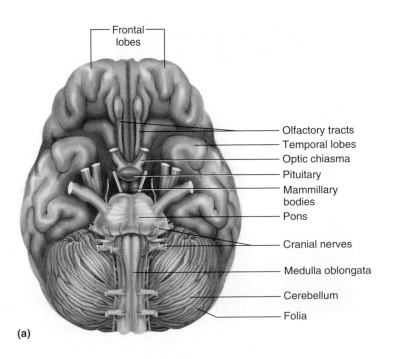

Frontal lobes

Olfactory tracts
Temporal lobes
Optic chiasma
Pituitary
Mammillary bodies
Pons
Cranial nerves
Medulla oblongata
Cerebellum
Folia

(a)

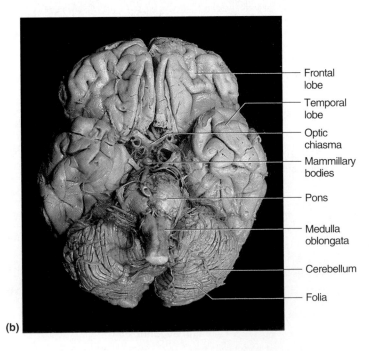

Frontal lobe
Temporal lobe
Optic chiasma
Mammillary bodies
Pons
Medulla oblongata
Cerebellum
Folia

(b)

FIGURE 21.9 Brain, Inferior View (a) Diagram; (b) photograph.

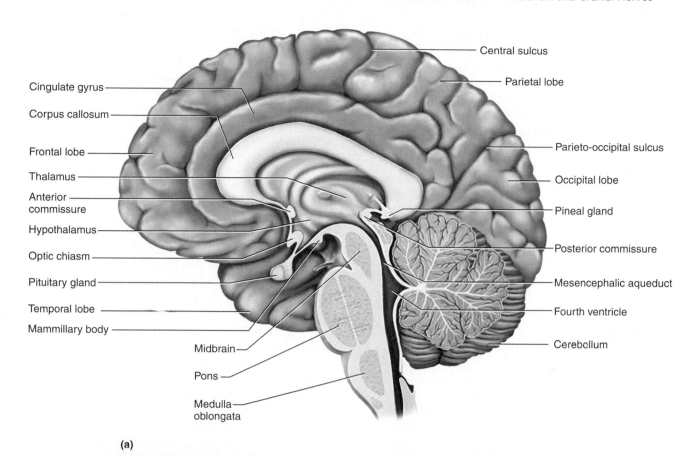

Central sulcus
Parietal lobe
Cingulate gyrus
Corpus callosum
Frontal lobe
Parieto-occipital sulcus
Thalamus
Occipital lobe
Anterior commissure
Pineal gland
Hypothalamus
Posterior commissure
Optic chiasm
Mesencephalic aqueduct
Pituitary gland
Fourth ventricle
Temporal lobe
Cerebellum
Mammillary body
Midbrain
Pons
Medulla oblongata

(a)

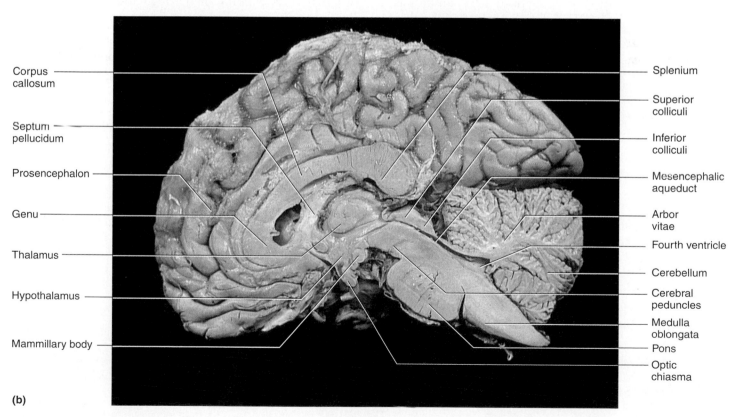

Corpus callosum
Splenium
Superior colliculi
Septum pellucidum
Inferior colliculi
Prosencephalon
Mesencephalic aqueduct
Genu
Arbor vitae
Fourth ventricle
Thalamus
Cerebellum
Cerebral peduncles
Hypothalamus
Medulla oblongata
Pons
Mammillary body
Optic chiasma

(b)

FIGURE 21.10 Brain, Midsagittal Section (a) Diagram; (b) photograph.

ventral surface of the brain. The **mesencephalic (cerebral) aqueduct** passes through the mesencephalon, with the **peduncles** being inferior and the **tectum** as a roof superior to the aqueduct. Locate these features in the material in the lab and in figure 21.10. Superior to the tectum are four hemispheric processes known as the **corpora quadrigemina,** which consist of the **superior colliculi** (areas of visual reflexes) and the **inferior colliculi** (areas of auditory reflexes). The midbrain also houses a center known as the **substantia nigra** (not seen in midsagittal sections) that, when not functioning properly, causes Parkinson's disease.

Hindbrain

The **hindbrain** consists of an anterior bulge known as the **pons,** a terminal **medulla oblongata,** and the highly convoluted **cerebellum** (see figure 21.10). The pons carries sensory information from the inferior regions of the body to the thalamus. The pons has important respiratory centers involved in controlling breathing rate.

The cerebellum is a location noted for muscle coordination, the maintenance of posture, the conceptualization of the passage of time, and other cognitive functions. The cerebellum consists of an outer **cerebellar cortex** and an inner, extensively branched pattern of white matter known as the **arbor vitae.** The **folia** of the cerebellum can be seen in this section. The triangular space anterior to the cerebellum is the fourth ventricle and can be seen in figure 21.10. Examine the features of the hindbrain as described here and seen in material in the lab.

Coronal Section of the Brain

The brain consists of **unmyelinated** gray matter and **myelinated** white matter. The **gray matter** of the brain is extensive and forms the **cerebral cortex.** Most of the active, integrative processes of the brain occur in the cerebral cortex (see figure 21.11). The cerebral cortex is approximately 4 mm thick and occupies the superficial regions of the brain. The cortex is where humans do most of their "thinking." It is the main metabolic area of the brain. Young adults have perhaps 50 billion neurons in the cerebral cortex. Deep to the cortex is the **white matter** of the brain, which consists of **tracts** that take information from deeper regions of the brain to the cerebral cortex for processing. Sensory information coming from the spinal cord moves through the inferior regions of the brain and through the white matter for integration in the cerebral cortex. White matter can also take information from one region of the cerebral cortex to another for integration or from the cerebral cortex back to the spinal cord and to other parts of the body for action.

Gray matter is not restricted to the cerebral cortex, however. Deep islands of gray matter in the brain compose the **basal nuclei (basal ganglia),** such as the **caudate nucleus, putamen,** and **globus pallidus.** Basal nuclei serve a number of functions in the brain, many of which involve subconscious processes, such as the swinging of arms while walking or the regulation of muscle tone. Basal nuclei are found not only in the cerebrum but in the thalamus and midbrain as well.

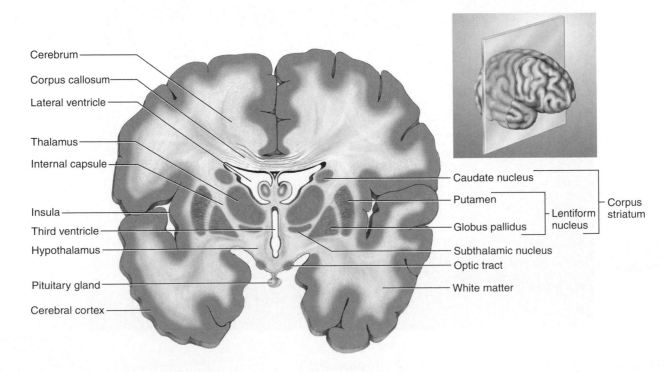

Cerebrum
Corpus callosum
Lateral ventricle
Thalamus
Internal capsule
Insula
Third ventricle
Hypothalamus
Pituitary gland
Cerebral cortex

Caudate nucleus
Putamen
Globus pallidus
Lentiform nucleus
Corpus striatum
Subthalamic nucleus
Optic tract
White matter

FIGURE 21.11 Brain, Coronal Section

Limbic System

Part of the limbic system can be seen in a coronal section. The limbic system is a complex region of the brain involved in mood and emotion and has centers for feeding, sexual desire, fear, and satisfaction. The inferior portion of the limbic system has neural fibers that come from the olfactory regions of the brain. These are best seen in a model of the limbic system or in a transected brain. Examine figure 21.12 for the major features of the limbic system.

Brainstem

The brainstem consists of the brain, excluding the cerebrum and cerebellum. Look at a model or section of brain that has had the cerebrum removed. Locate the corpora quadrigemina (see figure 21.13) along with the medulla oblongata and the pons.

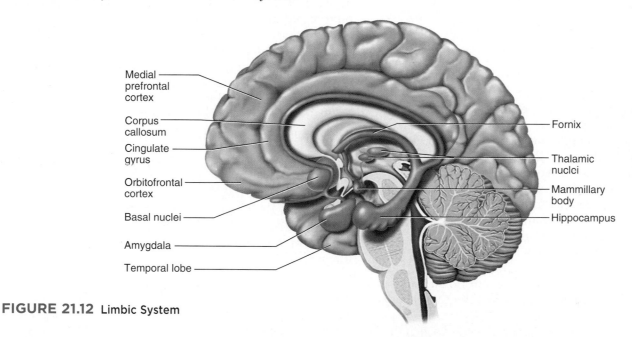

Medial prefrontal cortex
Corpus callosum
Cingulate gyrus
Orbitofrontal cortex
Basal nuclei
Amygdala
Temporal lobe
Fornix
Thalamic nuclei
Mammillary body
Hippocampus

FIGURE 21.12 Limbic System

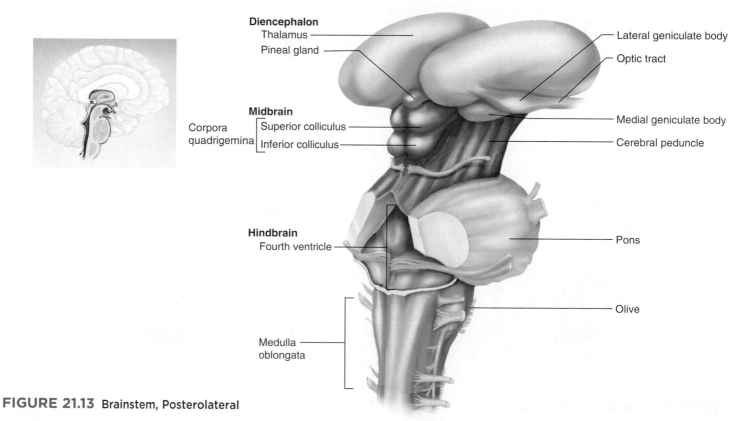

Diencephalon
Thalamus
Pineal gland
Lateral geniculate body
Optic tract

Midbrain
Corpora quadrigemina
Superior colliculus
Inferior colliculus
Medial geniculate body
Cerebral peduncle

Hindbrain
Fourth ventricle
Pons

Olive

Medulla oblongata

FIGURE 21.13 Brainstem, Posterolateral

Development of the Central Nervous System

The brain begins development, as does the rest of the nervous system, in the third week of pregnancy as a **neural groove** in the ectoderm. By the fourth week the brain has folded into a **neural tube** that contains the **central canal** (see figure 21.14). The posterior portion of the central canal becomes the central canal of the spinal cord. The anterior portion of the canal becomes the ventricles of the brain in adults. By the sixth week of development the cerebral hemispheres have begun to form and continue their development throughout pregnancy.

Cranial Nerves

The cranial nerves are part of the PNS but they are covered with the brain in lab as they are closely associated with it. There are 12 pairs of cranial nerves. The cranial nerves are listed by Roman numeral, and you should know the nerve by name and by number. Examine a model of the brain along with figure 21.15; the cranial nerves, except for nerve XII, are in sequence from anterior to posterior. When you study nerves you should examine the anatomy of the brain to see where the nerve emerges from the brain. If you know, for example, that the abducens is found between the pons and the medulla oblongata, then you can use that nerve as a way to locate other nerves. Nerves may be sensory, motor, or mixed (both sensory and motor).

The **olfactory nerve** runs along the anterior base of the brain at the inferior aspect of the frontal lobe. The **optic nerve** (*optic* refers to sight) is a sensory nerve from the eye to the base of the brain and forms the optic chiasma. Some tracts remain on one side of the brain, while others cross to the other side. The **oculomotor nerve** is anterior to the pons, more or less in the midline of the brain, while the **trochlear nerve** emerges at about a 45-degree angle from midline on the lateral aspect of the pons. The large **trigeminal nerve** is found at a 90-degree angle to the pons and is located on the lateral aspect of the pons. The **abducens nerve** is located at the midline junction of the pons and medulla oblongata, while the **facial nerve** is more lateral. Posterior to the facial nerve is the **vestibulocochlear nerve,** and the nerve directly behind that is the **glossopharyngeal nerve.** The **vagus nerve** (*vagus* = wandering) is a large nerve or large cluster of fibers, and on the lateral aspect of the medulla oblongata is the **accessory nerve.** More toward the midline is the **hypoglossal nerve,** the last of the cranial nerves. Locate these nerves in figure 21.15 and note their details in tables 21.2 and 21.3.

The cranial nerves are listed in table 21.4 as sensory nerves, motor nerves, or both sensory and motor nerves. There have been many mnemonic devices constructed to remember the sequence of cranial nerves. If you use the first letter of the name of the cranial nerve you can learn the sequence of these nerves. One such mnemonic is *Old Oliver Ogg Traveled Through Arches For Very Good Vacations And Holidays.* The first letters of the words in the mnemonic represent the first letters of the names of the nerves.

Dissection of the Sheep Brain

Take a sheep brain back to your table along with a dissecting tray and appropriate dissection tools. If the brains still have the **dura mater,** examine this tough connective tissue coat that occurs on the outside of the brain. Cut through this layer to examine the other meninges that occur underneath it. Deep to the dura mater is a filmy layer of tissue

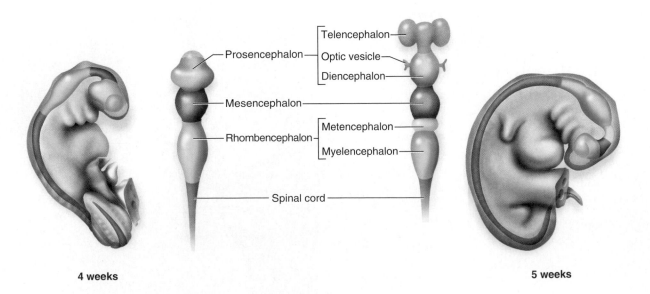

4 weeks   **5 weeks**

Prosencephalon — Telencephalon
Optic vesicle
Diencephalon
Mesencephalon
Metencephalon
Rhombencephalon — Myelencephalon
Spinal cord

FIGURE 21.14 Brain Development

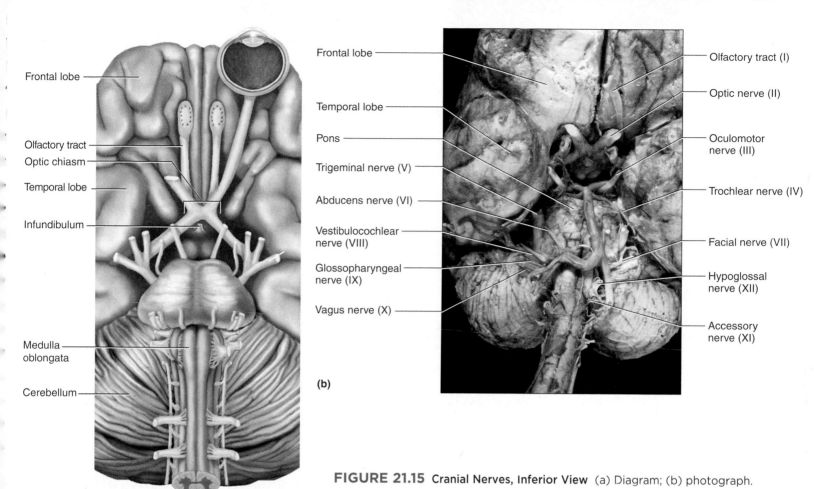

Frontal lobe

Olfactory tract

Optic chiasm

Temporal lobe

Infundibulum

Medulla oblongata

Cerebellum

(a)

Frontal lobe

Temporal lobe

Pons

Trigeminal nerve (V)

Abducens nerve (VI)

Vestibulocochlear nerve (VIII)

Glossopharyngeal nerve (IX)

Vagus nerve (X)

(b)

Olfactory tract (I)

Optic nerve (II)

Oculomotor nerve (III)

Trochlear nerve (IV)

Facial nerve (VII)

Hypoglossal nerve (XII)

Accessory nerve (XI)

FIGURE 21.15 Cranial Nerves, Inferior View (a) Diagram; (b) photograph.

TABLE 21.2	Cranial Nerves—Function	
Number	**Name**	**Function**
I	Olfactory	Receives sensory information from the nose
II	Optic	Receives sensory information from the eye, transmitting the sense of vision to the brain
III	Oculomotor	Transmits motor information to move the eye muscles particularly to the medial, superior, and inferior rectus muscles and to the inferior oblique muscle
IV	Trochlear	Transmits motor information to move the eye muscles, particularly the superior oblique muscle
V	Trigeminal	A three-branched nerve; transmits both sensory information from, and motor information to, the head
VI	Abducens	A motor nerve to move the eye muscles, particularly the lateral rectus muscle
VII	Facial	A large nerve that receives sensory information from the anterior tongue and takes motor information to the head muscles
VIII	Vestibulocochlear	Receives sensory information from the ear; the vestibular part transmits equilibrium information and the cochlear part transmits acoustic information
IX	Glossopharyngeal	A mixed nerve of the tongue and throat that receives information on taste
X	Vagus	Receives sensory information from the abdomen, thorax, neck, and root of the tongue; transmits motor information to the pharynx and larynx and controls autonomic functions of the heart, digestive organs, spleen, and kidneys
XI	Accessory	A motor nerve to the muscles of the neck that move the head
XII	Hypoglossal	A motor nerve to the tongue

TABLE 21.3 — Cranial Nerves—Location

Number	Name	Location
I	Olfactory	Begin in the upper nasal cavity and pass through the cribriform plate of the ethmoid bone. They synapse in the olfactory bulbs on each side of the longitudinal fissure of the brain. The fibers take information on the sense of smell and pass through the olfactory tracts to be interpreted in the temporal lobe of the brain.
II	Optic	Take sensory information from the retina at the back of the eye and transmit the impulses through the optic canal in the sphenoid bone. Some fibers cross at the optic chiasm and pass via the optic tracts to the occipital lobe, where vision is interpreted.
III	Oculomotor	Emerge from the surface of the brain near the midline and just anterior to the pons. They pass through the superior orbital fissure and innervate the inferior oblique muscle and the medial, superior, and inferior rectus muscles and carry parasympathetic fibers to the lens and iris.
IV	Trochlear	Seen at the sides of the pons at about a 45-degree angle from the midline of the brain. They pass through the superior orbital fissure to the superior oblique muscle.
V	Trigeminal	Seen at a 90-degree angle from the midline at the lateral sides of the pons. The trigeminal has three branches: (1) the ophthalmic branch passes through the superior orbital fissure; (2) the maxillary branch passes through the foramen rotundum of the sphenoid bone; (3) the mandibular branch passes through the foramen ovale of the sphenoid bone and enters the mandible by the mandible foramen and exits by the mental foramen.
VI	Abducens	Begins at the midline junction between the pons and the medulla oblongata and passes through the superior orbital fissure to carry motor information to the lateral rectus muscle of the eye.
VII	Facial	Begins as the first of a cluster of nerves on the anterolateral part of the medulla oblongata. It passes through the internal auditory meatus and through the inner ear to the stylomastoid foramen of the temporal bone to innervate facial muscles and glands. It receives sensory information from the anterior tongue. Sensory information of the tongue is interpreted in the temporal lobe of the brain.
VIII	Vestibulocochlear	Comes from the inner ear and passes through the internal auditory meatus. The conduction passes to the pons, and hearing and balance are interpreted in the temporal lobe.
IX	Glossopharyngeal	Passes through the jugular foramen to innervate muscles of the throat (pharyngeal branches) and the tongue (glossal branches). Motor portions of the nerve control some muscles of swallowing and salivary glands, while sensory nerves receive information from the posterior tongue and from baroreceptors of the carotid artery.
X	Vagus	Passes through the jugular foramen and along the neck to the larynx, heart, and abdominal region. The sensory impulses travel in this nerve from the viscera in the abdomen, the thorax, the neck, and the root of the tongue to the brain.
XI	Accessory	Multiple fibers arise from the lateral sides of the medulla oblongata and pass through the jugular foramen to numerous muscles of the neck.
XII	Hypoglossal	Begins at the anterior surface of the medulla and passes through the hypoglossal canal to innervate the muscles of the tongue

TABLE 21.4 — Type of Cranial Nerves

Number	Name	Name Mnemonic	Type	Mnemonic
I	Olfactory	Old	Sensory	Sally
II	Optic	Oliver	Sensory	Sells
III	Oculomotor	Ogg	Motor*	Many
IV	Trochlear	Traveled	Motor	Mangoes
V	Trigeminal	Through	Both	But
VI	Abducens	Arches	Motor	My
VII	Facial	For	Both	Brother
VIII	Vestibulocochlear	Very	Sensory	Sells
IX	Glossopharyngeal	Good	Both	Bigger
X	Vagus	Vacations	Both	Better
XI	Accessory	And	Motor	Mega
XII	Hypoglossal	Holidays	Motor	Mangoes

*Many of the motor nerves have sensory fibers that come from proprioreceptors in the muscles they innervate. Information about the tension of the muscle is sent back to the brain to make adjustments in contractile rate. Since the main function of these nerves is motor, they are listed as motor nerves, even though they have some sensory capabilities.

that contains blood vessels. This is known as the **arachnoid mater.** If you tease some of the membrane away from the brain you will see that it has a cobweblike appearance in the **subarachnoid space.** The subarachnoid space may contain some fluid, the CSF. Deep to this layer and adhering directly to the brain convolutions is the **pia mater,** the surface lining of the convolutions of the brain. Work in pairs during the dissection of the sheep brain. Examine the features presented in the beginning of this exercise and locate the structures that appear in figure 21.16.

Find the major lobes of the brain, the cerebellum, pons, and medulla oblongata. Sheep are quadrupeds, and the flexure of the brain does not occur in them as it does in humans. Sheep have a horizontal spinal cord, while humans have a vertical one. The cerebrum in sheep is smaller than that in humans. Examine the inferior surface of the sheep brain. You should see the olfactory bulbs and tracts, the optic nerve, and the optic chiasma easily from this view. The pituitary gland will probably not be

attached but you should locate the infundibulum, caudad to the optic chiasma. Locate these structures in figure 21.17. If the sheep brain is intact, you will need to decide which brain will be sectioned in the midsagittal plane and which will be sectioned in the coronal plane. For the midsagittal section, divide the brain along the length of the longitudinal fissure. Your cut should reflect a section illustrated in figure 21.18. Compared to humans, sheep have a larger corpora quadrigemina. The pons is larger in humans and the olfactory bulbs are larger in sheep. Locate the corpus callosum, lateral ventricles, third ventricle, hypothalamus, pineal gland, superior and inferior colliculi, cerebellum, arbor vitae, pons, medulla oblongata, mesencephalic aqueduct, and fourth ventricle.

The coronal section of a sheep brain is illustrated in figure 21.19. After making a coronal section about midway through the cerebrum, you should locate the cerebral cortex, cerebral medulla, lateral ventricles, corpus callosum, third ventricle, thalamus, and hypothalamus.

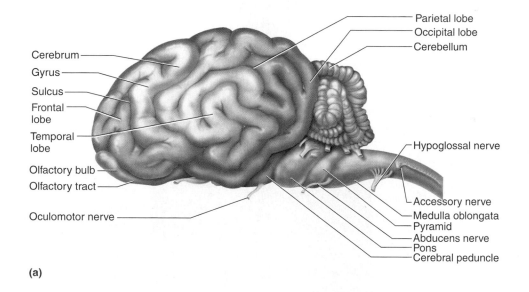

(a)

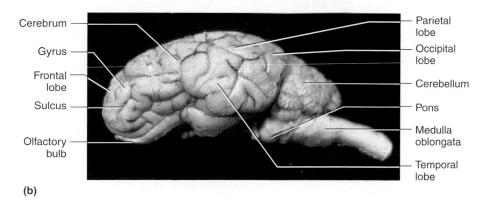

(b)

FIGURE 21.16 Brain of the Sheep, Lateral View (a) Diagram; (b) photograph.

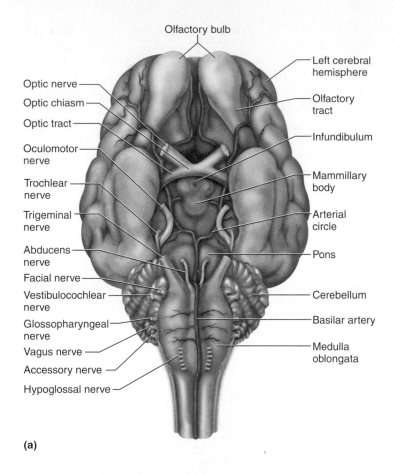

(a)

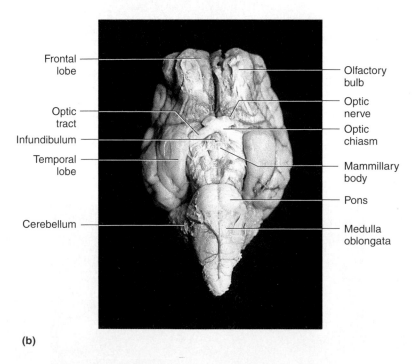

(b)

FIGURE 21.17 **Brain of the Sheep, Inferior View** (a) Diagram; (b) photograph.

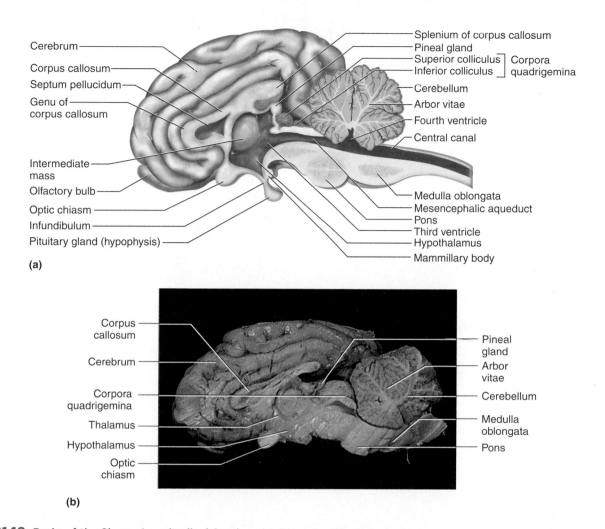

FIGURE 21.18 Brain of the Sheep, Longitudinal Section (a) Diagram; (b) photograph.

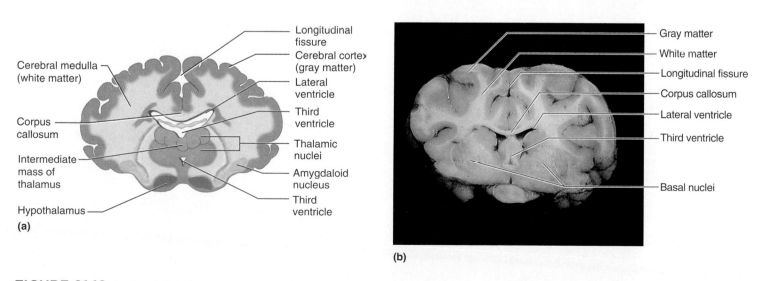

FIGURE 21.19 Brain of the Sheep, Coronal Section (a) Diagram; (b) photograph.

Notes

REVIEW SECTION

The Brain and Cranial Nerves

Name _____ *Date* _____

Lab Section _____ *Time* _____

Review Questions

1. Which of the meninges is between the outer and inner meninges?

2. Name the major veins that take blood from the brain.

3. The basilic artery in the brain receives blood from what two arteries?

4. What fluid is found in the ventricles of the brain?

5. Where does fluid flow from the mesencephalic aqueduct?

6. What is the difference between a gyrus and a sulcus?

7. What are the lobes of the cerebrum called?

8. What function does the precentral gyrus have?

9. What sense does the temporal lobe interpret?

10. What depression separates the temporal lobe from the parietal lobe?

11. What structure connects the cerebral hemispheres?

12. Name the major regions of the midbrain.

13. What function does the cerebellum have?

14. What function does the optic nerve have?

15. What function does the trochlear nerve have?

16. John pulled a "no-brainer" by hitting his forehead against the wall. What possible damage might he do to the function of his brain, particularly those functions associated with the frontal lobe?

17. If a stroke affected all the sensations interpreted by the brain just concerning the face and the hands, how much of the postcentral gyrus would be affected?

18. One convenient excuse that people often make for their inability to do something is to describe themselves as left-brained or right-brained individuals. Describe what effect the loss of an entire cerebral hemisphere would have on specific functions, such as spatial awareness or the ability to speak.

19. Aphasia is loss of speech. Different types of aphasia can occur. If the Broca area were affected by a stroke, would the content of the spoken word be affected, or would the ability to pronounce words be affected?

20. Label the following illustration using the terms provided.

corpus callosum   arbor vitae
mesencephalic aqueduct   hypothalamus
pons   medulla oblongata
pineal gland   pituitary gland
thalamus   fourth ventricle

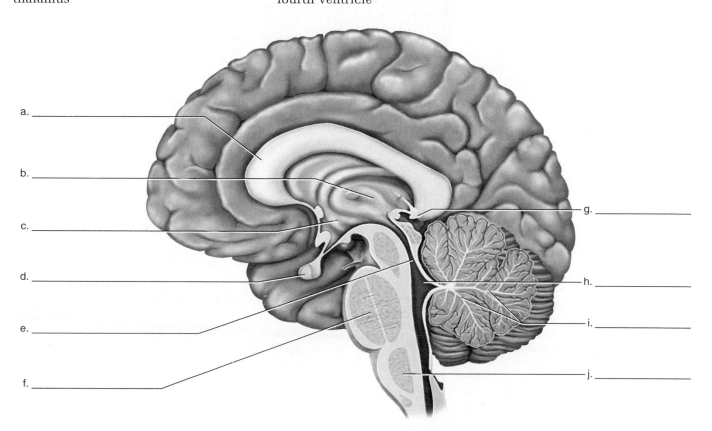

a. _____

b. _____

c. _____

d. _____

e. _____

f. _____

g. _____

h. _____

i. _____

j. _____

21. Label the following illustration using the terms provided.

olfactory nerves
trigeminal nerve
vagus nerve
medulla oblongata
hypoglossal nerves
pons
oculomotor nerve
cerebellum
optic nerve

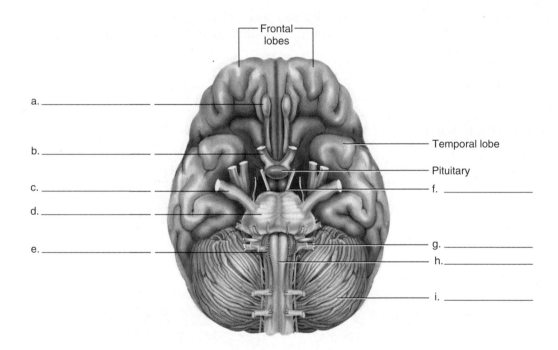

Frontal
lobes

a. _____ _____

b. _____ _____

Temporal lobe

Pituitary

c. _____ _____

f. _____

d. _____ _____

e. _____ _____

g. _____

h. _____

i. _____

LABORATORY

The Spinal Cord and Nerves

INTRODUCTION

The spinal cord is part of the central nervous system, which extends from the foramen magnum of the skull to about vertebra L1 or L2. The spinal cord is shorter than the vertebral column because the spinal cord stops growing early in life, yet the vertebral bodies continue to grow. The spinal cord receives sensory information from and transmits motor information to the spinal nerves, which radiate into the body as peripheral nerves. The peripheral nerves are part of the peripheral nervous system. We cover the spinal cord and peripheral nerves together because of their anatomical proximity and their functional relationship. Peripheral nerves **innervate** (functionally connect to) muscles and other parts of the body. In this exercise you learn the major features of the spinal cord in longitudinal aspect and in cross section and the major nerves and plexuses of the body. The spinal cord and somatic nerves are covered in the Saladin text in chapter 13, "The Spinal Cord, Spinal Nerves, and Somatic Reflexes."

OBJECTIVES

At the end of this exercise you should be able to

1. demonstrate the major regions in a cross section of the spinal cord;
2. describe the nature of the longitudinal aspect of the spinal cord;
3. list the major nerves that arise from each plexus;
4. name all the major nerves of the upper and lower extremities;
5. list the structures that carry impulses to and away from the spinal cord.

MATERIALS

Models or charts of the central and peripheral nervous systems

Prepared slide of a spinal cord in cross section

Model or chart of a spinal cord in cross section and longitudinal section

Cat
Materials for cat dissection

PROCEDURE

Longitudinal Aspect of the Spinal Cord

Examine a model or chart in the lab of a longitudinal view of the spinal cord and locate the major features illustrated in figure 22.1. The spinal cord terminates inferiorly as

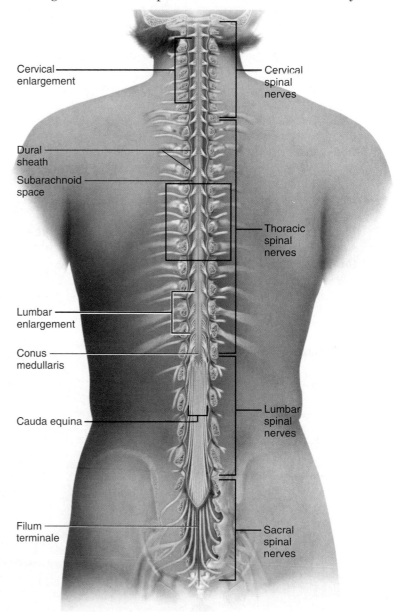

Cervical enlargement

Dural sheath

Subarachnoid space

Lumbar enlargement

Conus medullaris

Cauda equina

Filum terminale

Cervical spinal nerves

Thoracic spinal nerves

Lumbar spinal nerves

Sacral spinal nerves

FIGURE 22.1 Spinal Cord, Longitudinal View

the **conus medullaris** at approximately vertebra L1 and is attached to the coccyx by a continuation of the pia mater known as the **filum terminale.** The neural continuation of the spinal cord exists as an extension of spinal nerves from L2 through S5. These parallel nerves resemble a horse's tail and are called the **cauda equina.** The spinal cord has a couple of expansions in two locations. The **cervical enlargement** occurs at C3 through T2 and represents a bulge in the spinal cord that has increased innervation of the upper extremities. The **lumbar enlargement** occurs at about T7 through T11, and this expanse is due to the increased innervation of the lower extremity. Compare the material in the lab to figure 22.1.

Cross Section of Spinal Cord

Examine a model or chart in the lab of a cross section of the spinal cord. Locate the **gray matter,** which appears as an H pattern or a butterfly pattern in the middle of the spinal cord, with the **white matter** located on the periphery of the cord.

In a cross section of the spinal cord you should see the gray matter divided in two narrow horns and two rounded horns. The narrow horns are known as the **posterior gray horns** (dorsal gray horns), and these areas receive sensory information from the somatic nerves. The rounded horns are the **anterior gray horns** (ventral gray horns), and these send motor signals to the spinal nerves. In some parts of the spinal cord there are additional sections of gray matter known as **lateral gray horns.**

Each side of the gray matter is connected to the other by a crossbar known as the **gray commissure.** In the middle of the gray commissure is the **central canal,** which runs

the length of the spinal cord. The white matter of the cord is divided into **tracts,** or funiculi, which carry sensory information to the brain, called **ascending tracts,** or motor information from the brain, called **descending tracts.** You should also be able to see a depression in the posterior surface of the spinal cord. This is the **posterior median sulcus,** while the deeper depression on the anterior side is known as the **anterior median fissure.** Examine a model or chart in the lab and compare it to figure 22.2.

Histology of the Spinal Cord

Examine a prepared slide of the spinal cord under low power and locate the **anterior** and **posterior gray horns,** the **gray commissure,** the **central canal,** the **posterior median sulcus,** and the **anterior median fissure.** Examine the anterior gray horn of the spinal cord and look for the nerve cell bodies of the **multipolar neurons** there. These are motor neurons. Compare your slide to figure 22.3.

Meninges

The spinal cord is covered by **meninges,** as is the brain. The outer one is the **dura mater** (dural sheath). Between the dura mater and the vertebra is the **epidural space,** a region where anaesthetics are given during some deliveries to relieve pain during childbirth. The next deeper membrane is the **arachnoid mater (membrane).** Deep to the arachnoid membrane is the **pia mater,** the innermost of the meninges on the spinal cord proper. Between the pia mater and the arachnoid is the **subarachnoid space,** which contains the **cerebrospinal fluid.** Examine figures 22.2 and 22.4 for an illustration of the meninges of the spinal cord.

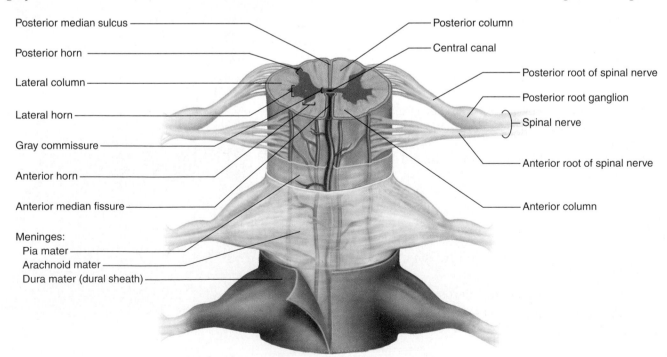

Posterior median sulcus
Posterior horn
Lateral column
Lateral horn
Gray commissure
Anterior horn
Anterior median fissure
Meninges:
  Pia mater
  Arachnoid mater
  Dura mater (dural sheath)

Posterior column
Central canal
Posterior root of spinal nerve
Posterior root ganglion
Spinal nerve
Anterior root of spinal nerve
Anterior column

FIGURE 22.2 Spinal Cord, Cross Section

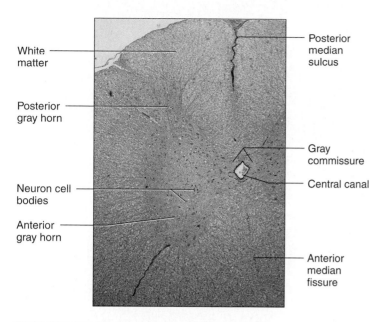

White matter

Posterior gray horn

Neuron cell bodies

Anterior gray horn

Posterior median sulcus

Gray commissure

Central canal

Anterior median fissure

FIGURE 22.3 Histology of the Spinal Cord (10×)

Nerves Associated with the Spinal Cord

The nerves that attach to the spinal cord are part of the peripheral nervous system. They are covered here because of their interplay with the spinal cord. The **posterior ramus** is a branch of sensory and motor nerve fibers that, with the **anterior ramus,** unite to form a **spinal nerve.** The spinal nerve enters the **intervertebral foramen** and divides into a **posterior root** and an **anterior root.** The posterior

root has a posterior root ganglion that contains the nerve cell bodies of the sensory nerves. The **posterior root** carries sensory information to the posterior gray horn of the spinal cord. The nerve cell bodies of the motor nerves are located in the anterior gray horn of the spinal cord and exit via the anterior root. Examine the material in the lab and compare it to figure 22.5.

Nerve Structure

Nerve fibers are clustered in parallel arrangements called **nerves** if they are in the peripheral nervous system and **tracts** if they are in the central nervous system. Nerves have a number of connective tissue wrappings that envelop the individual nerve fibers, clusters of fibers, and entire nerve. The sheath that wraps around single nerve fibers and their myelin sheaths is the **endoneurium,** and the sheath that wraps around groups of nerve fibers (nerve fasicle) is the **perineurium.** The wrapping that covers the entire nerve is called the **epineurium.** Examine figure 22.6 for these layers.

Spinal Nerves and Plexuses

There are 31 pairs of **spinal nerves** that exit from the spinal cord. The spinal nerves are **mixed nerves** carrying both sensory and motor information. The spinal nerves pass through the intervertebral foramina and are named according to their region of origin. There are **8 pairs of cervical nerves, 12 pairs of thoracic nerves, 5 pairs of lumbar nerves, 5 pairs of sacral nerves,** and **1 pair of coccygeal nerves.** Some of these nerves exit the spinal cord, and branches

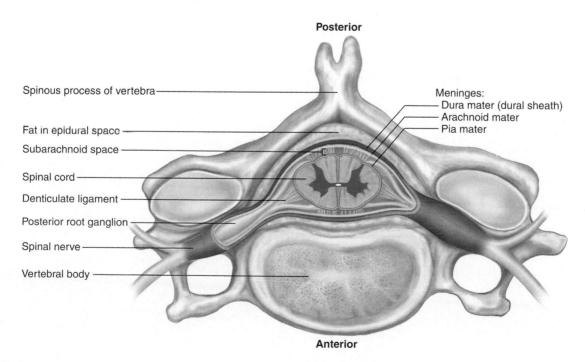

Posterior

Spinous process of vertebra

Fat in epidural space

Subarachnoid space

Spinal cord

Denticulate ligament

Posterior root ganglion

Spinal nerve

Vertebral body

Meninges:
Dura mater (dural sheath)
Arachnoid mater
Pia mater

Anterior

FIGURE 22.4 Spinal Meninges

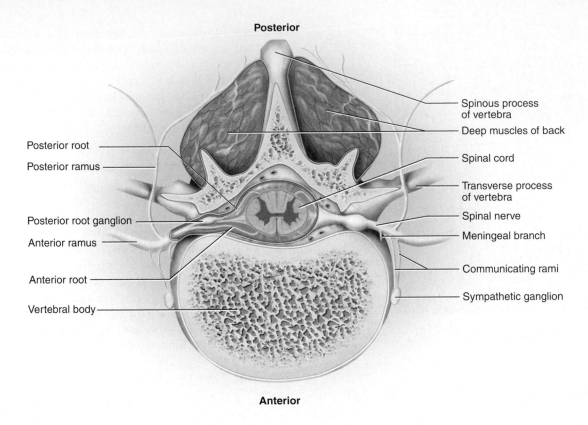

Posterior

Posterior root

Posterior ramus

Posterior root ganglion

Anterior ramus

Anterior root

Vertebral body

Spinous process
of vertebra

Deep muscles of back

Spinal cord

Transverse process
of vertebra

Spinal nerve

Meningeal branch

Communicating rami

Sympathetic ganglion

Anterior

FIGURE 22.5 Nerves Associated with the Spinal Cord

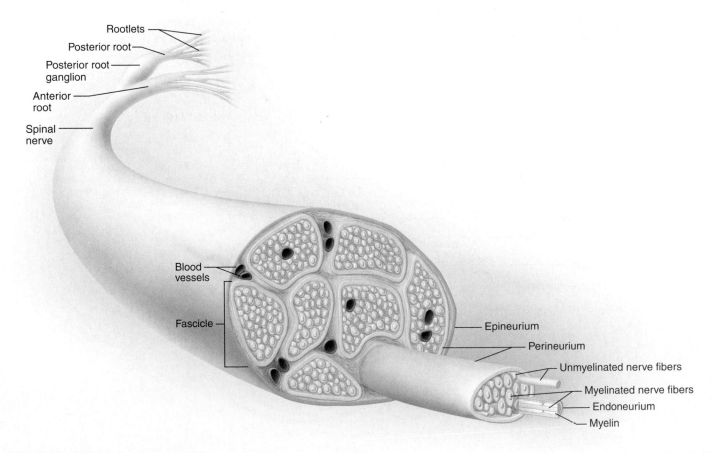

Rootlets

Posterior root

Posterior root
ganglion

Anterior
root

Spinal
nerve

Blood
vessels

Fascicle

Epineurium

Perineurium

Unmyelinated nerve fibers

Myelinated nerve fibers

Endoneurium

Myelin

FIGURE 22.6 Coverings of the Nerve

of spinal nerves travel to parts of the body individually, while others exit the spinal cord and form a branching network with other spinal nerves. These networks are called **plexuses,** and there are generally four recognized plexuses. Some spinal nerves, such as C5 and L4, contribute to more than one plexus. The composition of the plexuses are outlined in table 22.1 and illustrated in figure 22.7.

TABLE 22.1	Plexuses	
Name	**Spinal Nerves Contributing to Plexus**	**Major Nerves of Plexus**
Cervical	C1-5	Phrenic
Brachial	C5-T1	Radial, median, ulnar, musculocutaneous, axillary
Lumbar	L1-4	Femoral, obturator
Sacral	L4-S4	Sciatic (tibial and common fibular)

Communicating rami are branches in parts of the spinal cord that innervate sympathetic ganglia of the autonomic nervous system.

Cervical and Brachial Plexus Nerves

The cervical plexus has many nerves that innervate the head and neck. An important nerve that exits from each side of the cervical plexus is the **phrenic nerve.** This nerve runs to the diaphragm and is responsible for its contraction in breathing. Examine the nerves of the **cervical plexus** and **brachial plexus** in figures 22.8 and 22.9.

The nerves from the brachial plexus are those that primarily innervate the upper extremities. The innervation of the muscles of the upper extremity is covered in the exercises on the specific muscles. The major nerves are divided into anterior and posterior divisions. In the anterior division is the **musculocutaneous nerve,** innervating many of the muscles of the arm; the **median nerve,** running the length of each upper extremity and serving important hand and forearm flexors; and the **ulnar nerve,**

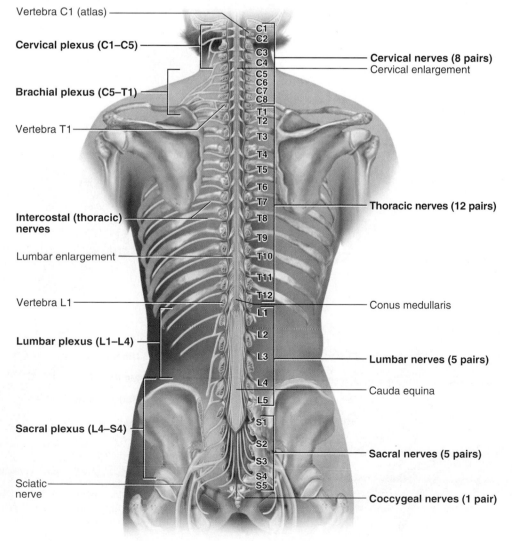

FIGURE 22.7 Nerve Plexuses

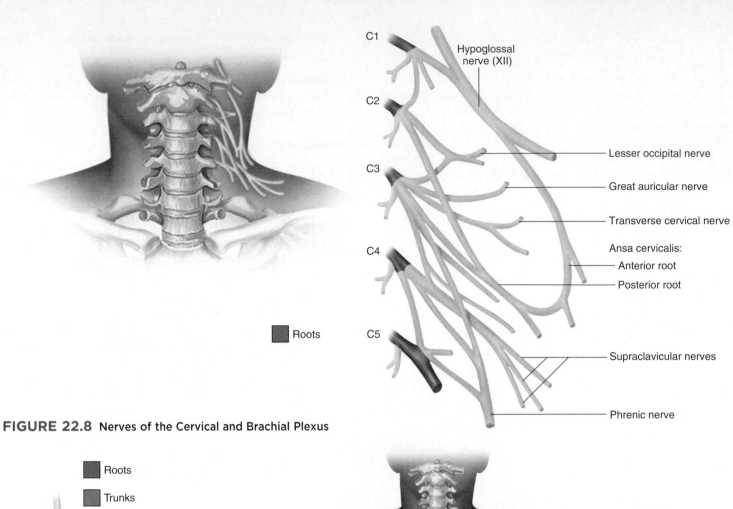

FIGURE 22.8 Nerves of the Cervical and Brachial Plexus

Labels in figure 22.8:

C1

Hypoglossal nerve (XII)

C2

Lesser occipital nerve

C3

Great auricular nerve

Transverse cervical nerve

Ansa cervicalis:
Anterior root
Posterior root

C4

Roots

C5

Supraclavicular nerves

Phrenic nerve

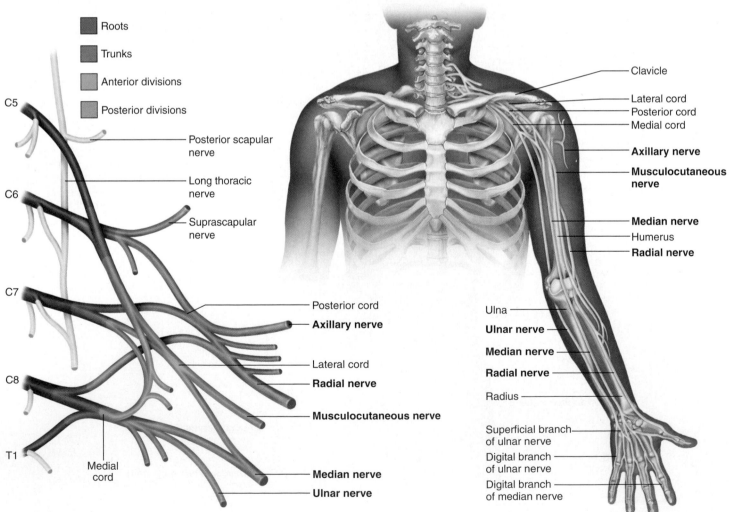

FIGURE 22.9 Nerves of the Cervical Plexus

Labels in figure 22.9:

Roots
Trunks
Anterior divisions
Posterior divisions

C5

Posterior scapular nerve

Long thoracic nerve

C6

Suprascapular nerve

C7

Posterior cord

Axillary nerve

Lateral cord

Radial nerve

C8

Musculocutaneous nerve

T1

Median cord

Median nerve

Ulnar nerve

Clavicle
Lateral cord
Posterior cord
Medial cord
Axillary nerve
Musculocutaneous nerve
Median nerve
Humerus
Radial nerve
Ulna
Ulnar nerve
Median nerve
Radial nerve
Radius
Superficial branch of ulnar nerve
Digital branch of ulnar nerve
Digital branch of median nerve

serving other forearm and hand flexors. The ulnar nerve crosses behind the medial epicondyle of the humerus and is commonly known as the "funny bone." In the posterior division is the **axillary nerve,** innervating the upper shoulder, and the **radial nerve,** predominantly innervating the triceps brachii and the extensors of the hand. Examine these nerves in the lab and compare them to figure 22.9.

Lumbar and Sacral Plexus Nerves

The **femoral nerve** is the most significant structure arising from the lumbar plexus. This large nerve passes posterior to the inguinal ligament and mostly innervates the muscles of the anterior thigh. The **obturator nerve** also arises from the lumbar plexus and innervates the adductor muscles. There are many nerves that come from the sacral plexus. Many innervate the pelvis and muscles that move the hip, thigh, and leg. Two of the nerves from this plexus, the **tibial** and **common fibular (peroneal) nerve,** unite to form the **sciatic nerve,** a large nerve of the posterior thigh, which innervates the leg and foot. Examine the material in the lab and compare it to figures 22.10 and 22.11.

Thoracic Nerves

There are numerous nerves not associated with a plexus. The **thoracic nerves** are a good example of this. Many of the thoracic nerves exit through the intervertebral foramina of the vertebral column and innervate the ribs, muscles, and other structures of the thoracic wall. These nerves can be seen in figure 22.12.

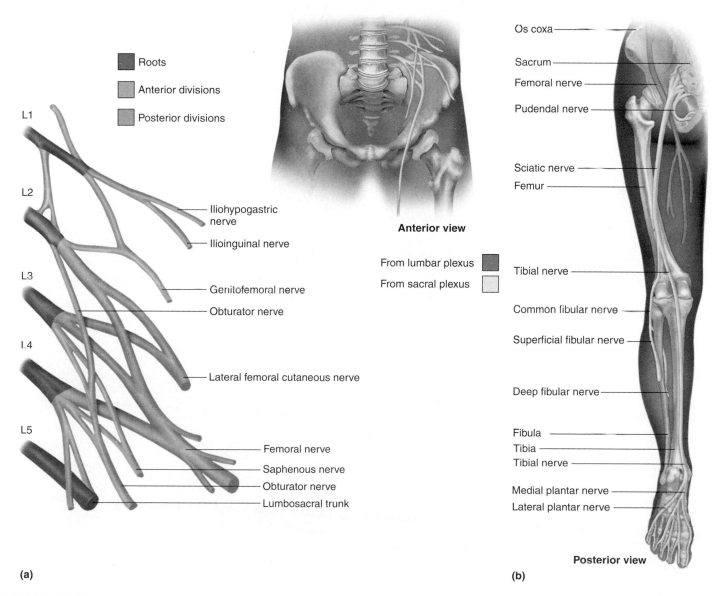

(a)

Roots
Anterior divisions
Posterior divisions

L1
L2

Iliohypogastric nerve
Ilioinguinal nerve

L3

Genitofemoral nerve
Obturator nerve

L4

Lateral femoral cutaneous nerve

L5

Femoral nerve
Saphenous nerve
Obturator nerve
Lumbosacral trunk

Anterior view

From lumbar plexus
From sacral plexus

(b)

Os coxa
Sacrum
Femoral nerve
Pudendal nerve

Sciatic nerve
Femur

Tibial nerve

Common fibular nerve

Superficial fibular nerve

Deep fibular nerve

Fibula
Tibia
Tibial nerve

Medial plantar nerve
Lateral plantar nerve

Posterior view

FIGURE 22.10 **Nerves of the Lumbar Plexus** (a) Lumbar plexus; (b) nerves of the lower extremity.

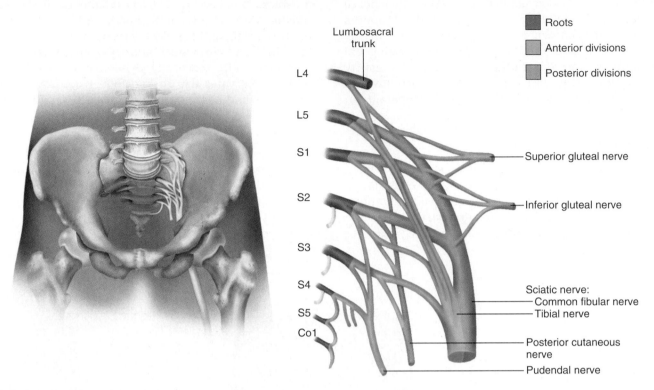

Roots

Anterior divisions

Posterior divisions

Lumbosacral trunk

L4

L5

S1 — Superior gluteal nerve

S2 — Inferior gluteal nerve

S3

S4

S5

Co1

Sciatic nerve:
— Common fibular nerve
— Tibial nerve

— Posterior cutaneous nerve

— Pudendal nerve

FIGURE 22.11 Nerves of the Sacral Plexus

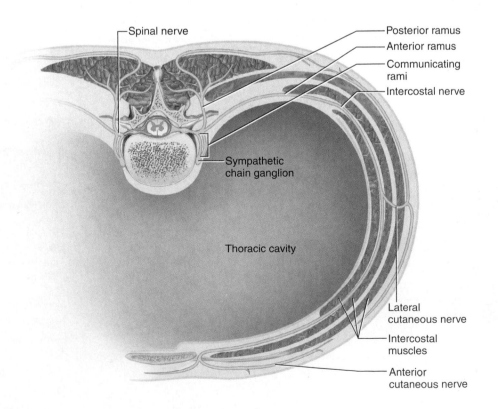

Spinal nerve

Posterior ramus

Anterior ramus

Communicating rami

Intercostal nerve

Sympathetic chain ganglion

Thoracic cavity

Lateral cutaneous nerve

Intercostal muscles

Anterior cutaneous nerve

FIGURE 22.12 Thoracic Nerves

Cat Dissection

Brachial Plexus

The dissection of the brachial plexus involves careful dissection in the axillary region. Your cat should be placed ventral side up as you begin the dissection. Remove any skin, if you have not done so already. Bisect the pectoralis major and pectoralis minor muscles and carefully fold them back to see the brachial plexus that is deep to these muscles. Remove adipose tissue and fascia to expose the blood vessels and the brachial plexus. Do not damage the blood vessels, as you will study these in future labs. Examine figure 22.13 as you study the cat.

The large and anterior nerve of the brachial plexus is the **musculocutaneous nerve.** It innervates the skin of the brachium and is the motor nerve of the biceps brachii muscle.

The **radial nerve** is the largest nerve of the plexus and is posterior to the musculocutaneous nerve. It is located on the posterior of the distal arm, where it innervates the muscles on the dorsal side of the forelimb, including many of the extensor muscles.

The **median nerve** is located alongside the brachial artery and is in the midline of the brachium and anterbrachium. It innervates many of the flexor muscles.

The **ulnar nerve** travels posterior to the medial epicondyle of the humerus to innervate the muscles on the ulnar side of the antebrachium.

Sacral Plexus

The sacral plexus is best seen from the dorsal view. Examine figure 22.14 and locate the large **sciatic nerve** by separating the biceps femoris on the dorsal thigh. The sciatic nerve is composed of two nerves, the medial **tibial nerve** and the lateral **common fibular nerve.** These can be seen at the distal part of the thigh.

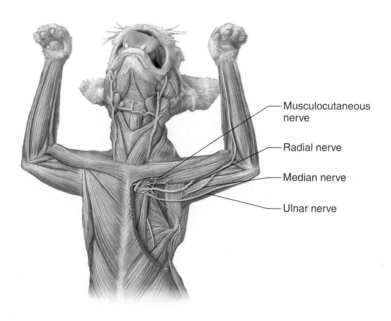

FIGURE 22.13 Brachial Plexus of the Cat

- Musculocutaneous nerve
- Radial nerve
- Median nerve
- Ulnar nerve

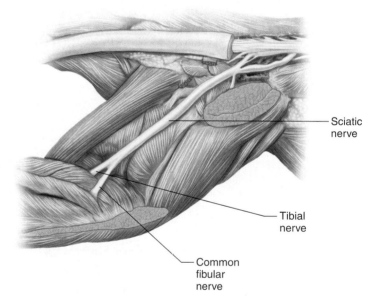

FIGURE 22.14 Sacral Plexus, Posterolateral View of Right Side

- Sciatic nerve
- Tibial nerve
- Common fibular nerve

Notes

REVIEW SECTION

The Spinal Cord and Nerves

Name _____ *Date* _____

Lab Section _____ *Time* _____

Review Questions

1. What type of signal (sensory/motor) travels through the

   a. anterior gray horn?

   b. posterior gray horn?

   c. ascending spinal tracts?

   d. descending spinal tracts?

2. What major nerves arise from the following plexuses?

   a. cervical

   b. brachial

   c. lumbar

   d. sacral

3. In function, how does the posterior spinal root differ from the anterior spinal root?

4. What causes the cervical enlargement of the spinal cord?

5. Where is the filum terminale found?

6. What is the conus medullaris?

7. What is the cauda equina?

8. What is the endoneurium?

9. In the spinal cord, which is deep to the other, the white matter or the gray matter?

10. What is the area of gray matter found between the lateral halves of the spinal cord?

11. The subarachnoid space is filled with what fluid?

12. How do tracts differ from nerves?

13. What is a mixed nerve?

14. The diaphragm contractions are regulated by what nerve?

15. The muscles of the arm, such as the biceps brachii, have what innervation?

16. The extensor muscles of the hand are controlled by what nerve?

17. The sciatic nerve is composed of two nerves. What are they?

18. A person has feeling from the deltoid and biceps brachii region but no feeling from the wrist extensors. Where on the spinal cord has injury occurred?

Notes

LABORATORY

Nervous System Physiology—Stimuli and Reflexes

INTRODUCTION

Nervous tissue shows two fundamental properties: excitability and conductivity. **Excitability** is the response of a nerve to some type of stimulus (chemical, mechanical, electrical), and **conductivity** is the transmission of a nerve impulse along the length of the neuron. Nerves receive information from a particular area (sense organ, regions in the CNS, etc.) and carry signals to either the brain for interpretation or an effector for some type of action. The nature of nerve conduction and reflexes are covered in the Saladin text in chapter 12, "Nervous Tissue," and chapter 13, "The Spinal Cord, Spinal Nerves, and Somatic Reflexes." In this exercise you study the basic properties of neuronal conduction and their sensitivity to various stimuli. In addition to studying the neurons, you experiment with nerves as they form reflex arcs in several parts of the body.

OBJECTIVES

At the end of this exercise you should be able to

1. describe the threshold nature of nerve responses;
2. list three things that cause a nerve to be stimulated;
3. name one substance that stimulates nerves and one that inhibits them;
4. describe reflex arcs;
5. list all the parts of a monosynaptic and polysynaptic reflex arc;
6. define hyporeflexic and hyperreflexic.

MATERIALS

Nerve Physiology Section

Frogs

Latex or plastic gloves

Glass rod with hook at one end

Hot pad or mitt

Bunsen burner

Matches or flint lighter

Frog Ringer's solution in dropper bottles (see Appendix C)

Microscope slide or small glass plate

Filter paper or paper towel

Dissection equipment for live animals

Scalpel or scissors

Cotton sewing thread

Stimulator apparatus with probe

Myograph transducer

Physiograph or physiology computer

Five small (50 mL) beakers

5% sodium chloride solution

0.1% hydrochloric acid solution (1 mL of concentrated HCl in 1 liter of water)

Procaine hydrochloride solution

Cotton applicator sticks

Gauze squares

Ice bath

Reflex Section

Patellar reflex hammer

Rubber squeeze bulb

Models or charts of spinal cord and nerves

PROCEDURE

Review the structure of the neuron in Laboratory Exercise 20 for descriptions of the dendrites, nerve cell body, and axon and the anatomy of the nerve in Laboratory Exercise 22. Make sure you read through all of this exercise prior to beginning the experiments. Wear latex gloves as a general precaution when working with fresh specimens, such as frogs.

Frog Nerve Conduction

In the first part of this lab exercise you will determine if nerves respond to only a specific stimulus or if they are more general and respond to many stimuli. You will also determine if certain materials or environmental conditions inhibit nerve response.

In this exercise you observe the process of nerve impulse conduction by experimenting on frogs or by watching a demonstration, depending on the wishes of your instructor. If you experiment on frogs, obtain a doubly pithed frog, dissection equipment, frog Ringer's solution, and various test solutions and take them to your table. Keep the frog nerve preparation moist with Ringer's solution during the entire experiment. Prepare the frog by cutting the skin away from the hip (see figure 23.1). Do not cut, pinch, or otherwise damage the sciatic nerve on the posterior side of the thigh.

Gently remove the nerve from between the muscles with a glass rod (see figure 23.1). Do not stretch the nerve; leave it intact alongside the muscles. You can attach the gastrocnemius to a myograph transducer, as you did in Exercise 19, or just examine the muscle for a response.

Nerve Response to Physical Stimuli

Flush the nerve with Ringer's solution and make sure it remains moist. Nerve stimulation can be determined by the corresponding muscular contraction. You measure the effects of the nerve stimulation by the contraction of the gastrocnemius muscle. If the nerve is stimulated, then the gastrocnemius muscle should contract.

Cut a small (10 cm) section of cotton thread and gently slip it under the sciatic nerve with a pair of fine forceps. Gently move the thread up toward the hip. When you reach the proximal location where the nerve descends into the muscle, tie off the nerve with thread and examine the effects. Loop the thread and ligate (tie off) the nerve

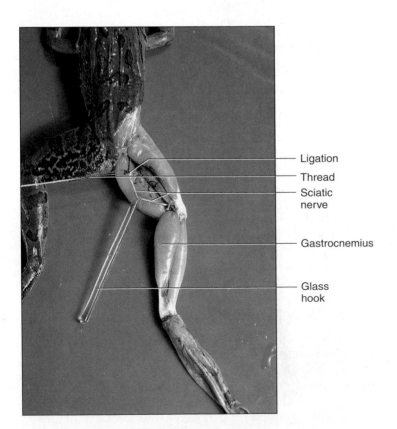

close to the sacrum (see figure 23.1) while watching the gastrocnemius muscle.

As the thread begins to tighten on the nerve, record the response in the space provided and in the Chapter Summary Data section.

? Response of the nerve to physical stimulation: _____1

_____

After you have ligated the nerve, cut it from the anterior side, leaving the nerve attached to the gastrocnemius muscle. Moisten the nerve with Ringer's solution and prepare for the next experiment.

Nerve Response to Electrical Stimuli

Place a stimulator probe connected to a stimulator underneath the sciatic nerve, lifting the nerve away from the gastrocnemius muscle. Keep the nerve moist as you determine the minimum voltage **(threshold voltage)** required for nerve conduction. Set the stimulator to a frequency of two pulses per second and a duration of 10 milliseconds. The voltage should be set at zero (with the knob on 0.1 volt). Slowly increase the voltage until you observe the gastrocnemius twitch. As soon as you see the gastrocnemius twitch at the *lowest voltage,* record this as threshold voltage in the space provided. Turn the voltage to zero, flush the nerve with Ringer's solution, and let it rest for a moment.

? Threshold voltage: _____2

Continue to increase the voltage until the muscle contracts maximally. Record this as the **maximum recruitment voltage.** This voltage is obtained when all the neurons of a particular nerve are stimulated.

? Maximum recruitment voltage: _____3

Nerve Response to Chemical Stimuli

In the following two experiments you test the response of the nerve to different chemical agents. Make sure you observe the nerve as soon as you apply the solution and rinse it as soon as the observation is made.

Acid Solution   Apply a 0.1% hydrochloric acid solution to a cotton applicator stick and gently touch the applicator to the nerve. Record the nerve response.

? Response to hydrochloric acid: _____4

Flush the nerve with Ringer's solution and let it rest for a moment.

FIGURE 23.1 Ligation of the Sciatic Nerve of the Frog

Ligation
Thread
Sciatic nerve
Gastrocnemius
Glass hook

Salt Solution Gently apply a 5% sodium chloride solution to the nerve with a new cotton applicator stick. Record the response. Flush the nerve with Ringer's solution and let it rest for a moment.

❓ Response to sodium chloride solution: _____ 5

Nerve Response to Anesthetics

Apply a solution of **procaine hydrochloride** (Novocain) to the nerve by soaking a small square of gauze or cotton with procaine solution and placing it on the nerve for a moment. As the gauze remains on the nerve, set up the stimulator apparatus and place the nerve over the stimulator probes. Remove the gauze and stimulate the nerve with a single pulse stimulus at the voltage that produced a maximum recruitment voltage in the previous experiment. If the nerve responds to the stimulus, leave the procaine hydrochloride on longer. When the nerve does not respond, remove the gauze, flush with frog Ringer's solution, and stimulate the nerve once every 30 seconds until it recovers from the local anesthetic. Keep the nerve moist at all times with frog Ringer's solution. Record the recovery time.

❓ Recovery time: _____ 6

Nerve Response to Changes in Temperature

Gently touch the nerve with a glass rod at room temperature. Record the response.

❓ Response to gentle touch: _____ 7

Place the glass rod in an ice bath. As it equilibrates in the ice water place a small chip of ice on the nerve and let it stay there for a moment. Gently touch the nerve with the cold rod and record the response. Flush the nerve with Ringer's solution and let it rest for a moment.

❓ Response to gentle touch with cold stimulation: ___ 8

Take another glass rod in a hot pad or mitts and heat one end of it in a Bunsen burner. Touch the nerve with the hot end of the glass rod. What is the response? Record your result.

❓ Response to gentle touch with hot stimulation: ___ 9

Clean Up Make sure you clean your station before continuing. Place the specimen in the appropriate container, and use care when cleaning sharp instruments, such as scalpels or razor blades.

Reflexes

A reflex is a motor response to a stimulus without conscious thought. Reflexes are involuntary, predictable responses to stimuli. Reflexes occur through **reflex arcs,** and these arcs have the following structure:

1. **Receptor** (structure that receives the stimulus and converts it to an action potential)
2. **Afferent** (sensory) **neuron** (the neuron taking the stimulus to the CNS)
3. **Integrating center** (brain or spinal cord)
4. **Efferent** (motor) **neuron** (the neuron taking the response from the CNS)
5. **Effector** (the structure causing an effect)

If the effector is skeletal muscle, it is a **somatic reflex.** If the effector is a gland, smooth muscle, or cardiac muscle, it is a **visceral,** or **autonomic, reflex.**

Most reflexes involve many neurons with many synapses and are called **polysynaptic reflex arcs.** A few reflexes involve only two neurons—a sensory neuron and a motor neuron with one synapse between them. These are **monosynaptic reflex arcs.** They are illustrated in figure 23.2.

Reflexes depend on a **stimulus,** or environmental cue; a **receptor** sensitive to the stimulus; an **afferent neuron** (or sensory neuron); an **efferent neuron** (or motor neuron); and an **effector.** The polysynaptic reflex has these structures as well as an **association neuron,** or **interneuron,** located between the afferent and efferent neurons. Interneurons take information to the brain via ascending tracts of the spinal cord.

Testing for reflexes is important for clinical evaluation of the condition of the nervous system. Decreased response or exaggerated response to a stimulus may indicate disease or damage to the nervous system. In this experiment you test several reflexes and determine if the response is **normal** (movement of an inch or two), **hyporeflexic** (showing less than average response), or **hyperreflexic** (showing an exaggerated response). If there is hyperreflexia this might indicate central nervous system damage. No reflex may indicate spinal cord damage, and hyporeflexia may be indicative of hypothyroidism.

Stretch Reflexes

For stretch reflexes, receptors are in the muscle spindle. Stretching the muscle causes an increase in action potentials in the sensory neuron. The impulse travels to the motor neuron, causing contraction of the muscle that is stretched. A typical example of a stretch reflex is the patellar reflex, in which striking the patellar ligament stretches the quadriceps muscles. Sensory neurons transmit this information to the spinal column, where motor neurons stimulate the quadriceps muscle to contract, thus extending the leg.

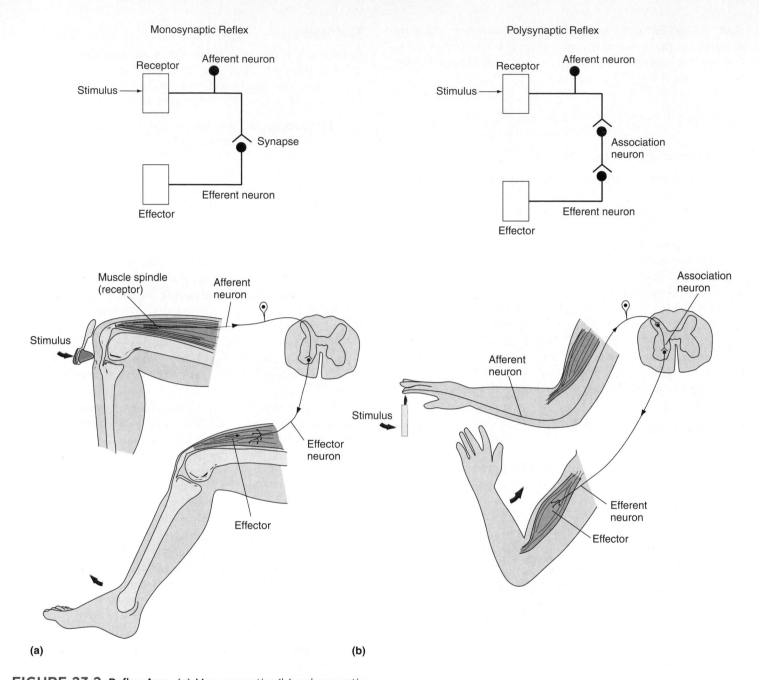

FIGURE 23.2 Reflex Arcs (a) Monosynaptic; (b) polysynaptic.

Patellar Reflex

The **patellar reflex** is a stretch reflex that tests the femoral nerve. Sensory neurons in the quadriceps muscle are stimulated by rapid lengthening of the muscle and conduct nerve impulses to the spinal cord. Sensory neurons synapse with motor neurons that innervate muscles, stimulating them to resist the stretch. It is the most frequent reflex test performed in clinical settings. Sit on the lab table with your leg hanging over the edge and have your lab partner tap you on the patellar tendon with a patellar reflex hammer. The percussion should

be placed about 3 to 4 cm below the inferior edge of the patella, and it should be firm but not hard enough to hurt. Look for extension of the leg as a response to the patellar reflex (see figure 23.3).

The tap stimulates the stretch receptors in the tendon and is representative of a monosynaptic reflex. Record the degree (hyperreflexic, normal, hyporeflexic) of the response.

? Patellar reflex: _____ 10

FIGURE 23.3 Patellar Reflex

FIGURE 23.5 Biceps Brachii Reflex

FIGURE 23.4 Triceps Reflex

Triceps Brachii Reflex

The **triceps brachii reflex** tests the radial nerve. Sit on a chair or lie down on your back on a clean lab table or cot and place your forearm on your abdomen. Have your lab partner tap the distal tendon of the triceps brachii muscle about 2 inches proximal to olecranon process. Look for the triceps muscle to twitch (see figure 23.4). Record your result.

❓ Triceps brachii reflex: _____ 11

Biceps Brachii Reflex

The **biceps brachii reflex** tests the musculocutaneous nerve. Sit comfortably and have your lab partner place his or her fingers on the biceps tendon just proximal to the antecubital fossa (see figure 23.5).

Your lab partner should tap his or her fingers with the reflex hammer, while they remain on the tendon, and look for the biceps brachii muscle contraction. Record your results.

❓ Biceps brachii reflex: _____ 12

Calcaneal (Achilles) Tendon Reflex

To test the **calcaneal tendon reflex** kneel on a chair with your foot dangling over the edge (see figure 23.6). Have your lab partner tap the calcaneal tendon to test the tibial nerve.

As your lab partner taps your calcaneal tendon, look for plantar flexion of the foot. You may see an initial movement of the foot due to the depression of the tendon by the reflex hammer, but there should be a slight pause and then another quick movement of the foot. Record your results.

? Calcaneal tendon reflex: _____ 13

Eye Reflexes

The automatic blinking of the eye is important to keep material, such as dust, away from the outer layer of the eye, the cornea. In the first part of this experiment have your lab partner try to make you blink by flicking his or her fingers near your eyes. They should not come close enough to touch your eyes. Can you prevent the blinking response? Record your answer.

? Control of blink reflex: _____ 14

Now have your lab partner take a *clean* rubber squeeze bulb (a large pipette bulb works well) and squirt a sharp blast of air across the surface of the eye (see figure 23.7).

Caution!   Use either new bulbs or ones free of debris to avoid damage to the eye.

Can you inhibit this response? Record your response.

? Control of corneal reflex: _____ 15

Plantar Response, or Babinski Reflex

Using the *metal end* of the patellar hammer, stroke the foot from the heel along the lateral, inferior surface and then toward the ball of the foot (see figure 23.8). The pressure should be firm but not uncomfortable.

FIGURE 23.6 Calcaneal Tendon Reflex

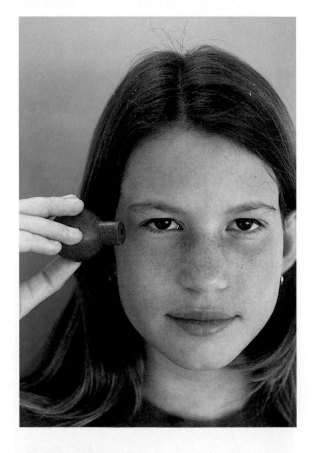

FIGURE 23.7 Corneal Reflex

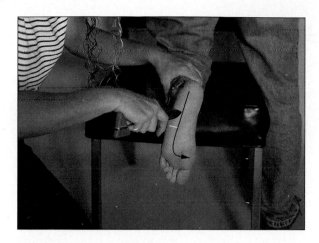

FIGURE 23.8 Babinski Reflex

Stroking the plantar surface normally results in flexion of the toes in adults. Damage to pyramidal tracts causes extension of the big toe (known as the Babinski reflex). This is important for determining spinal damage. Adults normally do not show a Babinski reflex, and the normal response in adults is to flex the foot and toes. The plantar response, or Babinski reflex, occurs in newborns and is normal. It is seen when the extension of the big toe or the fanning of the toes occurs when the plantar surface of the foot is stroked. Once myelination of the nerves occurs, the Babinski reflex disappears and is not normally found in the adult. Babinski responses in the adult are indicative of CNS problems.

Notes

REVIEW SECTION

Nervous System Physiology—
Stimuli and Reflexes

Name _____ Date _____

Lab Section _____ Time _____

Review Questions

1. What structure receives stimuli from the external environment and relays that stimuli to the afferent neuron?

2. What is another name for an efferent neuron?

3. What is a reflex?

4. What kind of reflex has only two neurons?

5. Polysynaptic reflexes have a neuron specific to them. What is the name of that neuron?

6. In numbers of synapses, what kind of reflex is a patellar reflex?

7. After patients leave the operating room they are transferred to an area called the "recovery room." Correlate the meaning of the word *recovery* in this context with what you have learned about the recovery of nerves in this exercise.

8. Draw a monosynaptic reflex arc in the space provided. Label your illustration using the terms provided.

   receptor

   motor neuron

   effector

   sensory neuron

   synapse

9. What action occurs with a hyperreflexic response? What action happens with a hyporeflexic response?

10. List the positive responses obtained in the frog experiment, and correlate this with the specificity of neuronal sensitivity.

11. What was the threshold voltage observed in the nerve response?

❓ Chapter Summary Data

Use this section to record your results from questions within the exercise.

1. _____   9. _____

2. _____   10. _____

3. _____   11. _____

4. _____   12. _____

5. _____   13. _____

6. _____   14. _____

7. _____   15. _____

8. _____

LABORATORY

Introduction to Sensory Receptors

INTRODUCTION

The gateway to understanding our world comes from our ability to sense the environment within our bodies and the environment around us. There are two main classes of sense—**general (somesthetic) senses** and **special senses.** General senses occur in many locations of the body and are found in places such as the skin, muscle, joints, and viscera. The senses of touch, pressure, changes in temperature, pain, blood pressure, and stretching are general senses. Special senses occur in specific locations, such as the eye, ear, tongue, and nose, and include taste, smell, sight, hearing, and balance.

Sense receptors are not uniformly distributed throughout the body. In some areas, specific sense receptors are absent, or few in number, while in other areas they are densely clustered. This pattern of uneven distribution is called **punctate distribution.**

Our perception of the environment is dependent on environmental **stimuli.** These stimuli are classified by type, or **modalities,** such as light, heat, sound, pressure, and specific chemicals. **Receptors** are the receiving unit in the body that respond to a stimulus. They transform the stimulus to neural signals transmitted by sensory nerves and neural tracts to the brain, which interprets the message. If any link in this sensory chain is broken, the perception of stimuli cannot occur.

Receptors respond to specific modalities, and each receptor can be classified according to the type of stimulus it responds to. **Photoreceptors** detect light (for example, the retina in the eye); **thermoreceptors,** located in the skin and other areas, detect changes in temperature; **proprioreceptors** detect changes in tension, such as those in tendons when a muscle contracts; **pain receptors,** or **nociceptors,** are present as naked nerve endings throughout much of the body; **mechanoreceptors** are receptive to mechanical stimuli (for example, touch receptors or receptors in the ear that respond to sound or motion); **baroreceptors** respond to changes in blood pressure; and **chemoreceptors** respond to changes in the chemical environment (for example, taste and smell). These receptors are discussed in the Saladin text in chapter 16, "Sense Organs."

The skin has several types of receptors and therefore makes a good starting point for understanding sense organs. There are receptors for pressure, pain, temperature, and light touch in your skin. You may want to review the major sensory receptors in the skin, such as Meissner's corpuscles, pacinian (lamellated) corpuscles, and pain receptors in Laboratory Exercise 7 before you begin this exercise.

OBJECTIVES

At the end of this exercise you should be able to

1. define modality and receptor;
2. list the major receptor types in the body;
3. define punctate distribution of sensory receptors;
4. distinguish between tonic and phasic receptors;
5. define adaptation in reference to a stimulus;
6. distinguish between relative and absolute determination of stimuli;
7. define referred pain.

MATERIALS

Blunt metal probes

Dishpan (or large finger bowls) of ice water (2 L)

Dishpan of room temperature water

Dishpan of warm water (45° C)

Towels

Small centimeter ruler

Black, washable, fine-tipped felt markers

Red, washable, fine-tipped felt markers

Blue, washable, fine-tipped felt markers

Two-point discriminators (or a mechanical compass)

Three lab thermometers

Hand lotion

Von Frey hairs (horsehair glued on a wooden stick)

Tweezers

PROCEDURE

Mapping Fine-Touch Receptors

Fine-touch receptors are of two types, Meissner's corpuscles and Merkel discs. You can map these receptors by testing the ability of your lab partner to distinguish fine touch.

1. Draw a square, 2 cm on a side, with a black, washable marker on the anterior surface of the forearm. If you apply a little hand lotion to the skin before doing this test, the ink comes off more easily after the experiment.
2. Use a Von Frey hair (a stiff bristle hair attached to a match stick) to map the number of areas in the square that can be perceived by your lab partner. Press only until the hair bends slightly to stimulate the touch corpuscles.
3. Record the location of each positive result with the marker. How many positive responses did you get in the square on the forearm? Record your result in the following space and in the Chapter Summary Data section at the end of the exercise.

? Number of anterior forearm responses: ——————— 1

4. Repeat the experiment on the *posterior* surface of the *arm* using another square 2 cm on a side. How does the number of receptors here compare to those on the forearm? Record your results.

? Number of responses on posterior side of the arm:

—————————————————————————— 2

Two-Point Discrimination Test

The sensitivity of touch is dependent on the number of fine-touch receptors per unit area of the skin. You can map the relative density of the receptors in the skin by performing a two-point discrimination test. The idea behind the test is to determine the *minimum distance* that your lab partner is able to recognize as two points. This is illustrated in figure 24.1.

1. Have your lab partner sit with eyes closed and his or her hand, palm up, resting on the lab counter.
2. Using the two-point discriminator, or a mechanical compass, touch your lab partner's fingertip simultaneously with both points of the instrument and see if he or she can sense one or two points. In order to establish an accurate reading make sure that you gently touch both points of the discriminator at the same time. If you touch with one point of the discriminator and then another, your lab partner may perceive two points in time and not in space.
3. A good way to establish accuracy is to occasionally touch just one of the points on your lab partner's fingertip. Another way is to vary the spread of the discriminator. You might begin with 3 cm and then adjust it to 0.5 cm followed by a 2 cm spread. Record your results.

? Minimum distance perceived as two points on

fingertip: ——————————————————— 3

4. Now move to the posterior of the arm. Establish the minimum distance that is perceived as two points by your lab partner. Record your results.

? Minimum distance perceived as two points on the

posterior arm: ——————————————— 4

? Is there a difference between the fingertip and the

posterior arm? ——————————————— 5

? If there is, how might the difference in distance perceived be explained in terms of the number of nerve

endings per unit area? ——————————— 6

5. Now try the palm and then the back of the shoulder or neck. Record the results.

? Minimum distance perceived as two points on palm:

—————————————————————————— 7

? Minimum distance perceived as two points on back of

shoulder or neck: ——————————————— 8

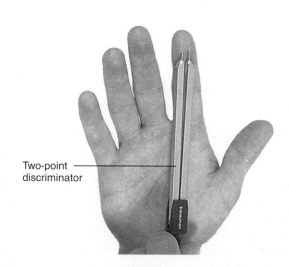

Two-point discriminator

FIGURE 24.1 Two-Point Discrimination

Mapping Temperature Receptors

The skin has receptors that are sensitive to cool or warm temperatures. In this part of the exercise you determine the relative numbers of these receptors.

1. Mark off a square that is 2 cm on a side on the anterior forearm of your lab partner using a washable marker.
2. Take two blunt metal probes and place the tip of one in an ice water bath and another in a warm water bath (45° C).
3. Let the probes reach the temperature of each bath, which should take a minute or two.
4. Have your lab partner close his or her eyes and rest an arm on the lab counter.
5. Remove one of the probes and quickly wipe it on a clean towel. Test the ability of your lab partner to distinguish between cool and warm by using the tip of the blunt probe on your lab partner's forearm. Systematically test areas in the square. When your lab partner perceives cold (not just touch) in a location, mark it with an "X." Retest the area with the warm probe and place an "O" in the location where warm is perceived.

? Number of cool receptors in the square: _____ 9

? Number of warm receptors in the square: _____ 10

? What is the ratio of "X" to "O" (cold/warm receptors)

in the square? _____ 11

Adaptation to Touch

Receptors can be classified by the length of time it takes for them to adapt to a stimulus. **Tonic receptors** continuously perceive stimuli (they do not adapt), while **phasic receptors** perceive the stimulus initially and then adapt. In this experiment you try to determine if the sense of fine touch is tonic or phasic.

1. Cut a small piece of paper, about 2 cm on a side, and crumple it into a small ball the size of a pea.
2. Have your lab partner close his or her eyes and place the hand, anterior side up, comfortably on the lab desk.
3. With a pair of tweezers, place the ball of paper on your lab partner's palm. Is the paper ball perceived after a few seconds?
4. Record your results in the following space and determine whether the sense of light touch is tonic or phasic.

? Results: _____ 12

Locating Stimulus with Proprioception

In this exercise you use washable markers of two colors. Location of the stimulus is dependent on both skin receptors and cerebellar function. Have your lab partner close his or her eyes and rest a forearm on the lab counter. Touch your lab partner's forearm with a felt marker and have him or her try to locate the same spot with a felt marker of another color. Test at least five locations on various parts of the forearm, and repeat each location at least twice. Now try the fingertip and palm of the hand and record the result.

? Maximum distance error on the forearm: _____ 13

? On the fingertip: _____ 14

? On the palm: _____ 15

Another method to test proprioreception is to close your eyes and *gently* try to touch the lateral corner of your eye with your fingertip. Have your lab partner watch you and determine the accuracy of your attempt. While your eyes are still closed, bring your hand far behind your head and then try to touch the bottom part of your earlobe or the exact tip of your chin. Record the error distance, if any, for each location.

? Corner of eye: _____ 16

? Earlobe: _____ 17

? Tip of chin: _____ 18

Temperature Judgment

In this exercise you examine the *adaptation* of thermoreceptors to temperature and the ability to determine temperature by **absolute value** or by **relative value.**

On the lab counter locate three dishpans or large finger bowls full of water. One bowl, located on the right, should be marked "Cold" (it should be about 10° C); another bowl, located on the left, should be marked "Warm" (it should be about 40–45° C); and the middle bowl should be marked "Room Temperature." Place one hand in the cold dish and the other in the warm dish and let them adjust to the temperature for a few minutes. If your hand begins to ache in the cold water, you may remove it for a short time, but try to keep it in the cold water for as long as possible during the adjustment time. After your hands have equilibrated, place them both in the room temperature water and describe to your lab partner the temperature (cold, warm, hot) of the water as sensed by each hand. How does the hand that was in cold water feel in the room temperature water, and how does the hand that was in warm water feel in the room temperature water? Record your results.

? Cold hand perception: _____ 19

? Warm hand perception: _____ 20

If the determination of temperature is absolute, the room temperature water should feel the same whether you are testing it with your cold hand or your warm hand. If it is relative, the room temperature water should feel warmer with your cold hand and colder with your warm hand.

? Is the determination of the temperature of water

absolute or relative? _____ 21

? How did your experiment prove this? _____ 22

Referred Pain

Referred pain is the perception of pain in one area of the body when the pain is somewhere else. An example of referred pain is the pain felt in the left shoulder and arm when a person is suffering from a heart attack or chest pain (angina pectoris) (see figure 24.2).

Frequently, neural impulses have convergent pathways where nerves receiving stimuli from the skin or muscle in one area follow the same general ascending tract that sensory information from an organ (such as the

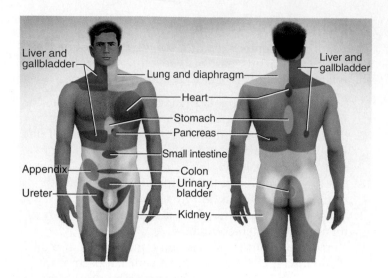

FIGURE 24.2 Referred Pain

kidney or gallbladder) do. If there are pain stimuli in that organ, an individual may feel the pain in the skin that shares the tract.

Place your elbow into a dish of ice water and leave it there for 2 painful minutes. Describe the sensation you feel and the location of the sensation. Record your results.

? Description of sensation: _____ 23

? Initial location of sensation: _____ 24

? Sensation after 2-minute period: _____ 25

REVIEW SECTION

Introduction to Sensory Receptors

Name _____ Date _____

Lab Section _____ Time _____

Review Questions

1. An area with a great number of fine touch receptors is the upper lip. What can you predict about the ability of the upper lip to distinguish two points?

2. Cool receptors in the skin are activated between 12° and 35° C. Warm receptors are activated between 25° and 45° C. When the temperature is below 12° C or above 45° C, pain receptors in the skin are activated. You or your lab partner may have had an experience with very cold conditions, such as when cleaning out a freezer or holding dry ice. What perception is sensed?

3. Adaptation is important to sensory stimulation. We are bombarded with stimuli during most of the day, and much of what we sense is filtered from conscious thought. How is adaptation used by pickpockets?

4. In terms of receptor density, describe why it is difficult to find the same location on the forearm when your eyes are closed.

5. In regard to sense organs, what is punctate distribution?

6. In reference to the sense organs, what is a modality?

7. What type of receptors are sensitive to the following modalities?

   a. light   d. sound

   b. touch   e. smell

   c. temperature

8. What type of receptor is responsive to extremely hot sensations?

9. Meissner's corpuscles respond to what type of sensation?

10. What type of receptor determines the weight of an object when you pick it up?

11. Which type of receptor (tonic/phasic) adapts to light in a darkened movie theater?

12. When you drink a burning hot liquid, the "chest pain" felt in the region of the sternum does not really occur there. What is this type of pain called?

? Chapter Summary Data

Use this section to record your results from questions within the exercise.

1. _____   14. _____
2. _____   15. _____
3. _____   16. _____
4. _____   17. _____
5. _____   18. _____
6. _____   19. _____
7. _____   20. _____
8. _____   21. _____
9. _____   22. _____
10. _____   23. _____
11. _____   24. _____
12. _____   25. _____
13. _____

LABORATORY

Taste and Smell

INTRODUCTION

Like most of our senses, we take for granted our senses of taste and smell. They are not fully appreciated unless they are lost. People who have lost the sense of smell find food difficult to eat, since they derive little or no pleasure from the act of eating. In this exercise you examine the structure and function of the organs involved in these two important senses.

Both taste and smell are examples of **chemoreception,** in which specific chemical compounds are detected by the sense organs and interpreted by various regions of the brain. The sense of taste and the sense of olfaction are covered in the Saladin text in chapter 16, "Sense Organs."

OBJECTIVES

At the end of this exercise you should be able to

1. list the two major chemoreceptors located in the region of the head;
2. diagram a taste bud;
3. trace the sense of smell from the nose to the integrative areas of the brain;
4. trace the sense of taste from the tongue to the integrative areas of the brain;
5. list the five tastes perceived by humans;
6. describe what happens in an olfactory reflex;
7. compare and contrast the senses of taste and smell.

MATERIALS

Sterile cotton-tipped applicators

Five dropper bottles containing one of the following:

　Solution of salt water (3%) labeled "Salty"

　Quinine solution (tonic water) labeled "Bitter"

　Vinegar solution (household vinegar or 5% acetic acid solution) labeled "Acidic"

　Sugar solution (3% sucrose) labeled "Sweet"

　Umami solution (15 g MSG in 500 mL water) labeled "Umami"

Biohazard bag

Prepared slides of taste buds

Microscopes

Roll of household paper towels

Small bowl of salt crystals (household salt)

Small bowl of sugar crystals (household granulated sugar)

Flat toothpicks

Small bottle (100 mL) of household ammonia

Several small vials (10–20 mL) screw-cap bottles with essential oil labeled "Peppermint," "Almond," "Wintergreen," and "Camphor" (keep vials in separate wide-mouthed jars to prevent cross-contamination of scents)

Four small vials colored red and labeled "Wild Cherry" filled with benzaldehyde solution

Four small vials of "Almond" essence

One vial of dilute perfume (one part perfume, five parts ethyl alcohol)

Marking pens

Noseclips and alcohol swabs

Selection of four or five fruit nectars (such as Kern's nectars), two cans each: apricot, coconut/pineapple, strawberry, mango, peach, apple

Small, 3 oz paper cups (89 mL), five per student (or student pair)

Napkins

PROCEDURE

Examination of Taste Buds

Examine the prepared slide of taste buds and compare them to figure 25.1. The taste buds are located on the sides of papillae on the tongue. The taste buds appear lighter than the surrounding tissue (like microscopic onions cut in long sections). Taste buds consist of **supporting cells** and **taste cells,** specialized epithelial cells, with hairs that project into the **taste pores** near the surface of the tongue. The basal portion of the taste cell synapses with sensory neurons.

Transmission of Sense of Gustation to the Brain

The sense of taste, or **gustation,** is picked up by the receptors in taste buds primarily in the tongue, although there are also receptors in the soft palate and pharynx. The sense of taste travels through the facial nerves, glossopharyngeal nerves, and vagus nerves to the medulla oblongata. From there, some fibers travel to either the hypothalamus or amygdala where autonomic reflexes (such as swallowing) occur. Other fibers travel to the thalamus and then to higher brain centers, such as the postcentral gyrus, where the sense of taste is determined. From the postcentral gyrus, fibers take neural impulses to the orbitofrontal cortex, where sight and smell are integrated with taste. The sense of taste is influenced by a food's smell, appearance, temperature, and texture, even the mood of the individual.

Taste Determination of Solid Materials

Gustatory receptors are stimulated by specific chemicals in solution. Fluid runs down the sides of the tongue papilla, where the hair cells of the receptors are located. Blot your tongue thoroughly with a paper towel. Make sure the surface is relatively dry. Have your lab partner select either the sugar crystals or the salt crystals and place a small scoop (with the end of a flat toothpick) on your tongue. Keep your mouth open and do not swirl saliva around. Can you determine what the sample is?

Now close your mouth and see if you can determine the nature of the sample.

Mapping the Tongue for Taste Receptors

In this section you determine if receptors for taste are distributed evenly over the tongue or are located in specialized areas. There are five primary tastes: sweet, sour, salty, bitter, and umami. Umami gives meat and cheese their tastes, and it can be referred to as the taste of "savory." We have varying degrees of sensitivity to these five tastes. Some of us find bitter tastes to be especially objectionable, while others do not seem to mind them as much. As you perform the taste experiments determine if members of your class are equally sensitive to the same tastes as you are. If you have food allergies you may wish to omit this part of the lab. People with allergies, migraines, or heart problems should avoid the tasting of umami, the taste of glutamate.

Select one of the dropper bottles labeled "Sweet," "Sour," "Salty," "Bitter," or "Umami." Using a sterile cotton-tipped applicator stick, apply one of the tasting solutions to the cotton tip. Saturate the cotton tip. Remember to keep track of the substance on your applicator stick. *Do not* reuse the applicator sticks. Dab the entire surface of your lab partner's tongue and determine by nod or hand signals when perception of the taste occurs. *Do not* have your lab partner tell you about the taste at first,

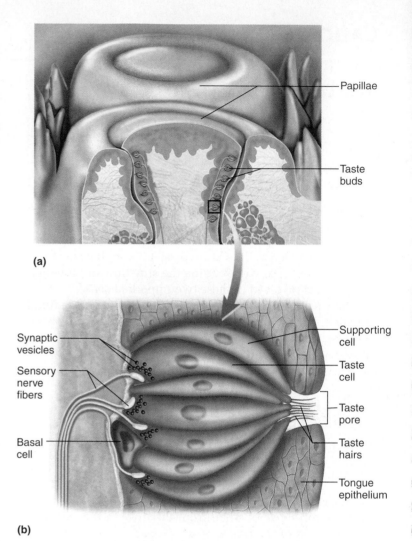

(a)

(b)

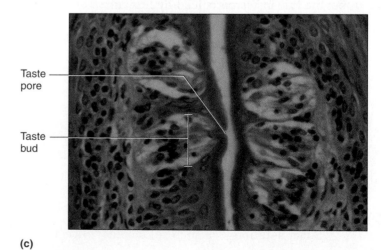

(c)

FIGURE 25.1 Taste Buds (a) Taste buds on sides of a tongue papilla; (b) details of taste buds; (c) photomicrograph (100x).

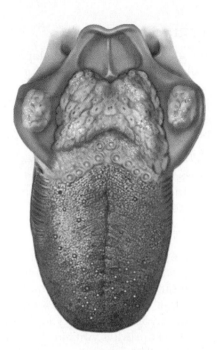

FIGURE 25.2 Map the Taste Receptors You Determine by Experiment Here

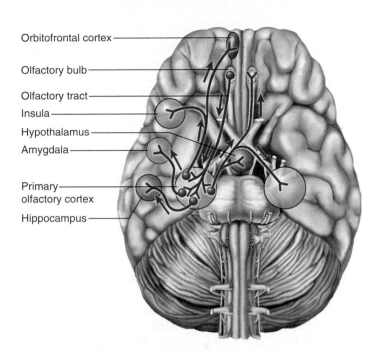

Orbitofrontal cortex

Olfactory bulb

Olfactory tract

Insula

Hypothalamus

Amygdala

Primary olfactory cortex

Hippocampus

FIGURE 25.3 The Transmission of Olfaction from the Olfactory Bulbs to the Brain

because the movement of the tongue will wash the solution to other areas of the tongue and you may get a false reading. You should test the front, lower sides, middle, and back of the tongue. Map the result of your taste test in figure 25.2. Certain regions of the tongue are more sensitive to a specific taste and these are variable among individuals. Throw the applicator stick in the biohazard bag. Repeat the test with a *new applicator stick* for each of the five solutions. *Do not contaminate* the solutions by reusing the applicator stick with the same or different solutions after it has been in your lab partner's mouth!

Transmission of Sense of Olfaction to Brain

Examine a model, chart, or diagram of an inferior view of the brain and locate the olfactory nerves, olfactory bulb, and olfactory tract. The sense of smell begins in the nose, where **olfactory cells** (specialized neurons) receive chemical input from the environment. These neurons are clustered in bundles that make up the **olfactory nerve** and they pass from the nasal mucosa to the **olfactory bulb** at the base of the frontal lobe (see figure 25.3). Some fibers travel to the temporal lobe, where the perception of smell occurs, while others travel to the hippocampus or amygdala, where the memory of smell is stored or the emotional response of smell occurs.

Olfactory Reflex

Take a small bottle of household ammonia and place it under the nose of your lab partner. Have your lab partner take a brief sniff from the bottle. If there is a visible move-

ment of the head in a posterior direction then your lab partner demonstrates an olfactory reflex to smell. Record the results of your experiment below.

Olfactory reflex:

? _____ 1a yes

? _____ 1b no

? What might be the adaptive benefit for people having

an olfactory reflex? _____ 2

Visual Cues in Smell Interpretation

In this section you examine the influence of visual cues on the interpretation of smell. You will seek to determine whether the color of a substance has any effect on what you perceive the smell to be. Have your lab partner show you a small vial labeled "Almond" and then smell it. Now examine and smell the small red vial labeled "Wild Cherry." Do you perceive these as two separate smells? Close your eyes and have your lab partner select a vial for you. Can you tell which one it is?

Olfactory Discrimination

Obtain four vials of different scents—peppermint, almond, wintergreen, and camphor. While keeping your eyes closed try to determine the name of each essential oil as your lab partner presents it to you. Record how many of the smells you got correct out of the four. If you are hypersensitive to smells, have your lab partner do this section of the experiment.

❓ Number correct: _____ 3

Adaptation to Smell

Adaptation to smell by the olfactory receptors occurs very rapidly, but the adaptation by the receptors is incomplete. Complete adaptation to smell probably occurs by additional CNS inhibition of the olfactory signals. In this section of the experiment you are trying to determine approximately how long olfactory adaptation takes. Close your eyes and plug one nostril. Inhale the scent from a vial of dilute perfume or one of the scents from the olfactory discrimination test, until the smell decreases significantly. Have your lab partner record the time when you begin the experiment and how long it takes for the perception of smell to decrease significantly. How long does this take?

❓ Length of time for significant reduction of the smell:

_____ 4

❓ What might be the evolutionary advantage of

adaptation to smell? _____ 5

Predict whether adaptation to one smell causes adaptation to another smell. Record your prediction that the smell of one material does or does not cause adaptation to another smell.

Now smell the wintergreen or peppermint vial. Does the adaptation of one smell cause the olfactory receptors to adapt to other smells?

Taste and Olfaction Tests

This experiment demonstrates the dependence of the sense of smell as a component of what we call *taste*. Obtain four or five small cups (3 oz) and, using a marking pen, label each with the name of the fruit juice or nectar it will contain. Have your lab partner select several types of fruit nectars and pour each into the proper cup.

 Caution!   If you have a particular food allergy notify your instructor. You may wish to omit part or all of this test.

Sit with your eyes and your nose closed (use noseclips or pinch off your nostrils with your finger and thumb) and try to determine the sample presented to you by your lab partner. Your lab partner should place the sample cup (sample unknown to you) in your hand and you should guess what fruit nectar is in the cup. After you have "tasted" the sample try to name it. Test all the samples with your nose closed. After you make your determination release your nostrils but still keep your eyes closed and taste the samples again. Record the results. How accurate is your comparison?

Trial Number	Sample	Accuracy or Detection (Yes/No)	
		Nose Closed	Nose Open
1	Apricot		
2	Mango		
3	Coconut/pineapple		
4	Strawberry		
5	Peach		
6	Other		

REVIEW SECTION

Taste and Smell

Name _____ Date _____

Lab Section _____ Time _____

Review Questions

1. Why does material have to be in solution for it to be sensed as taste?

2. What are the primary tastes?

3. Did everyone in your lab have the same sensitivity to taste, such as sweet, sour, or bitter?

4. Describe the pathway of smell from the olfactory receptors to the brain.

5. What structures are involved in taking the sense of taste from the taste buds to the brain?

6. Where are the taste buds located?

7. What is the exact region of the nasal cavity receptive to smell stimuli?

8. Can you determine the evolutionary advantage for having taste buds that determine unpleasant bitter compounds in many plant species?

9. Some individuals with severe sinus infections can lose their sense of smell. How can an infection that spreads from the frontal or maxillary sinus impair the sense of smell? What structure or structures might be affected?

10. Material must be in solution for it to be perceived by gustatory receptors. What process is used (olfaction, gustation) to perceive a lipid-based food, such as garlic or peppermint?

11. Some smells that we perceive as two separate smells are actually identical. What are the other cues that we use to distinguish these two "smells" as being distinct?

12. How does a cold (rhinovirus) influence our perception of the flavor of food?

13. Does adaptation to one smell influence the adaptation to another smell?

14. Compare your "taste map" of your tongue to the one from your lab partner. Are they exact? Did you have differences in sensitivity to specific tastes?

? Chapter Summary Data

Use this section to record your results from questions within the exercise.

1a. _____

1b. _____

2. _____

3. _____

4. _____

5. _____

LABORATORY

Eye and Vision

INTRODUCTION

In most people eyesight accounts for much of our accumulated knowledge. The importance of the eye can be inferred in that 5 of the 12 cranial nerves are dedicated, at least in part, to either receiving visual stimuli or coordinating the movement of the eyes. The anatomy and physiology of the eye are discussed in the Saladin text in chapter 16, "Sense Organs."

Anatomically, the eye consists of an anterior portion, visible as we look at the face of an individual, and a posterior portion, situated in the orbit of the skull. Light from the external environment travels through a number of transparent structures that bend and focus the light on the retina, the receptive layer of the eye that converts light energy to nerve impulses. Nerve impulses travel from the eyes in the optic nerves to the brain, where they are interpreted as sight in the occipital lobes of the brain. This exercise involves learning the structure of the eye and correlating those structures to the function of vision by performing basic physiology experiments.

OBJECTIVES

At the end of this exercise you should be able to

1. identify the major structures of the mammalian eye;
2. describe the six extrinsic muscles of the eye and their effect on the movement of the eye;
3. distinguish between the pupil and the iris of the eye;
4. describe the position of the choroid in reference to the retina and the sclera;
5. describe the function of the rods and the cones of the eye;
6. define the near point of the eye;
7. determine the visual field for both eyes;
8. demonstrate the Snellen vision tests and those for accommodation and astigmatism.

MATERIALS

Models and charts of the eye

Snellen charts

Astigmatism charts

Vision Disk (Hubbard) or large protractor

Microscopes

Prepared microscope slides of eye in sagittal section

Preserved sheep or cow eyes

Dissection trays

Dissection gloves (latex or plastic)

Scalpel or razor blades

Animal waste disposal container

Card with fine print (8-point font)

3-by-5-inch cards

Ishihara color book or colored yarn

Ruler (approximately 35 cm)

Ophthalmoscope and batteries

Penlight

Desk lamp with 60-watt bulb

Paper card with a simple colored image (red circle, blue triangle) printed on it

PROCEDURE

External Features of the Eye

The external anatomy of the eye and the accessory structures are illustrated in figure 26.1. Examine the eye of your lab partner and compare it to the figure. The **pupil** is located in the center of the eye and is surrounded by

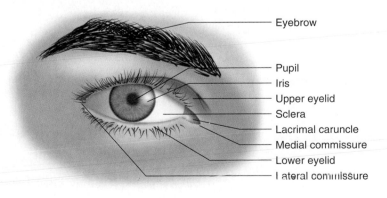

Eyebrow
Pupil
Iris
Upper eyelid
Sclera
Lacrimal caruncle
Medial commissure
Lower eyelid
Lateral commissure

FIGURE 26.1 External Anatomy of the Eye

the colored **iris.** The **sclera** is the white of the eye, and it is covered by a membrane, known as the **conjunctiva,** which continues underneath the eyelids. The eyelids join together at the **lateral commissure** and the **medial commissure.** There is a small piece of tissue near the medial commissure known as the **lacrimal caruncle.** Examine the **upper eyelid** and **eyelashes** and the **lower eyelid** and eyelashes, which prevent material from entering the eyes and (in the case of the eyelids) reduce visual stimulation when we sleep. Look also at the **eyebrow** located on the supraorbital ridge.

The sclera is a protective portion of the eye that is an attachment point for the muscles of the eye and helps maintain the intraocular pressure (the pressure inside the eye). The pressure maintains the shape of the eye and keeps the retina adhered to the back wall of the eye.

Numerous blood vessels traverse the sclera, and if they become dilated anteriorly they give the eye the appearance of being "bloodshot." The sclera is continuous with the transparent cornea in the front of the eye.

Attached to the sclera are the **extrinsic muscles** of the eye. There are six extrinsic muscles, which, in coordination, move the eye in quick and precise ways. Locate these muscles on a model and compare them to figure 26.2. These muscles and their action on the eye are listed in table 26.1.

A structure important in the maintenance of the exterior of the eye is the **lacrimal apparatus.** This consists of the **lacrimal gland** located superior and lateral to the eye (see figure 26.3). Tears bathe and protect the eye and clean dust from its surface. The fluid drains through the **nasolacrimal duct** into the nasal cavity.

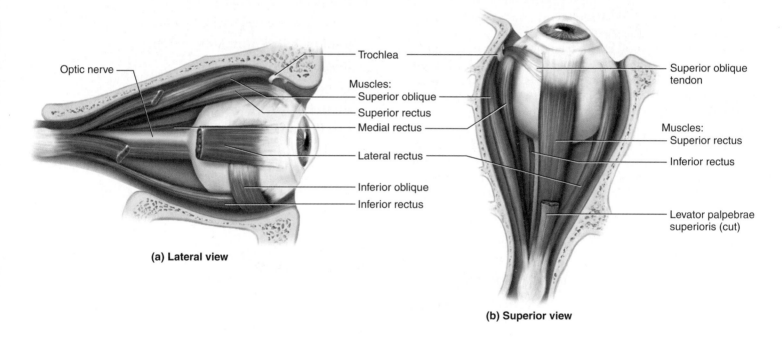

(a) Lateral view

(b) Superior view

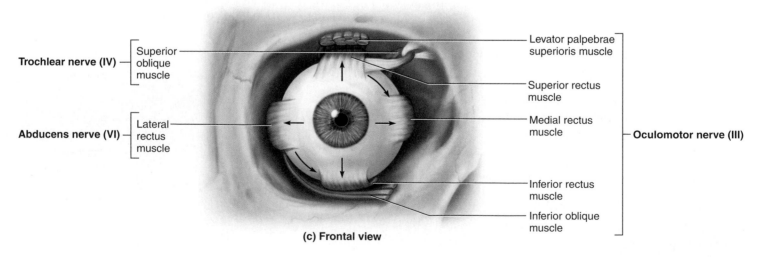

(c) Frontal view

FIGURE 26.2 Right Eye, External Features

TABLE 26.1	Extrinsic Muscles of the Eye	
Muscle Name	**Innervation**	**Direction Eye Turns**
Lateral rectus	VI (abducens)	Laterally
Medial rectus	III (oculomotor)	Medially
Superior rectus	III (oculomotor)	Superiorly
Inferior rectus	III (oculomotor)	Inferiorly
Inferior oblique	III (oculomotor)	Superiorly and laterally
Superior oblique	IV (trochlear)	Inferiorly and laterally

Interior of the Eye

From the front of the eye, the first layer covering the inside of the eyelid and extending across the sclera is the conjunctiva. The conjunctiva is composed of epithelial tissue and is an important indicator of a number of clinical conditions (for example, conjunctivitis). In the center of the eye is the transparent **cornea** (see figures 26.1 and 26.4). The cornea is the structure of the eye most responsible for the bending of light rays that strike the eye. The lens only fine-tunes images. It is composed of dense connective

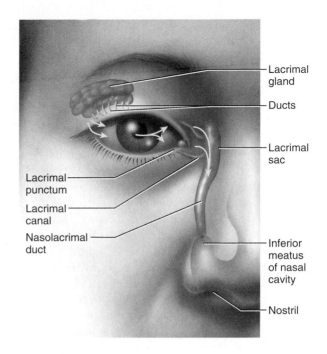

FIGURE 26.3 Lacrimal Apparatus

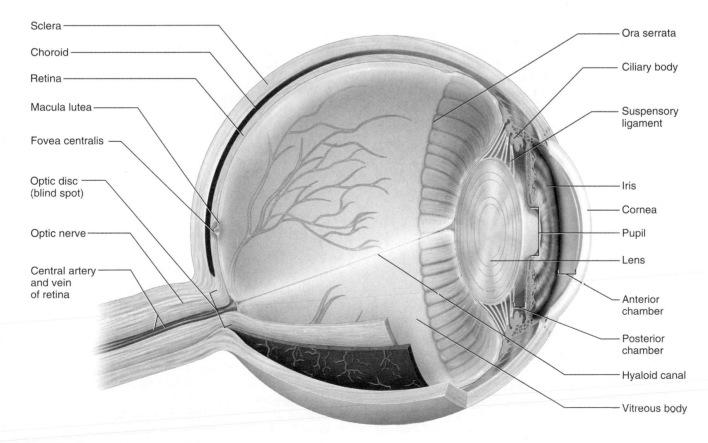

FIGURE 26.4 Sagittal Section of the Eye

tissue and is avascular. Why would the presence of blood vessels in the cornea be a visual liability?

Directly behind the cornea is the **anterior cavity,** subdivided into the **anterior chamber,** between the cornea and the **iris,** and the **posterior chamber,** between the iris and the **lens.** The anterior cavity is filled with **aqueous humor,** produced by the **ciliary body.** Only a few milliliters of aqueous humor are produced each day, and this amount is absorbed by a **venous sinus** (canal of Schlemm).

The iris is what gives us a particular eye color. People with blue and gray eyes are more sensitive to visible light than those with brown eyes, due to the protective pigment melanin found in brown eyes. The **circular muscles** of the iris constrict in bright light, reducing the diameter of the **pupil** (the space enclosed by the iris), and the **radial muscles** constrict in dim light, increasing the diameter of the pupil. Behind the pupil is the lens, made of a crystalline protein. The lens is more pliable in youth and stiffens as a person ages. Because of the loss of this elasticity, people in their forties usually begin to use reading glasses. The ciliary muscle in the ciliary body contracts and the suspensory ligaments that attach to the lens loosen, decreasing the pull on the lens. The lens becomes rounder, allowing for close focusing.

Behind the lens is the **posterior cavity,** or **vitreous chamber** (see figure 26.4). This cavity occupies most of the posterior portion, or fundus, of the eye. The posterior cavity is filled with **vitreous humor,** a clear, jellylike fluid that maintains the shape of the eyeball. Most of the posterior cavity is bounded by three layers, or **tunics,** of the eye. The outermost one, the sclera, is composed of dense irregular connective tissue. Inside of this is the **choroid,** a pigmented, vascular layer. The blood vessels found in this layer nourish the eye and the pigmentation prevents light from scattering and blurring vision. The layer closest to the vitreous humor is the **retina.** The retina consists of an outer **pigmented epithelium,** which absorbs light passing through the eye and prevents light scattering, and an inner **neural layer.**

The retina is a neural layer that converts light energy into nerve impulses. Light strikes the photoreceptor cells at the posterior portion of the retina, causing these cells to transmit signals to the **bipolar cells.** The bipolar neurons synapse with the **ganglion cells,** thus, the visual stimulation that occurs in the posterior portion of the eye is transmitted anteriorly (toward the vitreous humor) to the ganglion cells. The axons of the ganglion cells exit the eye as the **optic nerve.** From there the nerve impulse is transmitted to the lateral geniculate nucleus of the thalamus and then to the occipital region of the brain and integrated in the temporal lobe (see figure 26.5). Images from the left visual field cross over to the right side of the brain. The right side of the brain controls motor impulses for the left side of the body. The images in the right visual field cross over to the left side of the brain.

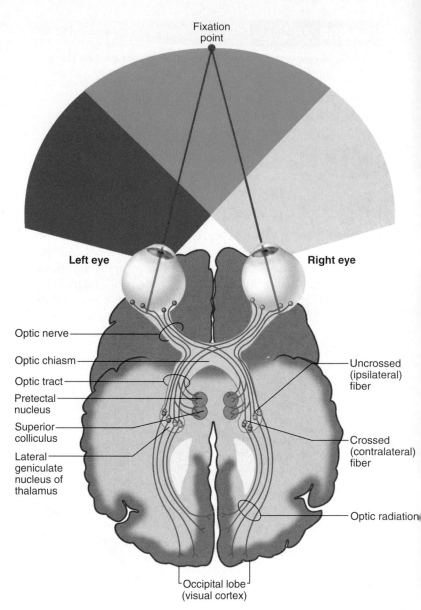

FIGURE 26.5 Visual Pathway to the Brain

Posterior Wall of the Eye

Obtain a microscope and a slide of the retina and compare what you see in the slide to figure 26.6. The neural layer of the retina is composed of three layers: **ganglionic, bipolar,** and **photosensitive.** The photosensitive layer is composed of **rods** and **cones.** Rods are important for determining the motion and general shape of objects and for sight in dim light. They do not function for color vision. Cones are involved in color vision and in visual acuity (determining fine detail). They function in bright light.

Examine a model or chart of the eye and locate the **macula lutea** at the posterior region of the eye. *Macula lutea* means "yellow spot," and in the center of this structure is the **fovea centralis,** a region where the concentration

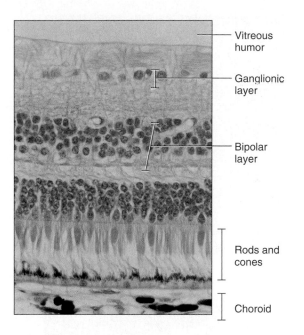

FIGURE 26.6 Retina (400×)

of cones is greatest. In the fovea the cone cells are not covered by the neural layers as they are in the other parts of the retina. When you focus on an object intently, you are directing the image to the fovea. Locate the fovea in figure 26.7 and in models available in the lab.

Dissection of a Sheep or Cow Eye

Rinse a sheep or cow eye in running water and place it on a dissection tray. Obtain a scalpel, scissors, or new razor blade and a blunt probe. Be careful with the sharp instruments and cut *away from* the hand holding the eye. Wear latex or plastic gloves while you perform the dissection. Carefully remove the fat and muscles from the eyeball. Using a scalpel, scissors, or razor blade, make a coronal section of the eye behind the cornea (see figure 26.8). Do not squeeze the eye with force or thrust the blade sharply because you may squirt yourself with vitreous humor. Cut through the eye entirely and note the jellylike material in the posterior cavity. This is the vitreous humor. Look at the posterior portion of the eye. Note the beige retina, which may have pulled away from the darkened choroid. The choroid in humans is very dark, but you may see an iridescent color in your specimen. This is the **tapetum lucidum,** which improves night vision in some animals. The tapetum lucidum produces the "eye shine" of nocturnal animals. Also examine the tough, white sclera, which envelops the choroid.

Now examine the anterior portion of the eye. Is the lens in place? The lens in your specimen probably will not be clear. Normally, the lens is transparent and allows for light penetration.

The lens is held to the ciliary body by the suspensory ligaments (see figure 26.4). These ligaments pull on the lens and alter its shape for close or distant vision. Locate the ciliary body at the edge of the suspensory ligaments.

Turn the eye over. Is there any aqueous humor left in the anterior cavity? Make an incision through the conjunctiva and cornea into the anterior cavity. Can you determine the region of the anterior chamber and the posterior chamber that make up the anterior cavity? Locate the iris and the pupil. Dispose of the specimen in a designated waste container and rinse your dissection tools.

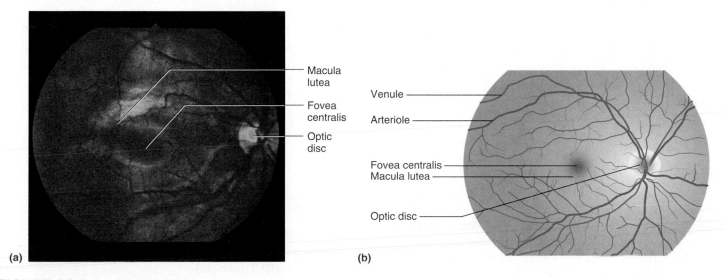

FIGURE 26.7 **Eye, Posterior View** (a) Photograph; (b) Diagram.

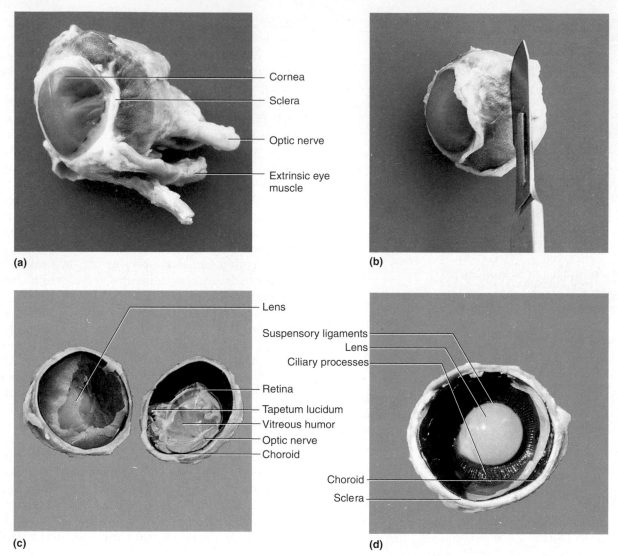

FIGURE 26.8 Dissection of a Sheep Eye (a) External features; (b) coronal section; (c) eye with vitreous humor; (d) anterior eye without vitreous humor.

Visual Tracking

Eye movements are precisely controlled by the six extrinsic muscles of the eye. To determine their effectiveness, have your lab partner follow your finger as you move it in front of the eyes. Is the movement of the eye smooth (a normal function) or do the eyes move in a jerky fashion (an abnormal condition)? Record your results here.

Eye movement: (smooth/jerky) (select one).

Determination of the Near Point

The minimum distance an object can comfortably be held in focus is called the **near point.** The eye's ability to focus is due to the elasticity of the lens. The pliability of the lens decreases with age. A 10-year-old is able to focus 8 to 10 cm away from the eye, yet a 65-year-old may not be able to focus closer than 80 to 100 cm. The decreased flexibility of the lens becomes noticeable around 40 to 50 years of age, when many people find reading glasses necessary

because of their aging lenses. You can measure your near point by holding a paper with fine print vertically at arm's length in front of you. Close one eye and slowly move the paper closer until either you see two objects or it becomes blurry. Have your lab partner measure the distance, in centimeters, from your eye to the paper. This is the near point distance. Measure the near point for both eyes in centimeters and record the data below and in the review section at the end of the exercise.

❓ Near point of right eye: _____ 1

❓ Near point of left eye: _____ 2

Measurement of the Distribution of Rods and Cones

To determine the distribution of rods and cones, have your lab partner look straight ahead. Slowly move a small colored object (a pen, piece of chalk, comb, etc.), without letting your lab partner see it, from the back of your lab partner's head, around the side, toward the front. With your lab partner still looking straight ahead, note the approximate angle when your lab partner is able to see the object (the use of rods). Your lab partner should keep his or her eyes directed forward. Continue to move the object forward slowly until recognition of the color of the object occurs. Note the approximate angle from the tip of the nose. You can record the approximate angle by using an apparatus called a Vision Disk or by placing a protractor above your lab partner's head while you make your measurements. Record these below and in the review section.

? Angle where object is perceived: _____ 3

? Angle where color is determined: _____ 4

What is the difference in the distribution of rods and cones in the eye?

? Rod distribution: _____ 5

? Cone distribution: _____ 6

When you are intently focusing on a subject, what cell type (rods or cones) are you using?

Measurement of Binocular Visual Field

Not all animals have the same **visual field.** Some prey species (such as deer or sheep) have extensive visual fields, with little binocular vision. On the other hand, many predators, birds, and arboreal animals have a more limited visual field, yet they have greater **binocular,** or **stereoscopic, vision.** Binocular vision allows arboreal animals to perceive depth—vital when judging how far away the next branch is! You can determine your visual field with the use of a Vision Disk or protractor. If you are using a Vision Disk, follow the instructions enclosed. If not, sit down and have your lab partner stand behind you. Close your right eye and look straight ahead with the left. While the right eye is closed, have your lab partner move an object (pen, paper disk, etc.) from behind your head from the left until you can just see the object. Measure the angle from the tip of the nose to where you saw the object. This is illustrated in figure 26.9 ("Angle A"). With the same eye directed ahead, continue moving the object until it is out of view (to the right of the nose somewhere). This is illustrated in figure 26.9 ("Angle B"). Determine the angle from the nose to where the object disappears. Add these two values and record the result below and in the review section.

? Angle where object is perceived: _____ 7

? Angle where object disappears: _____ 8

Now close your left eye and repeat the exercise on the right side. This is illustrated in figure 26.9 ("Angle C" and "Angle D"). Determine the total visual field for each eye by adding the sum of figure 26.9, "Angle A" and "Angle B," for the left eye and the sum of "Angle C" and "Angle D" for the right eye. For both eyes, add the sum of "Angle A" and "Angle C," and for the degree of overlap between the eyes, "Angle B" and "Angle D" (see figure 26.10). Record these data in the following spaces and in the review section.

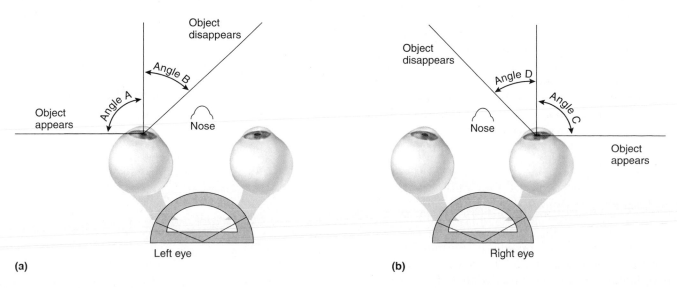

(a) Left eye; (b) right eye.

FIGURE 26.9 Measuring the Visual Field (a) Left eye; (b) right eye.

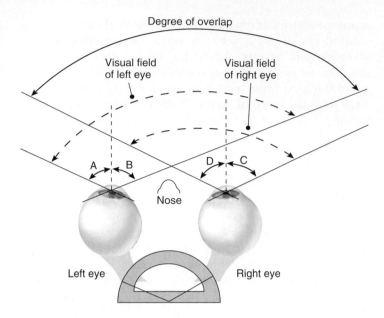

FIGURE 26.10 Visual Field

❓ Visual field of right eye: _____ 9

❓ Visual field of left eye: _____ 10

❓ Complete visual field of both eyes: _____ 11

❓ Degree of overlap: _____ 12

Measurement of Visual Acuity (Snellen Test)

Face the Snellen eye chart from 20 feet away and have your lab partner stand next to the chart. Cover one eye with a 3-by-5-inch card and have your lab partner point to the largest letter on the chart. Do *not* read the chart with both eyes open. Your lab partner should then progressively move down the chart and *note the line that has the smallest print in which you made no errors.* Your lab partner should record the numbers at the side of the line (such as 20/20). These numbers refer to your **visual acuity.** Switch the card to the other eye and repeat the test. Record your results below and in the review section.

❓ Left eye: _____ 13

❓ Right eye: _____ 14

A vision of 20/20 is considered normal. In 20/20 vision you can see the same details at 20 feet that most other people can see at the same distance. If your vision is 20/15, then you can see at 20 feet what most people can

see at 15 feet. If your vision is 20/100, then you see at 20 feet what most people see at 100 feet.

Astigmatism Tests

If the cornea or lens of the human eye were perfectly smooth, the incoming image would strike the retina evenly and there would be no blurry areas. In most people the cornea or lens is not perfectly smooth, and this is known as astigmatism. In a normal optical exam the distance correction is made first (to determine visual acuity) and then, with the corrective lenses in place, an astigmatism test is performed.

If you do not normally wear corrective lenses, then cover one eye and examine the astigmatism chart (as represented in figure 26.11) from 20 feet away. The chart consists of a series of parallel lines radiating from the center. Stare at the center of the chart and determine which of the sets of parallel lines, if any, appears light or blurry. Have your lab partner note the corresponding number on the chart and record the number below and in the review section.

❓ Astigmatism numbers: _____ 15

If you wear corrective lenses, then not only is the condition of nearsightedness or farsightedness corrected but astigmatism is corrected also. To test for astigmatism move about 12 feet from the chart, hold your glasses slightly away from your face, and then rotate them 90 degrees. If your glasses normally correct for astigmatism, the lines on the chart will be blurry as you rotate your glasses.

Ophthalmoscope

Clinical use of the ophthalmoscope is important not only for the diagnosis of variances in the eyeball but also as a potential indicator of diseases, such as **diabetes mellitus,** that may affect the eye. Familiarize yourself with the parts of the ophthalmoscope (see figure 26.12). Locate the ring (rheostat control) at the top of the handle. Depress the button and turn the ring until the light comes on. The head piece of the ophthalmoscope consists of a rotating disk of lenses. You can change the diopter (strength) of

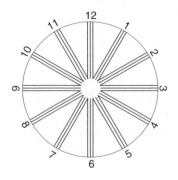

FIGURE 26.11 Astigmatism Chart

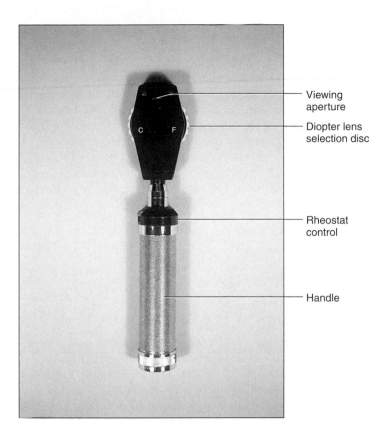

FIGURE 26.12 Ophthalmoscope

FIGURE 26.13 Use of Ophthalmoscope

the lenses by rotating the disc clockwise or counterclockwise. As the numbers get progressively larger, you are looking through more convex lenses. If you rotate the dial in the other direction, the lenses become more concave. At the zero reading there are no lenses in place.

Observation with the Ophthalmoscope

Caution! Examine the posterior region of the eye only for a short time. Extensive use of the ophthalmoscope can damage the eye.

Sit facing your lab partner. Select the left eye to examine. Have the ophthalmoscope setting at zero and use your left hand to hold the ophthalmoscope. Use your left eye to look through the ophthalmoscope and move in close to examine the eye of your lab partner (see figure 26.13). You may rest your hand on the cheek of your lab partner if you need to steady your hand. Look into the pupil and examine the back of the eye. If the back appears fuzzy, rotate the dial clockwise (positive diopters) and see if it comes into focus. If a positive setting provides a clear view of the eye, then your lab partner has **hyperopia** (hypermetropic vision), or farsightedness. In farsightedness the eyeball is too short and the image is focused posterior to the retina. Positive diopter lenses focus the image on the retina. If the image is indistinct, then adjust the dial counterclockwise to

obtain a negative diopter reading. If the image is clear with a negative number, then your lab partner has **myopia** (myopic vision), or nearsightedness. In nearsightedness the eyeball is too long and lenses with negative diopters focus the images farther back on the retina. This procedure is based on the assumption that your vision, as the examiner, is normal.

Pupillary Reactions

Have your lab partner sit in a dark room for a minute or two. Examine his or her eyes in dim light. Are the pupils dilated or constricted? Record the data below and in the review section.

? Pupil diameter in dim light: _____ 16

Shine a penlight in the right eye. Record what happens to the pupil diameter.

? Right pupil diameter: _____ 17

As you shine the light into the right eye, what occurs in the left pupil? This is called a **consensual reflex.** Record the effect.

? Left pupil diameter: _____ 18

Color Blindness

Color vision is dependent on three separate cone cell sensitivities. Cones may be red, green, or blue sensitive. Changes in the genes on the X chromosome are the most common cause of color blindness. Some individuals may be unable to see a particular color, while others may have a reduction in their ability to see a particular color. Color blindness is most common in males and relatively rare in females. This is due to the chromosomal makeup of the two sexes. The male chromosome makeup is XY. The Y chromosome does not carry the gene for color vision. If the X chromosome carries a gene for color blindness, then the male exhibits color blindness. On the other hand, if a female carries the gene for color blindness on the X chromosome, the chances are that she will have a normal gene on the other X chromosome. The normal gene is expressed, cone pigments are produced, and the female has normal color vision.

You can test for color blindness by matching various samples of colored yarn with a presented test sample or you can use Ishihara color charts, as represented in figure 26.14. If you use the color charts, flip through the book with a lab partner and record which charts were accurately viewed and which plates, if any, were missed. You can calculate the type of color blindness and the degree by following your instructor's directions or those that come with the color chart. Record your data below and in the review section.

❓ Your color vision: _____ 19

Afterimages

The photosensitive pigment of the rods is called **rhodopsin,** which is composed of light-sensitive **retinal** (a vitamin A derivative) and the protein **opsin.** When light strikes the retina, the purple-colored rhodopsin splits into its two component parts and becomes pale (a process known as "bleaching"). You can test the time for separation and reassembly of the photopigments by staring at a contrasting image on a card under a moderately

bright lightbulb (60 watt) for a few moments. Stare long enough to get an image (about 10 to 20 seconds) and then shut your eyes. Have your lab partner record the time. You should see a colored image against a dark background. This is known as a positive afterimage, due to the photoreceptors continuously firing. After a few moments you should see the reverse of the original image (dark against a light background). This is known as a negative afterimage. A negative afterimage reflects the bleaching effects of rhodopsin. Record the time change for the positive afterimage and the negative afterimage below and in the review section.

❓ Positive afterimage: _____ 20

❓ Negative afterimage: _____ 21

Determination of the Blind Spot

The **optic disc** is a region where the retinal nerve fibers exit from the back of the eye and form the **optic nerve.** The mass exit of the nerve fibers leaves a small circle at the back of the eye devoid of photoreceptors. This region, the optic disc, is also known as the **blind spot.** You can locate the blind spot by holding this lab manual at arm's length. Use your right eye (close your left eye) and stare at the following x. It should be in line with the middle of your nose. *Slowly* move the manual closer to you and stare only at the x. At a particular distance the dot disappears.

x   ●

You can test for the blind spot in your left eye as well. Make sure the dot is aligned with your nose (in the midsagittal plane).

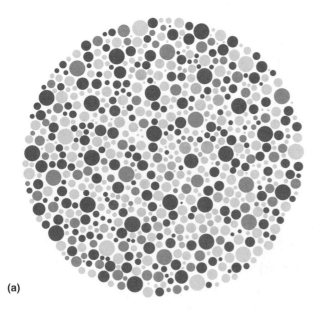
(a)

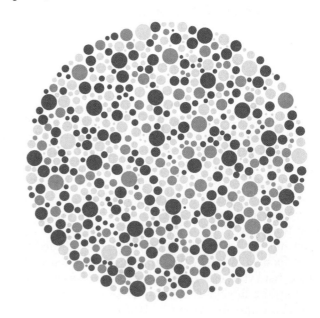
(b)

FIGURE 26.14 Ishihara Test for Color Blindness (a) Plate as seen by a person with normal color vision. (b) plate as seen by a person who has red-green color blindness.

REVIEW SECTION

Eye and Vision

Name _____ Date _____

Lab Section _____ Time _____

Review Questions

1. Eye shine in nocturnal mammals is different from the "red eye" seen in some flash photographs. Eye shine is the reflection from the tapetum lucidum. What might produce "red eye?"

2. Fill in the following illustration using the terms provided.

lens	retina	anterior cavity (anterior chamber)
sclera	optic nerve	choroid
ciliary body	cornea	pupil
		suspensory ligaments

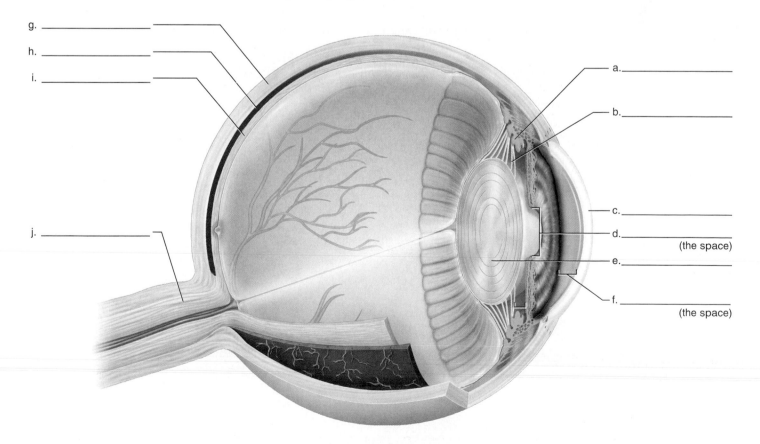

g. _____

h. _____

i. _____

a. _____

b. _____

c. _____

d. _____
(the space)

j. _____

e. _____

f. _____
(the space)

Sagittal section of the eye

3. Since the lens is made of protein, what effect might the preserving fluid used in lab have on the structure of the lens? How would this affect the clarity?

4. What is the consensual reflex of the pupil?

5. How does the vitreous humor differ from the aqueous humor in location and viscosity?

6. What layer of the eye converts visible light into nerve impulses?

7. What nerve carries the action potentials from the ganglion cells to the thalamus of the brain?

8. What is another name for the sclera?

9. How would you define an extrinsic muscle of the eye?

10. What gland produces tears?

11. What is the name of the transparent layer of the eye in front of the anterior chamber?

12. The iris of the eye has what function?

13. Where is vitreous humor found?

14. What is the middle tunic of the eye called?

15. Is the lens anterior or posterior to the iris?

16. Which retinal cells are responsible for vision in dim light?

17. How would you define the near point of the eye?

18. What do the numbers 20/100 mean for visual acuity?

19. What is an astigmatism?

20. In what area of the eye is the blind spot located?

21. Label the following illustration.

a. _____

b. _____
c. _____
d. _____
e. _____
f. _____
g. _____
h. _____

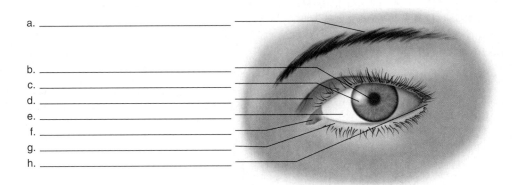

?Chapter Summary Data

Use this section to record your results from questions within the exercise.

1. _____   12. _____

2. _____   13. _____

3. _____   14. _____

4. _____   15. _____

5. _____   16. _____

6. _____   17. _____

7. _____   18. _____

8. _____   19. _____

9. _____   20. _____

10. _____   21. _____

11. _____

LABORATORY

Ear, Hearing, and Equilibrium

INTRODUCTION

The ear is a complex sense organ that performs two major functions, hearing and equilibrium. It consists of three regions, an outer ear, a middle ear, and an inner ear. The structure and function of the ear are covered in the Saladin text in chapter 16, "Sense Organs." Hearing is considered **mechanoreception,** because the ear receives mechanical vibrations (sound waves) and translates them into nerve impulses. This process begins with the vibrations reaching the outer ear and ends up being interpreted as sound in the temporal lobe of the brain. Equilibrium, on the other hand, involves receptors in the inner ear, along with other sensory perceptions, such as visual cues from the eye, and **proprioreception** in joints. There are two types of equilibrium sensed by the inner ear. These are **static equilibrium** and **dynamic equilibrium.** In static equilibrium an individual is able to determine his or her nonmoving position (such as standing upright or lying down). In dynamic equilibrium motion is detected. Sudden acceleration, abrupt turning, and spinning are examples of dynamic equilibrium.

OBJECTIVES

At the end of this exercise you should be able to

1. explain how mechanical sound vibrations are translated into nerve impulses;
2. list the structures of the outer, middle, and inner ear;
3. describe the structure of the cochlea;
4. perform conduction deafness tests, such as the Rinne and Weber tests;
5. compare dynamic and static equilibrium and the structures involved in their perception.

MATERIALS

Models and charts of the ear

Microscope

Slides of the cochlea

Tuning fork (256 Hz)

Rubber reflex hammer

Audiometer

Cotton balls

Model of ear ossicles

Meter stick

Ticking stopwatch

Bright desk lamp

Swivel chair

PROCEDURE

Anatomy of the Ear

The ear can be divided into three regions, the **outer (external), middle,** and **inner ear.** The outer ear consists of those auditory structures superficial to the **tympanic membrane (eardrum).** The middle ear contains the ear **ossicles (bones)** and **auditory (Eustachian) tube,** and the inner ear consists of the **cochlea, vestibule,** and **semicircular ducts** (see figure 27.1).

Structure of the Outer Ear

The outer ear consists of the **pinna (auricle),** which can further be subdivided into the **helix** and the **earlobe (lobule).** The helix is composed of stratified squamous epithelium overlying elastic cartilage. This cartilage allows the ears to bend significantly. Deep to the pinna the external ear forms the **auditory canal,** which penetrates into the temporal bone. The tympanic membrane is the border between the outer ear and the middle ear. It is composed of connective tissue covered by epithelial tissue. The membrane is sensitive to sound and vibrates as sound is funneled down the auditory canal.

Structure of the Middle Ear

The middle ear consists of a main cavity known as the **tympanic cavity;** three small bones, or **ossicles;** and the auditory tube (see figure 27.1). The ossicle attached to the tympanic membrane is the **malleus** (*malleus* = hammer). The malleus is attached to the **incus** (*incus* = anvil), which is attached to the **stapes** (*stapes* = stirrup).

Sound consists of pressure waves. As these waves strike the tympanic membrane, it vibrates. This vibration is conducted by the ossicles to the oval window. The process of moving from a large-diameter structure

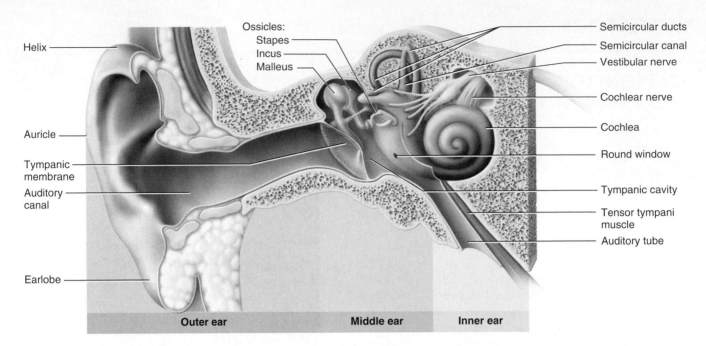

FIGURE 27.1 Anatomy of the Ear

(tympanic membrane) to a smaller-diameter structure (oval window) magnifies the sound about 18 times. See figure 27.2. Examine the ossicles in figure 27.1 and on models in the lab. In addition to the ossicles the **auditory tube** (Eustachian tube) occurs in the middle ear. This tube connects the middle ear to the nasopharynx (see figure 27.1) and equalizes pressure between the middle ear and the external environment when changes of pressure occur (such as during changes of elevation). The auditory tube can be a conduit for microorganisms that travel from the nasopharynx to the middle ear and lead to middle-ear infections, particularly in young children.

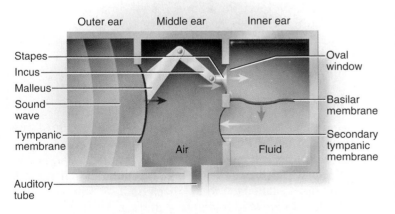

FIGURE 27.2 **Model of Hearing** Pressure waves of sound vibrate the tympanic membrane, which transfers sound to the oval window and finally to the basilar membrane, where vibration is converted to neural impulses.

Structure of the Inner Ear

The inner ear is encased in two labyrinthine structures and filled with two separate fluids. The outermost structure is the **bony labyrinth.** Inside the bony labyrinth is **perilymph,** a clear fluid external to the **membranous labyrinth.** The fluid enclosed by the membranous labyrinth is the **endolymph,** important in both hearing and equilibrium.

The inner ear is a complex structure composed of three regions, the cochlea, the vestibule, and the semicircular ducts (see figures 27.1 and 27.3). The cochlea is a spiral structure that resembles a seashell (*cochlea* = snail), while the semicircular ducts look like three loops. Between these two structures is the vestibule.

Cochlea   The cochlea (see figures 27.1, 27.3, and 27.4) is involved in hearing. As the sound waves travel down the auditory canal they cause the tympanic membrane to vibrate. This vibration rocks the ear ossicles, which are connected to the inner ear. As the stapes vibrates it moves back and forth in the **oval window,** and this causes fluid to move back and forth in the cochlea. The cochlea also has a **round window,** which allows the pressure wave from the ossicles to move fluid. Sound waves are measured by their amplitude (loudness) and wavelength (pitch). The wavelengths are measured in cycles per seconds, also known as hertz (Hz). The human ear is capable of hearing high-pitched sounds up to 20,000 Hz and sounds as low as about 20 Hz. High-pitched sounds with vibrations of short wavelength, up to 20,000 Hz in the human ear, stimulate the region of the cochlea closest to the middle ear. Low-pitched sounds with longer wavelengths, down

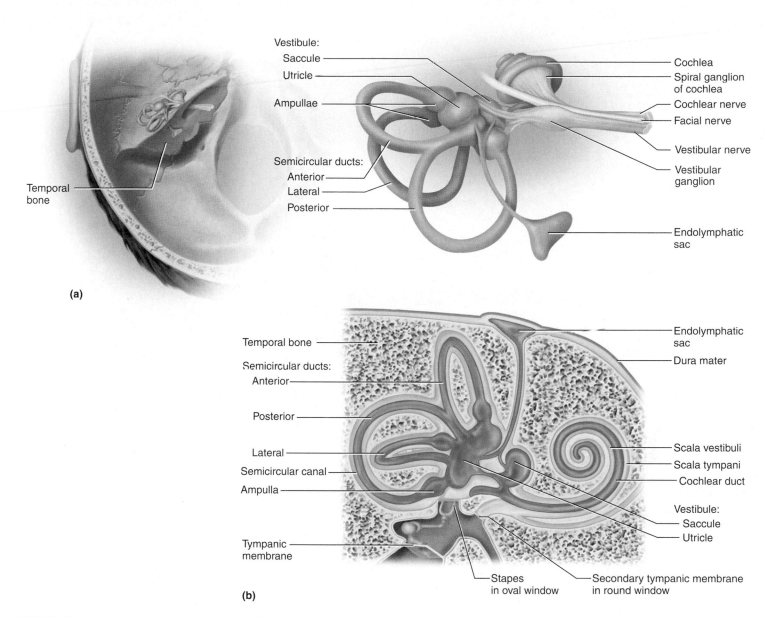

Vestibule:
Saccule
Utricle
Ampullae

Semicircular ducts:
Anterior
Lateral
Posterior

Temporal bone

Cochlea
Spiral ganglion of cochlea
Cochlear nerve
Facial nerve
Vestibular nerve
Vestibular ganglion
Endolymphatic sac

(a)

Temporal bone

Semicircular ducts:
Anterior
Posterior
Lateral
Semicircular canal
Ampulla

Tympanic membrane

Endolymphatic sac
Dura mater

Scala vestibuli
Scala tympani
Cochlear duct

Vestibule:
Saccule
Utricle

Stapes in oval window
Secondary tympanic membrane in round window

(b)

FIGURE 27.3 **Anatomy of the Inner Ear** (a) Major regions of the inner ear; (b) relationship of membranous labyrinth to bony labyrinth.

to 20 Hz, stimulate the region of the cochlea farther from the middle ear. In this way the cochlea can perceive sounds of varying wavelengths at the same time.

If you examine a microscopic section of the cochlea in cross section you will see a number of chambers. The chambers are clustered in threes. Find the **scala vestibuli, cochlear duct (scala media),** and **scala tympani** on the microscope slide. Compare these to figures 27.4 and 27.5.

Note the **spiral organ (acoustic organ,** or **organ of Corti)** in the area between the **vestibular membrane** and the **basilar membrane.** The spiral organ is seen here in cross section, but remember that it runs the length of the cochlea. The spiral organ is sensitive to sound waves.

As a particular region of the spiral organ is stimulated, the basilar membrane vibrates. This bends the cilia that connect the hair cells to the tectorial membrane. The hair cells send impulses to the cochlear branch of the **vestibulocochlear nerve.** These impulses travel to the **auditory cortex** of the **temporal lobe,** where they are interpreted as sound (see figure 27.6).

Vestibule Another part of the inner ear is the vestibule. The **vestibule** consists of the **utricle** and **saccule** (see figures 27.3 and 27.7). These two chambers are involved in the interpretation of static equilibrium and acceleration. The utricle and saccule have regions known as maculae, which

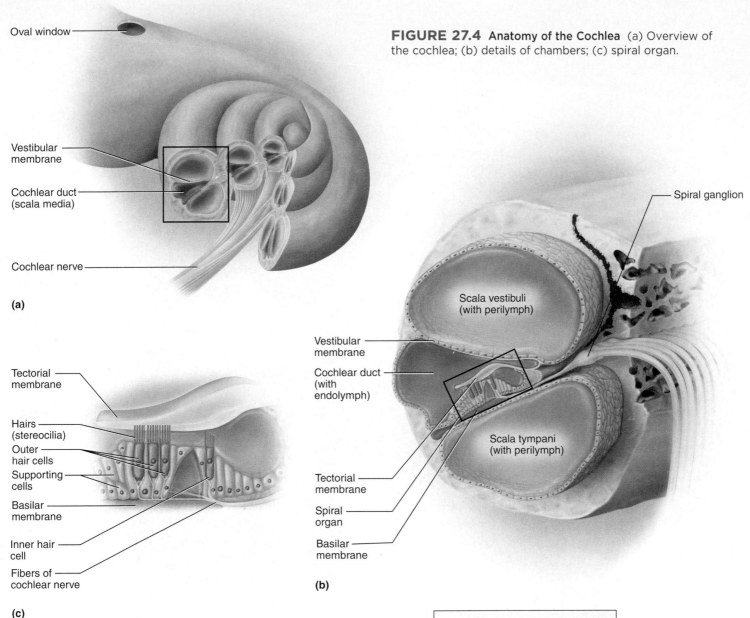

Oval window

Vestibular membrane

Cochlear duct (scala media)

Cochlear nerve

(a)

FIGURE 27.4 Anatomy of the Cochlea (a) Overview of the cochlea; (b) details of chambers; (c) spiral organ.

Spiral ganglion

Scala vestibuli (with perilymph)

Vestibular membrane

Cochlear duct (with endolymph)

Scala tympani (with perilymph)

Tectorial membrane

Spiral organ

Basilar membrane

(b)

Tectorial membrane

Hairs (stereocilia)

Outer hair cells

Supporting cells

Basilar membrane

Inner hair cell

Fibers of cochlear nerve

(c)

consist of cilia grouped together in a gelatinous mass called the otolithic membrane weighted with calcium carbonate stones called **otoliths.** These can be seen in figure 27.7. As the head is accelerated or tipped by gravity, the otoliths cause the cilia to bend, indicating that the position of the head has changed. Static equilibrium is perceived not only from the vestibule but from visual cues as well. When the visual cues and the vestibular cues are not synchronized, then a sense of imbalance or nausea can occur.

Semicircular Ducts The third part of the inner ear consists of the **semicircular ducts,** each located inside a semicircular canal, involved in determining dynamic equilibrium. There are three semicircular ducts, each at 90 degrees to one another (in the horizontal, sagittal, and coronal planes) (see figure 27.8). Each semicircular duct is

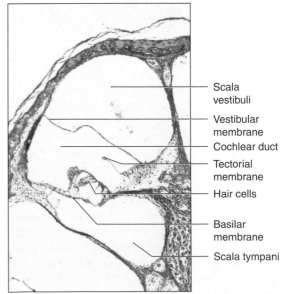

Scala vestibuli

Vestibular membrane

Cochlear duct

Tectorial membrane

Hair cells

Basilar membrane

Scala tympani

FIGURE 27.5 Photomicrograph of Cross Section of the Cochlea (100×)

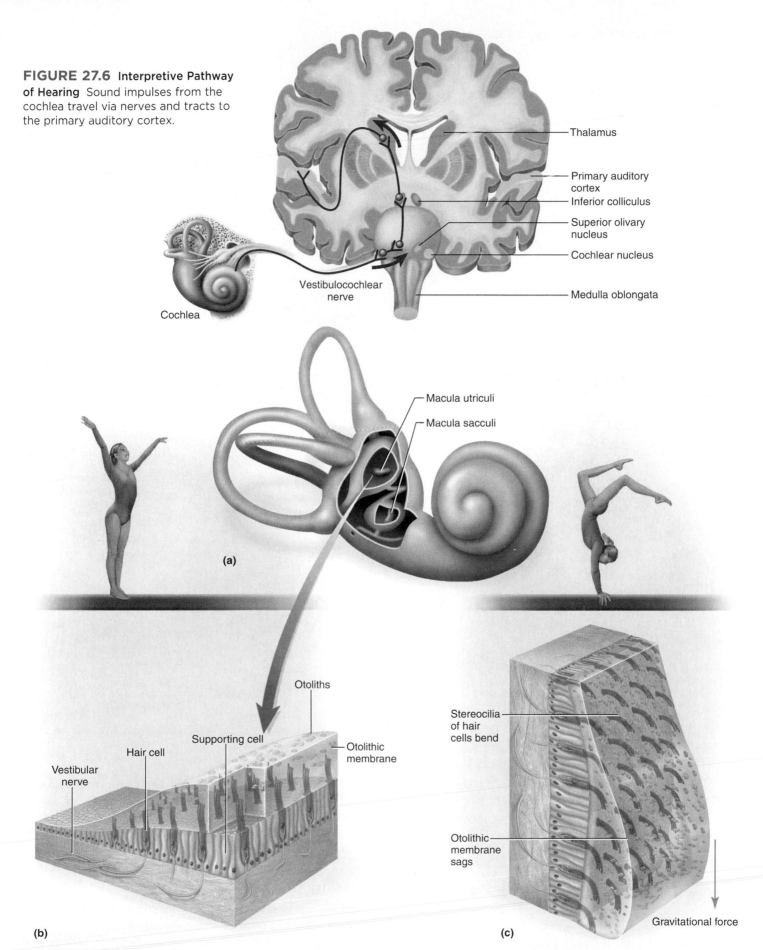

FIGURE 27.6 Interpretive Pathway of Hearing Sound impulses from the cochlea travel via nerves and tracts to the primary auditory cortex.

Thalamus

Primary auditory cortex

Inferior colliculus

Superior olivary nucleus

Cochlear nucleus

Medulla oblongata

Vestibulocochlear nerve

Cochlea

Macula utriculi

Macula sacculi

(a)

Otoliths

Supporting cell

Hair cell

Otolithic membrane

Vestibular nerve

(b)

Stereocilia of hair cells bend

Otolithic membrane sags

Gravitational force

(c)

FIGURE 27.7 Vestibule (a) Inner ear; (b) macula when head is upright; (c) macula when head is tilted.

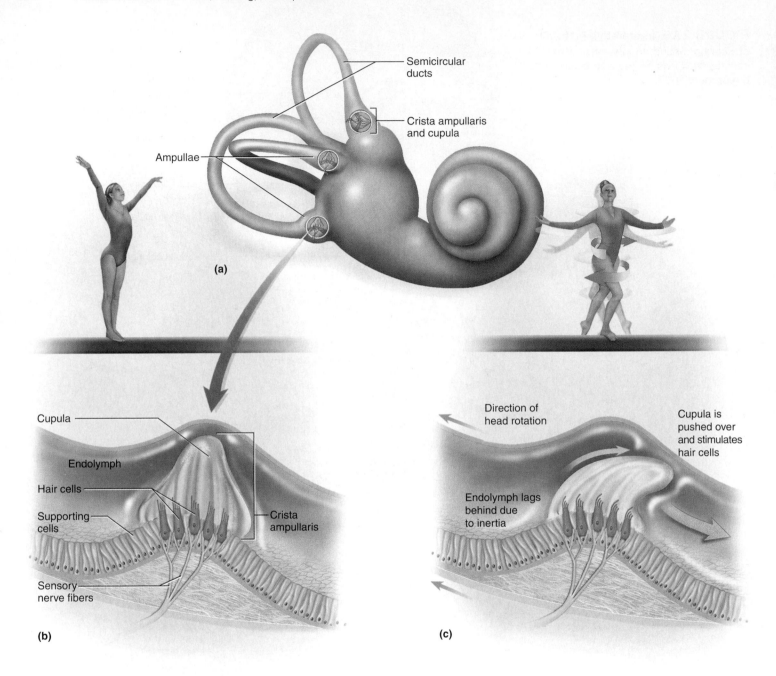

(a)

(b)

(c)

FIGURE 27.8 Semicircular Canals (a) Position of semicircular canals; (b) position of semicircular canals when still; (c) function of crista ampullaris and cupula during movement.

filled with endolymph and is expanded at its base into an **ampulla.** Inside each ampulla are clusters of hair cells and support cells (called the **crista ampullaris**). The cells have stereocilia and a kinocilium enclosed in a gelatinous material called the **cupula.** The endolymph in the membranous labyrinth has inertia; that is, it tends to remain in the same place. As the head is rotated, the endolymph pushes against the **stereocilia.** In this way **angular** or **rotational** acceleration can be determined as shown in figure 27.8 (c). The three semicircular ducts are at right angles to each other. If motion in the forward plane occurs (such as by

doing back flips), then the **anterior semicircular ducts** are stimulated. If you were to turn cartwheels, then the **posterior semicircular ducts** would be stimulated. If you were to spin around on your heels, then the **lateral semicircular ducts** would pick up the information.

Hearing Tests

Obtain a meter stick and a ticking stopwatch. Have your lab partner sit in a quiet room and slowly move the ticking watch away from one of his or her ears until the sound can no longer be heard. Record the distance in centimeters (when the sound can no longer be heard) in the space provided and in the review section. Check the other ear and record the data.

❓ Maximum distance sound perceived for right ear:

_____ 1

❓ Maximum distance sound perceived for left ear:

_____ 2

Audiometer Test

Ask your instructor to demonstrate how to use an audiometer to test hearing. Age-related hearing loss (presbycusis) (pres-bee-CUE-sis) is a progressive condition that affects many older adults, though hearing loss is greatly accelerated by loud music, moderate to extensive headphone use, and exposure to machines that operate at high decibel levels. Another common source of hearing loss is the use of recreational firearms without hearing protection.

Threshold for Hearing

The **threshold for hearing** is the lowest level at which your lab partner can hear the tone in at least 50% of the trials. To measure this value set the audiometer to 125 Hz and then lower the level in 10 dB increments from 50 dB to the point where your lab partner can no longer hear the tone. Record the threshold for hearing. You can test other frequencies as well. Hearing loss is significant when the hearing threshold is at 20 dB or more at any two frequencies in an ear or 30 dB at any particular frequency. Record the minimum sound levels.

Right Ear	Left Ear
125: _____	_____
250: _____	_____
500: _____	_____
1,000: _____	_____
2,000: _____	_____
4,000: _____	_____
8,000: _____	_____

Weber Test

Conduction deafness occurs when damage to the tympanic membrane or ear ossicles is present. You can test for this type of deafness with a tuning fork. Hold on to

FIGURE 27.9 Weber Test

the handle of a tuning fork at 256 Hz (middle C). Strike the tines (the forked portion) on a solid surface and gently place the end of the handle on your lab partner's chin or forehead (see figure 27.9).

If the sound is louder in one ear, then this may be a sign of conduction deafness. In this case the affected ear is not picking up sounds from the tympanic membrane and is more sensitive to vibrations coming through the skull bones. The sound will be louder in the ear that has conduction deafness. If the ear has nerve damage, then the sound will be louder in the normal ear. If you cannot hear the sound, you should place cotton balls in both ears. Record your results in the following space and in the review section.

❓ Sound perception (right/left or equal in both ears):

_____ 3

Rinne Test

Strike the tuning fork and place the handle on the mastoid process. When your lab partner indicates that the sound disappears, lift the tuning fork from the mastoid process and bring it to the outside of the pinna (see figure 27.10). Can the sound now be heard? If the sound cannot be heard after the tuning fork is placed near the outside of the ear, then there is damage to the tympanic membrane or ear ossicles (conduction deafness). This may also be due to obstructions in the ear canal (such as earwax). The sound is normally louder in air because of the amplification from the tympanic membrane to the oval window.

FIGURE 27.10 Rinne Test

Record your results in the following spaces and in the review section.

❔ Right ear: _____   4

❔ Left ear: _____   5

Sound Location

Have your lab partner sit with eyes closed. Strike the tuning fork with a rubber reflex hammer above his or her head. Have your lab partner describe to you where the sound is located. Strike the tuning fork behind the head, to each side, in front, and below the chin of your lab partner. Record the results in the following chart.

Location Where Sound Was Struck	Where Sound Was Perceived
Above head	
Behind head	
Right side	
Left side	
In front of head	
Below chin	

Postural Reflex Test

Postural reflexes are important for maintaining the upright position of the body. These reflexes are negative-feedback mechanisms. If, for example, you lean slightly to the left, your left foot abducts to regain the center of balance.

1. Select an area free from any obstacles.
2. While you read this manual, stand on your tiptoes. Your lab partner should give you a little nudge to the left or right (not enough to knock you off your feet) to push you off balance. The postural reflex is reflected in the movement of the foot on the opposite side of your lab partner. If you are nudged to the left, then your left foot should move to the side to correct against the direction of the contact. Did your postural reflex work?
3. Record your result in the following space and in the review section.

❔ Postural reflex: _____   6

Barany's Test

Barany's test examines visual responses to changes in dynamic equilibrium. As the head turns, one of the reflexes that occurs is movement of the eyes in the opposite direction of the rotation. Nerve impulses from the semicircular canals innervate the eye muscles and cause the eye movement. When rotating in one direction, the eyes move in the opposite direction, so that there is enough time for a visual image to be fixed. The volunteer for this test should not be subject to dizziness or nausea.

1. Place the subject in a swivel chair with four or five students close by, standing in a circle, in case the subject loses balance and begins to fall. The student chosen for this exercise should grab onto the chair firmly so as not to fall off. The subject should tilt his or her head forward about 30 degrees, which will place the lateral semicircular ducts horizontally for maximum stimulation.
2. One member of the group should spin the chair around about 10 revolutions while the subject keeps his or her eyes open.
3. Stop the chair and have the subject look forward. The twitching of the eyes is called **nystagmus** and is due to the stimulation of the endolymph flowing in the semicircular ducts. When the chair is stopped the fluid in the endolymph will have overcome the inertia and will continue to flow in the ducts. Did the movement of the eyes occur in the direction of the rotation or in the opposite direction? Record your results in the following space and in the review section.

❔ Direction of chair rotation relative to subject: _____   7

❔ Direction of eye rotation: _____   8

Romberg Test

The Romberg test involves testing the static equilibrium function of the body.

1. Place the subject in front of a blackboard or any surface on which you can see the shadow of the subject. Use a desk lamp, if needed.

2. Have the subject stand in that direction for 1 minute and determine if there are any exaggerated movements to the left or right. The subject should not rest against the wall.

3. Record your results in the following space and in the review section.

❓ Amount of lateral sway: _____ 9

You should then have your lab partner close his or her eyes (to reduce visual cues) and repeat the test. Is the swaying motion increased or decreased? Record your results in the following space.

❓ Amount of lateral sway with eyes closed: _____ 10

Notes

REVIEW SECTION

Ear, Hearing, and Equilibrium

Name _____ Date _____

Lab Section _____ Time _____

Review Questions

1. What are the three general areas or regions of the ear?

2. The pinna of the ear consists of what two main parts?

3. To what sense modality (or modalities) does the ear respond?

4. The ear performs two major sensory functions. What are they?

5. What structure separates the outer ear from the middle ear?

6. Name the three ear ossicles.

7. From the following choices, select the function of the cochlea.

   a. static equilibrium

   b. taste

   c. hearing

   d. dynamic equilibrium

8. What area is found between the scala vestibuli and the scala tympani?

9. What is the name of the nerve that carries signals from the cochlea and vestibule to the brain?

10. What units are used to measure sound energy?

11. What part of the inner ear is involved in transmitting signals of static equilibrium?

12. Name the parts of the ear that might be impaired if a person demonstrates conduction deafness.

13. What two diagnostic tests are used to determine conduction deafness?

14. What is the name of the canal that runs from the auricle to the tympanic membrane?

15. The auditory tube connects what two cavities?

16. What tube between the middle ear and the nasopharynx is responsible for the equalization of pressure when you change elevation?

17. What is the name of the space that encloses the ear ossicles?

18. Name the ear ossicles in sequence from the tympanic membrane to the oval window.

19. Name all the parts of the inner ear.

20. Background noise affects hearing tests. In the ticking watch test, what type of result, in terms of auditory sensitivity, would you have recorded if moderate background noise were present?

21. In the Weber test, the ear that perceives the sound as being louder is the deaf ear. Why is this the case?

22. Fill in the following illustration of the ear using the terms provided.

cochlea
ear ossicles
auditory canal
auditory tube
tympanic membrane
semicircular ducts

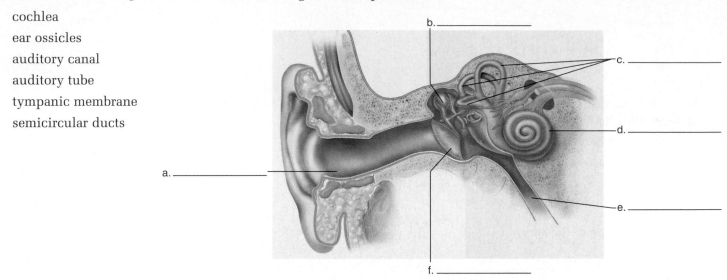

b. _____

c. _____

d. _____

a. _____

e. _____

f. _____

23. Label the following illustration of the cross section of cochlea using the terms provided.

hair cells
scala tympani
tectorial membrane
scala vestibuli
vestibular membrane
scala media (cochlear duct)

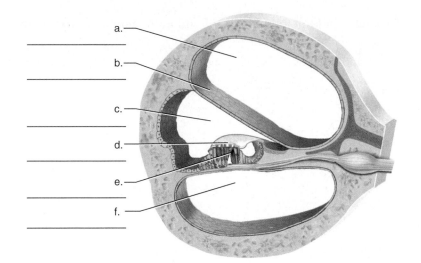

a. _____

b. _____

c. _____

d. _____

e. _____

f. _____

? Chapter Summary Data

Use this section to record your results from questions within the exercise.

1. _____   6. _____

2. _____   7. _____

3. _____   8. _____

4. _____   9. _____

5. _____   10. _____

LABORATORY

Endocrine System

INTRODUCTION

The **endocrine system** consists of glands that produce chemical messengers called **hormones,** which are picked up by blood capillaries. This type of release is called a "ductless secretion," which is characteristic of the endocrine system. On the other hand, the glands that secrete material in tubules or ducts (such as sweat and salivary glands) or directly to a surface (ovary) are known as **exocrine glands.** The secretions of the endocrine glands enter the interstitial fluid and then travel by blood vessels, which act as highways to carry the hormones throughout the body. Areas receptive to hormones are called **target cells,** and these may be organs or tissues. Hormones can have many effects. Some of these are growth; changes and development, or maturation; metabolism; sexual development; regulation of the sexual cycle; and homeostasis. Many organs of the body, such as the stomach, heart, and kidney, produce hormones and thus have endocrine functions. This exercise focuses on the glands that have a major endocrine component as their function. The endocrine system is presented in more detail in the Saladin text in chapter 17, "The Endocrine System."

OBJECTIVES

At the end of this exercise you should be able to

1. list the major endocrine organs of the human body;
2. name the hormones produced by these endocrine organs;
3. describe the detection of hormones by monoclonal antibodies;
4. list the hormones stored in the anterior and posterior pituitary glands;
5. discuss how the secretions of the endocrine glands differ from those of the exocrine glands;
6. identify endocrine organs in histological slides.

MATERIALS

Models and charts of endocrine glands

Microscopes

Microscope slides of thyroid, pituitary, adrenal gland, pancreas, testis, ovary

Urine samples from a nonpregnant and/or a pregnant individual

Ovulation test kit

Disposable gloves (latex or plastic) and biohazard container

PROCEDURE

Anatomy of the Major Endocrine Organs

Locate the major endocrine glands in figure 28.1, including the pineal gland, hypothalamus, pituitary gland, thyroid, parathyroids, thymus, pancreas, adrenals, and gonads (testes or ovaries). Once you have noted their location, you can proceed with a more detailed study.

Pineal Gland

The **pineal gland** develops from the diencephalon of the brain. The gland is named the pineal gland because it looks like a pine nut. Locate the gland in figure 28.1. The pineal gland secretes the hormone **melatonin.** The role of melatonin is not completely understood in humans. The production of melatonin from the pineal gland increases at night, so it has been named the "hormone of darkness." The level decreases during the day and these circadian rhythms (regular day/night cycles) have been suggested as contributing to sleep cycles. There is a condition known as seasonal affective disorder (SAD) that causes depression in people, typically during winter months. The increased levels of melatonin during this time of year have been implicated in SAD.

Hypothalamus and Pituitary Gland

The hypothalamus and the pituitary gland are so closely interrelated that it is appropriate to discuss them together. The hypothalamus is located in the inferior portion of the forebrain and is connected to the anterior pituitary by the hypophyseal portal system. Primary capillaries in the hypothalamus lead to the secondary capillaries in the anterior pituitary. The hypothalamus has neurons that extend into the posterior pituitary by the hypothalamo-hypophyseal tract, so the posterior pituitary can actually be considered an extension of the hypothalamus. The hypothalamus secretes both stimulating and inhibitory

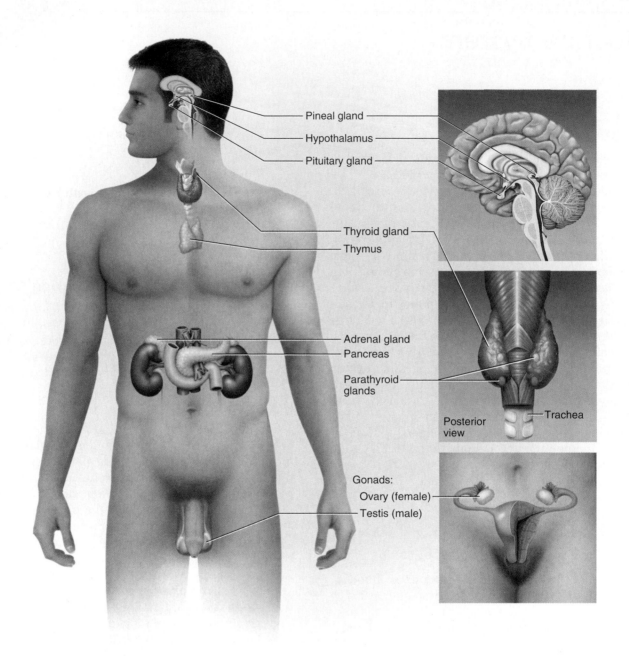

FIGURE 28.1 Major Endocrine Glands

hormones that either cause the release of hormones from their target areas or prevent the release of hormones from these areas. Locate the hypothalamus in figure 28.2.

The **pituitary gland,** or **hypophysis,** is seen in figures 28.1 and 28.2. It is divided into an **anterior lobe,** or **adenohypophysis;** an **intermediate lobe** (part of the anterior lobe); and a **posterior lobe,** or **neurohypophysis.** The pituitary is suspended from the hypothalamus by a stalk called the **infundibulum.**

Adenohypophysis

The adenohypophysis originates from the roof of the oral cavity during embryonic development. The result of this development can be seen if you examine a prepared slide of the pituitary gland. In the histological section, note that the adenohypophysis is composed of simple cuboidal epithelial cells (see figure 28.3). The cells of the anterior pituitary have the same embryonic origin, yet they have differentiated into several types of specialized cells.

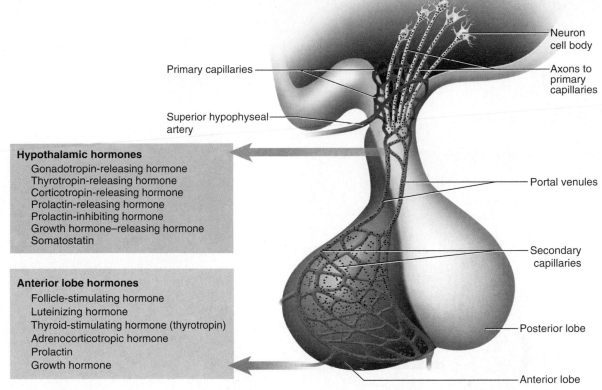

Hypothalamic hormones
Gonadotropin-releasing hormone
Thyrotropin-releasing hormone
Corticotropin-releasing hormone
Prolactin-releasing hormone
Prolactin-inhibiting hormone
Growth hormone–releasing hormone
Somatostatin

Anterior lobe hormones
Follicle-stimulating hormone
Luteinizing hormone
Thyroid-stimulating hormone (thyrotropin)
Adrenocorticotropic hormone
Prolactin
Growth hormone

(a)

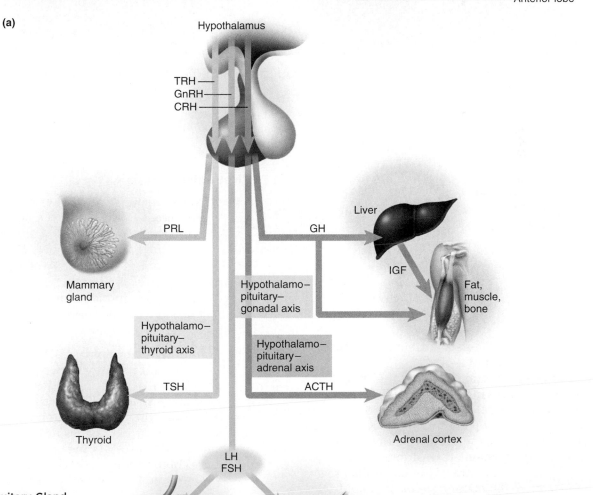

(b)

FIGURE 28.2 Pituitary Gland
(a) Overview of the pituitary gland;
(b) target areas of the anterior pituitary.

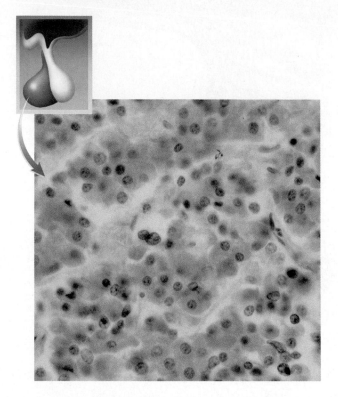

FIGURE 28.3 Histology of the Anterior Pituitary Gland (400×)

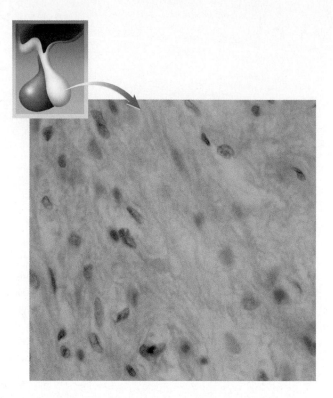

FIGURE 28.4 Histology of the Posterior Pituitary (400×)

These cells produce a number of hormones that have broad effects throughout the body.

Some of the hormones produced in the anterior pituitary and their functions are as follows:

- **Thyroid-stimulating hormone (TSH),** or thyrotropin, stimulates the thyroid gland to produce thyroid hormones.
- **Growth hormone (GH),** or somatotropin, promotes growth of most of the cells and tissues of the body.
- **Prolactin (PRL)** stimulates mammary glands to begin the production of milk.
- **Gonadotropins** stimulate ovaries and testes.
  —**Follicle-stimulating hormone (FSH)** stimulates the production of sex cells, causes ovarian follicles to mature (develop and produce egg cells), and stimulates testes to produce spermatozoa.
  —**Luteinizing hormone (LH)** stimulates ovaries to produce other hormones (for example, progesterone), causes the maturation of Graafian follicles and ovulation, and stimulates testosterone production in testes.
- **Adrenocorticotropin (ACTH)** controls hormone production in the adrenal cortex.

Neurohypophysis

The tissue of the neurohypophysis, or posterior pituitary, is very different from that of the adenohypophysis. The neurohypophysis is composed of nervous tissue that originated from the base of the brain. Compare the tissue of the neurohypophysis in a prepared slide to figure 28.4.

Hormones released from the posterior pituitary are synthesized by the **hypothalamus** and flow through axons to be stored in the posterior pituitary. These hormones and their actions include

- **Antidiuretic hormone (ADH)** stimulates the reabsorption and retention of water by the kidneys. It is also known as vasopressin because it causes arterioles to constrict, which elevates blood pressure.
- **Oxytocin** stimulates the contraction of the cells of the mammary glands, resulting in the release of milk, and causes uterine contractions. External influences on hormone actions can be seen in the release of oxytocin. Look at figure 28.5 and note that the sucking action of an infant on the mother's breast sends impulses to the hypothalamus. The numbers indicate the sequence of events. The hypothalamus stimulates the posterior pituitary to release oxytocin, which causes milk ejection.

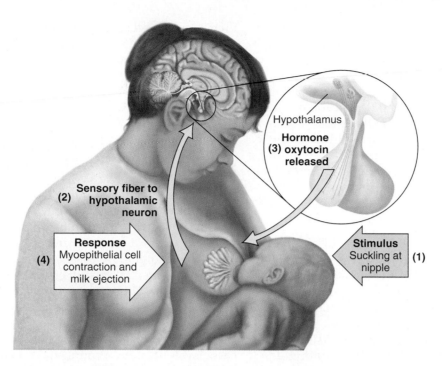

FIGURE 28.5 **External Influences on Hormonal Action** (1) External stimulus of suckling; (2) sensory neurons take impulse to brain, where (3) hypothalamus stimulates posterior pituitary to release oxytocin and (4) milk is released from the breast.

Thyroid Gland

Figure 28.6 illustrates the **thyroid gland,** named for its location inferior to the thyroid cartilage of the larynx. The thyroid gland has two lateral **lobes** and a medial **isthmus** that connects them.

The thyroid gland is easy to recognize histologically. Examine a slide of thyroid gland and compare it to figure 28.7. Locate the **follicle cells** that surround the colloid, a storage region for thyroid hormones. Colloid is mostly composed of thyroglobulin, a large molecular weight compound. Thyroid hormones are formed inside the larger thyroglobulin molecule. Also locate the **parafollicular cells,** which occur in the spaces between the follicles.

The thyroid gland secretes three specific hormones. Two of these are T_3, or **triiodothyronine** (a molecule that

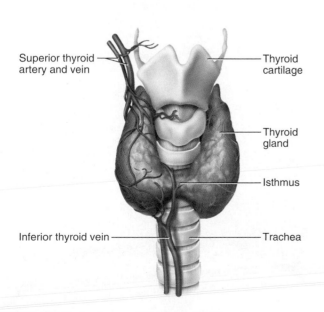

FIGURE 28.6 Anatomy of the Thyroid Gland

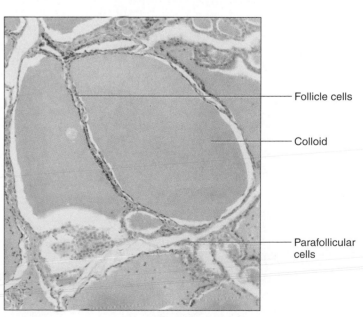

FIGURE 28.7 Histology of the Thyroid Gland (100×)

includes three iodine atoms), and T_4, or **thyroxine** (containing four iodine atoms per molecule). These hormones increase basal metabolic rates and are stored in the colloid. Another hormone of the thyroid gland is **calcitonin.** Calcitonin decreases blood calcium levels by causing excretion of calcium by the kidneys and deposition of calcium in bone by decreasing osteoclast activity and formation. The role of calcitonin is important in children and may protect individuals from hypercalcemia. Calcitonin is produced by the parafollicular, or C, cells. Calcitonin is antagonistic to parathyroid hormone (discussed next).

Parathyroid Glands

Attached to the posterior portion of the thyroid are usually two pairs of organs called the **parathyroid glands. Chief cells** in the parathyroid gland secrete **parathyroid hormone (PTH),** or parathormone, responsible for increasing calcium levels in the blood. This occurs by increasing calcium uptake in the intestines, increasing kidney reabsorption of calcium, and releasing calcium from bone. Examine the location of the parathyroid glands embedded in the posterior surface of the thyroid gland (see figure 28.8).

Thymus

The **thymus** is active in young individuals and plays an important part in immunocompetency. The thymus is located anterior and superior to the heart (see figure 28.9). It secretes a hormone, **thymosin,** that causes the maturation of T cells. These cells originate as stem cells in the bone marrow and migrate to the thymus. Under the influence of thymosin the cells mature to provide "cellular immunity" against antigens (foreign material, such as bacteria and viruses, which cause immune reactions). The T cells migrate from the thymus predominantly to the lymph nodes and the spleen to carry out their functions.

Pancreas

The **pancreas** is a mixed gland in that it has an exocrine function and an endocrine function. Locate the pancreas in figures 28.1 and 28.10. The exocrine function is digestive because pancreatic juice contains both sodium carbonate and digestive enzymes.

The endocrine function of the pancreas consists of the secretion of the hormones insulin and glucagon, which regulate blood glucose levels. When blood glucose levels drop, **glucagon** converts glycogen (a starch storage product) to glucose. Glucagon is produced in specialized cells **(alpha cells)** that occur in clusters called **pancreatic islets** (or the islets of Langerhans) (see figure 28.10). Pancreatic islets also produce **insulin,** which lowers the blood glucose level. Insulin, produced in **beta cells,** stimulates the conversion of glucose to glycogen. The pancreas also contains **delta cells,** which secrete somatostatin. Somatostatin is secreted after the ingestion of a meal and inhibits the secretion of insulin and glucagons from the pancreas. This may play a role in increasing efficiency in digestion. Locate the pancreatic islets on a microscope slide and compare them to those in figure 28.10.

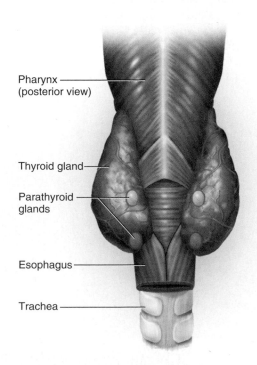

FIGURE 28.8 Parathyroid Glands

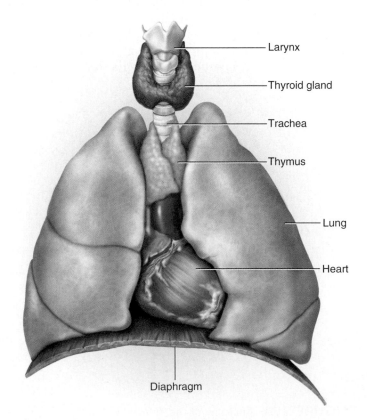

FIGURE 28.9 Thymus

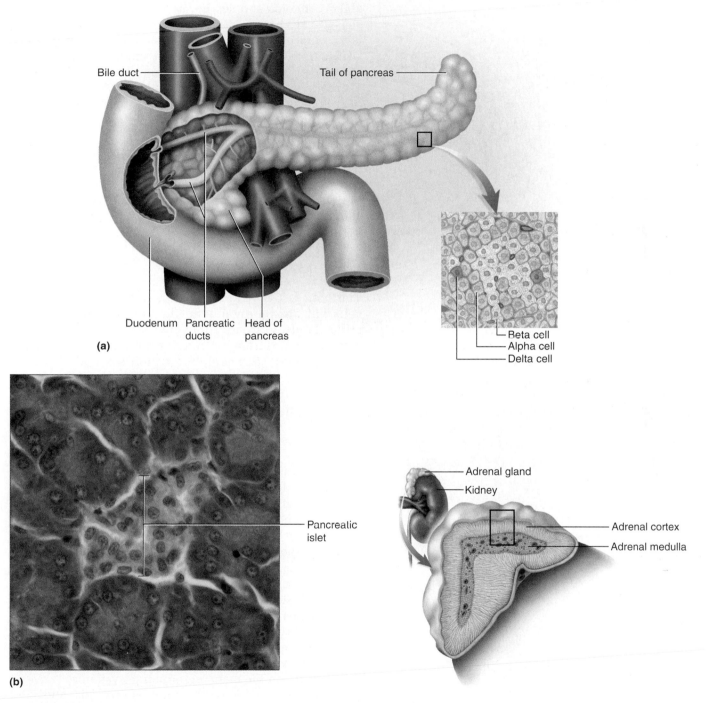

FIGURE 28.10 Pancreas (a) Gross anatomy; (b) histology (400×).

FIGURE 28.11 Adrenal Cortex and Medulla

Adrenal Glands

The **adrenal glands** (**ad** = next to, **renal** = kidney) are superior in position to the kidneys. Each adrenal gland is composed of an outer **cortex** and an inner **medulla.** Examine a microscope slide of an adrenal gland and locate the cortex and the medulla. Compare this to figure 28.11.

The hormones secreted from the adrenal cortex are called **corticosteroid hormones.** They are important in water and electrolyte balance (Na^+, K^+) in the body. They are also important for carbohydrate, protein, and fat metabolism, as well as stress management. The cortex can be divided into three regions. The outermost is the **zona**

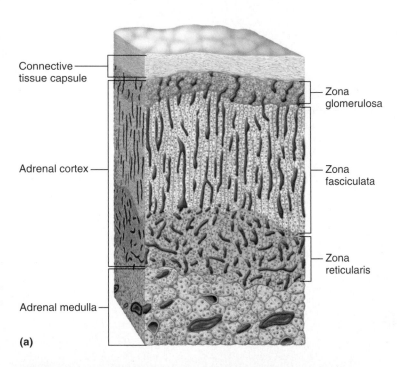

Connective tissue capsule

Adrenal cortex

Adrenal medulla

Zona glomerulosa

Zona fasciculata

Zona reticularis

(a)

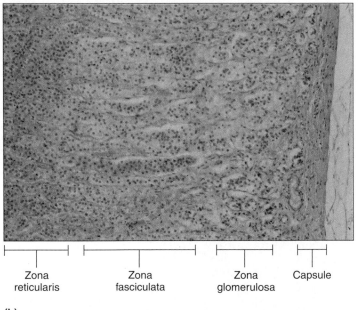

Zona reticularis

Zona fasciculata

Zona glomerulosa

Capsule

(b)

FIGURE 28.12 Histology of the Adrenal Cortex (a) Diagram; (b) photomicrograph (100×).

glomerulosa (see figure 28.12), which consists of clusters of cells that secrete **mineralocorticoids** (especially **aldosterone).** Mineralocorticoids affect salt and water balance. Inside this layer (closer to the medulla) is the **zona fasciculata,** which consists of parallel bundles of cells that secrete **glucocorticoids,** especially **cortisol (hydrocortisone).** Glucocorticoids regulate protein and fat catabolism and gluconeogenesis (production of glucose from amino

acids). They are also involved in reducing inflammation. The deepest cortical layer is the **zona reticularis,** which consists of a branched pattern of cells that produce both **glucocorticoids** and **sex hormones** (androgens and estrogens). The hormone dehydroepiandrosterone (DHEA) is an androgen, and both males and females produce DHEA from the adrenal glands in small amounts. DHEA is secreted from the adrenal glands but converted to testosterone or estrogens in other tissues. The major source of androgens in males is produced by the testes. Estrogen (estradiol) is secreted in small amounts in both sexes by the zona reticularis.

The hormones **epinephrine** and **norepinephrine** are produced in the adrenal medulla. Stimulation of the adrenal glands by the sympathetic nervous division causes the release of epinephrine and norepinephrine from the gland, which increases heart rate and prepares the body for "fight-or-flight" reactions.

Gonads

The gonads, testes and ovaries, not only produce sex cells but also hormones, so they are considered endocrine glands. Both are stimulated by follicle-stimulating hormone (FSH) from the anterior pituitary, which causes the production or maturation of the sex cells (spermatozoa or oocytes). They are also under the influence of luteinizing hormone (LH), which increases the level of hormone production.

Testes

In males the **testes** produce **testosterone,** a hormone responsible for secondary sex characteristics such as the development of facial and body hair, the expansion of the larynx (which produces a deeper voice), and the increased muscle and bone mass seen in males. Testosterone not only is responsible for secondary sex characteristics but also influences the development of the male genitalia (penis and scrotum) during embryological and fetal development. Testosterone also aids FSH in the production of **spermatozoa. Inhibin** is a hormone produced and secreted by the testes, and it is involved in negative feedback, providing regulation for testosterone production. The exocrine function of the testis is explored in Laboratory Exercise 46, on the male reproductive system.

Examine a microscope slide of a testis and find the **seminiferous tubules** and **interstitial cells.** Refer to figure 28.13 for assistance. The interstitial cells are found between the tubules and produce testosterone.

Ovaries

In females the ovaries produce oocytes (eggs) and have an endocrine function in that they produce **estrogen** and **progesterone** (see figure 28.14). Female hormones

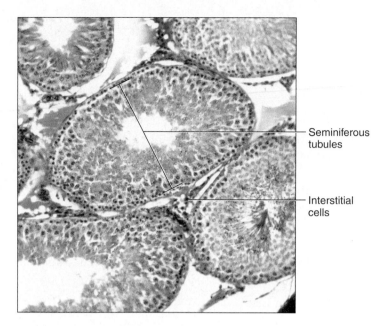

FIGURE 28.13 Histology of the Testis (100×)

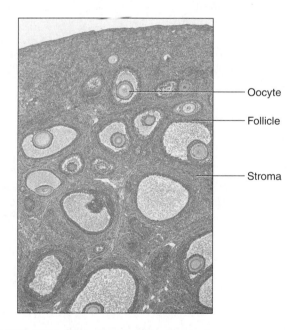

FIGURE 28.14 Histology of the Ovary (40×)

are also responsible for secondary sex characteristics in women such as the development of breasts, an additional subcutaneous adipose layer, and a higher voice. Estrogen and progesterone also influence the development of the endometrium, cause maturation of the oocytes, and regulate the menstrual cycle. *Estrogen* is a generic term for several hormones produced by females, including **estradiol.** Inhibin is also secreted by the ovary and regulates the levels of estrogen and progesterone.

Endocrine Physiology— Detection of Hormones

Hormones are secreted in small but detectable amounts in the bloodstream. Some of these hormones may be filtered in the urine and detected from a urine sample. **Luteinizing hormone (LH)** is one of these hormones. LH stimulates the final maturation of the oocyte and causes ovulation. Normally, small amounts of LH occur in a woman's body but an increase, or a spike, in the levels of LH occurs about 24–36 hours prior to ovulation (see figure 28.15). This increase lasts typically for 10–30 hours. In this part of the exercise you will use a test kit to determine the presence of LH. Your instructor may provide an unknown sample of urine, or female students in the middle of their ovarian cycle may be asked to volunteer to test for the presence of LH.

 Caution! Remember to treat bodily fluids as if they carried pathogens. Wear protective gloves and eyewear when handling fluids! Dispose of all contaminated disposable materials in the biohazard container. The reusable material should be placed in a 10% bleach solution.

The test kit uses **monoclonal antibody** techniques to determine the presence of LH. The development of a monoclonal antibody starts when a mouse (or another mammal) is injected with human LH. Since this hormone is foreign to the mouse, specific white blood cells, called beta lymphocytes, make antibodies against human LH. These can be removed from the mouse in a blood sample and fused with tumor cells. Tumor cells are used because they grow rapidly. The hybrid tumor-lymphocyte cells replicate, producing more cells that have the antibody against LH. The antibodies are isolated and absorbed by a solid substrate (the test strip). An enzyme that bonds

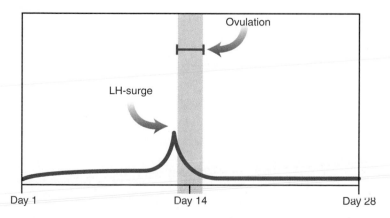

FIGURE 28.15 Surge of Luteinizing Hormone (LH) Prior to Ovulation Day 1 of the menstrual cycle begins with menstrual flow.

to the LH antibody is also attached to the substrate. This enzyme is attached to a dye marker and it changes color when attached to the antibody.

In summmary, the test strip contains antibodies to LH, an enzyme that will bind to the LH-antibody complex, and a dye that changes color when the enzyme binds to the LH-antibody complex. If a woman has significant amounts of LH, then the test strip produces a color (usually blue). The generic name for tests that use antibodies and enzymes to determine the presence of biological materials is *enzyme-linked immunosorbent assay (ELISA)*. In some cases, pregnancy, endometriosis, hyperthyroidism, and ingestion of some prescription drugs may produce false test results.

Select the unknown urine sample or have three women who are in the middle of their ovarian cycle perform the test. The test should be positive for women who are just about ready to ovulate, which is 2 weeks prior to the next anticipated menstrual cycle.

Obtain an ovulation test kit and follow the instructions provided in the kit or those from your instructor. Record whether the urine sample provided is an unknown or is from a female student in the class and whether it has significant levels of LH. Typically, the test involves collecting a urine

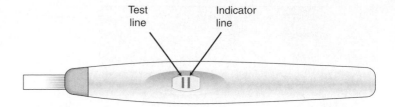

FIGURE 28.16 Ovulation Test Kit Test Line and Indicator Line should be the same from sample of Ovulating Female.

sample and dipping the absorbent sampler in the urine for a specific amount of time. The sample is then compared to a reference line to indicate ovulation, as seen in figure 28.16.

Sample: _____ (unknown/student)

Presence of LH: _____

Once you have studied the endocrine glands and material, complete table 28.1. You should fill in the blank spaces where needed.

TABLE 28.1		
Organ	**Hormones Produced**	**Effect of Hormones**
_____	Antidiuretic hormone	_____
Thyroid	Thyroxine	_____
_____	Corticosteroid hormones	Regulate electrolyte balance
Ovary	_____	Regulates ovarian cycle
_____	Melatonin	Regulates sleep cycles, inhibits release of reproductive hormones
_____	Luteinizing hormone	_____
Neurohypophysis	Oxytocin	_____
Parathyroid	_____	Increases calcium in blood
_____	Glucagon	Increases blood glucose levels
Pancreas	_____	Decreases blood glucose levels
_____	Calcitonin	_____

REVIEW SECTION

Endocrine System

Name _____ *Date* _____

Lab Section _____ *Time* _____

Review Questions

1. What is the general name of the types of organs that produce hormones?

2. What name is given to cells or tissues receptive to hormones?

3. Melatonin is secreted by what gland?

4. Where is ADH stored?

5. What is the effect of TSH, and where is it produced?

6. What connects the two lobes of the thyroid gland?

7. Does parathormone increase or decrease calcium levels in the blood?

8. What does glucagon do as a hormone, and where is it produced?

9. Which hormones in the adrenal gland control water and electrolyte balance?

10. What is the primary gland that secretes epinephrine?

11. Where is growth hormone produced?

12. What is another name for T_3?

13. Label the endocrine glands indicated in the following illustration.

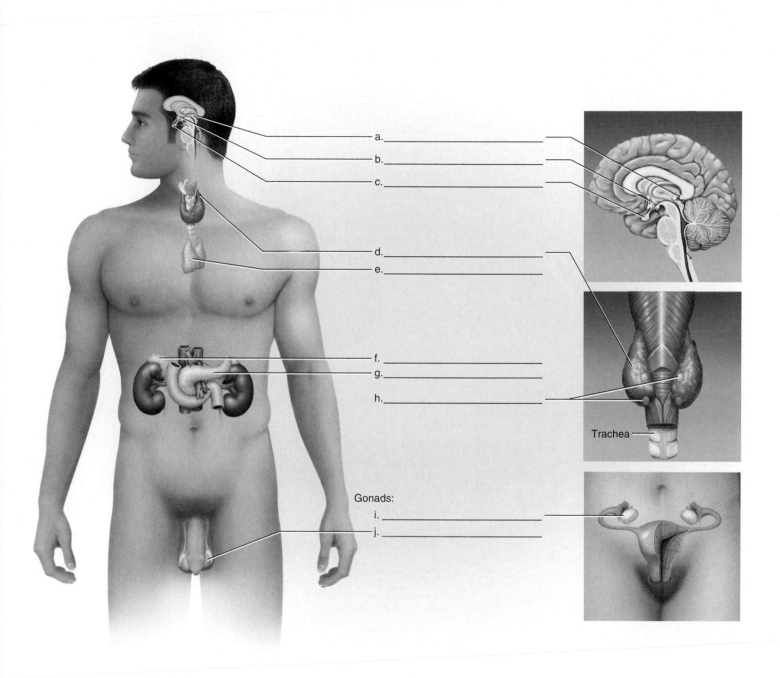

a. _____

b. _____

c. _____

d. _____

e. _____

f. _____

g. _____

h. _____

Trachea

Gonads:

i. _____

j. _____

14. Identify the parts of the pituitary as seen in the
     illustration, and list two hormones located in each.

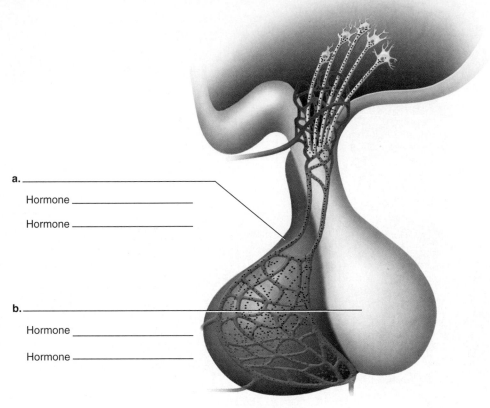

a. _____

  Hormone _____

  Hormone _____

b. _____

  Hormone _____

  Hormone _____

15. Identify the three layers of the adrenal cortex as illustrated, and list the hormones produced by each layer.

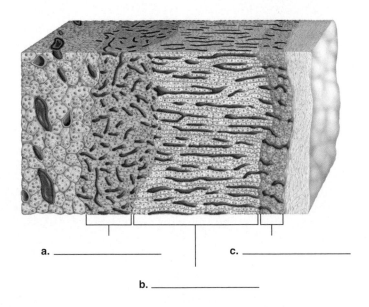

a. _____   c. _____

b. _____

16. Interstitial cells produce which hormone?

17. What structures are responsible for the production of estrogen?

18. Target cells must have the correct receptor site in order for homones to influence those cells. Cut out the following 12 outlines, which represent either hormones or an effect caused by a hormone. In the structures, hormones are circled above the endocrine glands that produce them. The effects caused by each hormone are not circled and the organ where the effect occurs is listed below the effects. Assemble the hormones and effects according to the sequence in which they fit. There may be hormones or cells that do not fit the sequence. Once the sequences are assembled, color each one with a different color. Two of the sequences represent an axis, which is a relationship among the hypothalamus, pituitary, and a more distant gland. There are three axes in the body: the hypothalamo-pituitary-gonadal axis (HPG), the hypothalamo-pituitary-thyroid axis (HPT), and the hypothalamo-pituitary-adrenal axis (HPA). Write *HPG, HPT,* or *HPA* on the corresponding sequence. Notice that, in some sequences, a hormone can have more than one effect (though not all the effects of the hormones are listed on this page). Select a hormone and target gland from your lab text and make a sequence of your own.

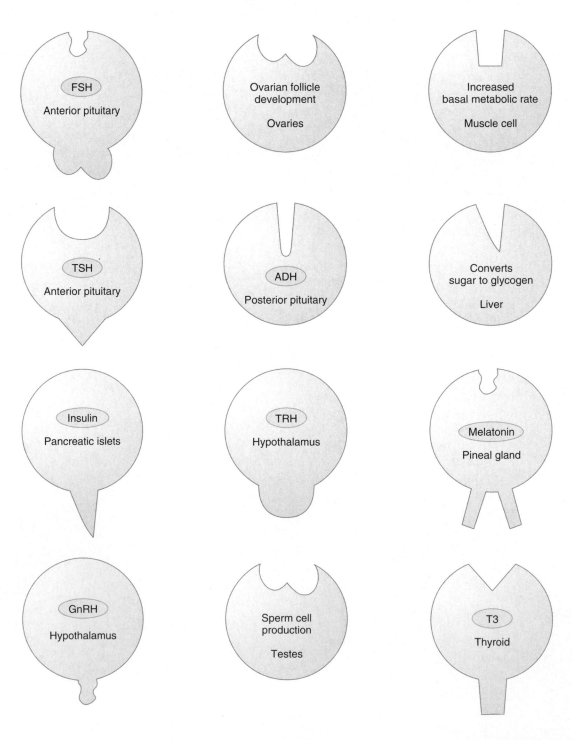

LABORATORY

Blood

INTRODUCTION

Blood consists of two parts, formed elements and plasma. Formed elements make up about 45% of the blood volume and can further be subdivided into erythrocytes (red blood cells), leukocytes (white blood cells), and platelets (thrombocytes). Platelets are not cells, and this is why the fraction of blood is called formed elements and not blood cells.

Plasma constitutes approximately 55% of the blood volume and contains water, lipids, dissolved substances, colloidal proteins, and clotting factors.

The study of blood cells is important because it is the fluid medium of the cardiovascular system and because it has significant clinical implications. Changes in the numbers and types of blood cells may be used as indicators of disease. In this exercise you examine the nature of blood cells, discussed in the Saladin text in chapter 18, "Blood."

Caution! The risk of blood-borne diseases has been significantly minimized by the use of sterilized human blood and nonhuman mammal blood in this exercise. However, use the same precautions you would if you were handling fresh and potentially contaminated human blood. Your instructor will determine whether to use animal blood (from nonhuman mammals), sterilized blood, or synthetic blood. In any of these cases, follow strict procedures for handling potentially pathogenic material.

1. Wear protective gloves during the procedures.
2. Do not eat or drink in lab.
3. If you have an open wound, do not participate in this exercise or make sure the wound is *securely* covered.
4. Your instructor may elect to have you use your blood. Due to the potential for disease transmission, such as AIDS or hepatitis in fresh blood samples, *make sure you keep away from other students' blood and keep them away from your blood.*
5. Do not participate in this part of the exercise if you have recently skipped a meal or if you are not well.
6. Place all disposable material in the biohazard bag.
7. Place all used lancets in the sharps container and all glassware to be reused in a 10% bleach solution.
8. After you have finished the exercise, clean and disinfect the countertops with a 10% bleach solution.

OBJECTIVES

At the end of this exercise you should be able to

1. discuss the composition of blood plasma;
2. distinguish among the various formed elements of blood;
3. determine the percent of each type of leukocyte in a differential white cell count;
4. describe the significance of an elevated level of a particular white blood cell and what it may indicate about a disease state or an allergic reaction.

MATERIALS

Prepared slides of human blood with Wright's or Giemsa stain

Compound microscopes

Lab charts or illustrations showing the various blood cell types

Dropper bottle of Wright's stain

Large finger bowl or staining tray

Toothpicks

Clean microscope slides

Coverslips

Squeeze bottle of distilled water or phosphate buffer solution

Pasteur pipette and bulbs

Sterile cotton balls

Adhesive bandages

Alcohol swabs

Sterile, disposable lancets

Latex gloves

Roll of paper towels

Hand counter

Biohazard bag or container

Container with 10% bleach

Sharps container

PROCEDURE

Plasma

Plasma is the fluid portion of blood and is over 90% water by volume. The remainder mostly consists of proteins, such as albumins, globulins, and fibrinogen. Albumins are produced by the liver and make up the majority of the plasma proteins. Some globulins are made by the plasma cells and make up the next largest amount of proteins. Fibrinogen is a clotting protein and this and other clotting factors are produced by the liver. Plasma also contains electrolytes (Na^+, K^+, and Cl^-), nutrients, hormones, and wastes.

Examination of Blood Cells

Your lab instructor will direct you to use a prepared slide of blood; fresh, nonhuman mammal blood; or your blood. If you are using a prepared slide of blood you can move to the section entitled "Microscopic Examination of Blood Cells." If you are to make a smear, read the following directions.

Preparing a Fresh Blood Smear

If you are using nonhuman mammal blood, wear latex gloves and withdraw a small amount of blood with a clean Pasteur pipette. Place a drop of blood on a clean microscope slide; then proceed to step 7 in the following procedure.

Withdrawal of Your Blood

If you are using your blood make sure you follow the safety precautions as discussed at the beginning of the exercise. Obtain the following materials: a clean, sterile lancet; an alcohol swab; an adhesive bandage; rubber gloves; sterile cotton balls; two clean microscope slides; and a paper towel.

1. Arrange the materials on the paper towel in front of you on the countertop.
2. Clean the end of the donor finger with the alcohol swab and let the hand from which you will withdraw blood hang by your side for a few moments to collect blood in the fingertips. You will be puncturing the pad of your fingertip (where the fingerprints are located).
3. Peel back the covering of the lancet and hold on to the *blunt* end as you withdraw it from the package. Do not touch the sharp end or lay the lancet on the table before puncturing your finger.
4. Jab your finger quickly and wipe away the first drop of blood that forms with a sterile cotton ball.
5. Place the second drop of blood from your finger on the microscope slide about 2 cm away from one of the ends of the slide (see figure 29.1) and place another cotton ball on your finger.

6. Put an adhesive bandage on your finger.
7. Whether you are using nonhuman blood or your blood, use another clean slide to spread the blood by touching the drop with the edge of the slide and pull the blood across the slide (see figure 29.1). This should produce a smooth, thin smear of blood. Let the blood smear dry completely. Throw the lancet in the sharps container. Never reuse the lancet or set it down on the table!
8. After the slide is dry, place it in a large finger bowl elevated on toothpicks or on a staining tray.
9. Cover the blood smear with several drops of Wright's stain from a dropper bottle. Let the stain remain on the slide for 1 to 2 minutes.
10. After this time add water or a prepared phosphate buffer solution to the slide. You can rock the slide gently with gloved hands or blow on it to stir the stain and water. A metallic green material should come to the surface of the slide. Let the slide remain covered with stain and water for 3 to 8 minutes.

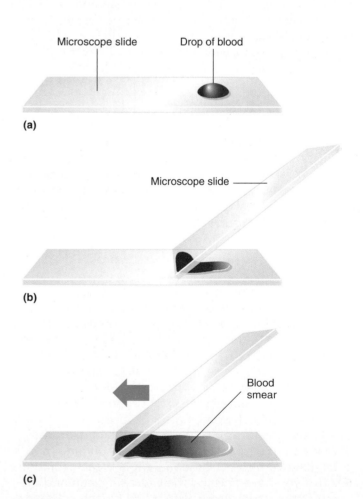

(a)

(b)

(c)

FIGURE 29.1 Making a Blood Smear (a) Placement of blood on slides; (b) touching glass slide to front of blood drop; (c) spreading blood across slide.

11. Wash the slide gently with distilled water until the material is light pink and then stand it on edge to dry. You may also stain blood using an alternate stain (Giemsa stain), following your lab instructor's procedure.

Place all blood-contaminated disposable material in the biohazard container. Place the slide used to spread the blood in the 10% bleach solution and place the lancets in the sharps container if you haven't done so already. Once the slide is completely dry it can be examined under the high-power or oil-immersion lens of your microscope.

Microscopic Examination of Blood Cells

Overview

As you examine the blood slide you should refer to figures 29.2 through 29.4. You will have to look at many blood cells to see all the cells included in this exercise.

Erythrocytes

Examine a slide of blood stained with either Wright's or Giemsa stain. Erythrocytes are the most common cells you will find on the slide. There are about 5 million erythrocytes per cubic millimeter. They do not have a nucleus but appear as pink, biconcave disks (like a donut with the hole partially filled in). This shape increases surface area and provides greater oxygen-carrying capabilities. Erythrocytes contain the pigment **hemoglobin,** which carries oxygen.

Erythrocytes are about 7.5 μm in diameter, on average, and have a life span of about 120 days, after which time they are broken down by the spleen or liver. The iron portion of the hemoglobin is used by the bone marrow, the proteins are broken down to amino acids and reabsorbed for general use by the body, and the heme is metabolized to bilirubin in the liver and secreted as part of the bile. Compare what you see under the microscope to figure 29.2.

Platelets

There are about 150,000 to 350,000 platelets per cubic millimeter of blood. **Platelets,** or **thrombocytes,** are small fragments of **megakaryocytes** involved in clotting. Examine the slide for small, purple fragments that may be single or clustered and compare them to the platelets in figure 29.2.

Leukocytes

There are far fewer leukocytes in the blood than erythrocytes. The number of leukocytes in a healthy adult is about 7,000 cells per cubic millimeter, with a range of 4,300 to 10,800 cells per cubic millimeter of blood. Leukocytes are formed in bone marrow. The life span of a leukocyte

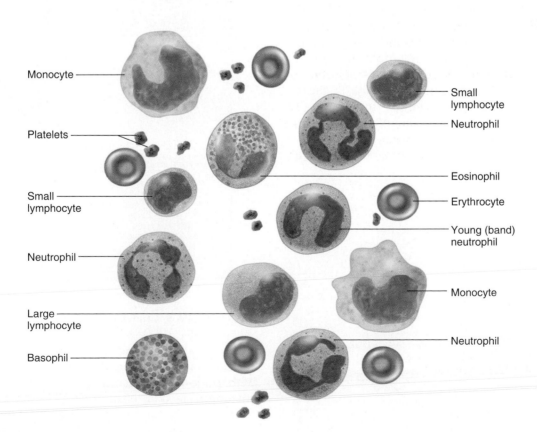

FIGURE 29.2 Formed Elements of Blood

varies from a few hours to several years. Many are capable of ameboid movement as they squeeze between cells of the vascular endothelium (a process termed *diapedesis*), engulfing foreign particles or cellular debris.

Leukocytes, or white blood cells, can be divided into two groups, **granular** and **agranular leukocytes.** Examine the slide under high power or oil immersion and identify the different leukocytes as presented in the following sections. Leukocytes have two major types of responses to immune reactions. One of these is **innate immunity,** which refers to a person having a reaction to microorganisms and foreign substances without prior exposure. Another kind of response is **adaptive immunity,** in which exposure to specific antigens initiates an immune reaction. Leukocytes have specific receptors and specific antibodies for specific antigens.

Granular Leukocytes   Granular leukocytes (granulocytes) are so named because they have granules in their cytoplasm. The three types of granular leukocytes are neutrophils, eosinophils, and basophils.

Neutrophils are the most common of all leukocytes. They are also known as polymorphonuclear (PMN) leukocytes due to the variable shape of their nuclei (which are lobed, not round). Neutrophils typically have a half life in the blood of about 6 hours and have ameboid capabilities. They move by diapedesis between blood capillary cells and in the interstitial areas between cells of the body. They move to connective tissue, where they typically live for an additional 1 to 4 days. Neutrophils move toward infection sites and destroy foreign material. They are the major phagocytic leukocyte. The granules of neutrophils absorb very little stain, but mature neutrophils can be distinguished from other leukocytes by their typically two- to five-lobed nucleus. They are about one and a half times the size of erythrocytes. Examine your slide for neutrophils, compare them to figure 29.3, and draw one in the following space.

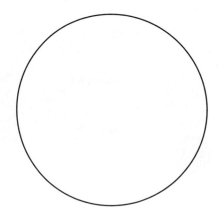

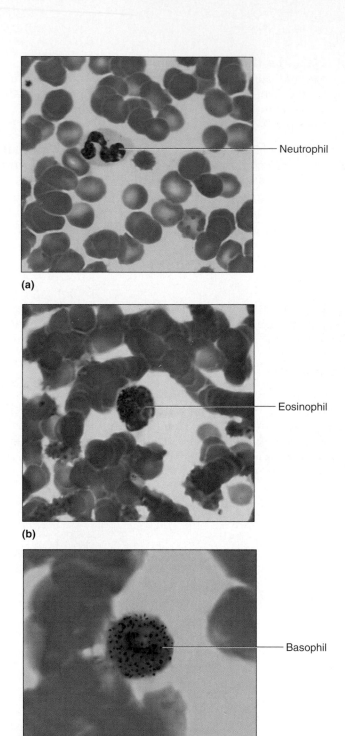

FIGURE 29.3 Granular Leukocytes (a) Neutrophil (1,000×); (b) eosinophil (1,000×); (c) basophil (1,500×).

Eosinophils typically have a two-lobed nucleus with pink-orange granules in the cytoplasm. The term **eosinophil** means "eosin-loving" (eosin is an orange-pink stain and is picked up by the granules). They are about twice the size of erythrocytes. Compare figure 29.3 to your slide as you locate the eosinophils and draw one in the following space.

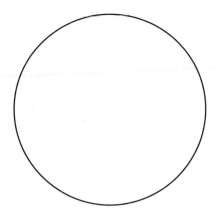

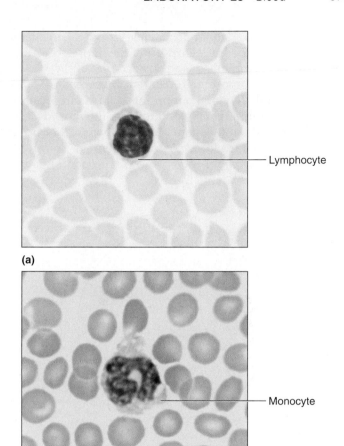

(a)

(b)

FIGURE 29.4 Agranular Leukocytes (1,000×)
(a) Lymphocyte; (b) monocyte.

Basophils are rare. The granules stain very dark (blue-purple), and sometimes the nucleus is obscured because of the dark-staining granules. The nucleus is S-shaped and the cell is about twice the size of erythrocytes. Their granules contain histamines (vasodilators) and heparin, an anticoagulant, which allows for movement of other leukocytes out of the capillaries to infection sites. They are involved in inflammatory and allergic reactions. You may have to look at 200 to 300 leukocytes before finding a basophil. Compare the basophil in your slide to figure 29.3 and draw one in the following space.

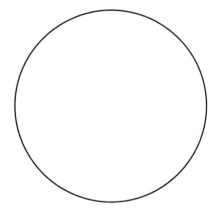

Agranular Leukocytes Agranular leukocytes are so named because they lack cytoplasmic granules. The nuclei are not lobed but may be dented or kidney bean–shaped. There are two types of agranular leukocytes, lymphocytes and monocytes.

Lymphocytes have a large, unlobed nucleus that usually has a flattened or dented area. The cytoplasm is clear and may appear as a blue halo around the purple nucleus. There are two groups of lymphocytes, the **B cells** and **T cells.** These cells cannot normally be distinguished from one another in standard histological preparations (for example, Wright's stain) and are considered lymphocytes in this exercise. Both B cells and T cells arise from fetal bone marrow. The **B cells** probably mature in the fetal liver and spleen and in the bone marrow of adults. B cell lymphocytes become **plasma cells** when activated and make **antibodies.** Plasma cells provide **antibody-mediated immunity** (plasma cells secrete antibodies that travel in the blood plasma).

T cells mature in the thymus and provide **cell-mediated immunity.** In cell-mediated immunity the cells (not antibodies in the blood plasma) move close to and destroy some types of bacteria or virus-infected cells. T cells also attack tumors and transplanted tissues. Most of the T cells are found in the lymph nodes, thymus, and spleen. They enter the bloodstream via the lymphatics. **Natural killer (NK) cells** are lymphocytes that attack bacteria, transplanted cells, and cancer cells. Compare your slide to figure 29.4 for lymphocytes and draw one in the following space.

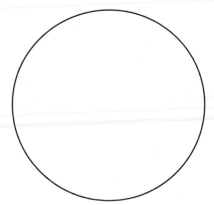

Monocytes are large and have a kidney bean–shaped or horseshoe-shaped nucleus. They are about three times the size of erythrocytes and are activated by T cells. Monocytes are important in that they are major phagocytic cells, and they are important in presenting foreign antigens to T lymphocytes. Monocytes move into tissues from the blood and become **macrophages.** Locate the large cells on your slide, compare them to figure 29.4, and draw one in the following space. You can review the formed elements in table 29.1.

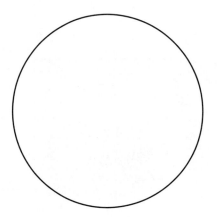

Differential Leukocyte Count

Once you have identified the various leukocytes on the blood slide, conduct a differential white blood cell count. There is great importance in performing differential leukocyte counts.

Changes in the relative percentages of leukocytes may indicate a type of disease.

1. Use a hand counter and count 100 leukocytes. Look into the microscope and methodically record the different types of leukocytes seen.
2. You should record how many of each type are found and keep track of the overall number with the use of a hand counter until 100 cells are counted.
3. Scan the slide in a systematic way, so that you don't count any cell twice. One method is illustrated in figure 29.5. Tally your results in the following spaces.

Neutrophils: _____

Eosinophils: _____

Basophils: _____

Lymphocytes: _____

Monocytes: _____

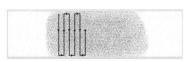

FIGURE 29.5 Counting Leukocytes

TABLE 29.1	Summary of Formed Elements in Blood		
	Size	**Number**	**Characteristics**
Erythrocytes	7.5 μm	5 million/mm³	Live 120 days on average No nucleus
Platelets	2-4 μm	130,000-360,000/mm³	Cell fragments
Leukocytes		7,000/mm³	Nucleus present
Granular leukocytes			
Neutrophils	9-12 μm	60-70% of leukocytes	Three– to five-lobed nucleus granules indistinct
Eosinophils	10-14 μm	2-4% of leukocytes	Two-lobed nucleus Orange granules
Basophil	8-10 μm	0.5-1% of leukocytes	S-shaped nucleus Large, dark granules
Agranular leukocytes			
Lymphocytes	5-17 μm	25-33% of leukocytes	Nucleus appearing dented Thin rim of cytoplasm in some
Monocytes	12-15 μm	3-8% of leukocytes	Large, kidney-shaped nucleus

Neutrophils represent about 60–70% of leukocytes. They increase in number in appendicitis or other acute bacterial infections.

Eosinophils represent about 2–4% of leukocytes. Eosinophils increase in number during allergic reactions and parasitic infections (for example, trichinosis).

Basophils represent about 0.5–1% of leukocytes. They increase in number during allergies and radiation.

Lymphocytes make up 25–33% of leukocytes. These cells increase in times of viral infection, such as infectious mononucleosis, and antibody-antigen reactions.

Monocytes make up about 3–8% of leukocytes. They increase in times of chronic infections, such as tuberculosis.

Fill in the following chart with what you know about the formed elements. Read all the information given before filling in the chart.

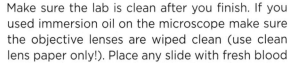

Clean Up Make sure the lab is clean after you finish. If you used immersion oil on the microscope make sure the objective lenses are wiped clean (use clean lens paper only!). Place any slide with fresh blood on it or any material contaminated with bodily fluid in the bleach solution. Place all gloves or contaminated paper towels in the biohazard container. All sharps material (broken slides or coverslips, lancets, etc.) should be placed in the sharps container. Clean the counters with a towel and a 10% bleach solution.

Characteristics of Formed Elements

Formed Element	Granules (If Present)	Shape of Nucleus (If Present)	Cause for Increase
Erythrocyte	No granules	No nucleus	_____
_____	Not obvious	_____	Mononucleosis
_____	Orange-staining	_____	Parasitic infections
_____	_____	two- to five-lobed	_____
_____	Not obvious	Kidney-bean	_____
Basophil	_____	_____	_____

Notes

REVIEW SECTION

Blood

Name _____ Date _____

Lab Section _____ Time _____

Review Questions

1. Formed elements have three main components. What are they?

2. What is the most common plasma protein?

3. What is another name for a platelet?

4. Which is the most common blood cell?

5. What is another name for a leukocyte?

6. What leukocyte is most numerous in a normal blood smear?

7. How many erythrocytes are normally found per cubic millimeter of blood?

8. What is an average number of leukocytes found per cubic millimeter of blood?

9. B cells and T cells belong to what class of agranular leukocytes?

10. What value is there to a change in the percentage of leukocytes to diagnostic medicine?

11. In counting 100 leukocytes you are accurately able to distinguish 15 basophils. Is this a normal number for the white blood cell count, and what possible health implications can you draw from this?

12. What is the function of platelets?

13. Label the blood cells in the following illustration.

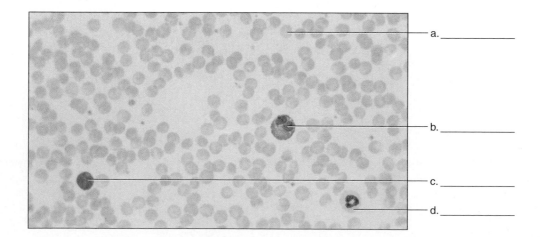

a. _____

b. _____

c. _____

d. _____

LABORATORY

Blood Tests and Typing

INTRODUCTION

Blood is a complex medium that has multiple functions in the body. Many clinical tests are performed on blood, including those to determine the number of erythrocytes or leukocytes in a known volume of blood, the blood type of an individual, and hemoglobin concentration. These topics are discussed in the Saladin text in chapter 18, "Blood." In this exercise you perform tests that allow you to evaluate different samples of blood. As in the previous exercise, handle the samples with care. This is important, so we've repeated the caution statement from that exercise here.

Caution!

In this excercise you may be using unsterilized human blood, sterilized human blood, nonhuman mammal blood, or synthetic blood. The risk of blood-borne diseases is significantly minimized by the use of sterilized human blood and nonhuman mammal blood; however, use the same precautions you would if you were handling fresh and potentially contaminated human blood. Your instructor will determine what type of blood to use. In all cases, follow strict procedures for handling potentially pathogenic material.

1. Wear protective gloves during the procedures.
2. Do not eat or drink in lab.
3. If you have an open wound, do not participate in this exercise or make sure the wound is *securely* covered.
4. Your instructor may elect to have you use your blood. Due to the potential for disease transmission, such as AIDS or hepatitis, in fresh blood samples, *make sure you keep away from other students' blood and keep them away from your blood.*
5. Do not participate in this part of the exercise if you have recently skipped a meal or if you are not well.
6. Place all disposable material in the biohazard bag.
7. Place all used lancets in the sharps container and all glassware to be reused in a 10% bleach solution.
8. After you have finished the exercise, clean and disinfect the countertops with a 10% bleach solution.

OBJECTIVES

At the end of this exercise you should be able to

1. perform specific diagnostic tests, such as hematocrit and hemacytometer tests;
2. determine the antigens (agglutinogens) present in a particular ABO blood type;
3. list the antibodies (agglutinins) present in a particular ABO blood type;
4. relate Rh-positive or Rh-negative blood to antigens present;
5. correlate hematocrit with erythrocyte counts.

MATERIALS

5 mL of fresh, nonhuman mammal blood (dog, sheep, or cow)

Sterilized human blood, labeled 1 to 4

Blood typing antisera (anti-A, anti-B, anti-D), test cards, and toothpicks

Rh warming tray

Heparinized blood microcapillary tubes

Capillary tube centrifuge

Hematocrit reader (Criticorp Micro-hematocrit tube reader, Damon Micro-capillary reader)

Seal-ease

300 mL Erlenmeyer flask

1000 µL micropipette or 1 mL pipette with pipette pump

0.9% saline solution

Hemoglobinometer

Hemacytometer, coverglass

Microscope

Protective gloves

Goggles

Tallquist paper

Tallquist chart

Cotton balls

Contamination bucket with autoclave bag

Container with 10% bleach solution

Sharps container

PROCEDURE

Blood Typing

An understanding of blood type is critical in clinical work. Proper matching of blood between donor and recipient is a vital process, and you will learn the essentials of this process in this exercise. There are many different typing systems that match blood, but the ABO and Rh systems are the two most common in clinical settings. Blood types are genetically determined. There are other typing systems, such as the Kell, Lewis, MNS, and Duffy blood groups. These are antigen/antibody groups used to type blood and they are frequently used in forensic studies.

The first blood typing system discovered was the ABO system. In this system a person has one of four blood types, type A, type B, type AB, or type O. These blood types are determined by the presence of **antigens.** Antigens are large molecules, such as glycoproteins, that occur on the outer membrane of a cell, with type A blood cells having the A antigen, type B blood having the B antigen, type AB blood having both the A and B antigens, and type O blood having neither.

Antigens can react to **antibodies** that occur in the plasma of another person. Antibodies are secreted by plasma cells and, in the case of blood, attach to the antigens. Antibodies cause plasma cells to stick together, causing a reaction known as agglutination. These antigen-antibody complexes are broken down by immune cells. In the case of blood, the antigens are called **agglutinogens** and the antibodies are called **agglutinins.** The agglutinins are found in the plasma portion of the blood.

In the ABO system, no prior exposure to the agglutinogen is needed for granulation or clumping to occur. For example, if a person has blood type A, that individual has agglutinins against blood type B. If the person with blood type A receives a transfusion of blood type B, then the anti-B agglutinins attack the agglutinogens in the blood introduced into the system. This causes a **transfusion reaction,** the agglutination and hemolysis of the transfused erythrocytes. If the reaction is severe enough, death may occur.

Table 30.1 gives details of the ABO blood system.

TABLE 30.1	ABO Blood System	
Blood Type	**Agglutinogens (Antigens)**	**Agglutinins (Antibodies)**
A	A	Anti-B
B	B	Anti-A
AB	A and B	None
O	None	Anti-A and anti-B

Procedure for Blood Typing

1. Read the entire blood typing procedure before you begin this part of the exercise. Wear protective gloves and eyewear.
2. Obtain a sample of sterilized blood as directed by your instructor. The vial should be labeled 1, 2, 3, or 4.
3. Record the sample number you are using in the space provided.
4. Place two drops of the sample blood on a blood test card or on a clean glass microscope slide (see figure 30.1).
5. Place antiserum-A on one drop and antiserum-B on the other drop.
6. Use separate toothpicks to stir each sample of blood and antiserum.
7. Keep the antisera separate from one another. Examine the sample for clumping or granulation within 2 minutes. Granulation may appear as fine red dust or larger particles. Make sure that you dispose of your toothpicks in the biohazard bag. If both blood samples coagulate, you have a sample of type AB blood. If neither sample coagulates, you have a sample of blood type O. If the blood with antiserum-A coagulates, you have a sample of blood type A, and if blood with antiserum-B coagulates, you have a sample of blood type B.
8. Compare your sample to figure 30.2.

Blood sample number: _____

Blood type: _____

Common ABO blood types for various groups in the United States are listed by percentages in table 30.2.

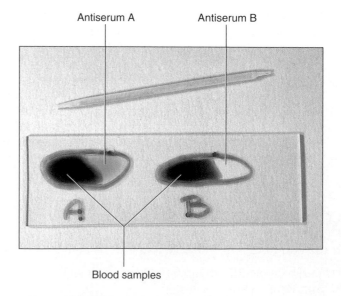

FIGURE 30.1 Blood Testing Procedures

TABLE 30.2	Blood Type Percentages			
Blood Type	**Caucasians**	**African Americans**	**Asian Americans**	**Native Americans**
A	41	27	28	16
B	9	21	23	4
AB	3	4	13	1
O	47	48	36	79
Total	**100**	**100**	**100**	**100**

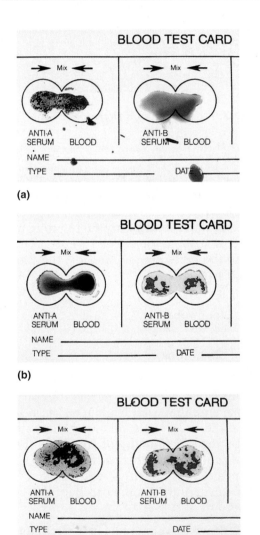

FIGURE 30.2 **Blood Types** (a) Type A; (b) type B; (c) type AB.

Determination of Rh Factor

The other significant blood group is the Rh system, which was named after Rhesus monkeys, the animal model used in discovering this blood group. The Rh factor is determined by the blood genotype DD or Dd. People with this genetic makeup are said to be Rh⁺ (Rh-positive). About 85% of the people in the United States are Rh-positive. People who have the genotype dd are said to be Rh⁻ (Rh-

negative) and they represent about 15% of the U.S. population. In an Rh-positive person the Rh antigen is present but the anti-Rh antibody is absent. In an Rh-negative person, the antigen is absent. Anti-Rh antibodies are normally not present in the blood. They develop in an Rh⁻ person after exposure to the Rh antigen.

Determination of the Rh type is a much more subtle test than ABO typing. The Rh antiserum (anti-D) is more fragile in shipping and storage and should not be considered clinically relevant in this exercise. Use the same precautions as outlined at the beginning of the exercise regarding bodily fluids.

1. Place one drop of blood on a slide and place a drop of antiserum-D (Rh antiserum) on the slide.
2. Use a new toothpick and stir the two drops together.
3. Place this mixture on a warming tray and gently rock the slide.
4. Examine the slide for clumping after a minute or two. If clumping occurs (which may appear as fine red granules), then the sample is Rh⁺ (the Rh antiserum reacted with the Rh antigen in the blood). If no granulation occurs, then the sample is Rh⁻. This test is more difficult to determine, so look for slight granulation. Rh antiserum should be used fresh.
5. Record your results.

Rh factor determination: _____

Rh determination is important if a pregnant woman is Rh⁻. If her unborn child is Rh⁺, then she may develop antibodies against the Rh antigen during pregnancy or delivery (a time when fetal and maternal blood frequently mix). If her second child is Rh⁺, then the antibodies that she produced during her first pregnancy may cross the placenta and cause severe reactions in the fetus. This is called **hemolytic disease of the newborn (HDN)**, or **erythroblastosis fetalis.** Rho-GAM (anti-Rh antibodies) prevents antibody formation in the mother against the Rh antigen and is frequently given to the mother during pregnancy and again after delivery. A woman who is Rh⁺ does not have the antibodies for the Rh factor and therefore will not produce antibodies against the developing fetus, regardless of the Rh factor of the fetus.

Blood Typing Problems

Extensive blood transfusions can lead to problems. Blood that is donated from one person carries antibodies, and these may react with the recipient's blood. There are other blood types in addition to the ABO and Rh systems and sensitivities may develop due to antibodies present in the recipient's blood.

Hematocrit

The **hematocrit** is the percentage of erythrocytes in the total blood volume. The percentage of erythrocytes can be calculated after centrifuging a sample of blood. The cells end up as a large sediment in the bottom of the capillary tube, leaving the lighter leukocytes and plasma on top. The hematocrit varies in individuals with values of 37% to 48% being normal in females and 42% to 52% being normal in males. An increase in the hematocrit above normal is known as **polycythemia** and can exceed 65%.

When blood is lost faster than it is replaced or when the production of erythrocytes is low, **anemia** occurs. In severe anemia, the hematocrit may drop to 15% or less. Anemia may also be due to low levels of hemoglobin in the blood. Hemoglobin is a complex oxygen-carrying molecule composed of an iron-containing **heme group** and the protein **globin.** Another form of anemia is sickle-cell anemia. This genetic condition causes erythrocytes to form long, curved shapes when a person is under oxygen stress (such as when you climb a set of stairs too fast). The erythrocytes are broken down, leaving the person anemic. You will next compare the hematocrit of sterilized human blood with that of another mammal.

Procedure for Determining the Hematocrit

Refer to the caution statement at the beginning of the exercise concerning working with blood.

1. While wearing protective gloves, fill two capillary tubes with blood. One should have commercially prepared, sterilized human blood and the other should have nonhuman mammal blood.
2. Fill each tube by touching the red end of the capillary tube to the blood sample. Let the blood flow up the tube by capillary action.
3. Once the tube is filled, place your finger on the other end of the tube (the nonred end) to prevent blood from flowing back out. Fill each tube to about three-quarters of the tube length. Both tubes should have about the same volume of blood in them.
4. Push the red end of the tube into Seal-ease® or modeling clay. Only the red end of the capillary tube should be sealed. Be careful—the capillary tubes are fragile and sharp! They can break and puncture the skin (see figure 30.3).
5. Place the capillary tube in the centrifuge with the clay plug against the outer rubber ring.

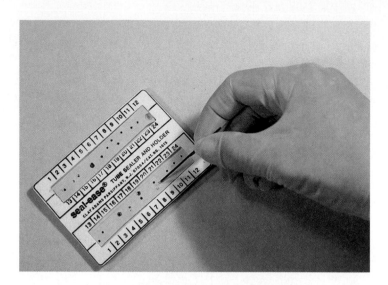

FIGURE 30.3 Preparation of a Hematocrit Tube

6. Place your tubes opposite each other and record the number of each tube. Make sure you keep track of the sample from each tube and don't mix them up! It is important that the capillary tubes are opposite one another to distribute the weight evenly in the centrifuge. If all the tubes are on one side, the centrifuge will be off balance.

Mammal blood tube #: _____

Sterilized blood tube #: _____

7. After many tubes have been placed into the centrifuge, make sure that each has the clay seal facing the rubber gasket. If the seals are facing the center of the centrifuge, then the blood will spray out of the tube when the centrifuge begins turning. Make sure the metal top is on the centrifuge and screwed in place. Close the cover and spin the capillary tubes for 3 to 5 minutes.

Once the centrifuge has come to a complete stop, remove each tube and calculate the percentage of red blood cells in the tube. This can be done with hematocrit readers. If you use a card reader, place the clay-blood interface at the zero line on the card. The capillary tube should be vertical; move the tube until the plasma-air interface is on the 100% line. Read where the red cells intersect the chart. This is the hematocrit.

Record the hematocrit in the following spaces and in the Chapter Summary Data section at the end of the exercise.

Sterilized blood: _____ 1

Mammal blood: _____ 2

? How do these values compare? _____ 3

? What commercial procedure might lead to a difference in the hematocrit between these two

samples? _____ 4

? Plasma with red coloration indicates hemolysis of the erythrocytes. Normally, the plasma should be straw-colored or light yellow. Does either of your samples

show hemolysis? _____ 5

The buff-colored layer between the erythrocyte layer and the plasma layer is the leukocyte layer.

Hemoglobin Determination

Anemia is a condition whereby the oxygen-carrying capacity of the blood is diminished, which may be due to a reduced number of erythrocytes in the blood and show up as a low hematocrit. In some cases the hematocrit is normal but there is a decrease in hemoglobin levels in the individual red blood cells.

Hemoglobinometer Method

Levels of hemoglobin can be examined with the use of a hemoglobinometer. This instrument uses differences in the absorption of green light by hemoglobin to measure its concentration in the blood. Different hemoglobinometers are available. The following test is for the Leica (AO) hemoglobinometer. You can determine the level of hemoglobin in the blood with the following procedure.

1. Make sure that the hemoglobinometer has fresh batteries installed and that the light turns on when the light switch is depressed.
2. Place two glass plates together, so that there is a chamber on the inside, and insert them into the metal clip.
3. Fill the space between the glass plates with blood.
4. Insert the metal clip with the glass plates into the hemoglobinometer.
5. Move the lever on the side of the hemoglobinometer until the green colors are the same shade.
6. Read the scale to determine the hemoglobin concentration.
7. Carefully remove the glass plates and clean them with a cotton ball soaked in alcohol. Dispose of the cotton ball in a biohazard container and place

the glass plates in a secure area where they can be used by other students or placed in a 10% bleach solution.
8. You may want to compare two different samples of blood, such as a sample of sterilized blood versus nonhuman mammal blood. Record your results.

Hemoglobin (Hb) levels are expressed as grams Hb/100 mL of blood. The average value for humans is 12 to 16 grams/100 mL of blood. In males the normal range is 13 to 18 g/100 mL and for females it is 12 to 16 g/100 mL.

Tallquist Method

Another method to determine the level of hemoglobin in the blood is by comparing the color of the oxyhemoglobin in a drop of blood on Tallquist paper to a comparative chart.

1. Place a piece of Tallquist paper on a paper towel on your desk.
2. Place a drop or two of blood on the paper and wait 15 seconds.
3. Compare the color on your paper to the Tallquist chart.

? Hemoglobin from sterile blood sample: _____ 6

? Hemoglobin from mammal blood sample: _____ 7

Blood Cell Counts

Another way to determine the number of blood cells is to count the number of cells in a known volume and subsequently determine the total number of cells per cubic millimeter (mm^3). Modern evaluation of erythrocyte and leukocyte counts is performed by injecting blood samples into an optical computer system, which automatically calculates the cell counts. Most hospitals use computer-driven optical systems to count cells (figure 30.4).

The average number of erythrocytes in blood is approximately 5 million per cubic millimeter. This is too large a number to count during the lab period, so another technique will be used. To determine the number of cells in blood, you will dilute the blood and count the cells that occur in a small volume. The process used in lab will involve the use of a **hemacytometer,** a glass slide that has very fine lines etched on its surface. When a coverslip is placed on the hemacytometer and a sample of blood is introduced, the blood fills a precise volume. This volume is 1/50th of a cubic millimeter.

Even with this small volume there are too many erythrocytes in the blood in a given volume to be counted accurately. The solution to this is to dilute the blood so that the cells can be counted individually. In the procedure that you will perform, you will take 1 part of blood and dilute it to 200 parts diluent (dilution solution).

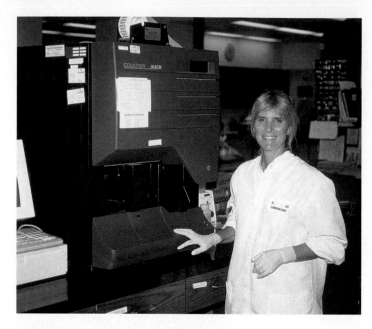

FIGURE 30.4 Optical Cell Counter

Procedure for Erythrocyte Counts

For males the average count is about 5.4 million erythrocytes per cubic millimeter of blood, with a range of 4.5 to 6.0 million cells per cubic millimeter. For females the average count is about 4.8 million erythrocytes per cubic millimeter of blood, with a range of 4.0 to 5.5 million cells per cubic millimeter.

You will perform this experiment twice: once using commercially prepared, sterilized human blood and once using fresh, nonhuman mammal blood. Obtain the following:

Micropipette or pipette and pipette pump

300 mL Erlenmeyer flask

0.9% saline solution

Sample of nonhuman mammal blood

Sample of sterilized blood

Hemacytometer and coverslip

Protective gloves and eyewear

Microscope

Cotton balls

Because of the volume of solution made, you may want to use one dilution of blood for the entire class, depending on the wishes of your instructor. It is particularly important to accurately measure the blood withdrawn and the amount of diluent in order to make an accurate erythrocyte count. For withdrawing blood from your samples you may use either a micropipette or a volumetric pipette. The following instructions provide directions for both. Make sure that the blood sample you use is well mixed, so that blood

cells are evenly distributed in the plasma. Wear protective gloves and eyewear during the entire procedure.

1. Add exactly 200 mL of 0.9% saline solution to an Erlenmeyer flask.
2. Use either a 1,000-microliter micropipette or a 1 mL pipette with a pipette pump. If you use a micropipette, depress the plunger until it stops and insert the tip into the blood sample. Slowly release the plunger to pull blood into the micropipette. If using the 1 mL pipette, use a pipette pump attached to the blunt end of the pipette and draw up exactly 1 mL of blood to the calibration line of the pipette. Never pipette anything by mouth! Use a pipette bulb or pump.
3. Transfer the blood to the Erlenmeyer flask. If you use a micropipette, depress the plunger until it stops and then push it a little farther. It you are using a volumetric pipette (one that has an expanded middle section), the blood should drain out of the pipette and you should NOT forcibly expel the remainder of the fluid in the pipette. Dispose of the pipette in a tray for that purpose or in a beaker of 10% bleach.
4. Gently swirl the flask to distribute the blood cells. Withdraw the diluted sample with a Pasteur pipette.
5. Place the hemacytometer coverslip (not a regular coverslip) over the hemacytometer chamber and fill the chamber with the Pasteur pipette. Do not overfill the chamber.
6. Place the hemacytometer under the microscope and examine the cells under high power (400–450×).
7. Count all of the erythrocytes in each of the five areas marked with an "R" (see figure 30.5). Some cells will be on the border of the counting grid. Count only those cells touching the left line and the upper line

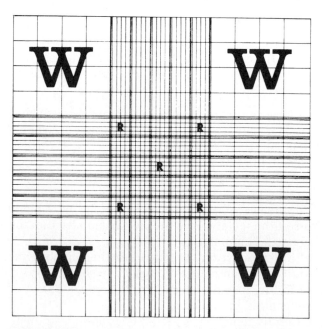

FIGURE 30.5 Erythrocyte Counts Using a Neubauer Hemacytometer

as part of the sample. Do not count the cells touching the right line or bottom line.

The blood was diluted 1:200, and the sum of the volume in the five areas represents 1/50th of a cubic millimeter. If you multiply the dilution factor (200) and the volume that would occur to fill 1 mm^3 (50), then you would multiply the number of erythrocytes counted by 10,000 to obtain the number of erythrocytes per cubic millimeter.

8. Record your results.

? Number of erythrocytes counted: _____8

? Number of erythrocytes per cubic millimeter: _____9

Use the sample of nonhuman mammal blood. Use the chamber on the opposite side of the hemacytometer and follow the same procedure you used for human blood. Record the results.

? Number of erythrocytes counted: _____10

? Number of erythrocytes per cubic millimeter: _____11

? How do the numbers of cells in sterilized blood and

nonhuman mammal blood compare? _____12

? Is there a correlation between the numbers of erythrocytes counted from these samples and the hematocrit

taken from the same samples? _____13

After you finish your counting, place the hemacytometer and coverslip *gently* into a 10% bleach solution.

Leukocyte Counts (Optional)

Leukocytes are counted in a similar fashion except that the four large corner regions of the hemacytometer are used and the sum of the cell count from these four areas is multiplied by 50. Your instructor may provide you with additional instructions for a leukocyte count, using a specific solution for leukocytes.

? Number of leukocytes counted: _____14

? Number of leukocytes per cubic millimeter: _____15

Normal blood levels for leukocytes are about 7,000 cells per cubic millimeter, with a range of 4,300 to 10,800 cells per cubic millimeter. Leukocytosis is an elevated white blood cell count typically above 11,000 cells per cubic millimeter. Leukocytosis indicates inflammation. Leukopenia, on the other hand, is a decreased white blood cell count, with less than 4,000 cells per cubic millimeter. This may be due to chemotherapy, radiation therapy, HIV infection, infectious hepatitis, cirrhosis of the liver, typhoid fever, or chronic infections, such as tuberculosis.

Clean Up Make sure the lab is clean before you leave. Wipe any blood from the microscope objective lenses with lens cleaner and lens paper. Make sure you do not leave glass or blood on the lab tables. Use a 10% bleach solution and a towel to clean off your lab table before you leave.

Notes

REVIEW SECTION

Blood Tests and Typing

Name _____ Date _____

Lab Section _____ Time _____

Review Questions

1. What is the name of a surface membrane molecule that causes an immune reaction?

2. What is the average range of hematocrit for a normal female?

3. What is the average range of hematocrit for a normal male?

4. What percent of the blood volume consists of formed elements?

5. A person with blood type B has what kind of agglutinins (antibodies)?

6. A person has antibody A and antibody B in his or her blood, with no Rh antibody. What blood type does this person have?

7. A total of 240 erythrocytes are counted in the hemacytometer chamber. What is the red blood cell count of this individual in terms of erythrocytes per cubic milliliter?

8. A person with blood type B negative is injected with type A positive blood. From an immunological (antigen/antibody) standpoint, what would happen after the injection?

9. How might changes in the pipette technique alter the final determined value of erythrocytes? What type of errors might you expect?

10. Using the following illustration, calculate the hematocrit of the individual. Determine if it falls within normal limits.

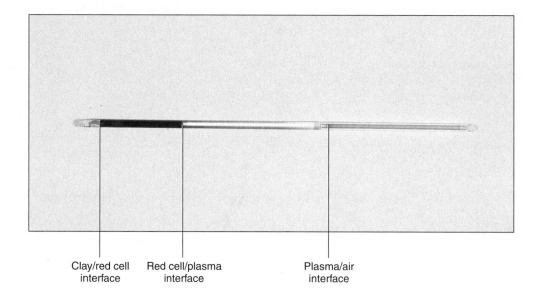

| Clay/red cell | Red cell/plasma | Plasma/air |
| interface | interface | interface |

11. Define anemia.

? Chapter Summary Data

Use this section to record your results from questions within the exercise.

1. _____   9. _____

2. _____   10. _____

3. _____   11. _____

4. _____   12. _____

5. _____   13. _____

6. _____   14. _____

7. _____   15. _____

8. _____

LABORATORY

Structure of the Heart

INTRODUCTION

In this exercise you study the structure of the heart, which is covered in the Saladin text in chapter 19, "The Heart." The heart is located deep in the thorax between the lungs in a region known as the **mediastinum.** The mediastinum contains the heart, the membranes surrounding the heart (the **pericardium**), and other structures, such as the esophagus and descending aorta. The mediastinum is located between the sternum, lungs, and thoracic vertebrae. Vessels that return blood to the heart are called **veins,** and those that carry blood from the heart are called **arteries.**

If you were to open the chest cavity, the first structure you would see is the **pericardial sac (parietal pericardium).** The pericardial sac encloses the heart and has two layers: a tough, outer connective tissue sheath called the fibrous layer and an inner layer called the serous layer. Deep to the pericardial sac is the **pericardial cavity,** which contains a small amount of **serous fluid.** This fluid reduces the friction between the outer surface of the heart and the pericardial sac. The heart wall has an outer layer known as the **epicardium,** or **visceral pericardium.** Locate these pericardial layers in figure 31.1.

In this exercise, you examine models of the heart, preserved sheep and human hearts, if available. As you look at the preserved material try to see how the structure of the heart relates to its function.

OBJECTIVES

At the end of this exercise you should be able to

1. list the three layers of the heart wall;
2. describe the position of the heart in the thoracic cavity;
3. describe the significant surface features of the heart;
4. describe the internal anatomy of the heart;
5. find and name the anatomical features on models of the heart and in the sheep heart;
6. describe the blood flow through the heart and the function of the internal parts of the heart;
7. discuss the functioning of the atrioventricular valves and the semilunar valves and their role in circulating blood through the heart.

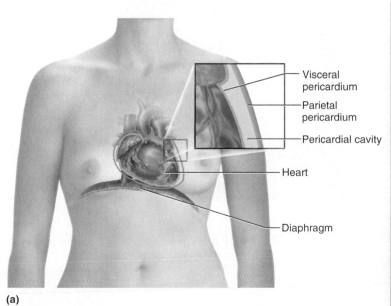

(a)

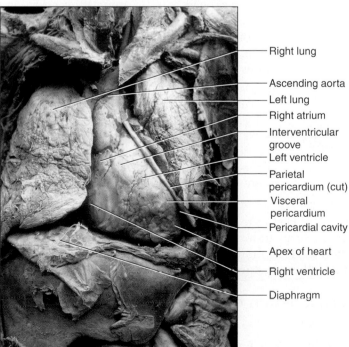

Right lung

Ascending aorta

Left lung

Right atrium

Interventricular groove

Left ventricle

Parietal pericardium (cut)

Visceral pericardium

Pericardial cavity

Apex of heart

Right ventricle

Diaphragm

(b)

Visceral pericardium

Parietal pericardium

Pericardial cavity

Heart

Diaphragm

FIGURE 31.1 Heart in Thoracic Cavity and Heart Coverings (a) Coronal section; (b) photograph of cadaver.

MATERIALS

Models and charts of the heart

Preserved sheep hearts

Preserved human hearts (if available)

Blunt probes (mall probes)

Dissection pans

Razor blades or scalpels

Sharps container

Disposable gloves

Waste container

Microscopes

Prepared slides of cardiac muscle

Fresh or thawed sheep heart

Clamp (hemostat) or string

PROCEDURE

Heart Wall

The heart wall is composed of three major layers. The outermost layer is the **epicardium,** or **visceral pericardium.** composed of epithelial and connective tissue. The middle layer is the **myocardium** and is the thickest of the three layers. It is mostly made of **cardiac muscle.** You may wish to review the slides of involuntary cardiac muscle and note the intercalated discs, branching fibers, and fine striations of cardiac tissue (as described in Laboratory Exercise 6). The cardiac muscle is arranged spirally around the heart, and this arrangement provides a more efficient wringing motion to the heart. The inner layer of the heart wall is known as the **endocardium,** a membrane consisting of endothelium (simple squamous epithelium) and connective tissue.

 Caution! Be careful when handling preserved materials. Ask your instructor for the proper procedure for working with preserving fluid and for handling and disposal of the specimen. Do not dispose of animal material in the sinks. Place it in an appropriate waste container.

It is best to examine heart models *before* dissecting a sheep heart unless your instructor directs you to do otherwise. Heart models are color coordinated and labeled to make the structures easier to locate.

Examination of the Heart Model

Overview

The heart is a four-chambered pump with two superior atria (Ay-tree-uh) and two inferior ventricles. Blood enters

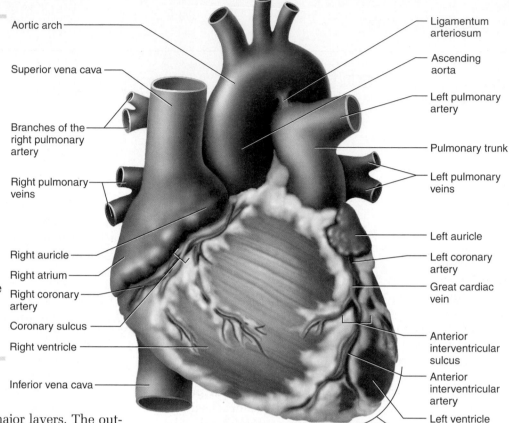

(a) Anterior View

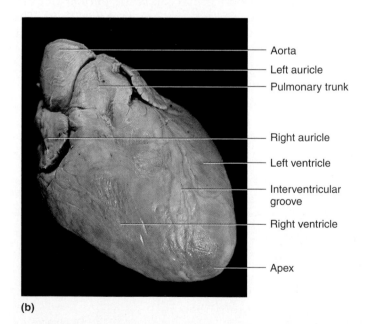

(b)

FIGURE 31.2 Surface Anatomy of the Heart, Anterior View (a) Diagram; (b) photograph.

the heart in the right atrium (see figure 31.2) and flows into the right ventricle. Once the blood is in the right ventricle the contraction of the ventricular wall sends the blood to the lungs. The blood is oxygenated in the lungs and returns to the heart by entering the left atrium. Blood moves from the left atrium to the left ventricle and is then pumped to the rest of the body.

Exterior of the Heart

Examine the heart model and notice that the heart has a pointed end, or **apex,** and a blunt end, or **base.** The apex of the heart is inferior, and the great vessels leaving the heart are located at the base (therefore, in the case of the heart, the base is superior to the apex). Compare the model to figure 31.2 and see how the **aorta** curves to the left in an anterior view of the heart and is posterior to the **pulmonary trunk.**

Locate the anterior features of the heart. The heart is composed of two large, inferior ventricles and two smaller and superior atria. The **left ventricle** extends to the apex of the heart and is delineated from the right ventricle by the **interventricular groove,** or sulcus. In this interventricular groove are some of the **coronary arteries** and **cardiac veins,** discussed later. The left ventricle is larger than the right ventricle. Note the two ear-like flaps on the anterior, superior atria. These are the **auricles.**

If you examine the heart from the posterior side you will see the atria more clearly. At the junction of the right atrium and the right ventricle is the **atrioventricular sulcus,** or groove. The **coronary sinus,** a large venous space that carries blood from the cardiac veins to the right atrium, is located in this sulcus. Locate the **superior vena cava** and the **inferior vena cava,** two vessels that also return blood to the right atrium. Locate the **pulmonary veins,** which carry blood from the lungs to the left atrium. Compare the heart model to figure 31.3.

The major vessels of the heart are illustrated in figures 31.2, 31.3, and 31.4. Locate the **pulmonary trunk, pulmonary arteries, ligamentum arteriosum** (between the pulmonary trunk and aortic arch), **ascending aorta, pulmonary veins, superior vena cava, inferior vena cava, coronary arteries,** and **cardiac veins.**

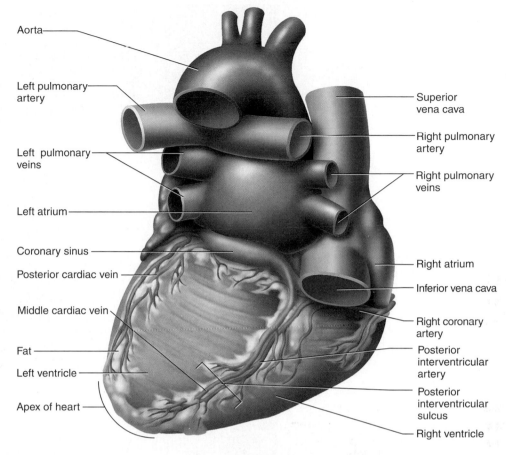

(a)

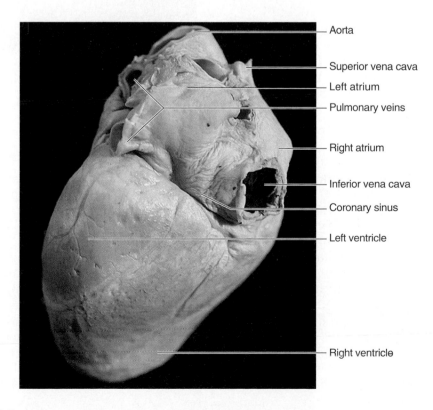

(b)

FIGURE 31.3 Surface Anatomy of the Heart, Posterior View (a) Diagram; (b) photograph.

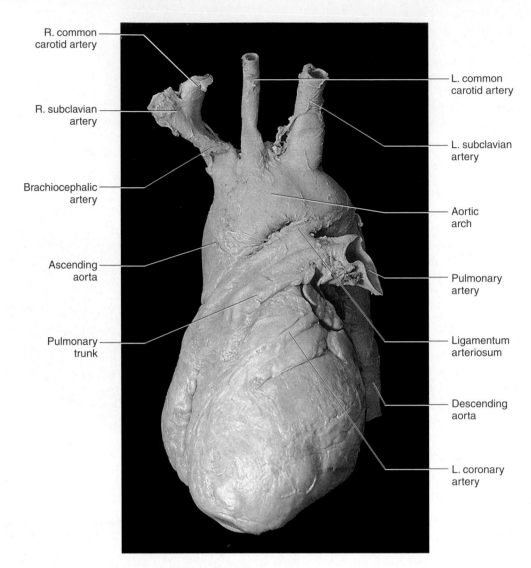

R. common carotid artery

R. subclavian artery

Brachiocephalic artery

Ascending aorta

Pulmonary trunk

L. common carotid artery

L. subclavian artery

Aortic arch

Pulmonary artery

Ligamentum arteriosum

Descending aorta

L. coronary artery

FIGURE 31.4 Vessels of the Heart, Anterior View

The heart tissue is nourished by coronary arteries. The **left coronary artery** arises from the ascending aorta (see figure 31.4) and then branches into the **anterior interventricular artery** and the **circumflex artery.** The **right coronary artery** also arises from the ascending aorta and branches to form the **posterior interventricular artery** and the **right marginal artery.** These major arteries of the heart supply blood to the myocardium. On the return flow from the heart muscle the **great cardiac vein** follows the depression of the interventricular groove and the atrioventricular groove to the coronary sinus. On the right side of the heart the **small cardiac vein** leads to the coronary sinus, which empties into the right atrium. Locate these vessels on the external surface of the model of the heart and compare them to figures 31.2, 31.3, and 31.4.

Interior of the Heart

Examine a model of the interior of the heart and locate the right and left ventricles. The muscular wall of the **right ventricle** is much thinner walled than the **left ventricle.** This is because blood from the right ventricle is pumped a short distance to the lungs while blood in the left ventricle is pumped more extensively through the body. The ventricles are separated by the **interventricular septum,** which forms a wall between the two ventricular chambers. Compare the model to figure 31.5.

Examine the **right atrium** and note how thin the wall is compared to the ventricles. The walls of the atria are thin because blood in the atria has to flow only a short distance to the ventricles. Examine the medial wall of the atrium, known as the **interatrial septum,** and locate a thin, oval

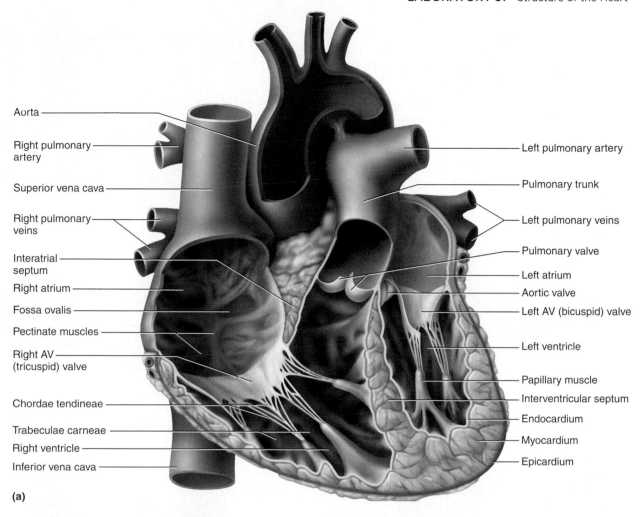

Aorta

Right pulmonary artery

Superior vena cava

Right pulmonary veins

Interatrial septum

Right atrium

Fossa ovalis

Pectinate muscles

Right AV (tricuspid) valve

Chordae tendineae

Trabeculae carneae

Right ventricle

Inferior vena cava

Left pulmonary artery

Pulmonary trunk

Left pulmonary veins

Pulmonary valve

Left atrium

Aortic valve

Left AV (bicuspid) valve

Left ventricle

Papillary muscle

Interventricular septum

Endocardium

Myocardium

Epicardium

(a)

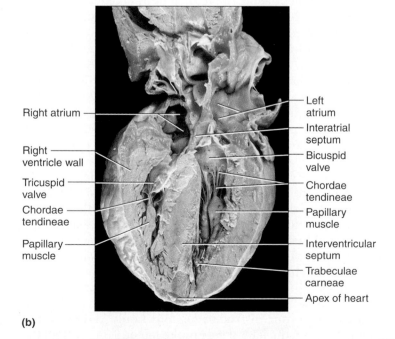

Right atrium

Right ventricle wall

Tricuspid valve

Chordae tendineae

Papillary muscle

Left atrium

Interatrial septum

Bicuspid valve

Chordae tendineae

Papillary muscle

Interventricular septum

Trabeculae carneae

Apex of heart

(b)

FIGURE 31.5 **Heart, Frontal Section** (a) Diagram; (b) photograph.

depression in the atrial wall. This depression is the **fossa ovalis.** In fetal hearts this is the site of the **foramen ovale (For-AYE-men o-VAL-eh),** but closure of the foramen usually occurs just after birth. Note the extensive **pectinate (PEK-tin-ate) muscles** on the wall of the atrium. These provide additional strength to the atrial wall. Blood in the superior vena cava, the inferior vena cava, and the coronary sinus returns to the right atrium. Examine the features of the right atrium in figures 31.5 and 31.6.

The blood from the right atrium flows into the right ventricle. Now examine the valve between the right atrium and right ventricle. This is the **right atrioventricular valve,** or **tricuspid valve,** and it prevents return of blood from the right ventricle into the right atrium during ventricular-contraction. Examine the valve for three flat sheets of tissue. These are the three cusps of the tricuspid valve. The tricuspid valve has thin, threadlike attachments called **chordae tendineae (COR-day TEN-din-ee).** These tough cords are attached to larger **papillary muscles,** extensions from the wall of the ventricle. The right ventricle wall has small extensions called **trabeculae carneae (trah-BEC-you-lee CAR-nee-ee).** The blood from the right ventricle flows into the pulmonary trunk toward the lungs.

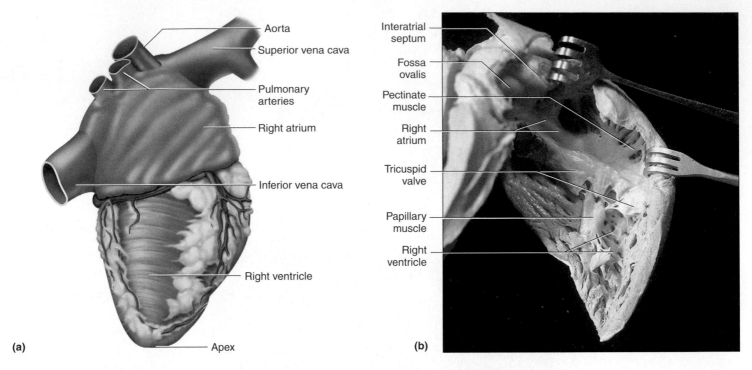

FIGURE 31.6 Right Side of the Heart (a) Diagram of surface view; (b) photograph of interior view.

Locate the **pulmonary semilunar valve.** It appears as three small cusps between the right ventricle and the pulmonary trunk and keeps blood from flowing in reverse from the pulmonary trunk into the right ventricle during ventricular relaxation. Examine the details of the right ventricle in models in the lab and in figure 31.5.

Blood from the pulmonary trunk flows into the pulmonary arteries prior to entering the lungs. Blood in the lungs releases carbon dioxide and picks up oxygen. The **pulmonary veins** carry oxygenated blood from the lungs into the **left atrium.** These vessels are located in the superior, posterior portion of the left atrium. Blood from the left atrium flows into the left ventricle. Locate the two large cusps of the bicuspid valve between the left atrium and left ventricle. The **bicuspid valve** is also known as the **mitral valve,** or **left atrioventricular valve.** It also has attached chordae tendineae and papillary muscles, which you should locate in the models in the lab. Note the thickness of the left ventricle wall compared to the wall of the right ventricle. Compare the left side of the heart to figure 31.5.

The **aortic semilunar valve** is located at the junction of the left ventricle and the ascending aorta. It has the same basic structure and general function as the pulmonary semilunar valve in that it prevents the flow of blood from the aorta into the left ventricle. Blood from the left ventricle moves into the aorta and subsequently to the rest of the body.

Dissection of the Sheep Heart

The sheep heart is similar to the human heart and usually is readily available as a dissection specimen. Dissection of anatomical material is valuable in that you can examine structures seen more accurately in preserved material than in models. Also, the preserved material has greater flexibility and is easily manipulated. There are some differences between sheep hearts and human hearts, especially in the position of the superior and inferior venae cavae. In sheep these are called the anterior and posterior venae cavae, but we refer to them using the human terminology.

If your sheep heart has not been dissected, then you will need to open the heart. If your sheep or other mammalian heart has been previously dissected, then you can skip the next two paragraphs.

Place the heart under running water for a few moments to rinse off the preserving fluid. Examine the external features of the heart, as seen in figure 31.7. The sheep heart may still be in the pericardial sac. If this is the case, remove the sac before proceeding. Note the fat layer on the heart. The amount of fat on the human or sheep heart is variable. Locate the **left ventricle,** the **right ventricle,** the **interventricular groove (sulcus),** the **right atrium,** and the **left atrium.** Note the **auricles** that extend on the anterior surface of the atria. Carefully remove the adipose tissue from the major vessels of the heart.

Using a sharp scalpel or razor blade, make an incision along the right *side* of the heart (lateral side) from the apex of the heart to the lateral side of the right atrium. If you are unsure about how to proceed during any part of the dissection, ask your instructor for directions. Make another long cut from the lateral side of the left atrium through the lateral side of the left ventricle. You will make a coronal section of the heart if you cut through the **interventricular septum.** Once you have opened the heart, compare the structures of the sheep heart to figure 31.8.

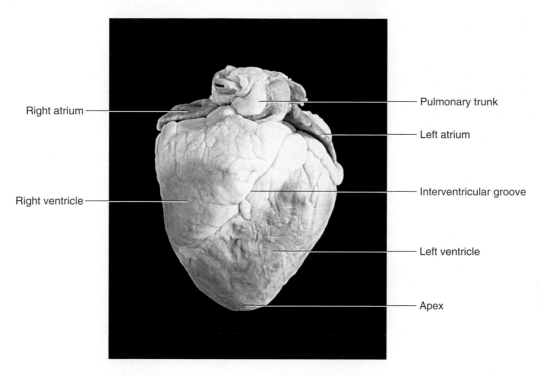

FIGURE 31.7 Sheep Heart, Anterior View

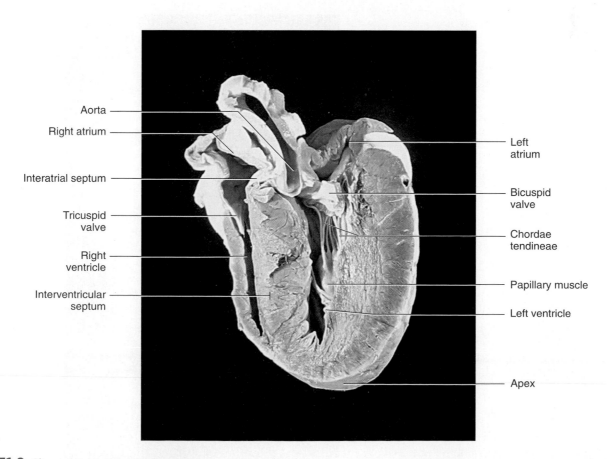

FIGURE 31.8 Sheep Heart, Coronal Section

You can locate the vessels of the heart by gently inserting a blunt metal probe into the vessels and determining which chamber the vessel goes to or comes from. Place the heart in anatomical position and insert the probe into the large, anterior vessel that exits toward the specimen's left side. Be careful and do not tear the heart valves. The blunt end of the probe should enter the right ventricle. This vessel is the **pulmonary trunk.** The pulmonary trunk may still have the **pulmonary arteries** attached. Locate the large vessel directly behind the pulmonary trunk (see figures 31.7 and 31.8). This is the **ascending aorta.** If the vessels are cut farther away from the heart, you can see the **aortic arch.** Insert the probe into this vessel and into the **left ventricle.**

Turn the heart to the posterior surface and locate the **superior (anterior) vena cava** and **inferior (posterior) vena cava.** Insert the probe into the superior and inferior vena cavae, pushing the probe into the **right atrium.** If you find only one large opening in the atrium, you may have cut through either the superior or inferior vena cava during your initial dissection. The probe can be felt through the wall more easily here than in a ventricle because the atrial walls are thinner than those of the ventricles. On the left side the **pulmonary veins** may appear as four separate veins, or you may see a large hole on each side of the **left atrium** if the vessels were cut close to the atrial wall. Locate the same structures in the sheep heart as you found on the model and compare them to figure 31.9.

Cut into the right atrium and use your blunt probe to locate the opening of the **coronary sinus** in the posterior, inferior portion of the atrium. It is small and somewhat difficult to find.

Examine the opening between the right atrium and the right ventricle to locate the **tricuspid valve.** You may have dissected through one of the cusps as you opened the heart. Locate the major features of the **right ventricle.** Find the **chordae tendineae** and the **papillary muscles.** You can find the **pulmonary trunk** by inserting a blunt probe into the superior portion of the right ventricle. Make an incision in the pulmonary trunk near the right ventricle to expose the three thin cusps of the **pulmonary semilunar valve.** Note how the cusps press against the wall of the pulmonary trunk when the probe is pushed against them in a superior direction. These cusps close when blood begins to flow back into the right ventricle as the ventricle relaxes.

Locate the **left atrium, bicuspid valve,** and **left ventricle.** In the left ventricle you should find the papillary muscles, chordae tendineae, and **trabeculae carneae.** You can find the aorta and aortic semilunar valve by inserting a blunt probe toward the superior end of the left ventricle toward the middle of the heart.

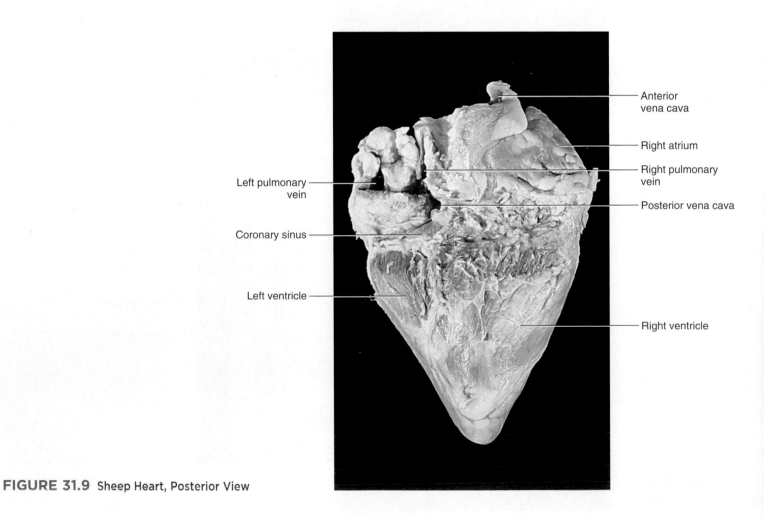

Left pulmonary vein

Coronary sinus

Left ventricle

Anterior vena cava

Right atrium

Right pulmonary vein

Posterior vena cava

Right ventricle

FIGURE 31.9 Sheep Heart, Posterior View

Clean Up When you have finished your study of the sheep heart make sure you clean your dissection equipment with soap and water. Be careful with sharp blades. Place the sheep heart either back in the preserving fluid or in the appropriate waste container as directed by your instructor.

Heart Valves

The semilunar valves and atrioventricular valves prevent the backflow of blood. Your instructor may demonstrate the procedure, or you can do it yourself by using a fresh or thawed sheep heart. Flush any remaining blood from the heart before you locate the valves.

Make an incision into the right atrium, exposing the **tricuspid valve.** Pour water into the right ventricle and notice how the water flows past the tricuspid valve and into the right ventricle. Clamp off the pulmonary trunk or tie it with string to prevent blood from flowing out of the pulmonary trunk. *Gently* squeeze the right ventricle and notice how the right atrioventricular valve closes and prevents blood from backing up into the right atrium.

What is the adaptive value for the closing of atrioventricular valves?

Cut the **pulmonary trunk** close to the right ventricle and slowly pour water into it as if you were trying to fill the right ventricle. The **pulmonary semilunar valve** should fill with water and close the entrance of the right ventricle, preventing backflow. This normally occurs as the right ventricle begins diastole. Compare these valve closures to figure 31.10.

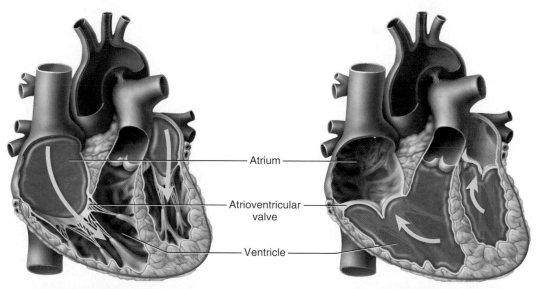

	Atrium
	Atrioventricular valve
	Ventricle

(a)   **Atrioventricular valves open**   **Atrioventricular valves closed**

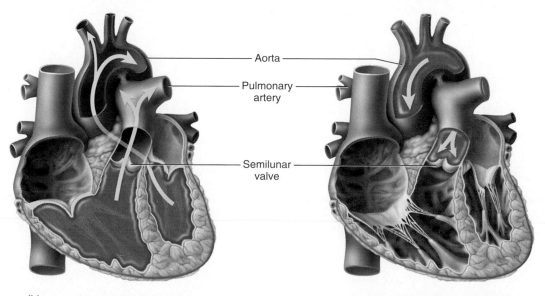

	Aorta
	Pulmonary artery
	Semilunar valve

FIGURE 31.10 Closure of the Heart Valves (a) Atrioventricular valve; (b) semilunar valve.

(b)   **Semilunar valves open**   **Semilunar valves closed**

Notes

REVIEW SECTION

Structure of the Heart

Name _____ Date _____

Lab Section _____ Time _____

Review Questions

1. The heart is located between the lungs in an area known as the:

2. Name the outer (superficial) layer of the pericardium.

3. What is the innermost layer of the heart wall called?

4. Name the depression between the two ventricles seen on the anterior surface of the heart.

5. Are auricles extensions of the atria or the ventricles?

6. What three vessels take blood to the right atrium?

7. Where do the great cardiac vein and the small cardiac vein take blood?

8. Is the apex of the heart superior or inferior to other parts of the heart?

9. What blood vessels nourish the heart tissue?

10. What structure separates the left atrium from the right atrium?

11. The bicuspid valve is located between what two chambers of the heart?

12. What is the function of the aortic semilunar valve?

13. Name the structure found between the atrioventricular valve and the papillary muscle.

14. What is another name for the tricuspid valve?

15. What is the cell type that makes up most of the myocardium?

16. The walls of the left ventricle are thicker than those of the right ventricle. What explanation can you give for this?

17. How does cardiac muscle resemble skeletal muscle?

18. In terms of function, how is cardiac muscle different from skeletal muscle?

19. What is the function of the semilunar valves?

20. Label the following illustration using the terms provided.

right atrium

right ventricle (wall)

left ventricle (wall)

tricuspid valve

aorta

left atrium

apex

chordae tendineae

interventricular septum

bicuspid valve

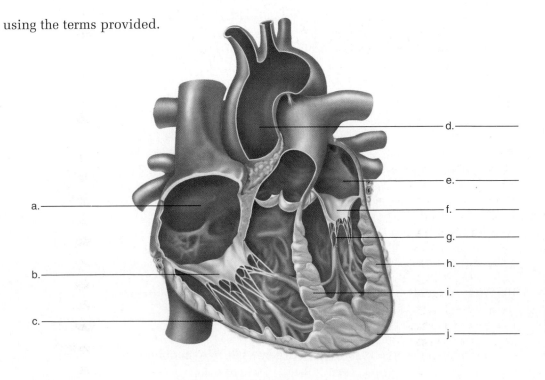

LABORATORY

Electrical Conductivity of the Heart

INTRODUCTION

Before you begin the study of the function of the heart, review the anatomy of the heart in Laboratory Exercise 31. The topics for this exercise are covered in the Saladin text in chapter 19, "The Heart."

The living heart is a phenomenal organ. It contracts an average of 72 times per minute for the lifetime of an individual. The heart has been described as a muscular pump or, more accurately, two pumps acting in unison. As a pump, the heart has a contraction mode and a relaxation mode. **Systole** is contraction of the heart muscle and can be described more specifically as **atrial systole** and **ventricular systole** (contraction of the atria and ventricles, respectively). **Diastole** is the relaxation of the heart muscle, and there is **atrial diastole** and **ventricular diastole.**

Electrical activity of the heart stimulates the heart muscle to contract. In this exercise you observe the electrical activity of the heart and correlate it to the mechanical functioning of the heart. You learn the conductive structures of the heart, make a recording of the electrical activity of the heart, and correlate the recording with the activity of the heart.

The initiation of the electrical impulse in the heart begins at the **sinoatrial (SA) node,** which is commonly known as the **pacemaker.** The sinoatrial node is located in the superior portion of the right atrium. Cells of the sinoatrial node use the sodium-potassium pump to generate a difference in electrical voltage across the cells' membranes. The cells are said to be **polarized** because the inside of the cell membrane has more negative ions than the outside of the membrane. This separation of charged particles is reflected in the voltage difference across the membrane. The sinoatrial node spontaneously **depolarizes,** which causes a change in the total amount of charge across the cell membrane. This change in membrane potential occurs approximately 72 times per minute (the average heart rate). Conduction from the sinoatrial node travels across the atria, causing the muscles of the atria to contract. The impulse that spreads out across the atria reaches the **atrioventricular node (AV) node.** The impulse has a slight delay (about 0.1 second) in the node before being conducted further. This delay allows the atrial cardiac muscle to contract prior to ventricular firing (see figure 32.1).

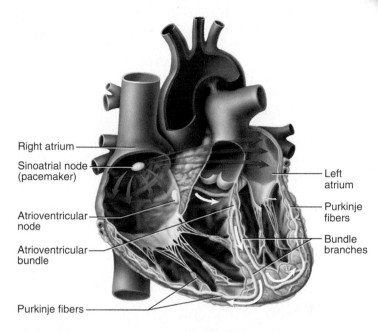

Right atrium

Sinoatrial node
(pacemaker)

Atrioventricular
node

Atrioventricular
bundle

Purkinje fibers

Left
atrium

Purkinje
fibers

Bundle
branches

FIGURE 32.1 Conduction System of the Heart Conduction follows the path indicated by arrows.

The electrical impulse then travels from the AV node to the **atrioventricular bundle (bundle of His),** to the **right** and **left bundle branches,** and finally to the **Purkinje (conduction) fibers.** The Purkinje fibers stimulate the cardiac muscle of the ventricles to contract. The ventricles are thus stimulated from the apex toward the base, and the contraction proceeds from the inferior end of the ventricles toward the atria. Locate the structures of the heart's conduction system in figure 32.1. Contraction of the heart muscle occurs just after depolarization of the heart muscle cells. Relaxation of the heart occurs just after repolarization of the heart muscle cells.

Two concepts are important in understanding the electrical activity of the heart. One is that the heart produces low-voltage **electrochemical impulses** in a similar way to the production of impulses in the nervous system. The "average" potential difference of **−90 millivolts** is a little less than one-tenth of a volt; therefore, the body is operating electrically at slightly less than 0.1 volt. The other important concept in understanding the electrical activity of the heart is that these impulses can travel through the saline medium of the body and be picked up by **sensors (electrode plates)** attached to the skin. The cells of the body are bathed in a saline solution of a little less than 1% salt, which is an excellent conducting medium for electrical impulses. The electrical impulse generated by the atria and ventricles depolarizing and repolarizing can be recorded using an electrocardiograph machine attached to the electrode plates.

Early work done by Willem Einthoven established the modern techniques of recording the electrical activity of the heart by an **electrocardiograph (ECG)** machine. It is an instrument that measures slight changes in the voltage related to cardiac activity. The electrocardiograph produces a chart paper recording or a data file called an **electrocardiogram** (**ECG,** or **EKG**). Time is measured along the horizontal axis (or x-axis, with each millimeter equal to 0.04 second), and voltage difference is measured along the vertical axis (or y-axis, with 1 millimeter equal to 1 millivolt). The ECG has a **baseline** known as the **isoelectic line** and deflections from that line record the electrical activity of the heart. Einthoven established three standard leads that record the heart's electrical events with the heart sitting in the middle of a theoretical shape known as Einthoven's triangle (figure 32.2).

Lead I connects the right arm and the left arm (RA-LA). It measures the potential voltage across the horizontal axis of the heart.

Lead II connects the right arm and the left leg (RA-LL). This is the lead that records the potential voltage from the base to the apex of the heart.

Lead III connects the left arm and the left leg (LA-LL). This is the lead that records the potential voltage along the left side of the heart.

In many college physiology labs, the ECG electrode plates are attached to four areas. These are the medial side of the **left** and **right ankles** and the anterior surface of the **left** and **right wrists.** The attachment of the electrode plate to the right ankle serves as an **electrical ground** and is not used for measurement. The ECG is measured as the potential voltage difference between selected electrode plates. There are numerous "V" electrodes, which are chest electrodes. You will not use the V electrodes in this exercise unless directed to do so by your instructor.

Before you record an ECG, examine the ECG in figure 32.3 and read the accompanying description. The ECG typically has three major events. The first event is the **P wave,** which is a small bump called a **deflection wave.** The P wave represents **atrial depolarization.** The **QRS complex** represents the **ventricular depolarization.** Because the ventricles are more massive than the atria, the electrical events produced as the ventricles depolarize are much larger than the P wave generated by the atria. The **T wave** represents the **ventricular repolarization.** The **atrial repolarization** occurs during ventricular depolarization and is masked by the larger QRS complex.

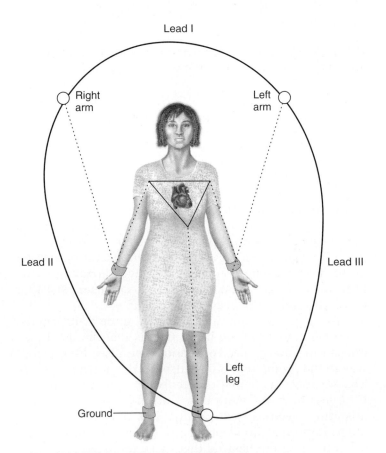

FIGURE 32.2 Leads of a Standard ECG

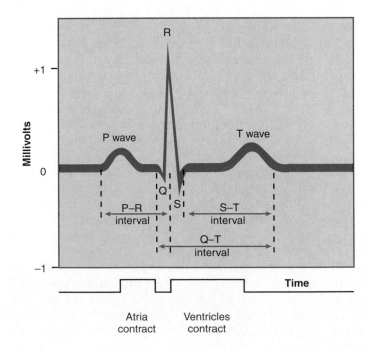

FIGURE 32.3 ECG

OBJECTIVES

At the end of this exercise you should be able to

1. distinguish between systole and diastole;
2. list the structures in the conductive system of the heart;
3. describe how the electrocardiogram can be used to monitor electrical events in the heart;
4. correlate the electrical potentials of the heart to mechanical activity in a cardiac cycle;
5. describe the sequence of electrical conductivity of the heart;
6. associate the P wave, QRS complex, and T wave of an ECG with electrical events that occur in the heart;
7. relate how the electrical activity of the heart is transferred to the ECG;
8. calculate heart rate, PR interval, QRS interval, and QT interval from an ECG recording;
9. describe the variance in an ECG from normal if a person has a heart block;
10. describe the changes in an ECG immediately after exercise;
11. determine the mean electrical axis of the heart.

MATERIALS

Cot or covered lab table

Alcohol swabs

Electrode jelly, paste, or saline pads

BIOPAC Section

EL 503—general purpose disposable electrode pads

SS2L—lead cables

CBLSERA—CBL serial for MP30

MP30—data acquisition unit and power supply (AC100A)

Compatible computer

ECG Machine Section

Electrocardiograph machine

PROCEDURE

Recording ECGs

There are a few ways to record ECGs. One is with a conventional ECG machine that puts out a paper chart strip, another is with a physiograph, and a third is with a computer program and hardware, such as BIOPAC. This exercise will describe the BIOPAC procedure and the ECG machine procedure.

BIOPAC Procedure

You will collect data for the electrical activity of your heart by attaching receiving electrodes to your skin. In this part of the exercise you will be using lead II. The resulting electrocardiogram is displayed on the computer monitor and saved as a file in your BIOPAC folder. If you are going to determine the mean electrical axis of the heart later in this exercise you will need to use BIOPAC Lesson 6.—L06 Electrocardiography 2. The following instructions are for determining a simple ECG and are found in BIOPAC Lesson 5. These are the steps you will take:

1. Setup

a. Make sure the computer is ON.
b. Connect the MP30 unit to an electrical outlet with the power supply.
c. With the MP30 unit OFF, plug in the SS2L lead cables, into the CBLSERA cable, which should be connected to CH2 on the MP30 unit.
d. Turn the MP30 unit ON.
e. Have your lab partner remove all jewelry and lie comfortably. Make sure that you have three of the disposable electrode pads.
f. Clean the inner side of each ankle of your lab partner with an alcohol swab and attach one of the disposable EL 503 electrode pads to the inner left leg just proximal to the medial malleolus, as seen in figure 32.2. You may want to place a small amount of electrode gel in the center of the pad in order to get proper electrical conduction. Attach another electrode pad to the right leg at the same height. Clean the distal, anterior, right forearm just proximal to the wrist with an alcohol swab and attach the third electrode pad there. You should have three electrode cables, a white, black, and red one.
g. Place the **white lead** on the right forearm. The clip has a squeeze lever and attaches to the metal button of the electrode pad. The clip connects on only one surface. If you cannot get it to attach one way, try flipping the connector over.
h. Place the **black lead** on the electrode pad of the right leg. This is a ground.
i. Place the **red lead** on the left leg.
j. Open the BIOPAC Student Lab (BSL) Software on the computer desktop and select BIOPAC Lesson 5.—L05 Electrocardiography 1.
k. Type in your (folder) name. When done, click OK. If you have a folder on this computer station a window should appear with the message "A folder with this name already exists. Would you like to use it or create a new folder?" Choose USE IT.

2. Calibration
Make sure that your test subject is comfortable and relaxed. It is often beneficial to talk to the subject calmly as you prepare the test. Any extraneous

movement may cause electrical interference. People who are nervous or jittery do not produce good ECGs.

Click on CALIBRATE in the upper left corner of the computer screen. The calibration will stop automatically after approximately 8 seconds. You should see a small ECG tracing on the computer screen with a flat baseline. If not, check your connections and click REDO CALIBRATION.

3. Record Data   Have your lab partner relax, click on the RECORD button, and let several cycles occur. After about 15 seconds, click on SUSPEND. If the recording does not have a flat baseline, then click on the REDO button. If your recording looks similar to figure 32.3, then you have made a successful recording.

4. Analyze Data   Find the **Review Saved Data** folder from the lessons menu and look for your file.

You can determine the length of various sections of your ECG by selecting the **Delta T** mode. This mode determines the time from one part of the ECG tracing that you select to another part of the tracing that you select. You can use the **I-beam** tool to select the appropriate area of your ECG. Chose the magnifying tool in the lower right side of the computer screen to give you a close-up view of the area you are going to choose.

Beats per minute—bpm Select the area from one R peak to another R peak using the I-beam tool. bpm will show up in the channel measurement box. Record your bpm in table 32.1.

PR interval Use the I-beam and highlight a section of the ECG from the beginning of the P wave to the beginning of the QRS complex. Record this time (use the Delta T mode) in table 32.1.

QRS complex Use the same procedure as you did with the PR interval to determine the time for the QRS complex. The length of the QRS complex begins at the first deflection of the Q wave and ends when the S wave returns to the baseline. Record this time in table 32.1.

QT interval This interval is determined with the beginning of the Q wave to the end of the T wave. Record this time in table 32.1.

Procedure for Conventional ECG Machine

The ECG machine should be plugged in and turned on. Make sure that the ECG machine is on "standby" when you attach the electrodes. Accidental grounding of the electrodes may cause damage to the equipment. Older ECG machines should warm up for a few minutes prior to running the experiment. Remove metal watches or any other jewelry that might interfere with the electrical signal between the heart and the extremities. Have your lab partner scrub the inside of the ankle, about 2 cm above the medial malleolus, with an alcohol swab to remove dust and skin oil. Clean the anterior sides of the wrist in the same way. Saline pads or electrode jelly may be applied to these areas. Attach the electrodes firmly, using the straps provided, but not so tight that you cut off blood flow. The four connections are

> **LA** attaches to the left arm.
>
> **RA** attaches to the right arm.
>
> **RL** attaches to the right leg.
>
> **LL** attaches to the left leg.

Your instructor will explain how to operate your electrocardiograph. Have your lab partner rest comfortably on a cot or table for a few moments before monitoring the ECG. Some machines operate automatically and record all three leads. If you need to record leads manually, then turn the knob on the ECG machine from "standby" to "run" and record lead I for about 10 cycles. Turn the knob to "standby" after recording each lead. Then record lead II and lead III for about 10 cycles each. Tear the ECG from the machine and make sure you write your lab partner's name on the paper and write the appropriate lead on the paper. It is important that you provide a relaxed atmosphere for your lab partner to produce a good ECG record-

TABLE 32.1	ECG Data			
Record your value for the various ECG data either from the BIOPAC ECG or the ECG machine and determine if your data falls within the normal values.				
bpm	Yours _____	Above 100 Tachycardia	Below 60 Bradycardia	Within norm? _____
PR interval	Yours _____	Normal 0.16–0.18 sec	Longer than 0.18 sec Partial AV heart block	Within norm? _____
QRS complex	Yours _____	Normal 0.08 sec	Longer than 0.08 sec Rt/lft bundle branch block	Within norm? _____
QT interval	Yours _____	Normal 0.3–0.4 sec	Shortens with increased heart rate	Within norm? _____

ing. People who are fidgeting or nervous generate extraneous interference and do not produce a level baseline recording. You can demonstrate this electrical "noise" by the following activity.

Interference in an ECG While your lab partner is still attached to the ECG machine, set the dial to lead II. Run three to four cardiac cycles and then have your lab partner clench his or her fists. Note the electrical interference that occurs as the skeletal muscles depolarize and repolarize. Emotional state also plays a part in electrical activity. If a patient is nervous or excitable, this will show in the ECG. Background electrical noise in the ECG varies from person to person, but you should be able to get a good ECG in the lab.

ECG and Exercise Once you have compared your ECG to normal, you should exercise vigorously for a few minutes. You can do this by running outside of the lab room or doing jumping jacks for a minute or two. Once you feel that your heart is pumping faster and harder, quickly lie down on the cot and re-measure your ECG.

How does your ECG compare to normal in terms of distance between the P waves?

How does the height of the QRS complex compare to the resting ECG?

The time between the T wave and the next P wave is the resting interval of the heart. As you exercise, this interval shortens.

ECG Evaluation Select one of the leads (lead II usually works best) for the following evaluations.

Beats per Minute The standard rate of paper travel in an ECG machine is 25 millimeters/second. Therefore, each millimeter is equal to 0.04 second. The small squares are millimeters and the darker lines indicate 5 mm or 0.5 cm. You can calculate the resting heart rate by counting how many millimeters ("Y" mm) occur between two peaks (such as R to R). This value is going to be multiplied by 0.04 second in the following equation:

_____ mm/beat × 0.04 second/mm = "Y" seconds/beat

$$Y = _____$$

There are 60 seconds/minute and you want to calculate the beats/minute, so you can use the following equation to find out the beats per minute.

$$\frac{60 \text{ seconds/minute}}{\text{"Y" seconds/beat}} = \text{your number of beats/minute}$$

Calculated beats/minute: _____.

Record your bpm in table 32.1.

PR Interval Select a section of the ECG from the beginning of the P wave to the beginning of the QRS complex. Record this interval in table 32.1.

QRS Complex The length of the QRS complex begins at the first deflection of the Q wave and ends when the S wave returns to the baseline. Record this time in table 32.1.

QT Interval This interval is determined with the beginning of the Q wave to the end of the T wave. Record this time in table 32.1.

Analysis of the ECG ECG recordings are important in assessing the health of the heart. **Conduction problems, myocardial infarcts (heart attacks),** and **heart blocks** are a few of the problems that may show up as variances on an ECG. The interpretation of ECGs in this exercise is for educational purposes only. Clinical evaluations of ECGs should be done only by trained health-care workers. Locate the P wave, the QRS complex, and the T wave. The whole cardiac cycle should take, on average, 0.7 to 0.8 second.

Irregularities in Heart Rate An excessively high heart rate is termed **tachycardia.** In young adults a heart rate above 100 beats per minute is considered tachycardia. However, 100 beats per minute in small children may be perfectly normal. An excessively low heart rate is termed **bradycardia.** In young adults a rate below 60 beats per minute is considered bradycardia unless they are highly trained aerobic athletes.

PR Intervals PR intervals (also known as PQ intervals) are normally about 0.16 second. The PR interval is the time between the beginning of atrial depolarization and the beginning of ventricular depolarization. If the PR interval is longer than 0.2 second (5 mm on the chart paper), this might indicate a **heart block.** Heart blocks result from reduced conduction between the atria and the ventricles. This may be caused by damage to the AV node or decreased transmission in the AV bundle. In a **complete heart block** the atria do not stimulate the ventricular depolarization at all; therefore, the atria fire independently from the ventricles. In a complete heart block the P waves may be spaced at 0.8 second apart, the SA node depolarization rate, but the ventricles fire at a much lower rate (1.5 to 2.0 seconds apart). Are the times measured on your recording within normal limits?

QRS Complex The **QRS complex** is 0.08 to 0.10 second, on average. If the QRS complex spans longer than 0.12 seconds this may indicate a **right** or **left bundle branch block.** In this condition the two ventricles contract at slightly different times, increasing the length of the QRS complex.

QT Interval   The **QT interval** is 0.3 second, on average. The interval is shorter as the heart rate increases, and the interval becomes longer as the heart rate slows down.

Cardiac Arrhythmias   One of the major diagnostic uses of the ECG is to detect arrhythmias, abnormal rhythms, of the heart. Arrhythmias range from mild variations due to emotions, or stimulants to severe abnormalities causing life-threatening conditions. Compare your ECG to figure 32.4. The normal ECG is represented in the figure along with abnormal patterns. These are only a few examples of irregular ECGs.

Mean Electrical Axis of the Heart

If you have recorded an ECG from lead I and lead III, you can determine the electrical axis of the heart. If you are using BIOPAC you must use BIOPAC Lesson 6.—L06 Electrocardiography 2. If you are using a standard strip chart ECG you will need to have lead I and lead III in hand. It is possible to calculate the relative position of the heart in the thoracic cavity by using the electrical data as was first analyzed by Einthoven. The mean electrical axis of the heart is essentially parallel to the interventricular septum. If you examine the three standard leads, they form a triangle known as Einthoven's triangle (see figure 32.2). Normally, the electrical axis of the heart runs from the superior right side to the inferior left side, meaning that the heart is in the middle of the chest but the apex points to the inferior and left side. There are many reasons for deviations from the normal pattern, including loss of electrical activity in a portion of the heart (a bundle branch block), death of heart tissue, or an enlarged ventricle. If you are using a standard ECG you can calculate the mean electrical axis using the following information. You may be using a program that automatically calculates the mean electrical axis; however, the interpretation of the axis will be more meaningful if you also read the following description.

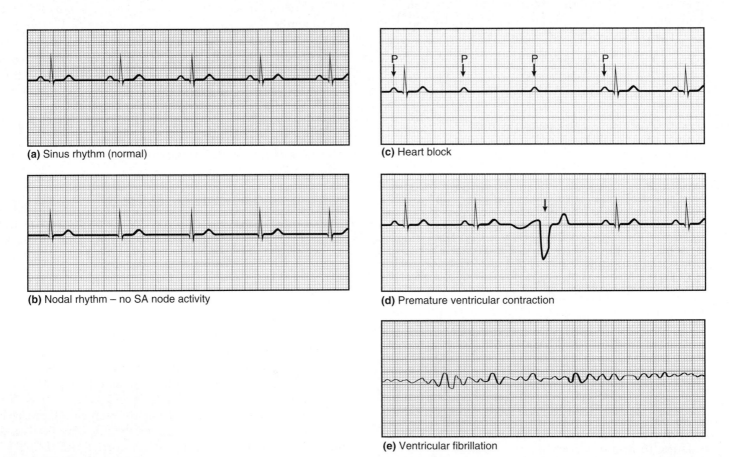

(a) Sinus rhythm (normal)

(b) Nodal rhythm – no SA node activity

(c) Heart block

(d) Premature ventricular contraction

(e) Ventricular fibrillation

FIGURE 32.4 Normal and Pathological Electrocardiograms (a) Normal sinus rhythm; (b) nodal rhythm generated by the AV node in the absence of SA node activity—note the lack of P waves; (c) heart block, in which some P waves are not transmitted through the AV node and do not generate QRS complexes; (d) premature ventricular contraction (PVC), or extrasystole—note the inverted QRS complex, misshapen QRS and T, and absence of a P wave preceding this contraction; (e) ventricular fibriliation, with grossly irregular waves of depolarization.

To determine the mean electrical axis of the heart, you must acquire data from leads I and III. Using your ECG find the QRS complex in lead I. Count the number of millivolts (this is equal to the number of millimeters on chart paper) that the QRS complex projects above the isoelectric line (the baseline of the recording). If you are using BIOPAC L06, you can determine the amplitude of the wave by selecting the **delta mode.** Select the region of the graph using the I-beam tool. Subtract from that the sum of the number of millivolts of both the Q and the S waves that project below the isoelectric line. Mark this number on the scale for lead I and draw a vertical line perpendicular to the lead I axis on the figure 32.5 (if you obtained 15 mv this would the red line in figure 32.6).

Repeat this procedure for lead III, but this time mark the area on the lead III scale and draw a line perpendicular to it on figure 32.5 (if you obtained 5 mv this would look like the blue line on figure 32.6). The lines for lead I and lead III should intersect.

Use a ruler and draw a line from the center mark of the triangle through the point where lead I and III lines

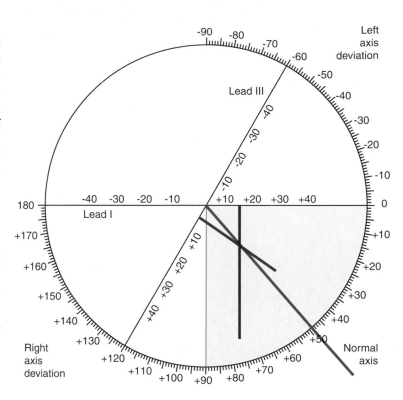

FIGURE 32.6 Electrical Axis of the Heart Sample shown to determine mean electrical axis.

intersect and record the number that represents the mean electrical axis on the edge of the circle that the line passes through. This is seen as the green line on figure 32.6 and shows a mean electrical axis of +49. As this is in the shaded area of 0 to +90, it is in the normal range. Record your value.

Mean electrical axis: _____

Is yours within the normal range? _____

If the value is greater than +90, there is a right axis deviation. This may be due to right ventricular enlargement, displacement of the heart to the right side, or damage to the left side of the heart.

If the value is less than 0 (some physicians say less than −30), there is a left axis deviation. This may be due to left ventricular enlargement, left bundle branch block, displacement of the heart to the left side, or damage to the right side of the heart.

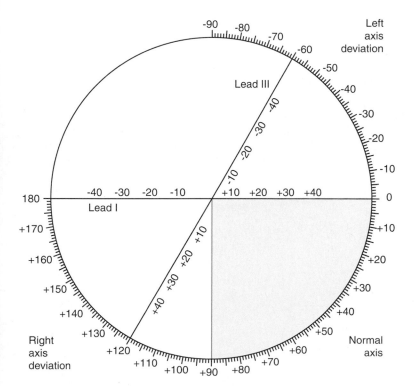

FIGURE 32.5 Electrical Axis of the Heart

Notes

REVIEW SECTION

Electrical Conductivity of the Heart

Name _____ Date _____

Lab Section _____ Time _____ ____

Review Questions

1. The sinoatrial node has a common name. What is it?

2. Which two chambers of the heart (atria or ventricles) contract last in a normal cardiac cycle?

3. What two chambers are stimulated immediately after the SA node depolarizes?

4. After the AV node depolarizes, what structures conduct the impulse to the myocardium of the ventricles?

5. What are the main events recorded by an ECG?

6. What electrical event in the heart does the QRS complex represent?

7. Ventricular repolarization is represented by what part of an ECG?

8. What ECG wave is represented by the atrial depolarization?

9. Why is the ECG event indicating atrial repolarization not seen in an ECG?

10. What does a heart block do to impulse transmission in the heart?

11. Fibrillation is uncoordinated cardiac muscle contraction. Predict what an ECG would look like if there were no uniform conduction of electrical activity in the heart. Draw what it might look like.

12. What consequence does fibrillation have for cardiac muscle contraction and for the pumping efficiency of the heart? Which is more serious—atrial or ventricular fibrillation?

13. If a myocardial infarct (heart attack) destroyed a portion of the right or left bundle branches, what potential change might you see in an ECG?

14. Tape or paste your ECG in the following space. Label the P wave, the QRS complex, and the T wave.

15. What was your value for the mean electrical axis of the heart? _____

    Was this within normal limits? _____

LABORATORY

Functions of the Heart

INTRODUCTION

One of the fundamental physiological activities of the body is to maintain adequate blood pressure. If blood pressure is too low, cells may not function correctly or they may die. If blood pressure is too high, damage may occur to the organs of the body or excess fluid may be expressed from the capillaries. One way to control blood pressure is through changes in **cardiac output.** When more oxygen is required cardiac output increases with an elevated heart rate or an increase in the volume ejected per contraction, or both. The heart has an **intrinsic control,** which is a pacemaker located in the heart wall that determines heart rate, and it has **extrinsic controls,** from organs other than the heart such as hormones or input from the nervous system that control the heart rate or contraction strength.

In this exercise you explore some of the functions of the human heart, with particular emphasis on changes in heart rate and contraction strength. These topics are covered in the Saladin text in chapter 19, "The Heart." You examine the resting rate of human hearts, the natural rate of heart contraction in a frog, and changes that occur when various solutions are added to the frog heart muscle. The heart of a frog is physically somewhat different from a human heart. The frog has two atria and only a single ventricle; however, the physiological response of the frog to various cardiac-influencing substances is similar to that of humans.

Due to the decline in some frog populations (the "leopard frogs" in North America), bullfrogs may be preferable as study specimens. Bullfrogs are not only plentiful but frequently a pest species in many areas and, as of today, may be used without significant impact on their population levels.

Cardiac Muscle Characteristics

Cardiocytes (cardiac muscle cells), like neurons, undergo a normal **polarization** process. This produces a **resting membrane potential.** The mechanism by which electrochemical impulses occur in heart muscle depends on three ion channels, **fast sodium channels, slow calcium channels,** and **potassium channels.** In cardiocytes polarization occurs by activation of the **sodium-potassium pump.** The result is an increase in sodium and potassium ions outside the cell membrane. This produces a resting membrane potential.

When the cardiocyte is stimulated, the fast sodium channels open, causing depolarization. This leads to a twitch of the muscle fiber. In skeletal muscle the twitch is quick (2 msec). In cardiac muscle the twitch is much longer (250 msec), caused by a **plateau** in the repolarization of the cell. This may be due to the closing of slow calcium channels or the slow movement of calcium to the cytosol. The plateau can be seen in figure 33.1. Depolarization travels across the cardiocyte. **Repolarization** occurs relatively quickly by reestablishing the resting membrane potential, and the heart muscle is ready for its next contraction.

Cardiac Pacemaker

Pacemaker cells in the **sinoatrial (SA) node** of the heart wall are specialized muscle cells that spontaneously and periodically depolarize, sending action potentials across the heart wall, causing the heart to contract. In neurons, by contrast, the resting membrane potential is stable. In the sinoatrial node the depolarization is thought to occur due to the leaking of sodium ions into the pacemaker cells without the corresponding outflow of potassium ions. When threshold is reached, fast calcium-sodium channels open and calcium and sodium ions flow into the cell. When 0 mv is reached, the **potassium channels** open and potassium ions leave the cell.

Once this happens, a **field effect** occurs, depolarizing cardiocytes adjacent to the pacemaker. This depolarization spreads throughout the atria and eventually to the ventricles, stimulating the cardiac muscle to contract. Action potentials from the sinoatrial node open **voltage-gated slow calcium channels** in the cardiocytes. **Ligand-gated calcium channels** in the sarcoplasmic reticulum open and calcium moves into the cell, binding to troponin, as it does in skeletal muscle.

Because the heart muscle cells are linked together by intercalated discs with gap junctions between cells, they act as one electrical unit. An impulse generated in the nodal tissue spreads to the surrounding muscle, causing it first to depolarize (an electrical event) and subsequently to contract (a mechanical event). Review the conduction pathway of the heart in Laboratory Exercise 32.

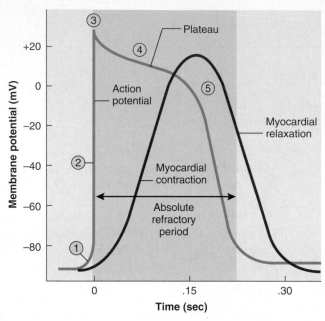

① Voltage-gated Na^+ channels open.

② Na^+ inflow depolarizes the membrane and triggers the opening of still more Na^+ channels, creating a positive feedback cycle and a rapidly rising membrane voltage.

③ Na^+ channels close when the cell depolarizes, and the voltage peaks at nearly +30 mV.

④ Ca^{2+} entering through slow Ca^{2+} channels prolongs depolarization of membrane, creating a plateau. Plateau falls slightly because of some K^+ leakage, but most K^+ channels remain closed until end of plateau.

⑤ Ca^{2+} channels close and Ca^{2+} is transported out of cell. K^+ channels open, and rapid K^+ outflow returns membrane to its resting potential.

FIGURE 33.1 Cardiac Muscle Action Potential

Normally, the resting heart rate of mammals is lower than the rate of excised hearts. In human hearts that have been removed for heart transplant surgery, the cardiac rate is about 100 beats per minute, which is the native sinoatrial depolarization rate.

This slowing of the heart rate from 100 beats per minute to an average of 70 to 80 beats per minute is due to the action of the **parasympathetic nervous system.** The **vagus nerve** conveys parasympathetic fibers that innervate the SA node. **Acetylcholine (ACh)** is a neurotransmitter released by this nerve, and it promotes potassium (K^+) leakage from the nodal cells to the interstitial fluid, increasing the flow of potassium ions to the outside of the membrane, thus **hyperpolarizing** the cells. By increasing the difference in voltage between the inside and the outside of the cells, it takes longer for the cells to reach threshold and depolarize. In this way the heart rate is slowed. Acetylcholine is a **parasympathomimetic substance,** one that mimics the effects of innervation by the parasympathetic nervous system.

The heart rate can be elevated by increasing the firing rate of the SA node. An increase in the heart rate from the normal resting rate to above 100 bpm is due to the progressive *inhibition* of the **parasympathetic nervous system** and subsequent *stimulation* of the **sympathetic nervous system.** As the vagus nerve secretes less acetylcholine at the node, the firing rate increases. Above 100 beats per minute, the heart rate is controlled by the sympathetic nervous system. Nerves from the cervical region of the spinal cord release **norepinephrine,** which binds to **beta-adrenergic receptors.** This causes calcium channels to open, *decreasing* the threshold and causing more rapid firing of the SA node. Like norepinephrine, **epinephrine** increases the heart rate and strength of contraction by making the calcium channels more permeable. Epinephrine is a **sympathomimetic drug,** one that mimics the effect of sympathetic nervous innervation. Other chemicals function as stimulants. Caffeine is a central nervous system stimulant, and it has an effect on heart muscle. It increases heart rate and the strength of contraction. High concentrations of caffeine cause **arrhythmias,** or irregular heart rates.

The addition of calcium chloride to the heart tissue increases the heart rate and contraction strength, but it can also produce pacemakers in regions of the heart other than the SA node. These are called **ectopic foci.**

The electrical activity of the nodes of the heart occurs in specialized *muscle* cells, not in nervous tissue. The electrochemical transfer of impulses by cardiocytes is known as **myogenic conduction.** The muscle tissue may be influenced by the nervous system, but the activity is initiated and generated in specialized muscle fibers. In addition to the influence of the nervous system, certain hormones, some drugs, and ion concentration have an effect on the heart rate.

OBJECTIVES

At the end of this exercise you should be able to

1. describe the basic contraction characteristics of the heart;
2. list the effects of various substances, such as epinephrine, calcium chloride, and acetylcholine, on the heart rate;
3. outline the mechanisms by which these substances change heart rate;

4. measure the resting pulse rate of the heart;
5. correlate the sounds the heart makes with the action of the heart;
6. explain how the valves of the heart increase the efficiency of the contraction.

MATERIALS

Pulse Rate and Heart Sounds

Clock or watch with second hand

Stethoscope

Alcohol wipes

Frog Heart Rate and Contraction Strength

Frog Preparation

Live frog

Clean Dissection Instruments

Scissors

Razor blades or scalpels

Dissection tray

Forceps

Dissection pins

Small heart hook (fishhook, copper wire, or Z wire)

Nylon thread

Frog anchoring board

Ring stand

Goggles

Animal waste container

BIOPAC Equipment

Force transducer assembly—SS12LA

HDW100A tension adjuster

Data acquisition unit MP30

BIOPAC *BSL PRO* A04 Frog Heart (FrogHeart.gtl)

50-gram weight with hook

Physiograph Equipment

Myograph transducer

Physiograph or Duograph

Cable

Channel amplifier

Solutions

Frog Ringer's solution—room temperature, 37° C, and iced

2% calcium chloride solution

0.1% acetylcholine chloride solution

0.1% epinephrine solution

Saturated caffeine solution

PROCEDURE

Pulse Rate

Measure the pulse rate (heart rate) and listen to the heart sounds of your lab partner.

1. Place your fingers on the radial artery or the carotid artery of your lab partner and count the number of beats that occur in 1 minute and record the resting pulse rate.

   Resting pulse rate: _____ beats/minute

2. You can calculate the average pulse rate for your class by having all the members of your class record their pulse rate in beats per minute on the chalkboard or whiteboard. Add all the pulse rates together and divide by the total number to determine the average. Compare the value you calculate from your class with the average heart rate.

   Average pulse rate for class: _____ bpm

   There are differences in pulse rate, depending on whether you are sitting, lying, or standing. The normal resting pulse rate is measured when you are sitting. One of the functions of changes in pulse rate is to alter blood pressure. The carotid sinus is one of the areas of the body that senses changes in blood pressure.

3. After you measure the pulse rate while sitting, lie down on a table or cot and measure the pulse rate. Record any change in the pulse rate in the following space.

   Pulse rate while lying down: _____ bpm

4. Stand up and record the change in the pulse rate in the following space.

   Pulse rate while standing: _____ bpm

   Can you explain any changes in the rate from the one you recorded when you were sitting?

Heart Sounds

The **lubb/dupp** sounds of the heart reflect the closure of the heart valves. The first heart sound is the lubb sound, and it occurs due to the closing of the atrioventricular valves. The second heart sound is the dupp sound, and it is due to the closing of the semilunar valves. You may hear an additional whooshing sound as you listen through the stethoscope. This is attributed to the imperfect closure of the valves, known as a **heart murmur.**

1. Clean the earpieces of a stethoscope with alcohol swabs and examine the stethoscope. The earpieces should point in an anterior direction as you place them into your ears.
2. Listen for heart sounds generated by your lab partner by placing the diaphragm of the stethoscope inferior to the second rib on the left side of the body. The pulmonary semilunar valves are best heard in this location.
3. Place the diaphragm of the stethoscope on the right side of the body, in the intercostal space between the second and third ribs, for the location of the aortic semilunar valve.
4. Use figure 33.2 as a guide and listen to heart sounds in these locations.

Examination of Frog Heart Rate and Contraction Strength in Frogs

Either you have a frog prepared for you or your instructor may wish for you to double pith a frog for this exercise. If

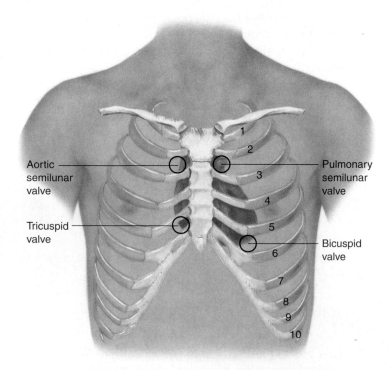

FIGURE 33.2 Auscultation Areas Circles are locations where specific valvular sounds can best be heard.

you need to pith a frog, follow the instructions outlined in Laboratory Exercise 11 and then follow the procedure described next. You may be using a BIOPAC setup or a physiograph setup. In either case, the frog will be prepared the same way.

Frog Preparation

Procedure

1. Once the frog is pithed, pin the frog down to a board.
2. Cut through the pectoral skin with a scissors and through the sternum. The heart should be beating inside the pericardial sac. Moisten the heart with frog Ringer's solution.
3. Carefully cut the pericardial membrane of the heart away from the dorsal surface of the frog.
4. Cover the heart with a moist paper towel and prepare either the BIOPAC setup or the physiograph setup.

Biopac Section

Biopac Setup for MP30   You can access detailed instructions or customize the experiment at www.biopac.com.

1. Make sure the computer is ON.
2. Connect the MP30 unit to an electrical outlet with the power supply.
3. With the MP30 unit OFF, connect MP30 to the computer.
4. Plug SS12LA into Channel 1.
5. Turn the MP30 unit ON.

Computer Setup

1. Start the BSL *PRO* software. An untitled window should appear.
2. Go to the File menu and choose "Graph Template (*GTL) > File Name: FrogHeart.gtl."
3. Physically attach the HDW100A tension adjuster to a ring stand.
4. Attach the force transducer assembly to the tension adjuster. Make sure that the place where the S-hook attaches is facing down. The adjuster should be set so that, when the frog heart is hooked up, the string will be vertical.
5. On the computer, select 0–50 grams force range.
6. Select the Setup Channels from the menu, choose the wrench icon from the Channel 1 menu, and locate the wrench icon and scaling window.
7. With only the S-hook on the tension adjuster, click Cal 1.
8. Attach a 50-gram weight to the adjuster and click Cal 2.
9. Click OK and exit the scaling window.

Running the Experiment

1. Place the frog on the board underneath the force transducer assembly and remove the wet paper towel from the frog heart.

2. Attach a metal hook or wire to the membrane attached to the dorsal portion of the heart. You can pierce the apex of the heart with a wire or hook, but you must be very careful not to puncture the ventricle.

3. Attach the hook to the force transducer assembly (see figure 33.3).

4. Adjust the tension, so that the nylon thread is taut but not so much that you tear the hook from the heart.

5. Click the "start" button and record 10 cycles of the heart.

6. Save your data by selecting the **File menu.** Choose **Save As** and select the file type: **BSL Pro files (*.ACQ) File name: (your name).** Choose **Save.**

7. Alter the temperature or add chemicals as described after "Physiograph Section."

Physiograph Section

Procedure

1. Attach a metal hook or wire to the membrane at the dorsal portion of the heart. You can pierce the apex of the heart with a wire or hook, but you must be very careful not to puncture the ventricle.

2. Attach the hook to a thread and tie the thread to the myograph transducer, adjusting the tension until you get a deflection in the myograph leaf. Your setup should look similar to figure 33.3.

3. Set the chart speed to 0.5 cm/second and record the contraction rate and strength of contraction of the heart muscle. The force that is generated by the heart is measured by the height of the stylus on the chart paper. The rate can be determined by the distance between the peaks on the chart paper.

4. Alter the temperature or add chemicals as described in the following section.

Data Acquisition

Keep the heart moistened with frog Ringer's solution. Unless directed to do otherwise, use the Ringer's solution that is at room temperature.

Changes in Heart Rate with Temperature You have already recorded the heart rate at room temperature.

1. Run tests by flooding the heart with iced Ringer's solution. Note the change in heart rate.

   Heart rate with iced Ringer's solution: _____ bpm

2. You can further study the effect of temperature on enzymes by flooding the frog heart with frog Ringer's solution at 37° C. Record the contraction rate.

   Heart rate at 37° C: _____ bpm

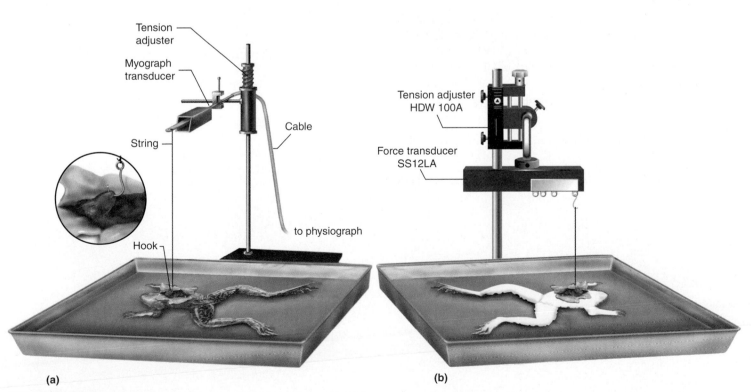

(a)   **(b)**

FIGURE 33.3 Frog Setup (a) For physiograph; (b) for BIOPAC.

The **Q10** is an enzymatic relationship that states that, for every 10° rise in temperature (from 0° C to 50° C), there is a doubling of enzyme activity. Does this correlate with your findings on the temperature relationships with the heart? _____

Calcium

Effect of Various Chemicals on Heart Rate and Contraction Strength   Add several drops of 2% calcium chloride solution to the heart muscle. Wait until you get a change in the heart rate before you record the data. Turn on the recording device (BIOPAC or physiograph) and produce a tracing. Calculate the heart rate and record the number in the following space.

Heart rate: _____ bpm

What happens to the contraction strength?

Rinse the heart with room temperature frog Ringer's solution.

Acetylcholine

Add several drops of 0.1% acetylcholine solution to the outer surface of the heart. You may have to wait a few minutes to observe any changes in heart rate. Record the heart rate for about 10 seconds after waiting 2 minutes. Wait 3 additional minutes and then record the rate for another 10 seconds. Record the rate in the following space.

Heart rate after 2 minutes: _____ bpm

Heart rate after 3 additional minutes: _____ bpm

When you have determined a change in the heart rate and recorded your results, flood the heart with frog Ringer's solution.

Epinephrine

Add several drops of 0.1% epinephrine solution to the outer surface of the heart. Examine the rate over the next 2 to 3 minutes, and, when you notice a change, record the heart rate for about 10 seconds. Write your results in the space provided.

Heart rate: _____ bpm

What is the change in the contraction strength?

After you record your results, flush the heart with room temperature frog Ringer's solution.

Caffeine

Add several drops of saturated caffeine solution to the outer surface of the heart and note the time when the rate changes. Once the rate changes, record the heart rate and strength of contraction of the heart muscle for about 10 seconds. Enter the rate in the following space.

Heart rate with caffeine: _____ bpm

Strength of contraction (increase/decrease):

 Clean Up   When you are finished, dispose of the frog in the appropriate animal waste container. Clean your desk and equipment. Dispose of scalpel blades in the sharps container.

REVIEW SECTION

Functions of the Heart

Name _____ Date _____

Lab Section _____ Time _____ _____

Review Questions

1. Decreasing heart rate is under the control of what nervous division?

2. What is the resting heart rate of the average person?

3. Are there more sodium ions inside or outside a cardiac muscle cell during the resting membrane potential?

4. What happens to sodium ions when a membrane depolarizes?

5. What region in the heart depolarizes spontaneously?

6. What happens to the heart when an action potential is generated in the SA node?

7. The movement of electrochemical impulses in the myocardium is called _____ conduction.

8. What effect do calcium slow channels have on shortening or lengthening the contraction time of the heart muscle?

9. Beta-adrenergic blockers bind to norepinephrine sites, preventing these neurotransmitters from having an effect. What effect would the use of "beta blockers" have on heart rate?

10. How much of a change in the heart rate of the frog did you see after the addition of calcium chloride?

What was the change in the contraction strength?

What process might account for this in terms of cardiac muscle interactions with calcium?

11. What heart sound is produced by the closure of the atrioventricular valves in the heart?

12. A heart murmur is normally caused by what event?

13. When would a murmur occur in the lubb/dupp cycle if the AV valves were not closing properly?

14. Pilocarpine stimulates the release of acetylcholine from the vagus nerve, thereby increasing parasympathetic stimulation. What impact would this drug have on heart rate?

LABORATORY

Introduction to Blood Vessels and Arteries of the Upper Body

INTRODUCTION

Blood vessels are the conduits that carry oxygen, nutrients, and other materials to the cells, as well as remove wastes from the tissue fluid near the cells. These vascular conduits consist of numerous vessels, such as large **arteries;** smaller **arterioles; capillaries,** which exchange materials with the cells; and **venules,** which return blood to the **veins,** which carry blood back to the heart.

Arteries are defined as blood vessels that carry blood away from the heart. Arteries have thicker walls than veins, and the thickness of arterial walls reflects the higher blood pressure found in them. These topics are covered in the Saladin text in chapter 20, "Blood Vessels and Circulation."

Arteries are frequently named for the region of the body they pass through. The brachial artery is found in the arm, while the femoral artery is located in the thigh. In this exercise you learn the basic structure of blood vessels and locate the arteries of the upper body.

OBJECTIVES

At the end of this exercise you should be able to

1. draw a cross section of the wall of a generalized blood vessel and label the three layers;
2. distinguish between arteries and veins;
3. describe the sequence of major arteries that branch from the aortic arch;
4. list the arteries of the upper extremity;
5. trace the arteries that supply blood to the head;
6. describe the major organs that receive blood from the upper arteries.

MATERIALS

Microscopes

Prepared slides of arteries and veins

Models of the blood vessels of the body

Charts and illustrations of the arterial system

Cats

Materials for cat dissection

    Dissection tray or pan

Scalpels

Pins

Blunt probes

Gloves

Scissors or bone cutter

Forceps

First aid kit

Sharps container

Animal waste container

PROCEDURE

Cross Sections of Arteries and Veins

Obtain a prepared microscope slide of a cross section of artery and vein. Both arteries and veins have walls that consist of three layers. The outer layer is known as the **tunica externa (tunica adventitia)** and consists of a connective tissue sheath. The middle layer, or **tunica media,** is composed of smooth muscle in both arteries and veins. The tunica media is much thicker in arteries, and there are pronounced elastic fibers in the wall of the tunica media in arteries. The innermost layer (near the blood) is the **tunica interna (tunica intima)** and consists of a thin layer of connective tissue and a thin layer of simple squamous epithelium, known as **endothelium.** In arteries there is an **inner elastic lamina,** a thin layer of elastic fibers. The **vaso vasorum** is a blood vessel that takes blood to the outer portion of an artery. The inner part of the artery is fed by the blood carried in the artery.

Valves are features of veins not found in arteries. The valves in veins prevent backflow in the vessel. Veins carry blood under low pressure and the valves aid in maintaining a one-way flow of blood. You can easily distinguish veins from arteries in a prepared slide where the vessels are cut in cross section. Veins have a large lumen (though the vein is frequently collapsed in prepared sections) and a thin tunica media relative to its overall size. You may also find nerves in the prepared slide. These are also circular structures, but they are solid, not hollow. Examine a prepared slide of an artery and a vein and locate these layers. Compare your slide to figure 34.1.

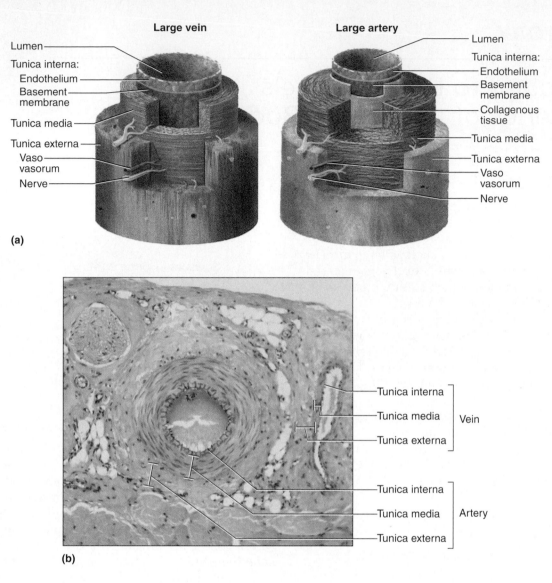

FIGURE 34.1 Artery and Vein Cross Section (a) Diagram; (b) photomicrograph of artery, vein, and nerve (40×).

Arteries

Examine the models, charts, and illustrations of the cardiovascular system in the lab and locate the major arteries of the upper body. Read the following descriptions of the arteries. Name the vessels that take blood to an artery and those that receive blood from the artery in question. An overview of the major arteries of the body is shown in figure 34.2.

Aortic Arch Arteries

Locate the heart and find the large **aorta** that exits from the left ventricle. This is the **ascending aorta,** a large vessel about the size of a garden hose. The ascending aorta is relatively thick-walled due to the high pressure of the blood coming from the heart. The first two arteries that arise from the ascending aorta are the **coronary arteries,** covered in Laboratory Exercise 31. The ascending aorta curves to the left side of the body and forms the **aortic arch.** In humans, there are three main arteries that receive blood from the aortic arch. The first major artery to receive blood is on the right side of the body and is called the **brachiocephalic trunk (artery).** This artery shortly divides into arteries that feed the right side of the head (the **right common carotid artery**) and the right upper extremity (the **right subclavian artery**) (see figure 34.3). The aortic arch has two other arteries. On the left side is the **left common carotid artery,** which takes blood to the left side of the head, and the **left subclavian artery,** which takes blood to the left upper extremity. Another artery that branches from the subclavian is the vertebral artery. The vertebral artery and the internal carotid artery take blood to the brain.

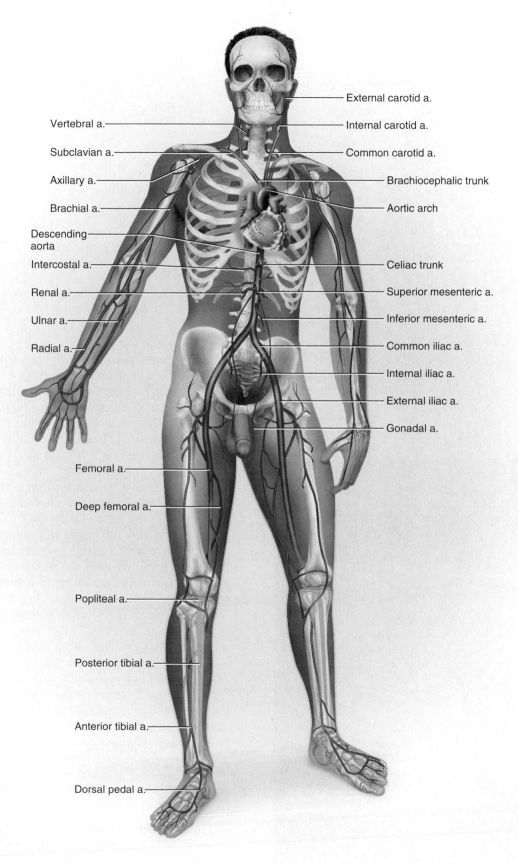

FIGURE 34.2 Major Arteries of the Body "a." refers to *artery*.

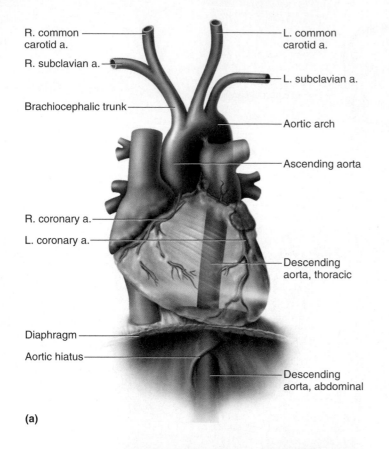

R. common carotid a.

R. subclavian a.

Brachiocephalic trunk

R. coronary a.

L. coronary a.

Diaphragm

Aortic hiatus

L. common carotid a.

L. subclavian a.

Aortic arch

Ascending aorta

Descending aorta, thoracic

Descending aorta, abdominal

(a)

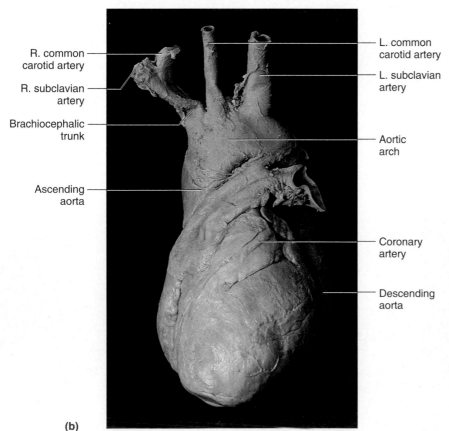

R. common carotid artery

R. subclavian artery

Brachiocephalic trunk

Ascending aorta

L. common carotid artery

L. subclavian artery

Aortic arch

Coronary artery

Descending aorta

(b)

FIGURE 34.3 Arteries of the Aortic Arch (a) Diagram; (b) photograph.

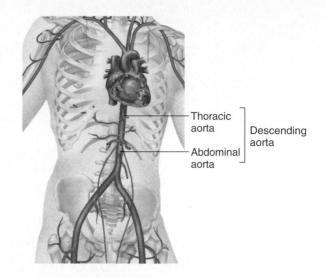

Thoracic aorta

Abdominal aorta

Descending aorta

FIGURE 34.4 Thoracic and Abdominal Aorta

As the aortic arch turns inferiorly behind the posterior part of the heart, it becomes the **descending aorta,** which is composed of two segments. Above the diaphragm the descending aorta is known as the **thoracic aorta** and below the diaphragm it is known as the **abdominal aorta** (see figure 34.4). The thoracic aorta has numerous branches that run between the ribs called **intercostal arteries** (see figure 34.5).

Pulmonary Circulation The pulmonary circulation involves the heart pumping blood to the lungs for oxygenation, removing carbon dioxide, and returning that blood to the heart. Blood in the **right ventricle** exits the heart by the **pulmonary trunk.** The blood enters the left and **right pulmonary arteries.** These arteries are unusual in that they carry deoxygenated blood. Most arteries carry oxygenated blood. From the pulmonary arteries blood enters the **lobar arteries** of the lungs and eventually flows to the capillaries, where the blood is oxygenated and carbon dioxide is released. The return flow from the lungs eventually enters the **pulmonary veins.** There are two pulmonary veins from each lung and these enter the heart at the **left atrium.** Examine models or charts in lab and compare them to figure 34.6.

Arteries That Feed the Upper Extremities

If you return to the brachiocephalic trunk, you can see that it is a short tube that takes blood from the aortic arch and then branches into the right common carotid artery and the right subclavian artery. The right subclavian artery has one branch that takes blood to the brain,

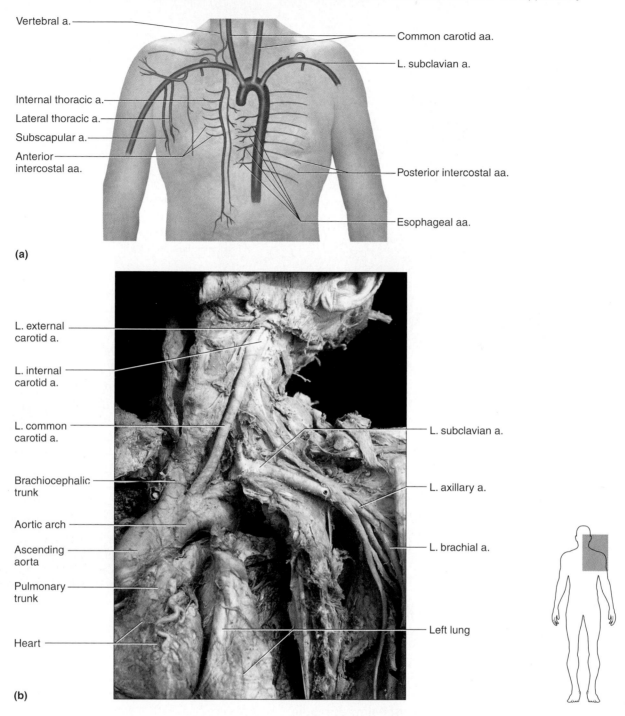

Vertebral a.

Common carotid aa.

L. subclavian a.

Internal thoracic a.

Lateral thoracic a.

Subscapular a.

Anterior intercostal aa.

Posterior intercostal aa.

Esophageal aa.

(a)

L. external carotid a.

L. internal carotid a.

L. common carotid a.

Brachiocephalic trunk

Aortic arch

Ascending aorta

Pulmonary trunk

Heart

L. subclavian a.

L. axillary a.

L. brachial a.

Left lung

(b)

FIGURE 34.5 Branches of the Subclavian Artery (a) Diagram; (b) photograph of cadaver.

while another branch becomes the **axillary artery.** The left subclavian artery also has a branch that takes blood to the brain and another branch that becomes the left axillary artery (see figures 34.5 and 34.7).

The axillary artery on each side of the body turns into the **brachial artery,** which has a pulse that can be palpated by finding a groove on the medial, distal region of the arm between the biceps brachii and brachialis muscle. This location is the site for the placement of the diaphragm of a stethoscope during blood pressure

measurement. The brachial artery supplies blood to the triceps brachii muscle and the muscles of the anterior arm. The brachial artery bifurcates (splits in two) to form the **radial artery** on the lateral side of the forearm and the **ulnar artery** on the medial side of the forearm. These arteries supply blood to the forearm and part of the hand. The radial artery can be palpated just lateral to the flexor carpi radialis muscle at the wrist, a common site for the measurement of the pulse. The radial artery and ulnar artery become united again as the **palmar arch arteries.**

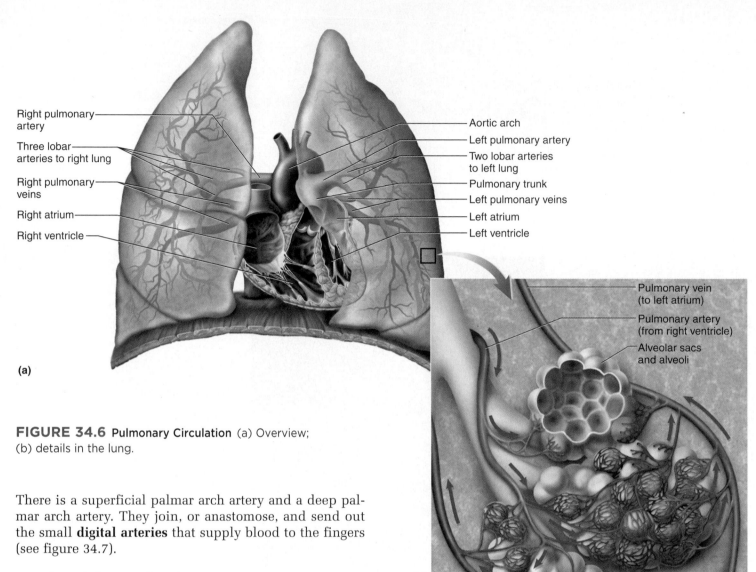

Right pulmonary artery

Three lobar arteries to right lung

Right pulmonary veins

Right atrium

Right ventricle

Aortic arch

Left pulmonary artery

Two lobar arteries to left lung

Pulmonary trunk

Left pulmonary veins

Left atrium

Left ventricle

(a)

Pulmonary vein (to left atrium)

Pulmonary artery (from right ventricle)

Alveolar sacs and alveoli

(b)

FIGURE 34.6 Pulmonary Circulation (a) Overview; (b) details in the lung.

There is a superficial palmar arch artery and a deep palmar arch artery. They join, or anastomose, and send out the small **digital arteries** that supply blood to the fingers (see figure 34.7).

Arteries of the Head and Neck

The **vertebral arteries** receive blood from the subclavian arteries and take it to the brain by traveling through the transverse foramina of the cervical vertebrae and then into the foramen magnum of the skull. The other main arteries that take blood to the brain are the **common carotid arteries,** which occur on the anterior sides of the neck. Each common carotid artery branches just below the angle of the mandible to form the **external carotid artery,** which takes blood to the face, and the **internal carotid artery,** which passes through the carotid canal and takes blood to the brain. The external carotid artery on each side has several branches, including the **facial artery,** the **temporal artery,** the **maxillary artery,** and the **occipital artery** (see figure 34.8).

The brain receives blood from four major arteries. These are the left and right **vertebral arteries** and the left and right **internal carotid arteries.** The vertebral arteries pass through the foramen magnum and unite to form the **basilar artery** at the base of the brain. These were discussed in Laboratory Exercise 21 and illustrated in figure 21.3.

Cat Dissection

Prepare for the cat dissection by obtaining a dissection tray, a scalpel, scissors or bone cutter, string, forceps, and a plastic bag with a label. Remember to place all excess tissue in the appropriate waste container and not in a standard wastebasket or down the sink!

Be careful as you make an incision into the body cavity of the cat. Cut with the scalpel blade facing to one side or, even better, use a lifting motion and cut upwards through the tissue. If you use a downwards, sawing motion you will damage the internal organs of the cat.

Open the thoracic and abdominal regions of the cat if you have not done so already. Make an incision into the body wall of the cat in the belly, slightly lateral to midline. Cut into the muscle and then grasp the tissue with a forceps and lift the body wall away from the viscera. Continue cutting anteriorly through the belly. Stop at the

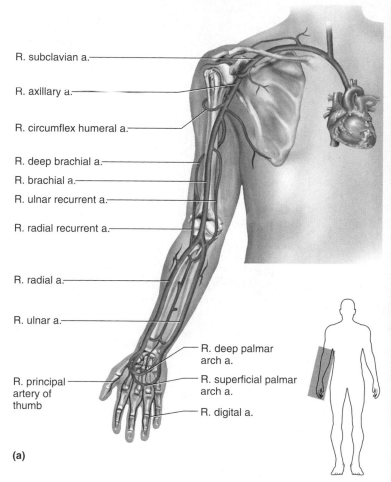

R. subclavian a.

R. axillary a.

R. circumflex humeral a.

R. deep brachial a.

R. brachial a.

R. ulnar recurrent a.

R. radial recurrent a.

R. radial a.

R. ulnar a.

R. deep palmar arch a.

R. principal artery of thumb

R. superficial palmar arch a.

R. digital a.

(a)

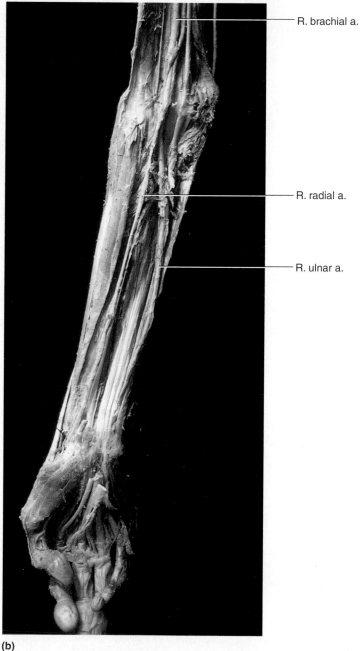

R. brachial a.

R. radial a.

R. ulnar a.

(b)

FIGURE 34.7 Arteries of the Upper Extremity (a) Diagram; (b) photograph.

level of the diaphragm, and then continue cutting a little off center as you cut through the costal cartilages. Use a scalpel or scissors to cut through the thoracic region.

You can now make a transverse incision in the lower region of the belly, so that you have an inverted T-shaped incision (see figure 34.9). Lift the tissue near the ribs and expose the diaphragm. Gently and carefully snip or cut the diaphragm away from the ventral region by cutting as close to the ribs as possible.

Next make an incision by cutting transversely across the upper part of the chest. Be careful not to cut the blood vessels of the neck exposed during the original skinning process. Open the body cavity by pulling the two flaps laterally, exposing the thoracic and abdominal cavities. You may want to cut through the ribs at their most lateral point to facilitate the opening of the thoracic cavity. Use a pair of bone cutters or heavy scissors to cut through the ribs of the cat, so that the thoracic cavity is easier to open.

Carefully expose the thoracic region of the cat and locate the two lungs and the heart. The heart is enclosed in the pericardial sac. Look for the fatty material, called the greater omentum, that covers the large liver, stomach, and intestines in the abdominal region. As you dissect the arteries of the cat, pay careful attention to the veins and nerves that travel along with the arteries. *Do not dissect other organs.* You will study these systems later (for example, the respiratory, digestive, and urogenital systems).

As you open the abdominal cavity you will see the space that contains the internal organs, known as the coelom, or body cavity. The lining on the inside of the body wall is the parietal peritoneum, and the lining that wraps around the outside of the intestines is the visceral peritoneum.

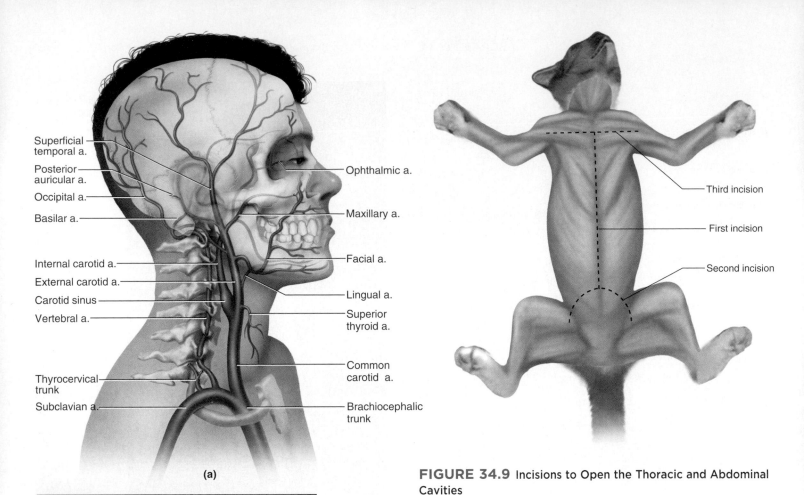

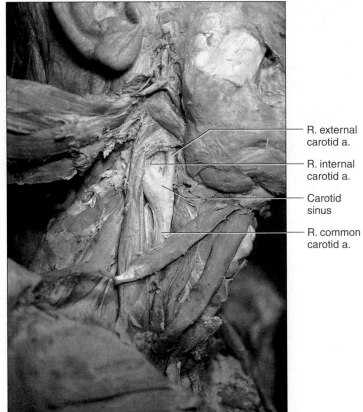

FIGURE 34.8 Arteries of the Head (a) Diagram; (b) photograph of cadaver.

FIGURE 34.9 Incisions to Open the Thoracic and Abdominal Cavities

Arteries of the Cat

The blood vessels in the cat have been injected with colored latex. If the cat is doubly injected the arteries are red and the veins are blue. If the cat has been triply injected the hepatic portal vein is typically injected with yellow latex. Be careful locating the arteries in the cat. You can tease away some of the connective tissue with a dissection needle to see the vessel more clearly. Note the difference in the aortic arch arteries in the cat, compared to that of the human. Look for the **coronary arteries,** which take blood from the ascending aorta to the heart tissue.

You can begin your study of the arteries of the cat by examining the ascending aorta. The **ascending aorta** bends and forms the **aortic arch** and then moves posteriorly to form the **descending aorta.** The descending aorta is composed of the **thoracic aorta** above the diaphragm and the **abdominal aorta** below the diaphragm. The pattern of arteries that leave the aortic arch is somewhat different in cats than in humans. In the cat typically only two large vessels leave the aortic arch, the large **brachiocephalic trunk** and the **left subclavian artery.** The brachiocephalic trunk further divides into the **left common carotid artery,** the **right common carotid artery,** and the **right subclavian artery.** Examine figure 34.10 for the pattern in the cat. How is the pattern in the cat different from that in the human?

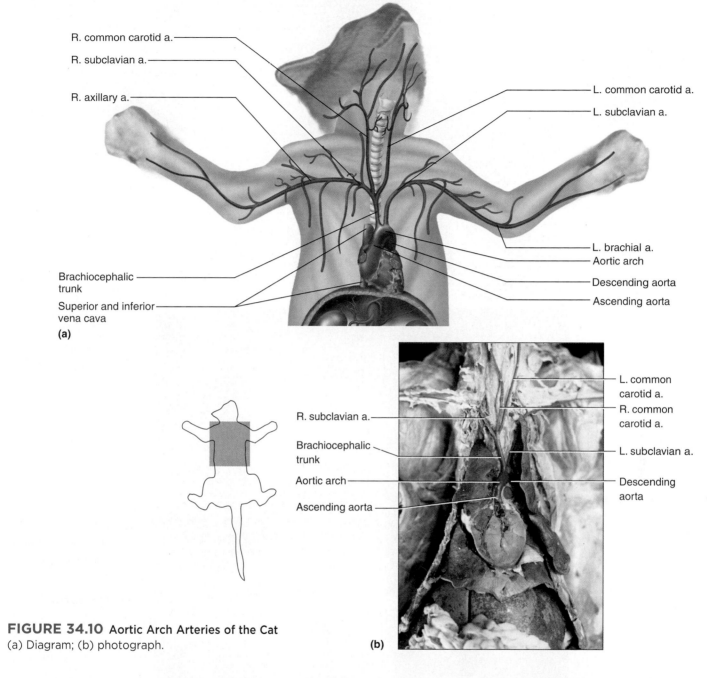

R. common carotid a.

R. subclavian a.

R. axillary a.

L. common carotid a.

L. subclavian a.

L. brachial a.

Aortic arch

Brachiocephalic trunk

Descending aorta

Superior and inferior vena cava

Ascending aorta

(a)

R. subclavian a.

Brachiocephalic trunk

Aortic arch

Ascending aorta

L. common carotid a.

R. common carotid a.

L. subclavian a.

Descending aorta

FIGURE 34.10 Aortic Arch Arteries of the Cat
(a) Diagram; (b) photograph.

(b)

Several arteries arise from the **subclavian artery.** The **vertebral artery** takes blood to the head, and the **subscapular artery** takes blood to the pectoral girdle muscles, along with the **ventral thoracic artery** and **long thoracic artery.** The ventral thoracic and long thoracic arteries supply the latissimus dorsi and pectoralis muscles with blood. The left and right subclavian arteries become the left and right axillary arteries, which in turn take blood to the **brachial arteries.** The brachial artery takes blood to the **radial** and **ulnar arteries.** Additional arteries that branch from the subclavian artery are the **thyrocervical arteries,** which take blood to the region of the shoulder and neck (see figure 34.11).

The **common carotid artery** takes blood along the ventral surface of the neck to feed the small **internal carotid artery,** which supplies the brain. The common carotid artery also empties into the **external carotid artery,** which takes blood to the external portions of the head. The **occipital artery** receives blood from the common carotid artery and supplies the muscles of the neck.

The thoracic aorta not only takes blood to the abdominal aorta but also supplies the rib cage with blood from the **intercostal arteries. Esophageal arteries** also branch from the thoracic aorta, supplying the esophagus with blood. The thoracic aorta continues as the abdominal aorta as it passes through the diaphragm.

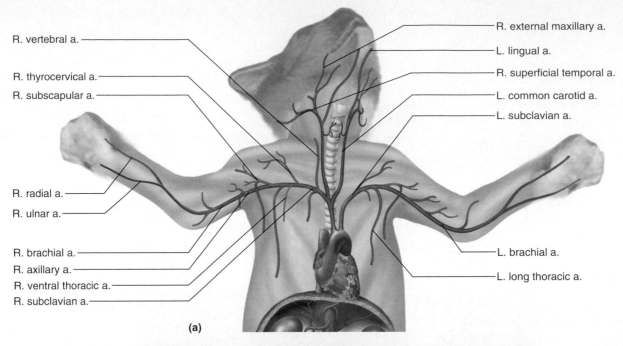

R. vertebral a.

R. thyrocervical a.

R. subscapular a.

R. radial a.

R. ulnar a.

R. brachial a.

R. axillary a.

R. ventral thoracic a.

R. subclavian a.

R. external maxillary a.

L. lingual a.

R. superficial temporal a.

L. common carotid a.

L. subclavian a.

L. brachial a.

L. long thoracic a.

(a)

(b)

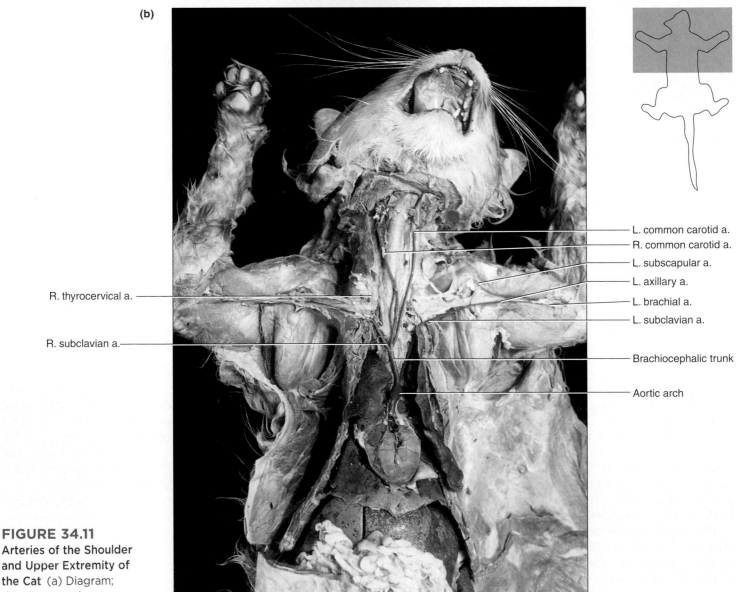

L. common carotid a.

R. common carotid a.

L. subscapular a.

L. axillary a.

L. brachial a.

L. subclavian a.

Brachiocephalic trunk

Aortic arch

R. thyrocervical a.

R. subclavian a.

FIGURE 34.11
Arteries of the Shoulder and Upper Extremity of the Cat (a) Diagram; (b) photograph.

REVIEW SECTION

Introduction to Blood Vessels
and Arteries of the Upper Body

Name _____   Date _____

Lab Section _____   Time _____

Review Questions

1. Label the following illustration with the major arteries of the body. Try to complete the illustration first and then review the material in this exercise to determine your accuracy.

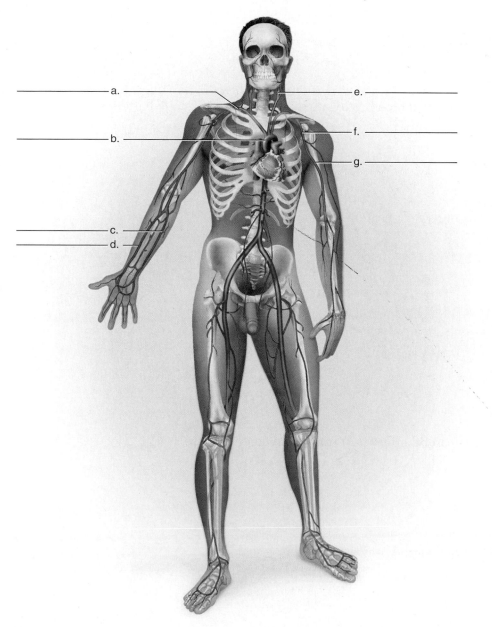

a. _____   e. _____

b. _____   f. _____

                g. _____

c. _____

d. _____

2. Blood from the left subclavian artery flows into what vessels as it moves toward the left arm?

3. Blood in the radial artery comes from what blood vessel?

4. An aneurysm is a weakened, expanded portion of an artery. Ruptured aneurysms can lead to rapid blood loss. Describe the significance of an aortic aneurysm versus a digital artery aneurysm.

5. The pulmonary arteries carry deoxygenated blood from the heart to the lungs. Umbilical arteries carry a mixture of oxygenated and deoxygenated blood. Why are these blood vessels called arteries?

6. What is the name of the outermost layer of a blood vessel?

7. What type of blood vessels have valves?

8. Blood from the common carotid artery next travels to what two vessels?

9. Blood from the right brachial artery travels to what two vessels?

10. Where does blood in the right subclavian artery come from?

11. The internal carotid artery takes blood to what organ?

12. From what blood vessel does the descending aorta get blood?

13. What is the general name of a large vessel that takes blood away from the heart?

14. Blood in the left common carotid artery receives blood from what vessel?

15. Name three blood vessels that exit from the aortic arch.

16. Working in pairs, have your lab partner select an artery for you to name. Quiz each other on the material learned in this exercise.

Notes

LABORATORY

Arteries of the Lower Body

INTRODUCTION

The arteries of the lower body receive blood directly or indirectly from the descending aorta. As with the arteries of the upper extremities, these vessels are frequently named for their location, such as the iliac artery, the femoral artery, and the mesenteric arteries, or for the organs they serve, such as the common hepatic artery and the splenic artery. These arteries are presented in the Saladin text in chapter 20, "Blood Vessels and Circulation." In this exercise you continue your study of arteries. It is very important that you are careful when dissecting the arteries of the abdominal region of the cat. Do as little damage as possible to the other structures as you examine the arteries in this region.

OBJECTIVES

At the end of this exercise you should be able to

1. describe the sequence of major arteries that originate from the abdominal aorta or its derivatives;
2. list the arteries of the lower extremities;
3. describe the major organs that receive blood from the lower arteries;
4. name the vessels that take blood to a particular artery and the vessels that receive blood from a particular artery.

MATERIALS

Microscope

Prepared slide of arteriosclerosis

Models and charts of the blood vessels of the body

Cadaver (if available)

Materials for Cat Dissection

    Cats

    First aid kit

    Sharps container

    Animal waste disposal container

PROCEDURE

Examine the models and charts in the lab and the accompanying illustrations to locate the major arteries in the abdomen and lower body. Read the following descriptions of the vessels and locate the individual arteries.

Abdominal Arteries

The **abdominal aorta** is the portion of the descending aorta inferior to the diaphragm. It passes through a hole in the diaphragm known as the **aortic hiatus.** The first major branch of the abdominal aorta is the **celiac artery** (also known as the **celiac trunk**). The celiac trunk splits into three separate arteries: the **splenic artery,** which takes blood to the spleen, pancreas, and part of the stomach; the **left gastric artery,** which takes blood to the stomach and esophagus; and the **common hepatic artery,** a vessel that takes blood to the liver. Locate the celiac artery and the branches of the celiac in figure 35.1.

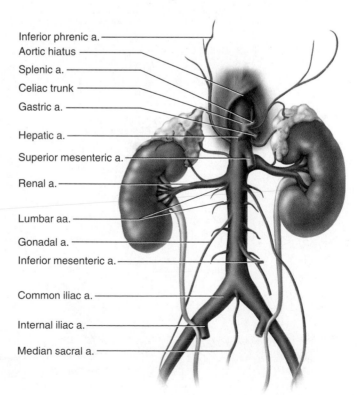

Inferior phrenic a.
Aortic hiatus
Splenic a.
Celiac trunk
Gastric a.
Hepatic a.
Superior mesenteric a.
Renal a.
Lumbar aa.
Gonadal a.
Inferior mesenteric a.
Common iliac a.
Internal iliac a.
Median sacral a.

FIGURE 35.1 Arteries of the Anterior Abdominal Region

Inferior to the celiac artery is the **superior mesenteric artery.** This vessel takes blood from the abdominal aorta and continues through the mesentery until it reaches the small intestine and proximal portions of the large intestine, including the cecum, ascending colon, and part of the transverse colon. The major branches of the superior mesenteric artery are the **intestinal arteries,** the **ileocolic arteries,** and the right and middle **colic arteries.** The next vessels are the paired **suprarenal arteries,** which take blood to the adrenal glands. Inferior to the suprarenal arteries are the left and right **renal arteries.** These arteries take blood to the kidneys. Locate these vessels in the lab and in figures 35.2 and 35.3.

The **gonadal arteries** branch inferior to the renal arteries and descend to either the testes or ovaries. The **inferior mesenteric artery** is the next vessel to leave the aorta, and it takes blood to the lower portion of the large intestine, including part of the transverse colon, the descending colon, and the rectum. These can be located in figures 35.2 and 35.3.

The abdominal aorta terminates by dividing into the two **common iliac arteries.** The common iliac arteries take blood to the **external** and **internal iliac arteries.** Each internal iliac artery takes blood to the pelvic region, including branches that feed the rectum; pelvic floor; external genitalia; groin muscles; hip muscles; uterus; ovary; and vagina. These arteries are illustrated in figures 35.3 and 35.4.

Each external iliac artery branches from a common iliac artery. As the external iliac artery leaves the body cavity, it continues into the thigh as the **femoral artery.** The femoral artery has a superficial branch that feeds the thigh and a deeper branch, called the **deep femoral artery,** that takes blood to the muscles of the thigh, knee, and femur. The femoral artery divides into the **posterior tibial artery** and **anterior tibial artery,** which take blood to the knee and leg, and the **fibular (or peroneal) artery,** which supplies the muscles on the lateral side of the leg. The tibial and fibular arteries anastomose in the foot and supply blood to the **plantar arteries,** the **dorsal pedal artery,** and the **digital arteries.** Locate these arteries in figure 35.4.

Arteriosclerosis

Arteriosclerosis is a condition known commonly as hardening of the arteries. Cholesterol, fatty acids, and other molecules can be deposited in the arterial wall, deep to the endothelium. This can produce a hard region known as **plaque.** Examine a microscope slide or an illustration

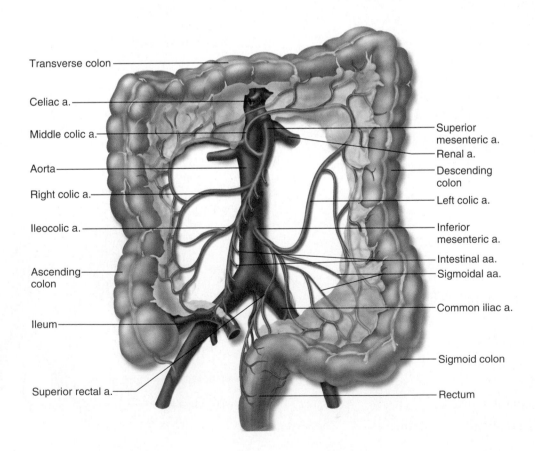

FIGURE 35.2 Middle Abdominal Arteries

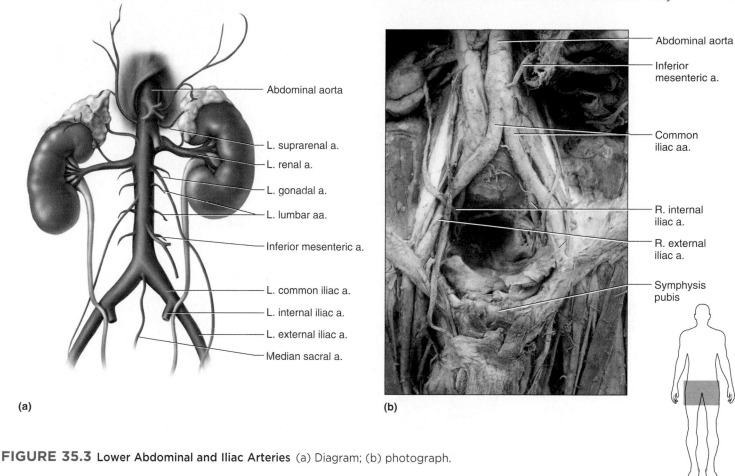

(a)

(b)

FIGURE 35.3 Lower Abdominal and Iliac Arteries (a) Diagram; (b) photograph.

of arteriosclerosis and note the development of plaque under the endothelial layer. Compare it to figure 35.5. Draw what you see in the space provided.

Drawing of an artery with arteriosclerosis:

Cat Dissection

We will use human terminology for the cat unless a vessel is specific to the cat and not found in humans. Take the cat back to your table along with dissection supplies. Remember to place all excess tissue in the appropriate waste container and not in a standard classroom wastebasket or down the sink!

Follow the dissection procedures discussed in the previous exercise. Make sure you do not cut the blood vessels away from the organs that will be studied later.

Abdominal Arteries

The determination of the abdominal arteries is best done by looking for both the exit of these arteries from the aorta and their entrance into the organs supplied by these vessels. The abdominal cavity of the cat should already be opened. If not, make an incision in the lower abdominal wall and carefully cut toward the sternum. Do not cut deeply into the body cavity or you will puncture or cut into the viscera. Locate the major organs of the abdominal region, such as the stomach, spleen, liver, and small and large intestines. Examine the specimen from the *left side,* gently lifting the stomach, spleen, and intestines toward the ventral right side as you look for the abdominal arteries. The **descending aorta** is located on the left side of the body, while the **inferior vena cava** is located on the right side. Locate the short **celiac trunk (celiac artery)**, which quickly divides into three major vessels, the **splenic artery,**

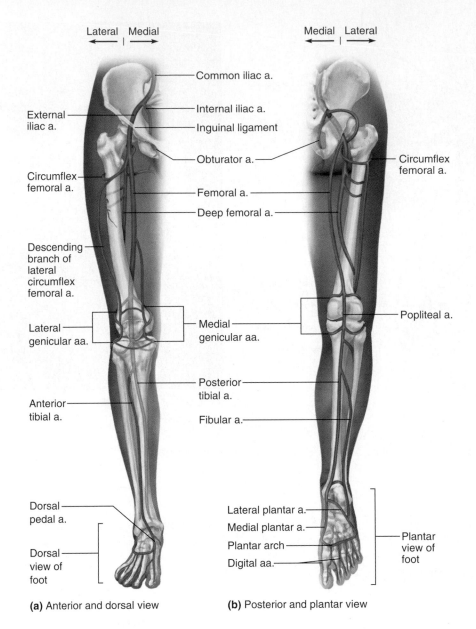

Lateral   Medial

Medial   Lateral

Common iliac a.

Internal iliac a.

Inguinal ligament

External iliac a.

Obturator a.

Circumflex femoral a.

Circumflex femoral a.

Femoral a.

Deep femoral a.

Descending branch of lateral circumflex femoral a.

Popliteal a.

Lateral genicular aa.

Medial genicular aa.

Posterior tibial a.

Anterior tibial a.

Fibular a.

Dorsal pedal a.

Lateral plantar a.

Medial plantar a.

Plantar arch

Plantar view of foot

Dorsal view of foot

Digital aa.

(a) Anterior and dorsal view

(b) Posterior and plantar view

FIGURE 35.4 Arteries of the Lower Extremity

the **left gastric artery,** and the **hepatic artery.** The splenic artery takes blood to the spleen, the left gastric artery supplies the stomach, and the hepatic artery reaches the liver. Compare your dissection to figure 35.6.

The **superior mesenteric artery** is the next major vessel to take blood from the abdominal aorta. The superior mesenteric artery travels to the small intestine and the proximal portion of the large intestine. It branches into the **jejunal arteries,** the **ileal arteries,** and the **colic arteries.** Compare the superior mesenteric artery illustrated in figures 35.6 and 35.7 with your specimen.

The paired **adrenolumbar arteries** branch on each side of the abdominal aorta and take blood to the adrenal glands, parts of the body wall, and the diaphragm. Just caudal to the adrenolumbar arteries are the paired **renal arteries,** which take blood to the kidneys. The **gonadal arteries** are the next set of paired arteries, and these take blood to the testes in male cats, or ovaries in female cats (see figure 35.6).

A single **inferior mesenteric artery** takes blood from the abdominal aorta to the terminal portion of the large intestine. The inferior mesenteric artery can be found caudal to

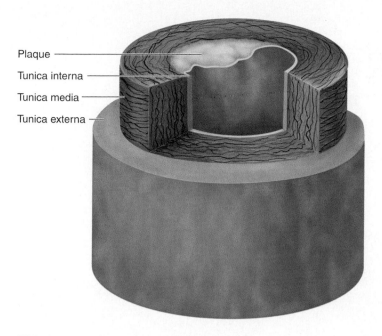

Plaque

Tunica interna

Tunica media

Tunica externa

FIGURE 35.5 Arteriosclerotic Plaque

the gonadal arteries. Locate the lower abdominal arteries in your specimen and compare them to figure 35.6.

The pelvic arteries are somewhat different in cats than in humans. The abdominal aorta splits caudally into the **external iliac arteries,** and a short section of the aorta continues on and then divides to form the two **internal iliac arteries** and the **caudal artery.** These are illustrated in figures 35.8 and 35.9. There is no common iliac artery in cats as there is in humans. In cats the caudal artery takes blood to the tail. The internal iliac artery takes blood to the urinary bladder, the rectum, the external genital organs, the uterus, and some thigh muscles. As the external iliac artery exits from the pelvic region and enters the thigh, it becomes the **femoral artery,** which supplies blood to the thigh, leg, and foot. The **deep femoral artery** supplies blood to some of the thigh muscles. The femoral artery becomes the **popliteal artery** behind the knee, which further divides to form the **tibial arteries** (see figures 35.8 and 35.10).

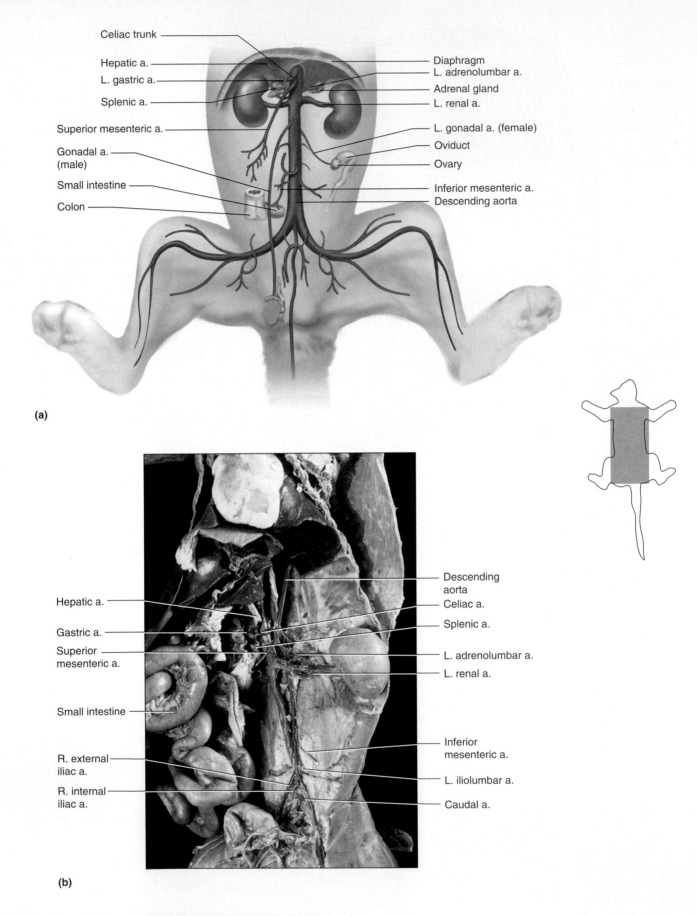

FIGURE 35.6 Posterior Arteries of the Cat (a) Diagram; (b) photograph.

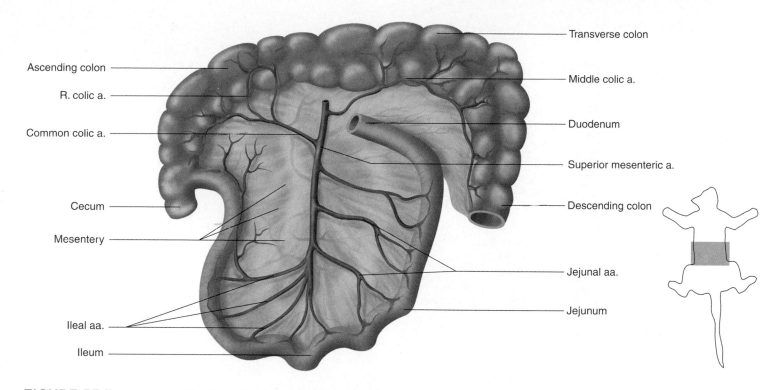

FIGURE 35.7 Branches of the Superior Mesenteric Artery of the Cat

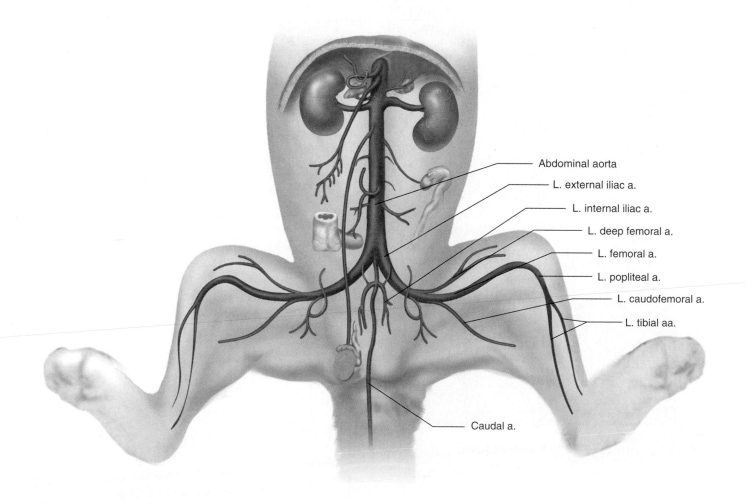

FIGURE 35.8 Arteries of the Pelvis and Lower Extremity of the Cat

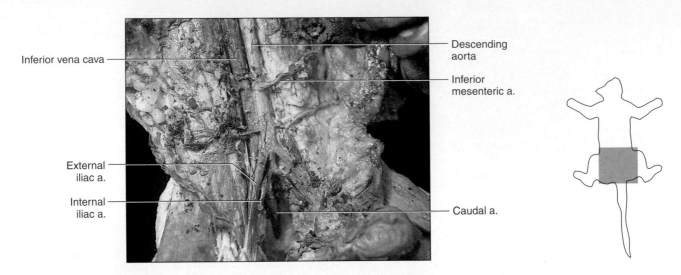

FIGURE 35.9 Pelvic Arteries of the Cat

Inferior vena cava

Descending aorta

Inferior mesenteric a.

External iliac a.

Internal iliac a.

Caudal a.

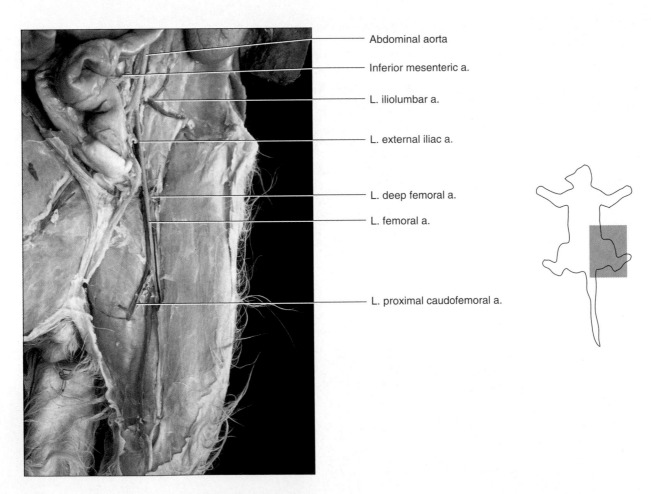

Abdominal aorta

Inferior mesenteric a.

L. iliolumbar a.

L. external iliac a.

L. deep femoral a.

L. femoral a.

L. proximal caudofemoral a.

FIGURE 35.10 Arteries of the Left Lower Extremity of the Cat

REVIEW SECTION

Arteries of the Lower Body

Name _____ Date _____

Lab Section _____ Time _____

Review Questions

1. Blood from the popliteal artery comes directly from what artery?

2. Blood from the celiac artery flows into three different blood vessels. What are these vessels?

3. The superior mesenteric artery takes blood to what major abdominal organs?

4. In what part of the arterial wall does cholesterol plaque develop?

5. How do the lower pelvic arteries in humans differ from those in cats?

6. Name the section of descending aorta inferior to the diaphragm.

7. What artery takes blood directly to the femoral artery?

8. In humans, where does blood in the external iliac artery come from?

9. Blood in the inferior mesenteric artery travels to what organs?

10. What is arteriosclerosis?

11. Name the vessel that takes blood to the adrenal glands in the cat.

12. What vessels take blood to the kidneys?

13. The ovaries or testes receive blood from which arteries?

14. Label the following illustration using the terms provided.

inferior mesenteric artery   celiac trunk
popliteal artery   tibial artery
femoral artery   internal iliac artery
common iliac artery   superior mesenteric artery
gonadal artery

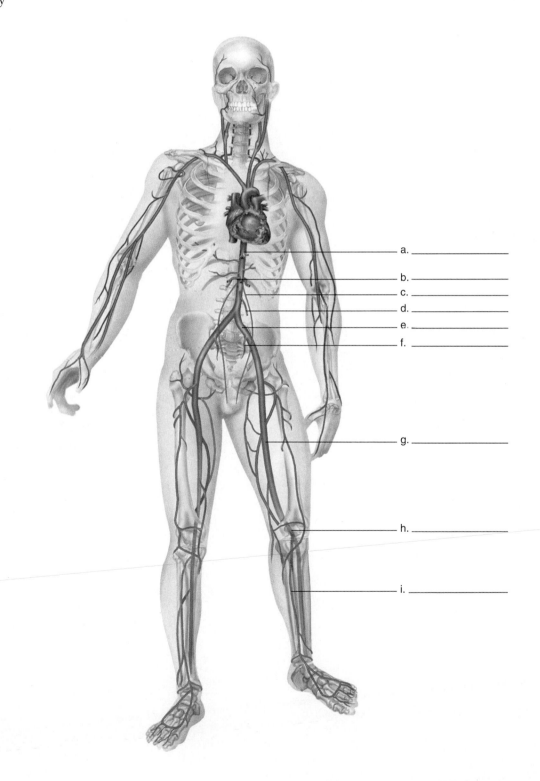

a. _____

b. _____
c. _____
d. _____
e. _____
f. _____

g. _____

h. _____

i. _____

Notes

LABORATORY

Veins and Special Circulations

INTRODUCTION

Veins are defined as blood vessels that carry blood toward the heart. Veins are thinner walled than arteries and contain valves for a one-way flow of blood to the heart. A description of the differences in the walls of arteries and veins was covered in Laboratory Exercise 34. Veins can be superficial or deep. The deep veins of the body frequently travel alongside the major arteries and take on the arterial names, while superficial veins have names specific to each vessel. Examples of veins that carry the arterial names are the femoral vein, brachial vein, and subclavian vein. In this exercise you begin the study of veins by examining charts or models of the entire venous system and then studying the veins of specific regions of the body. Veins are covered in the Saladin text in chapter 20, "Blood Vessels and Circulation."

OBJECTIVES

At the end of this exercise you should be able to

1. describe the sequence of major veins that drain the upper extremities;
2. describe the sequence of major veins that drain the lower extremities;
3. trace the blood flow from the brain to the heart;
4. distinguish a portal system from normal venous return flow;
5. describe the major digestive organs and vessels that supply blood to the hepatic portal vein;
6. identify major veins in models, a cadaver, or a cat.

MATERIALS

Models and charts of the blood vessels of the body

Cadaver (if available)

Cats

Materials for cat dissection

    Dissection trays

    Scalpel and two or three extra blades

Gloves (household latex gloves work well for repeated use)

Blunt (mall) probe

Forceps and sharp scissors

First aid kit in lab or prep area

Sharps container

Animal waste disposal container

PROCEDURE

Examine the models and charts in the lab and locate the major veins of the body. Read the following descriptions of the veins and find them as they are represented in lab. As you locate a specific vein, name the vessel that takes blood to the vein and those that receive blood from the vein. Veins in this exercise are studied in the direction of their flow from the cells of the body to the heart. In this way they resemble tributaries of rivers as smaller veins flow into larger veins.

An overview of the major veins of the body is shown in figure 36.1. Compare that figure as well as the material in lab to the following list of some of the major veins of the body.

_____ Radial vein

_____ Ulnar vein

_____ Brachial vein

_____ Axillary vein

_____ Subclavian vein

_____ Brachiocephalic vein

_____ Superior vena cava

_____ Inferior vena cava

_____ Internal jugular vein

_____ Femoral vein

_____ Great saphenous vein

_____ Common iliac vein

_____ External iliac vein

_____ Internal iliac vein

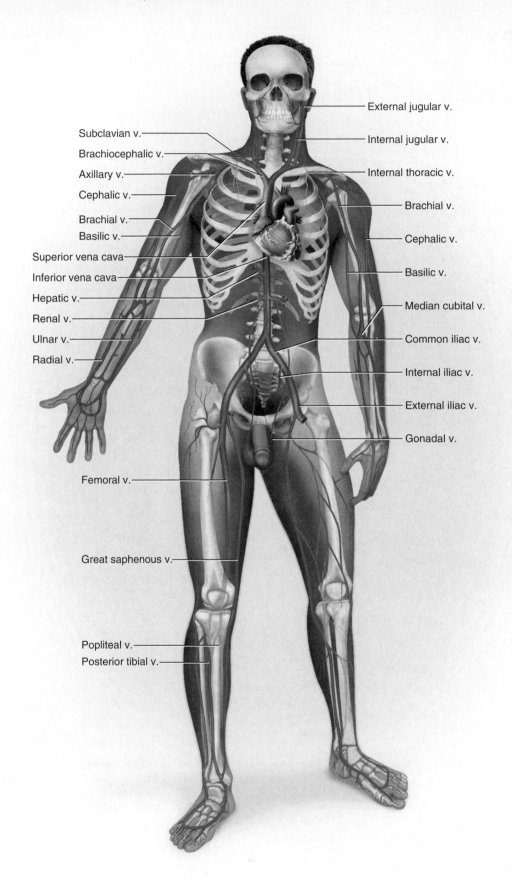

External jugular v.

Subclavian v.
Brachiocephalic v.
Axillary v.
Cephalic v.
Brachial v.
Basilic v.
Superior vena cava
Inferior vena cava
Hepatic v.
Renal v.
Ulnar v.
Radial v.

Internal jugular v.
Internal thoracic v.
Brachial v.
Cephalic v.
Basilic v.
Median cubital v.
Common iliac v.
Internal iliac v.
External iliac v.
Gonadal v.

Femoral v.

Great saphenous v.

Popliteal v.
Posterior tibial v.

FIGURE 36.1 Major Systemic Veins

Veins of the Upper Extremities

Examine the models and charts in the lab and locate the veins of the upper extremities. The fingers are drained by the small **digital veins,** which lead to the **palmar arch veins.** The major superficial veins of each upper extremity are the **basilic vein,** on the anterior, medial side of the forearm and arm, and the **cephalic vein,** on the anterior, lateral side of the forearm and arm. The two vessels have many anastomosing branches (cross-connections) between them. One of the significant anastomosing veins is the **median cubital vein,** which crosses the anterior cubital fossa and is a common site for the withdrawal of blood. Locate these superficial veins in figure 36.2.

The deep veins of the forearm are the **radial vein** and the **ulnar vein,** each of which can be found traveling near the artery of the same name. The deep veins of the arm are the **brachial veins,** next to the brachial artery. The brachial veins are formed by the union of the radial and ulnar veins and merge superiorly with the **basilic vein** to form the short **axillary vein.** The axillary vein connects with the **cephalic vein** to form the **subclavian vein.** The blood of the upper extremities is carried to the heart by the left and right subclavian veins, which flow to the **brachiocephalic veins,** then to the **superior vena cava,** and finally to the **right atrium** of the heart. Locate the deep veins of the upper extremities in figure 36.3.

Veins of the Head and Neck

Examine the models and charts in the lab and locate the veins that drain blood from the head. Blood from the head returns to the heart by a number of routes. Each **internal jugular vein** receives blood from the brain and passes through a jugular foramen of the skull. It then passes along the lateral aspect of the neck as it moves toward the **brachiocephalic vein.** The brachiocephalic vein is formed by the union of the **internal jugular vein** and the **subclavian vein.** The superficial regions of the head (musculature and skin of the scalp and face) are drained by the **external jugular vein.** The external jugular veins join with the subclavian veins prior to reaching the brachiocephalic veins.

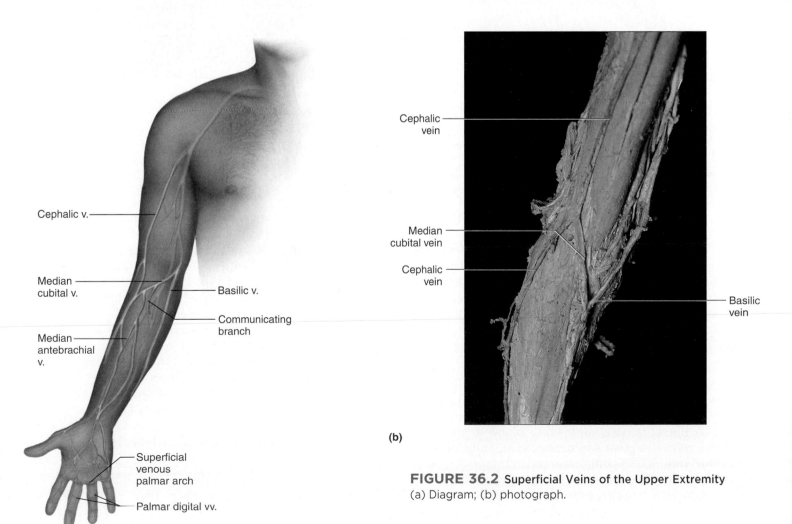

(a)

Cephalic v.

Median cubital v.

Median antebrachial v.

Basilic v.

Communicating branch

Superficial venous palmar arch

Palmar digital vv.

(b)

Cephalic vein

Median cubital vein

Cephalic vein

Basilic vein

FIGURE 36.2 Superficial Veins of the Upper Extremity (a) Diagram; (b) photograph.

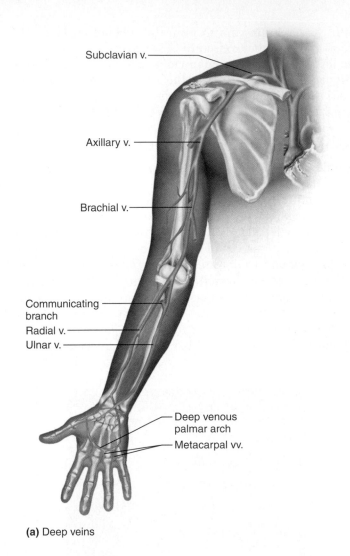

(a) Deep veins

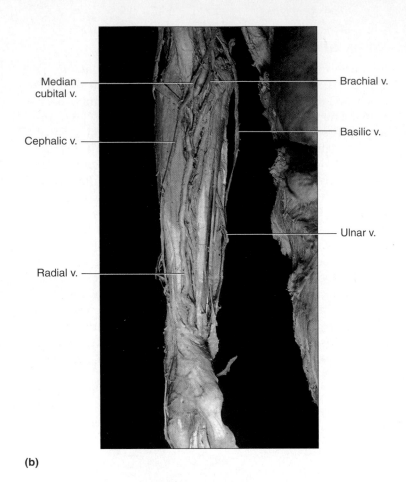

(b)

FIGURE 36.3 Deep Veins of the Upper Extremity
(a) Diagram; (b) photograph.

The left and right brachiocephalic veins take blood to the superior vena cava. Another vessel that takes blood from the brain is the **vertebral vein,** which, like the vertebral arteries, travels through the transverse foramina of the cervical vertebrae. The vertebral veins take blood to the **subclavian veins,** which in turn flow to the **brachiocephalic veins.** Locate these major vessels of the head and neck in figure 36.4.

Veins of the Lower Extremities

The blood vessels that drain the lower extremities operate under relatively low pressure and must take blood back to the heart against gravity. Digital veins in the feet take blood to the **plantar venous arch** and **dorsal venous arch** veins. The dorsal venous arch takes blood to the **anterior tibial vein,** the **great saphenous vein,** and the **small saphenous vein.** The plantar venous arch takes blood to the small saphenous vein and the **posterior tibial vein.** The anterior and posterior tibial veins unite posterior to the knee and form the **popliteal vein,** which joins with the small saphenous vein to form the **femoral vein.** The longest vessel in the human body is the **great saphenous vein,** just underneath the skin on the medial aspect of the lower

extremity beginning near the medial malleolus and traversing the lower extremity to the proximal thigh. This vessel frequently is embedded in adipose tissue below the skin in humans, yet it is still considered a superficial vein. Two other vessels in the thigh are the **femoral vein** and the **deep femoral vein,** which take blood from the thigh. The **femoral vein** travels alongside the femoral artery. As the great saphenous vein reaches the inguinal region, it joins with the femoral vein, which takes blood from the thigh region and passes under the inguinal ligament. Once the **femoral vein** crosses under the ligament it becomes the **external iliac vein.** Examine these vessels in the lab and compare them to figure 36.5.

Veins of the Abdomen and Pelvis

The **external iliac vein** joins with the **internal iliac vein,** which drains the region of the pelvis, and they form the **common iliac vein.** The common iliac veins unite and form the **inferior vena cava,** which travels superiorly along the right side of the vertebrae, taking blood to the right atrium of the heart. Locate these veins and compare them to figure 36.6.

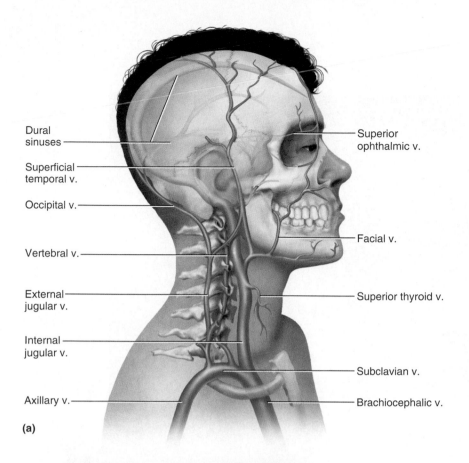

Dural sinuses

Superficial temporal v.

Occipital v.

Vertebral v.

External jugular v.

Internal jugular v.

Axillary v.

Superior ophthalmic v.

Facial v.

Superior thyroid v.

Subclavian v.

Brachiocephalic v.

(a)

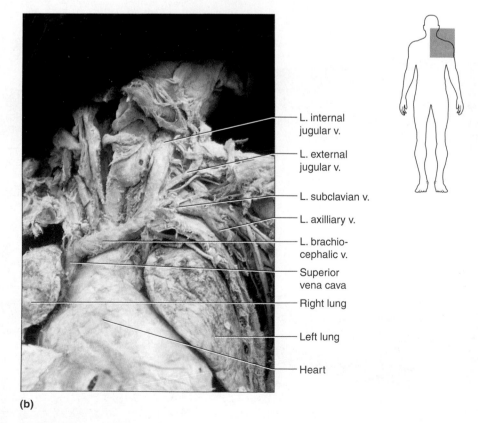

L. internal jugular v.

L. external jugular v.

L. subclavian v.

L. axilliary v.

L. brachio-cephalic v.

Superior vena cava

Right lung

Left lung

Heart

(b)

FIGURE 36.4 Veins of the Head and Neck (a) Diagram of right side; (b) photograph of left veins.

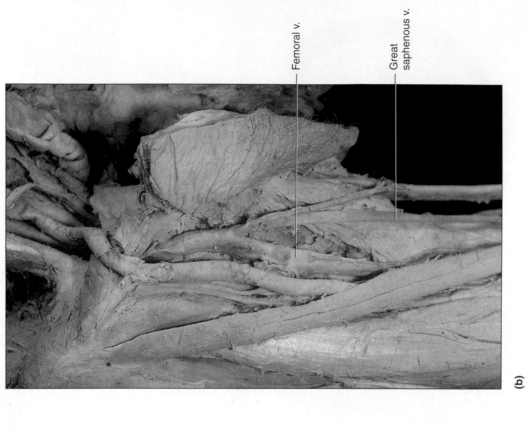

(b)

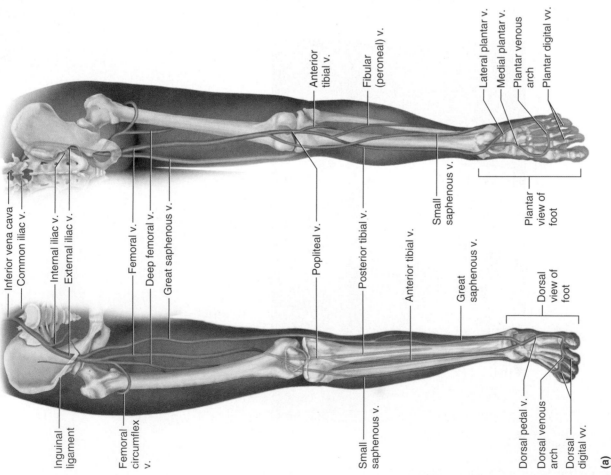

(a)

FIGURE 36.5 Veins of the Lower Extremities (a) Diagram of entire lower extremity; (b) photograph of proximal right thigh.

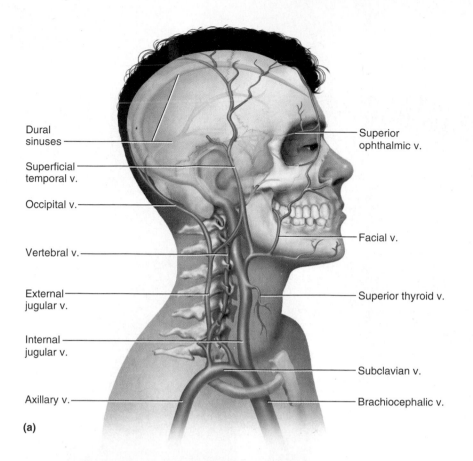

Dural
sinuses

Superficial
temporal v.

Occipital v.

Vertebral v.

External
jugular v.

Internal
jugular v.

Axillary v.

Superior
ophthalmic v.

Facial v.

Superior thyroid v.

Subclavian v.

Brachiocephalic v.

(a)

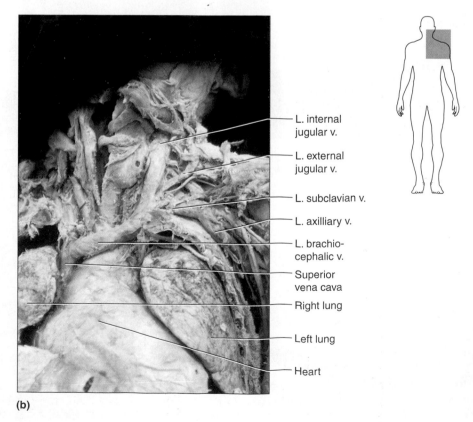

L. internal
jugular v.

L. external
jugular v.

L. subclavian v.

L. axilliary v.

L. brachio-
cephalic v.

Superior
vena cava

Right lung

Left lung

Heart

(b)

FIGURE 36.4 Veins of the Head and Neck (a) Diagram of right side; (b) photograph of left veins.

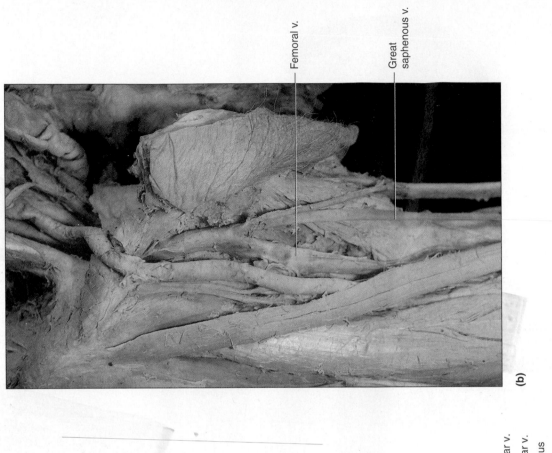

(b)

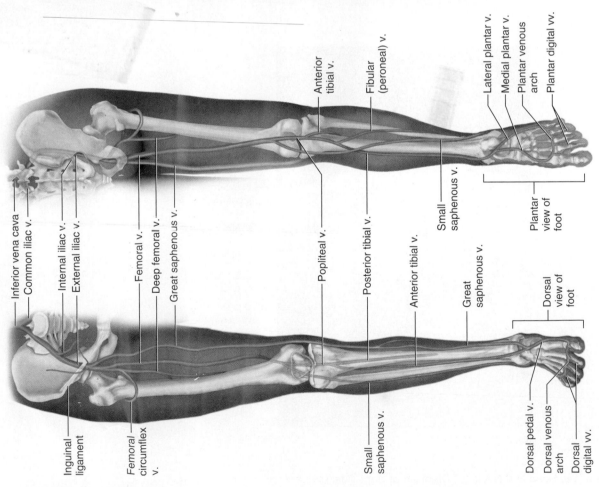

(a)

Femoral v.

Great saphenous v.

Anterior tibial v.

Fibular (peroneal) v.

Lateral plantar v.

Medial plantar v.

Plantar venous arch

Plantar digital vv.

Plantar view of foot

Inferior vena cava

Common iliac v.

Internal iliac v.

External iliac v.

Femoral v.

Deep femoral v.

Great saphenous v.

Popliteal v.

Posterior tibial v.

Anterior tibial v.

Small saphenous v.

Inguinal ligament

Femoral circumflex v.

Small saphenous v.

Great saphenous v.

Dorsal view of foot

Dorsal pedal v.

Dorsal venous arch

Dorsal digital vv.

FIGURE 36.5 Veins of the Lower Extremities (a) Diagram of entire lower extremity; (b) photograph of proximal right thigh.

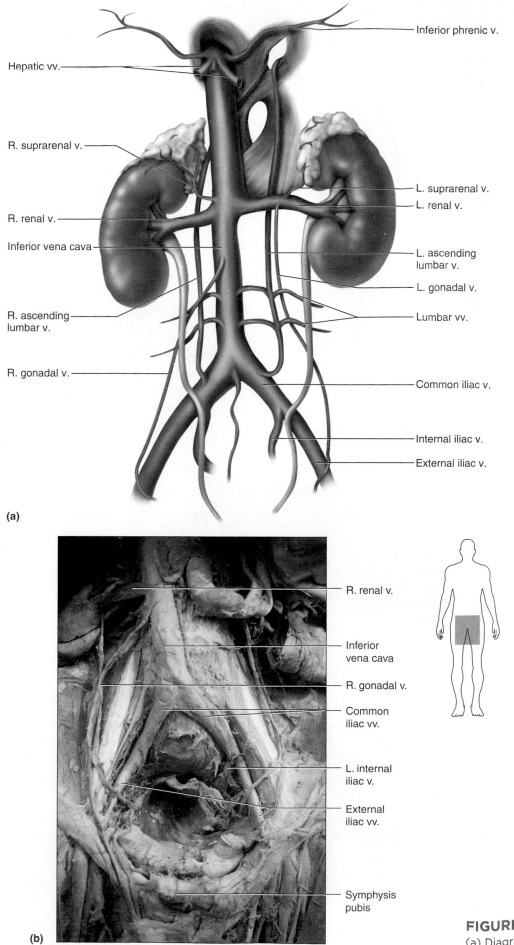

Inferior phrenic v.

Hepatic vv.

R. suprarenal v.

R. renal v.

Inferior vena cava

R. ascending lumbar v.

R. gonadal v.

L. suprarenal v.

L. renal v.

L. ascending lumbar v.

L. gonadal v.

Lumbar vv.

Common iliac v.

Internal iliac v.

External iliac v.

(a)

R. renal v.

Inferior vena cava

R. gonadal v.

Common iliac vv.

L. internal iliac v.

External iliac vv.

Symphysis pubis

(b)

FIGURE 36.6 Veins of the Pelvis
(a) Diagram; (b) photograph.

Some abdominal veins take blood directly to the inferior vena cava and some pass through the liver before reaching the inferior vena cava. Those that take blood directly to the inferior vena cava are the **renal veins,** the **suprarenal veins,** and the **lumbar veins.** The **right gonadal vein** takes blood directly to the inferior vena cava, but the **left gonadal vein** takes blood to the left renal vein prior to flowing into the inferior vena cava. These vessels can be seen in figure 36.6.

Hepatic Portal Circulation

The veins that flow into the liver before returning to the heart are part of the **hepatic portal system.** Most veins take blood from capillaries and venules and return the blood to the heart. In **portal systems** the flow is different. Blood flows from a *capillary bed* through veins to another *capillary bed.* In the **hepatic portal system,** blood from the capillaries of major digestive organs, as well as the spleen, travels to the capillaries of the liver. The **inferior mesenteric vein** drains the distal part of the large

intestine, receives blood from the **gastroepiploic vein,** and empties into the **splenic vein.** The splenic vein joins with the **superior mesenteric vein,** which drains the small intestine and the proximal portion of the large intestine. The splenic vein and the superior mesenteric vein unite to form the **hepatic portal vein,** which takes blood to the capillaries of the liver, where the blood is cleaned by macrophages and nutrients are either stored in the liver or passed into the bloodstream. The liver receives the blood from the digestive organs and processes it before sending it through the **hepatic vein** to the inferior vena cava and toward the heart. Examine models or charts in lab and compare them to figure 36.7 for the main vessels of the hepatic portal system and identify the following veins:

_____ Inferior mesenteric vein

_____ Superior mesenteric vein

_____ Splenic vein

_____ Gastroepiploic vein

_____ Hepatic portal vein

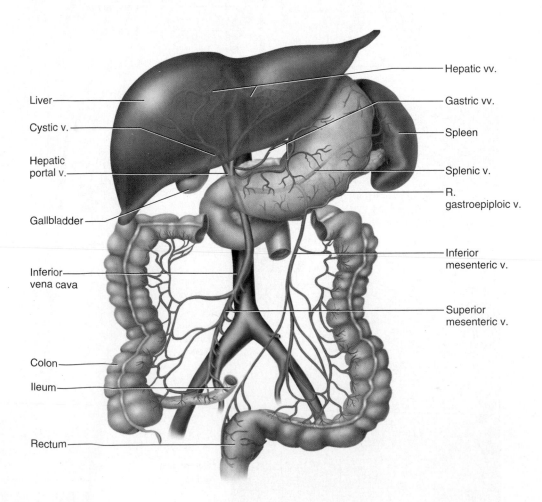

Liver
Cystic v.
Hepatic portal v.
Gallbladder
Inferior vena cava
Colon
Ileum
Rectum

Hepatic vv.
Gastric vv.
Spleen
Splenic v.
R. gastroepiploic v.
Inferior mesenteric v.
Superior mesenteric v.

FIGURE 36.7 Hepatic Portal System

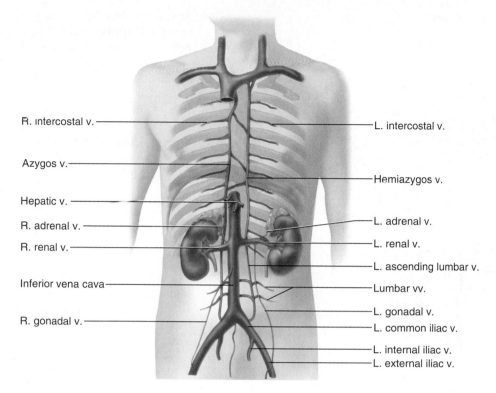

R. intercostal v.

Azygos v.

Hepatic v.

R. adrenal v.

R. renal v.

Inferior vena cava

R. gonadal v.

L. intercostal v.

Hemiazygos v.

L. adrenal v.

L. renal v.

L. ascending lumbar v.

Lumbar vv.

L. gonadal v.

L. common iliac v.

L. internal iliac v.

L. external iliac v.

FIGURE 36.8 Veins of the Abdomen and Thorax

Thoracic Veins

The major veins of the thoracic region consist of the **intercostal veins,** which drain the intercostal muscles, and the **azygos vein** and **hemiazygos vein,** which drain blood from the thoracic region. Locate these vessels in the lab and compare them to figure 36.8.

Fetal Circulation

The pathway of fetal blood is somewhat different from that of the adult in that the lungs are nonfunctional in the fetus. Oxygen and nutrients move from the maternal side of the **placenta** to the fetal bloodstream, while carbon dioxide and metabolic wastes move from the fetal bloodstream to the placenta. In the fetus, the blood flows from the placenta through the **umbilical vein,** which is located in the **umbilical cord.** From there it travels through the **ductus venosus,** which serves as a shunt to the **inferior vena cava** of the fetus. The maternal blood, relatively high in oxygen and nutrients, mixes with the deoxygenated, nutrient-poor fetal blood and thus the fetus receives a mixture of blood. Examine figure 36.9 for an overview of fetal circulation.

The blood from the **inferior vena cava** travels to the **right atrium** of the heart. While in the right atrium, the blood can travel either to the **right ventricle** (which pumps blood to the lungs) or through a hole in the right atrium called the **foramen ovale.** Since the lungs are nonfunctional in the fetus, the foramen ovale serves as

a bypass route away from the lungs and to the chambers of the heart that will pump blood to the rest of the body. Blood in the right ventricle is pumped to the **pulmonary trunk,** where another shunt vessel, the **ductus arteriosus,** carries blood to the **aortic arch** bypassing the lungs. The lungs do receive some blood, but it is for the nourishment of the lung tissue as opposed to gas exchange. Locate the structures of the fetal heart in figure 36.9.

Blood from the heart exits the left ventricle and passes through the aorta to the **systemic arteries.** Blood travels down the common iliac arteries to the lower extremities. Some blood moves into the **internal iliac arteries** and into the **umbilical arteries,** which carry blood to the **placenta** (see figure 36.9). At birth, the pressure changes in the newborn's heart cause the closing of a flap of tissue over the **foramen ovale,** leaving a thin spot in the interatrial septum known as the **fossa ovalis.** Lack of closure of this foramen can lead to a condition known as "blue baby."

Cat Dissection

Prepare for the cat dissection by obtaining a cat and dissection equipment. Remember to place all excess tissue in the appropriate waste container and not in a standard classroom wastebasket or down the sink!

Begin your dissection by reviewing the arterial system previously studied in Laboratory Exercises 34 and 35. As stated in previous exercises, the blood vessels in the cat have been injected with colored latex. If the cat has been

Fetus

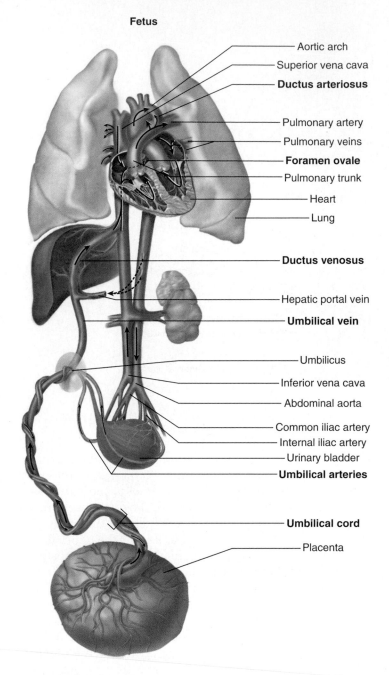

- Aortic arch
- Superior vena cava
- **Ductus arteriosus**
- Pulmonary artery
- Pulmonary veins
- **Foramen ovale**
- Pulmonary trunk
- Heart
- Lung
- **Ductus venosus**
- Hepatic portal vein
- **Umbilical vein**
- Umbilicus
- Inferior vena cava
- Abdominal aorta
- Common iliac artery
- Internal iliac artery
- Urinary bladder
- **Umbilical arteries**
- **Umbilical cord**
- Placenta

FIGURE 36.9 Fetal Circulation in the Human

doubly injected the arteries are red and the veins are blue. If the cat has been triply injected the hepatic portal vein is also injected, typically with yellow latex. The terms used in this lab manual will be human terms unless the vessels are specific to the cat.

Veins of the Upper Extremities

Begin your dissection by examining the veins of the upper extremities. The basilic vein is not found in cats, but you should be able to find the **ulnar vein;** it joins the **radial vein** to form the **brachial vein.** The brachial vein is next to the brachial artery. Locate the **median cubital vein** and the **cephalic vein.** The **axillary vein** and **subscapular vein** (from the shoulder) join to form the **subclavian vein,** which takes blood to the **brachiocephalic vein.** The cephalic vein continues into the transverse scapular vein that takes blood to the external jugular vein. Locate these veins in figure 36.10.

Veins of the Head and Neck

The veins in this region of the cat have a few variations from those in the human. In the cat the **external jugular vein** is larger than the **internal jugular vein.** The reverse is true in humans. What anatomical difference between the cat and human might explain the difference in the volume of blood carried by these two vessels?

The **transverse jugular vein,** present in cats but absent in humans, connects the two external jugular veins. The jugular veins, along with the **costocervical veins** and the **subclavian veins,** unite to form the **brachiocephalic veins** (see figure 36.10). The vertebral veins take blood from the head of the cat and empty into the subclavian veins.

Thorax

Anterior to the diaphragm are the veins of the thorax. The **internal mammary vein** (internal thoracic vein) may be difficult to locate because it is frequently cut while opening up the thoracic cavity. The internal mammary vein receives blood from the ventral body wall. Another vein of the thoracic region is the **azygos vein.** This vessel is found on the right side of the body and takes blood from the esophageal, intercostal, and bronchial veins. Compare the veins in your cat to figure 36.10.

Veins of the Lower Extremities

Locate the superficial **great saphenous vein** in the cat on the medial side of the lower extremity. The great saphenous vein joins the femoral vein just above the knee. Note the long **tibial veins** of the leg that join and form the **popliteal veins.** Find the **femoral vein,** found with the femoral artery. The vessels of the distal part of the lower extremity take blood to the femoral vein, which turns into the **external iliac vein** as it passes into the body cavity at the level of the inguinal ligament. Find these veins in the cat and compare them to figure 36.11.

Veins of the Pelvis

The **external iliac vein** and the **internal iliac vein** join to form the **common iliac vein.** The two common iliac veins lead to the **inferior vena cava** along with the **caudal vein,** which takes blood from the tail. Compare your cat to figures 36.11 and 36.12. Be careful as you dissect these structures in the cat. There are numerous ducts that cross the external iliac veins, such as the ureter, and, in males, the ductus deferens. Look for these structures and *do not cut them.* You will study them later.

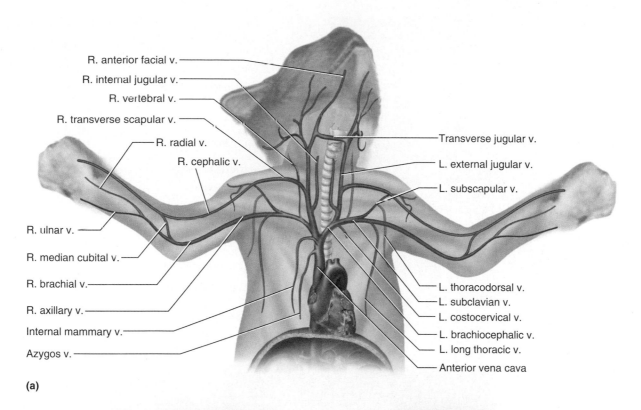

R. anterior facial v.
R. internal jugular v.
R. vertebral v.
R. transverse scapular v.
R. radial v.
R. cephalic v.
R. ulnar v.
R. median cubital v.
R. brachial v.
R. axillary v.
Internal mammary v.
Azygos v.

Transverse jugular v.
L. external jugular v.
L. subscapular v.
L. thoracodorsal v.
L. subclavian v.
L. costocervical v.
L. brachiocephalic v.
L. long thoracic v.
Anterior vena cava

(a)

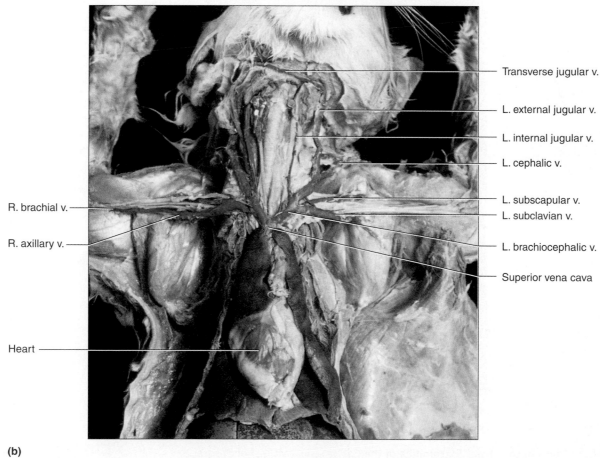

Transverse jugular v.
L. external jugular v.
L. internal jugular v.
L. cephalic v.
L. subscapular v.
L. subclavian v.
L. brachiocephalic v.
Superior vena cava

R. brachial v.
R. axillary v.

Heart

(b)

FIGURE 36.10 **Anterior Veins of the Cat** (a) Diagram; (b) photograph.

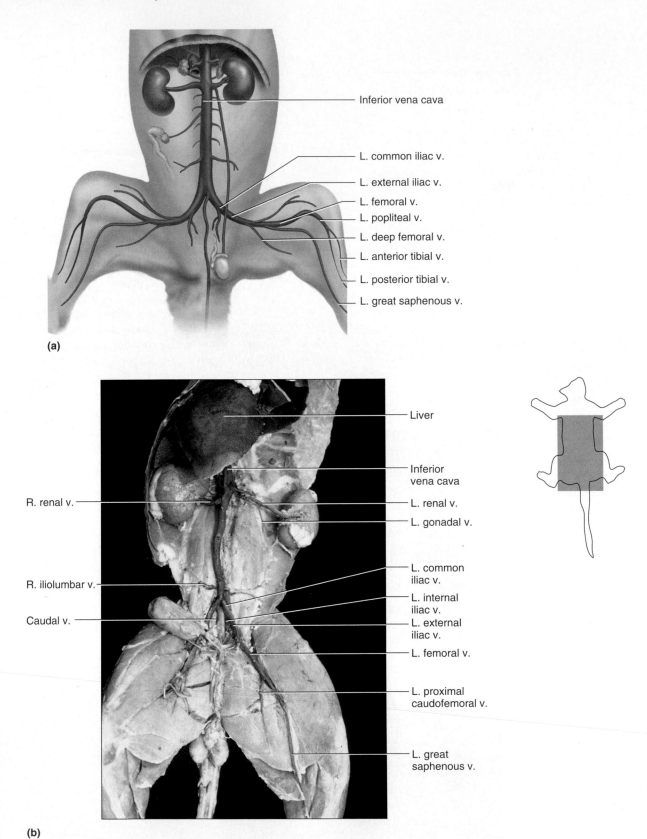

(a)

Inferior vena cava

L. common iliac v.

L. external iliac v.

L. femoral v.

L. popliteal v.

L. deep femoral v.

L. anterior tibial v.

L. posterior tibial v.

L. great saphenous v.

(b)

Liver

Inferior vena cava

R. renal v.

L. renal v.

L. gonadal v.

R. iliolumbar v.

L. common iliac v.

L. internal iliac v.

Caudal v.

L. external iliac v.

L. femoral v.

L. proximal caudofemoral v.

L. great saphenous v.

FIGURE 36.11 Veins of the Lower Extremities of the Cat (a) Diagram; (b) photograph.

Abdominal Veins

Make sure you identify the major organs in the digestive tract before you proceed with the study of the veins of this region. Pay particular attention to the stomach, liver, spleen, kidneys, adrenal glands, small intestine, and large intestine. The **inferior vena cava** is a large vessel that travels up the right side of the body in the cat. You may have to gently move some of the digestive organs to the side to find the inferior vena cava. Examine where the common iliac veins join to form the inferior vena cava. The **caudal vein** joins the inferior vena cava at this point. It takes blood from the tail of the cat. This vein is absent in humans. The **gonadal veins** receive blood from the testes or ovaries. The **right gonadal vein** takes blood to the inferior vena cava, and the **left gonadal vein** takes blood to the **left renal vein,** which subsequently leads to the inferior vena cava. The paired **renal veins** receive blood from each kidney and empty into the inferior vena cava. The **iliolumbar veins** take blood from the body wall and return it via the inferior vena cava. The body wall is also drained by the **adrenolumbar veins,** which receive blood from the adrenal glands in addition to the body wall. Compare the veins in your cat to figure 36.12.

Hepatic Portal System

The vessels of the **hepatic portal system** take blood from a number of abdominal organs and transfer it to the liver, an organ where significant metabolic processing occurs. As you examine the lower portion of the digestive tract (transverse colon and descending colon), locate the **inferior mesenteric vein.** The **superior mesenteric vein** receives blood from the duodenum, the pancreas, the jejunum, the ileum, and part of the large intestines. It also receives blood from the **inferior mesenteric vein.** The **gastrosplenic vein** joins the superior mesenteric vein and takes blood to the **hepatic portal vein,** along with other digestive veins. The gastrosplenic vein receives blood from the spleen and the stomach. The hepatic portal vein thus receives blood from the spleen and the digestive organs and transports that blood to the liver. After the liver receives the blood, it is transferred to the inferior vena cava by the short **hepatic vein.** This vein is usually difficult to dissect because it is at the dorsal aspect of the liver and joins the inferior vena cava at about the junction of the diaphragm. Examine the hepatic portal system of the cat while trying to maintain the integrity of the digestive organs. Compare your dissection to figure 36.13.

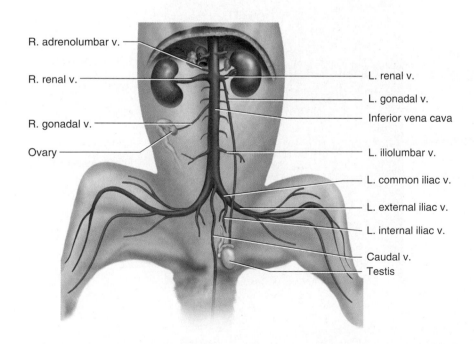

FIGURE 36.12 Veins of the Pelvis and Abdomen of the Cat

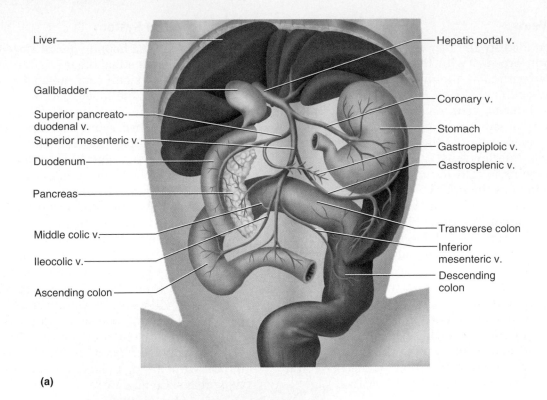

Liver

Gallbladder

Superior pancreato-
duodenal v.

Superior mesenteric v.

Duodenum

Pancreas

Middle colic v.

Ileocolic v.

Ascending colon

Hepatic portal v.

Coronary v.

Stomach

Gastroepiploic v.

Gastrosplenic v.

Transverse colon

Inferior
mesenteric v.

Descending
colon

(a)

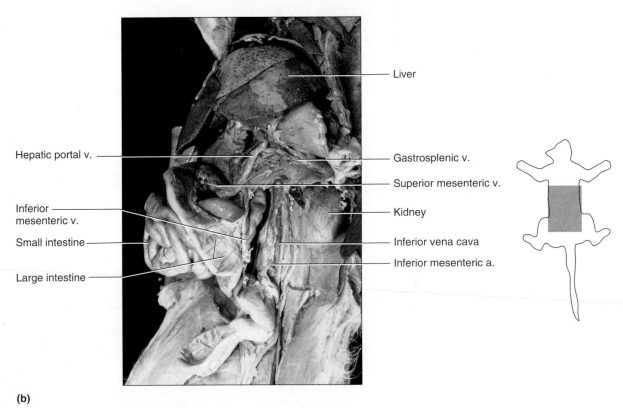

Hepatic portal v.

Inferior
mesenteric v.

Small intestine

Large intestine

Liver

Gastrosplenic v.

Superior mesenteric v.

Kidney

Inferior vena cava

Inferior mesenteric a.

(b)

FIGURE 36.13 **Hepatic Portal System of the Cat** (a) Diagram; (b) photograph.

REVIEW SECTION

Veins and Special Circulations

Name _____   _____   *Date* _____

Lab Section _____   *Time* _____

Review Questions

1. Label the following illustration using the terms provided.

    brachial vein

    internal jugular vein

    brachiocephalic vein

    basilic vein

    common iliac vein

    superior vena cava

    great saphenous vein

    inferior vena cava

    cephalic vein

    femoral vein

    external jugular vein

    internal iliac vein

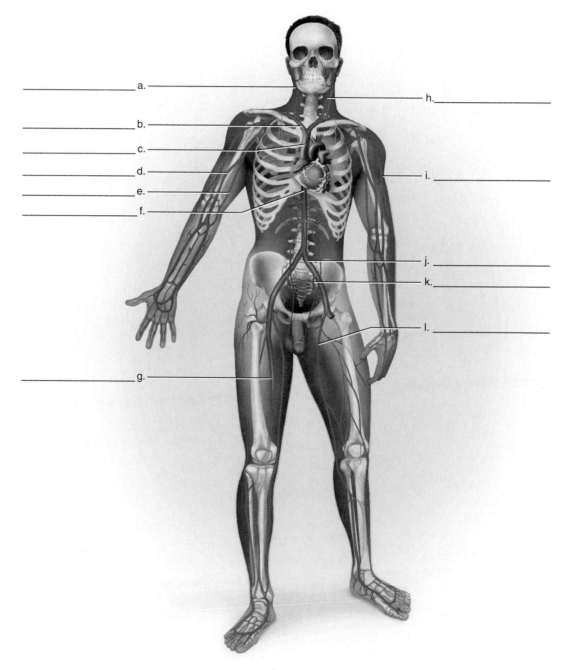

2. The brachiocephalic veins take blood to what vessel?

3. Is the radial vein a superficial or deep vein?

4. The common iliac vein receives blood from two vessels. What are these two vessels?

5. Blood from the right axillary vein next travels to what vessel?

6. What vessels take blood to the left femoral vein?

7. What area do the right and left external jugular veins drain?

8. What is the functional nature of a "portal system," and how does it differ from normal venous return flow?

9. What major vessels take blood to the hepatic portal vein?

10. Which veins (superficial/deep) have names that do not correspond to arterial names?

11. Where is the median cubital vein found?

12. What vessel receives blood from the ulnar vein?

13. Where is the cephalic vein located?

14. Blood in the superior vena cava next flows to what area?

15. The internal jugular vein takes blood from what area?

16. What veins pass through the transverse vertebral foramina?

17. In what region of the body is the great saphenous vein found?

18. Where does blood flow after it leaves the femoral vein?

19. Blood in the small intestine travels to the hepatic portal vein by what vessel?

20. In the fetal heart, what is the name of the shunt between the pulmonary trunk and the aortic arch?

21. Name the opening between the atria in the fetal heart.

22. Why might a clot in the lungs (a pulmonary embolism) occur after a deep vein thrombosis in the leg?

LABORATORY

Functions of Vessels, Lymphatic System

INTRODUCTION

Fluid in the circulatory route may travel by several different pathways. Typically, blood is pumped from the **heart** through **arteries** and distributed through the **arterioles** to the **capillaries,** where nutrients, water, and oxygen are exchanged with the cells of the body. The blood returns via **venules** to **veins** and back to the heart under relatively low pressure. The valves in veins keep blood flowing in the direction of the heart. Their role is examined in this exercise. Arteries and veins transport blood cells and plasma. Blood cells and plasma proteins typically stay in the vessels, yet a significant amount of fluid from the plasma leaks from the capillaries and bathes the cells of the body. This fluid flows between the cells of the body and is known as **interstitial (tissue) fluid.** It provides nutrients to the cells and receives dissolved wastes from the cells along with cellular debris. Capillaries absorb most of the interstitial fluid, with the remainder picked up by **lymph capillaries,** which return the fluid, now known as lymph, to the lymphatic vessels. Lymphatic vessels take the lymph back to the venous system. Fluid leaked from the capillaries returns to the cardiovascular system. This process is illustrated in figure 37.1.

The lymphatic system is also instrumental in the absorption of lipids from the digestive system. Lipids in the small intestine are converted to **chylomicrons** (phospholipids and other molecules) in the cells that line the digestive tract. These chylomicrons are conducted into the lymphatic system, which then transports the material through the thoracic duct and into the subclavian vein where it enters the cardiovascular system.

As the fluid flows through the lymphatic vessels, macrophages in **lymph nodes** phagocytize the cellular debris and foreign material (bacteria, viruses) that may have entered into the lymphatic system. Lymph nodes also activate B cells and T cells. In this exercise you examine the nature of the lymphatic system. These topics are covered in the Saladin text in chapter 20, "Blood Vessels and Circulation," and in chapter 21, "The Lymphatic and Immune Systems."

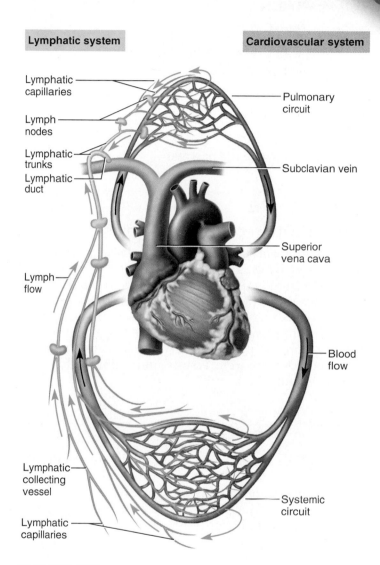

Lymphatic system

Cardiovascular system

Lymphatic capillaries

Lymph nodes

Lymphatic trunks

Lymphatic duct

Lymph flow

Lymphatic collecting vessel

Lymphatic capillaries

Pulmonary circuit

Subclavian vein

Superior vena cava

Blood flow

Systemic circuit

FIGURE 37.1 Flow of Interstitial Fluid and Lymph

OBJECTIVES

At the end of this exercise you should be able to

1. describe the function of valves in veins;
2. identify the valves in lymphatic vessels;

3. locate the major histological features of a lymph node;
4. demonstrate on a model the location of the tonsils and other lymph organs;
5. describe the function of the spleen and thymus.

MATERIALS

Microscopes

Microscope slides of lymphatics with valves

Torso models

Live frog

Dissection scopes

Paper towels

Small squeeze bottle of water

PROCEDURE

Demonstration of Valves in Veins

One way to examine the nature of **valves** in veins is to let your arm hang at your side until the veins become engorged with blood. As you hold your arm in this position, stroke the superficial veins from distal to proximal with the index finger of your free hand, applying uniform and constant pressure (see figure 37.2). Maintain pressure on the vein at its proximal end and see if the vein fills with blood. Record your results, indicating if the veins fill with blood.

Results: _____

Try the experiment again, but keep pressure on the distal part of the vein with your index finger as you push the blood toward the heart with your thumb (see figure 37.3). Release the vein from the proximal area and see if blood fills the vein. Record your results.

Results: _____

Do the valves prevent the blood from flowing in a superior or inferior direction?

Types of Arteries

Conducting (elastic) arteries are those close to the heart. Their appearance is made distinctive by the presence of significant amounts of elastic fibers in the tunica media. **Distributing (muscular) arteries** are farther away from the heart and have more smooth muscle in the tunica media.

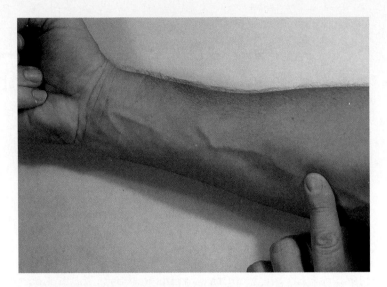

FIGURE 37.2 Testing Valves in Veins (Part 1)

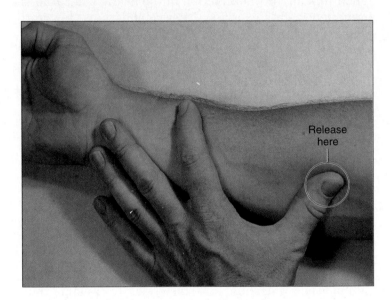

FIGURE 37.3 Testing Valves in Veins (Part 2)

Movement Through Capillaries

Blood capillaries have a relatively simple structure in that they are composed of **endothelium.** The endothelium consists of a single layer of **simple squamous epithelium** that forms a tube slightly larger than the erythrocytes that pass through it. You can observe capillaries in living tissue by examining the blood flow through the webbed foot of a live frog. Obtain a *living* frog and wrap its body in a wet paper towel. Leave the head and the foot exposed. *Do not let the frog dry out while you are examining it.* Hold the foot of the frog under a dissecting microscope and fan out the webbing of the foot. Make sure to hold the frog securely. Observe the flow of the blood through the capillaries, keeping the foot moist with water. Describe the flow of the blood through the capillaries of the foot in terms of speed and size of the capillary relative to the diameter of an erythrocyte.

Your description:

Lymphatic System

The lymphatic system is difficult to study in preserved specimens because it collapses at death. The lymphatic system is composed of **lymphatic capillaries, lymph nodes, lymphatic vessels, lymph organs,** and **lymphatic (lymphoid) tissue.** The lymphatic capillaries have blunt ends and flaplike valves that conduct lymph away from interstitial areas. Examine figure 37.4 and describe how the valves might operate.

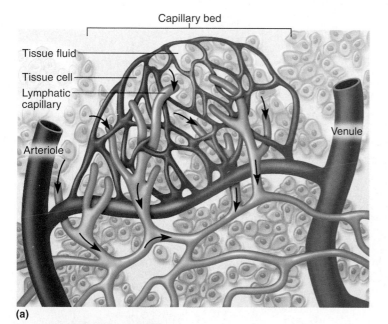

(a)

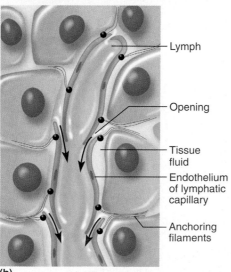

(b)

FIGURE 37.4 Lymphatic Capillary

Labels in figure (a): Capillary bed, Tissue fluid, Tissue cell, Lymphatic capillary, Arteriole, Venule

Labels in figure (b): Lymph, Opening, Tissue fluid, Endothelium of lymphatic capillary, Anchoring filaments

Valve operation: _____

Lymph Node

Look at a slide of a **lymph node.** The lymph node is enclosed by a sheath of tissue called the capsule. Locate the dark purple **lymphatic nodules** in the lymph node. The outer portion of the lymph node is called the **cortex** and the inner part is the **medulla.** The cortex is the outer region of the node that has numerous lymphatic nodules. In the medulla, medullary cords and sinuses cleanse the lymph of foreign particles and cellular debris. **Afferent lymphatic vessels** take lymph to the node, and **efferent lymphatic vessels** take lymph away from the node. Compare your slide to figure 37.5.

Lymphatic Vessels

Use figure 37.6 to examine the drainage pattern of lymph in the body. Compare this illustration to models or charts in lab. The **lymphatic vessels** lead to regions of the body where lymph nodes are found. Nodes are clustered in the groin (inguinal region), axilla, antecubital fossa, popliteal region, neck, thorax, and abdomen. The **thoracic duct** drains most of the body, taking lymph to the left subclavian vein, where the fluid is returned to the cardiovascular system. The **right lymphatic duct** drains the right side of the head and neck, the right thoracic region, and the right upper extremity. The right lymphatic duct returns lymph to the right subclavian vein. Identify the **cisterna chyli,** an enlarged portion of the thoracic duct in the abdominal region.

Examine a prepared slide of a lymphatic vessel and note the presence of the **valve.** Valves prevent the backflow of lymph away from the vascular system into which they drain. Compare your slide to figure 37.7.

Lymph Organs

Tonsils Tonsils are lymph organs in the oral cavity and nasopharynx. Look at illustrations in your text and models in the lab and locate the three pairs of tonsils. These are the **palatine tonsils** along the sides of the oral cavity near the oropharynx, the **lingual tonsils** at the posterior portion of the tongue, and the **pharyngeal tonsils** in the nasopharynx. Adenoids are enlarged pharyngeal tonsils. The tonsils have crypts lined with **lymphatic nodules** that serve as a first line of defense against foreign matter that is inhaled or swallowed. Compare these to figure 37.8.

Spleen The **spleen** is an important lymph organ that removes aging erythrocytes and foreign particles from the blood. The spleen is located on the left side of the body adjacent to the stomach. It is a highly vascular organ that cleanses the blood and produces lymphocytes. The spleen contains **red pulp,** which filters blood, and **white pulp,** which produces lymphocytes. Locate the spleen in models and charts in the lab and compare it to figure 37.9.

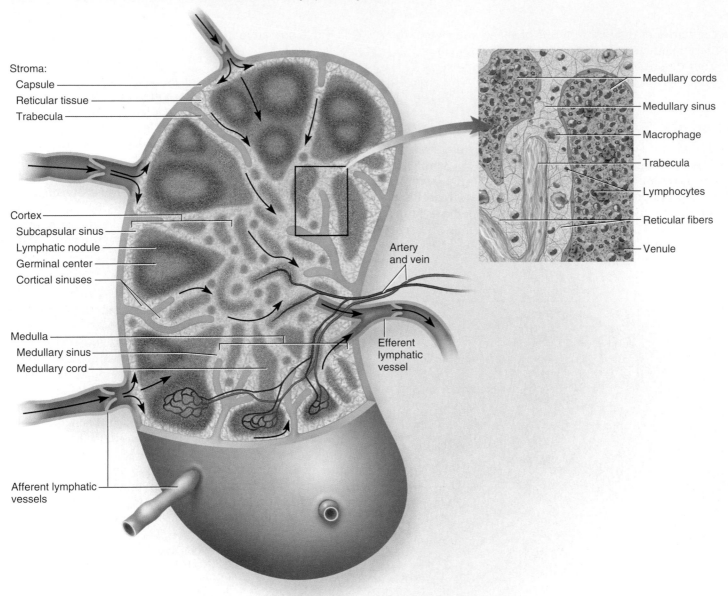

Stroma:
  Capsule
  Reticular tissue
  Trabecula

Cortex
  Subcapsular sinus
  Lymphatic nodule
  Germinal center
  Cortical sinuses

Medulla
Medullary sinus
Medullary cord

Afferent lymphatic
vessels

Artery
and vein

Efferent
lymphatic
vessel

Medullary cords
Medullary sinus
Macrophage
Trabecula
Lymphocytes
Reticular fibers
Venule

(a)

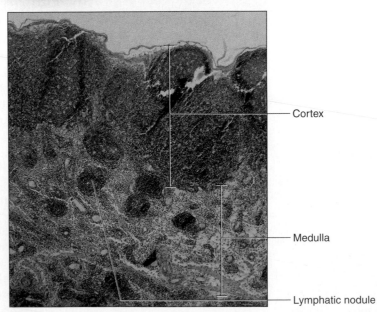

Cortex

Medulla

Lymphatic nodule

(b)

FIGURE 37.5 Section of
a **Lymph Node** (a) Diagram;
(b) photomicrograph (40×).

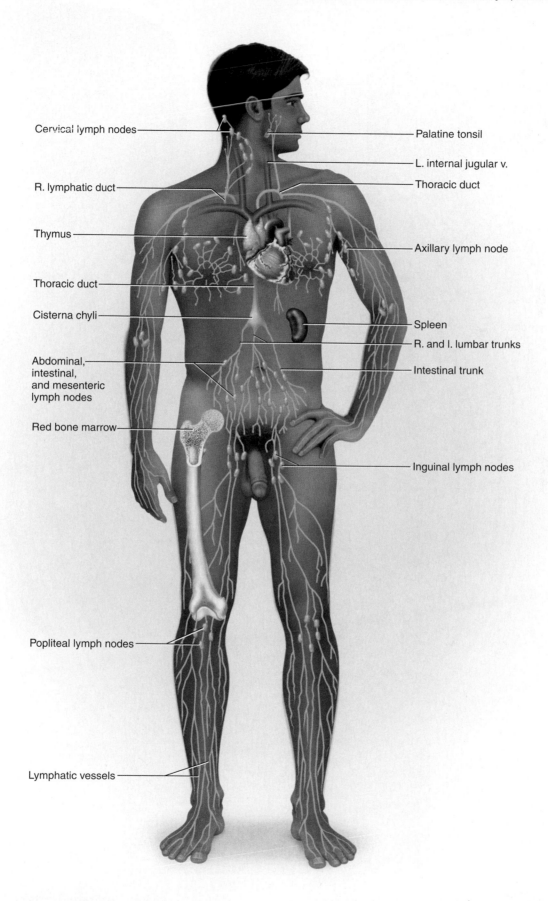

Cervical lymph nodes

Palatine tonsil

L. internal jugular v.

R. lymphatic duct

Thoracic duct

Thymus

Axillary lymph node

Thoracic duct

Cisterna chyli

Spleen

R. and l. lumbar trunks

Abdominal, intestinal, and mesenteric lymph nodes

Intestinal trunk

Red bone marrow

Inguinal lymph nodes

Popliteal lymph nodes

Lymphatic vessels

FIGURE 37.6 Lymphatic Vessels of the Body

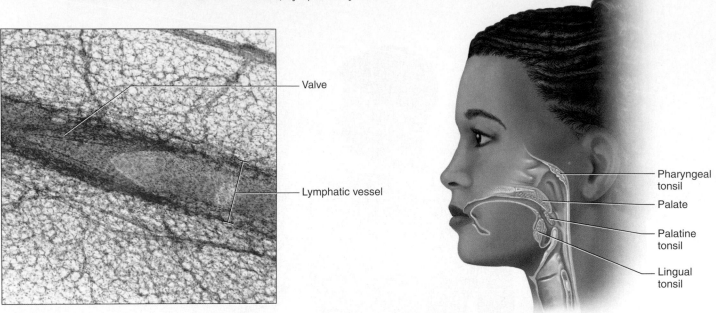

FIGURE 37.7 Lymphatic Vessel with Valve

FIGURE 37.8 Tonsils

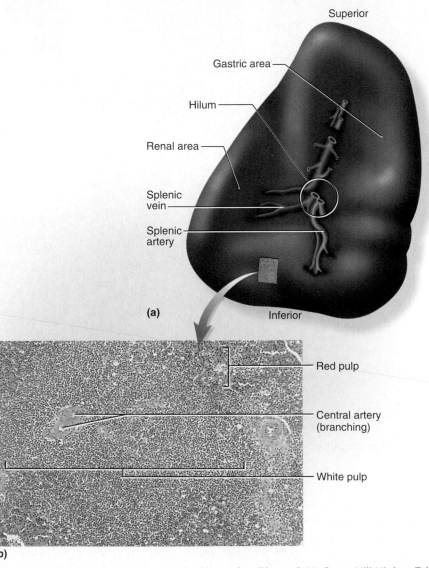

FIGURE 37.9 Spleen (a) Medial surface; (b) histology showing red and white pulp. (Photo © McGraw-Hill Higher Education, Inc/ Dennis Strete, photographer)

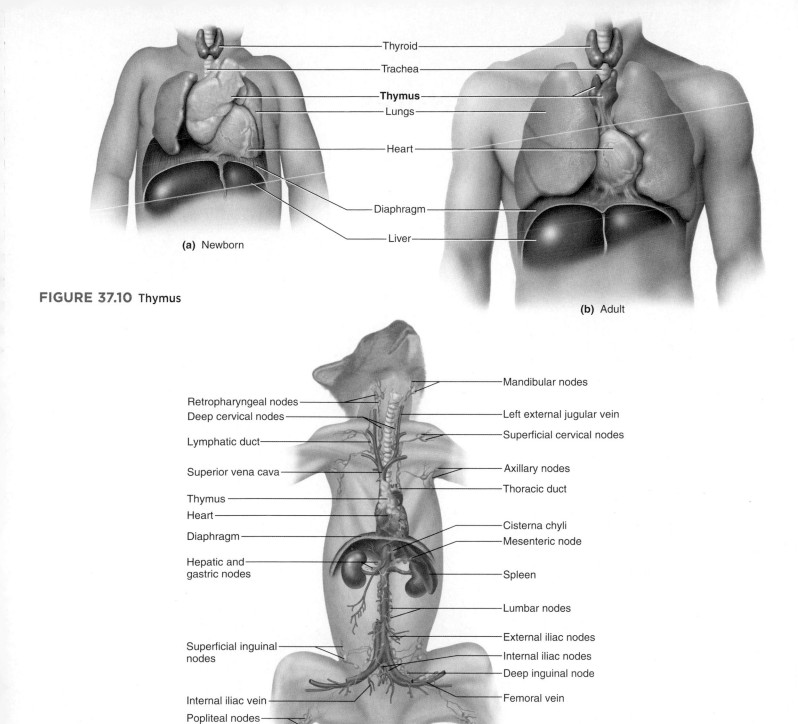

FIGURE 37.10 Thymus

(a) Newborn

Thyroid
Trachea
Thymus
Lungs
Heart
Diaphragm
Liver

(b) Adult

Retropharyngeal nodes
Deep cervical nodes
Lymphatic duct
Superior vena cava
Thymus
Heart
Diaphragm
Hepatic and gastric nodes
Superficial inguinal nodes
Internal iliac vein
Popliteal nodes

Mandibular nodes
Left external jugular vein
Superficial cervical nodes
Axillary nodes
Thoracic duct
Cisterna chyli
Mesenteric node
Spleen
Lumbar nodes
External iliac nodes
Internal iliac nodes
Deep inguinal node
Femoral vein

FIGURE 37.11 Lymphatic System of the Cat

The appendix is another lymph gland consisting of lymphoid tissue. It destroys intestinal bacteria and produces lymphocytes.

Thymus The **thymus** overlays the vessels superior to the heart. It is an important site in determining immune competence. Lymphocytes travel from the bone marrow to the thymus, where they undergo maturation essential to immune responses. Once these cells mature they are known as **T cells.** Examine the charts and models in the lab and compare them to figure 37.10.

Cat Dissection

Locate the **spleen** in the cat, which is an elongated organ near the stomach on the left side of the abdominal cavity. It is a brown organ, and you should find the splenic artery and splenic vein associated with the organ.

You should also look for the **thymus,** which is anterior to the heart. It may be small in older cats or may have been removed in the study of the heart.

There are many locations where you can find **lymph nodes,** and these are illustrated in figure 37.11. Use a forceps or a probe and tease away loose connective tissue to locate the nodes. They should appear as small lumps of beige tissue.

Notes

REVIEW SECTION

Functions of Vessels, Lymphatic System

Name _____ Date _____

Lab Section _____ Time _____

Review Questions

1. What type of cell makes up the endothelium of capillaries?

2. What is the name of the vessels that carry lymph from the lymphatic capillaries to the veins?

3. What is the name of the inner region of a lymph node?

4. Once tissue fluid enters the lymphatic vessels, what is it called?

5. What kind of vessel takes lymph away from a lymph node?

6. The adenoids are enlarged _____ tonsils.

7. Which tonsils are found on the sides of the oral cavity?

8. What tonsils are located at the back of the tongue?

9. Blood is recycled by which lymph organ in the adult?

10. What part of the spleen is involved in producing lymphocytes?

11. Where do T cells mature?

12. From what you know of the functions of lymph nodes, make a prediction of the difference between lymph entering a node and lymph leaving a node. What materials may be missing from the lymph leaving the node?

13. Elephantiasis is a disease that may be caused, in some cases, by a parasitic worm blocking the lymphatic vessels. Examine the following illustration and predict where the lymphatic blockages occur.

14. In the analysis of breast cancer, lymph nodes of the axillary region are removed and a biopsy is performed. The removal of the nodes is done to determine if cancer has spread from the breast to other regions of the body. What effect would the removal of lymph nodes have on the drainage of the pectoral region?

15. Superficial veins contain valves. Inactive people may have problems with their veins in that blood pools in the veins. Can you propose a mechanism by which blood from the veins may be returned to the heart (other than standing on your head!)?

16. Lymphatic vessels have a one-way flow from the extremities to the heart. Damage to the lymphatic system can lead to edema, an increase in tissue fluid. From the standpoint of reducing edema, how does the use of medical leeches (segmented worms that drain tissue fluid) work for a region that has suffered trauma?

LABORATORY

Blood Vessels and Blood Pressure

INTRODUCTION

Maintenance of blood pressure is important for the health of the heart and for proper functioning of various organs. High blood pressure increases the workload of the heart by increasing the force with which the heart must pump to provide blood to the body. High blood pressure can harm the arteries, causing fatty deposits in the walls of arteries, a condition called **atherosclerosis** (ATH-er-oh-Skler-OH-sis), which narrows the arteries and reduces blood flow to organs, such as the brain, heart, and kidneys. These deposits can harden, causing arteriosclerosis, or hardening of the arteries.

Many things influence blood pressure. As people age, their arteries typically become less elastic and blood pressure increases. Kidney disease and other illnesses can cause hypertension. Poor diets, such as those rich in fats or salt; lack of exercise; and elevated caffeine, nicotine, or alcohol intake can also increase blood pressure, as can obesity. Men generally have higher blood pressure until about the age of 55 and then women show an increase in hypertension after that age.

Hypotension can also be dangerous. Low blood pressure may occur when you stand up quickly. This is known as orthostatic hypotension. Severe dehydration, genetics, and early pregnancy can all lead to low blood pressure. A decrease in pressure may lead to fainting or dizziness.

Historically, blood pressure has been indirectly measured with a **sphygmomanometer** (blood pressure cuff), which measures pressure in millimeters of mercury rising in a glass column. The greater the pressure, the higher the rise of mercury. Mercury-filled sphygmomanometers are still used but aneroid sphygmomanometers (using air) and differential pressure transducers are also used.

Blood pressure is not uniform throughout the body but is influenced by gravity; therefore, the pressure in the arteries of the head and neck is less than the blood pressure from the heart. The pressure in the arteries of the leg is greater than the blood pressure in the heart. Blood pressure is measured in the **brachial artery,** which is at the level of the heart and has the approximate pressure as that leaving the heart. Ausculatory measurement of blood pressure involves listening to sounds as blood passes through the brachial artery. Normally, no sound is heard through a stethoscope as blood passes through the brachial artery. The blood passes smoothly through the vessel, like water through a garden hose. When pressure is applied by pinching off a hose the turbulence creates sound. When pressure is applied to the arm due to the constriction from the blood pressure cuff, the turbulence of the blood passing through the vessel creates sound. The significant difference between the flow of blood in the body versus water in a hose is that the blood pulses through the arteries as the heart contracts during ventricular systole. The sounds made by the flow of blood in a partially constricted artery are known as **Korotkoff sounds.** Completely open arteries or completely closed arteries do not produce Korotkoff sounds.

Hypertension is elevated blood pressure. Normal blood pressure is approximately 120/80 mmHg. Hypertension may be due to a number of factors; however, the majority of hypertensive cases occur for unknown reasons. Hypertension is typically a blood pressure in excess of 140/90 mmHg for young adults, though the greatest concern is in the diastolic reading. The diastolic pressure should be less than 90 mmHg. These topics are covered in the Saladin text in chapter 20, "Blood Vessels and Circulation." In this exercise you learn how to determine blood pressure and some factors affecting blood pressure.

OBJECTIVES

At the end of this exercise you should be able to

1. determine the pulse rate of an individual;
2. define hypertension in terms of millimeters of mercury pressure;
3. distinguish between systolic pressure and diastolic pressure;
4. properly take and record blood pressure;
5. list three factors that affect blood pressure.

MATERIALS

Blood Pressure Experiment

Stethoscope

Alcohol wipes or isopropyl alcohol and sterile cotton swabs

Washable felt pen

Sphygmomanometer

Watch or clock with accuracy in seconds

Flow Experiment

Corn syrup

1 yard of 1/4-inch plastic flexible tubing

1 yard of 1/2-inch plastic flexible tubing

Small funnel to fit small tubing

Large funnel to fit large tubing

Stopwatch

Two ring stands

Beakers

Clamps

Wash basin

Roll of paper towels

PROCEDURE

Measurement of Pulse Rate in Beats per Minute (bpm)

Initially record a baseline pulse rate. The pulse rate is measured in beats per minute (bpm) by locating regions of the body where the pulse can be palpated—the radial artery or carotid artery. If you are recording your lab partner's pulse, make sure you use the tips of your fingers to measure the pulse rate, not your thumb. If you use your thumb you may feel the pulse in your thumb and not the pulse of the person you wish to record. Determine beats per minute by counting the pulse for 1 minute.

Pulse rate of lab partner: _____ bpm

Measurement of Arterial Blood Pressure

1. To determine blood pressure, locate the pulse of the brachial artery on your lab partner. This can be done by placing two fingers on the medial side of the biceps brachii muscle near the antecubital fossa. You can place a small "X" with a felt pen on this location on the arm (see figure 38.1).

2. Place the blood pressure cuff around the arm of your lab partner at the level of the heart. Make sure the inflatable portion of the cuff is on the anterior medial side. Some cuffs have a metal bar that provides a loop through which a part of the cuff goes through. This bar should *not* be located on the medial side of the arm. The reason for this is that, if the bar is located on the medial side of the arm, it may not constrict the brachial artery effectively as the cuff is inflated (see figure 38.2).

3. Clean the ear pieces of the stethoscope with alcohol wipes, place the diaphragm of the stethoscope on the "X" where you located the brachial pulse, and have your lab partner rest his or her forearm on the lab counter. You should *not* hear any sound at this time.

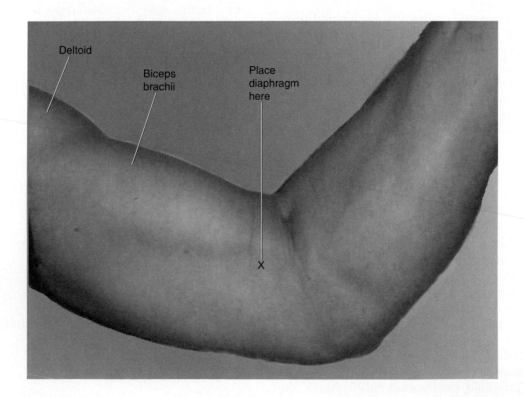

FIGURE 38.1 Location of Stethoscope Diaphragm, Left Arm

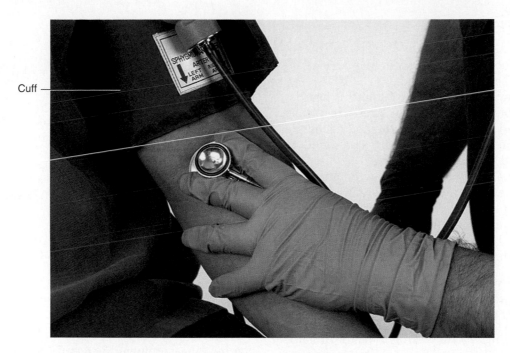

FIGURE 38.2 Proper Placement of Cuff

4. Hold on to the rubber squeeze bulb with the attached rubber tubing leading away from you. Turn the metal dial clockwise until it is closed. You can now begin to inflate the cuff (see figure 38.3).

5. Pump the cuff up to about 80 mmHg. Look at the mercury in the glass tube or the dial on an aneroid instrument. If it seems to bounce up and down a little, then listen closely for the sounds in the stethoscope. (The ear pieces are inserted into the ears, facing forward.) If you do not hear any sound, then inflate the mercury or the dial on an aneroid instrument to 100 or 120 mmHg. If you see the mercury pulsing, listen again for the sound. Make sure the diaphragm of the stethoscope is in the right place.

6. Once you are sure you have heard the sound of the heart, remove the cuff and place it on the other arm. Excessive constriction of the arm may elevate your lab partner's blood pressure.

7. Inflate the cuff on the other arm, but make sure you exceed the level where you see motion in the needle or pulsing in the mercury (usually around 150 mmHg). Do not leave the cuff inflated for a long period.

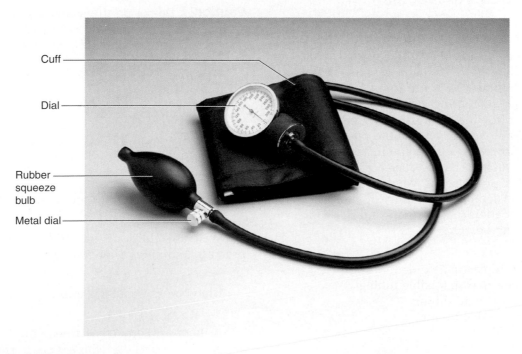

Cuff

Dial

Rubber squeeze bulb

Metal dial

FIGURE 38.3 Parts of a Sphygmomanometer

8. Release the knob slowly, so that the mercury or needle slowly begins to descend.

9. Record the number when the first sound is heard. This sound represents the **systolic pressure** of the ventricles, which is the pressure the heart generates that exceeds the pressure of the cuff. The sound may muffle a bit and then come in strong.

10. Continue to let air out of the cuff slowly until the sound starts to muffle. Listen very carefully until the sound completely disappears. The exact level at which the sound disappears is the **diastolic pressure.**

11. Record the level of blood pressure as the systolic pressure over the diastolic pressure:

$$\frac{\text{Systolic pressure}}{\text{Diastolic pressure}} = \underline{\hspace{3cm}}$$

If you are unsure of exactly where the pressure is, you can fine-tune your readings by increasing the cuff pressure at each approximate end to obtain an accurate reading. Do not subject your lab partner to more than two or three attempts. Ask your instructor for help if you have difficulty determining blood pressure.

Pulse Pressure

You can now determine the pulse pressure of your lab partner. The pulse pressure is determined by subtracting the diastolic pressure from the systolic pressure. It is normally around 50 mmHg.

Pulse pressure: _____

Influence of Blood Vessel Diameter on Resistance

In this part of the lab you do a little investigative work to try to determine the influence that diameter has on the resistance to blood flow in blood vessels. The blood flow in a vessel increases with increasing pressure and decreases with increasing resistance. This can be expressed in the following formula: **Flow = pressure / resistance.** Therefore, if you increase the pressure of the system the rate of flow increases. If you increase the resistance in the blood vessel the flow decreases.

Using the items provided by your instructor, design an experiment that demonstrates how an increase or a decrease in diameter alters the resistance and subsequently the flow rate in blood vessels.

1. corn syrup
2. 1 yard of ¼-inch plastic flexible tubing
3. 1 yard of ½-inch plastic flexible tubing
4. small funnel to fit small tubing
5. large funnel to fit large tubing
6. stopwatch
7. ring stand

8. clamps
9. wash basin
10. beakers

Flow is altered in humans by cardiac output, peripheral resistance, precapillary sphincters, and general vasoconstriction.

Factors Affecting Blood Pressure
Body Position

Measure the blood pressure of your lab partner while he or she lies down. Record this value.

Measurement lying down: _____

Have your lab partner stand suddenly, and quickly take a new measurement. *Be careful not to drop the sphygmomanometer!* Record the results.

Blood pressure on immediately standing: _____

How can you account for these two readings?

Exercise

Have your lab partner run around the building for awhile or do strenuous physical activity for a couple of minutes. Measure the blood pressure immediately after the cessation of exercise and record your results.

Caution! Do not do strenuous exercise if you are not feeling well, have a history of heart trouble, or have been advised by a physician to avoid exercise.

Blood pressure after exercise: _____

Wait for 1 minute and record your blood pressure in the following space.

Blood pressure after 1 minute: _____

Wait an additional minute and record your blood pressure 2 minutes after exercise.

Blood pressure after 2 minutes: _____

Is there a difference between individuals in class who are in shape and those who are out of shape?

REVIEW SECTION

Blood Vessels and Blood Pressure

Name _____ _____ Date _____

Lab Section _____ Time _____

Review Questions

1. The letters bpm stand for what phrase in cardiac measurement?

2. What does a sphygmomanometer measure?

3. To measure blood pressure, what artery would you most commonly use?

4. If you have a blood pressure of 140/80, what does the 80 represent?

5. What is the clinical threshold for high blood pressure in young adults?

6. When the first sound is heard during measurement with a blood pressure cuff, what is measured—systolic or diastolic pressure?

7. Emotions have an effect on blood pressure. Predict the blood pressure of an individual who recently had a heated argument with a roommate about rent money.

8. Illness can also affect blood pressure. Illness tends to increase stress responses. Predict the blood pressure of an individual with a sinus headache and postnasal drip.

9. Nicotine and caffeine both elevate blood pressure. Explain how an increase in blood pressure could have a negative effect on the cardiac output.

10. Record your blood pressure.

11. How did the change in blood pressure after exercise vary from those people who were in shape versus those who were out of shape?

12. According to the potential risk factors for hypertension, which, if any, do you have?

LABORATORY

Structure of the Respiratory System

INTRODUCTION

The respiratory system provides oxygen to the cells of the body and removes carbon dioxide. Cells use oxygen as the terminal electron acceptor, and carbon dioxide is a waste product of cellular respiration.

Atmospheric oxygen moves into the lungs and diffuses into the circulatory system. This vital exchange occurs due to the respiratory system, discussed in the Saladin text in chapter 22, "The Respiratory System." The oxygen subsequently reaches the individual cells of the body while the metabolic waste product, carbon dioxide, is released from the intercellular environment and travels via the blood to the lungs, where it is released by exhalation. Too much carbon dioxide in the blood increases the hydrogen ion concentration, producing acidosis of the blood. The respiratory system is therefore integrated with the other body systems. As you study the anatomy of the respiratory system in this exercise, be aware of the role that other systems play in respiration.

OBJECTIVES

At the end of this exercise you should be able to

1. list the organs and significant structures of the respiratory system;
2. define the role of the respiratory system in terms of the overall function of the body;
3. explain the physical reason for the tremendous surface area of the lungs;
4. identify the cartilages of the larynx;
5. distinguish among a bronchus, bronchiole, and respiratory bronchiole.

MATERIALS

Lung models or detailed torso model, including midsagittal section of head

Model of larynx

Microscopes

Prepared microscope slides of lung tissue

Prepared microscope slide of "smoker's lung"

Charts and illustrations of the respiratory system

Cats

Materials for cat dissection

    Dissection trays

    Scalpel and two or three extra blades

    Protective gloves

    Blunt (mall) probe

    Forceps and sharp scissors

    First aid kit in lab or prep area

    Sharps container

    Animal waste disposal container

PROCEDURE

Overview of the Respiratory System

Look at the charts and models of the respiratory system for a general orientation and compare them to figure 39.1. Locate the following structures:

    nose

    nasal cavity

    pharynx

    larynx

    trachea

    bronchus

    left lung

    right lung

Nose and Nasal Cartilages

Examine a midsagittal section of a model or chart of the head and look for the **nose, nasal cartilages, external nares** (nostril), and **nasal septum.** The nasal septum separates the nasal cavities and is composed of the perpendicular plate of the ethmoid bone, the vomer, and the nasal cartilage. Examine these features in figures 39.2 and 39.3.

The entrance of the external nares is protected by guard hairs. The region of the nose, just posterior to the

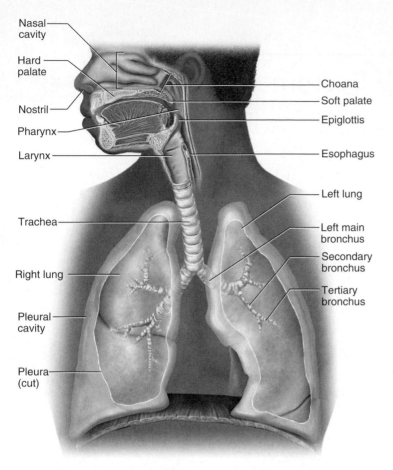

FIGURE 39.1 Overview of the Anatomy of the Respiratory System

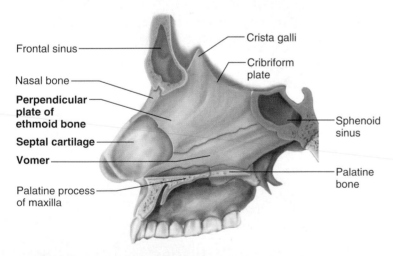

FIGURE 39.2 Structures That Make Up the Nasal Septum

external nares, is the **nasal vestibule,** lined with stratified squamous epithelium. Behind the vestibule is the nasal cavity lined with a **mucous membrane** that moistens air entering the respiratory system. The membrane consists of **respiratory epithelium,** composed of **pseudostratified ciliated columnar epithelium** with **goblet cells.** This mucous membrane overlays a superficial venous plexus that warms the air. The lateral walls of the cavity have three

protrusions that push into the nasal cavity. These are the **nasal turbinates** (TUR-bih-nates). The turbinates consist of the nasal conchae and the mucous membrane covering. They cause the air to swirl in the nasal cavity and come into contact with the mucous membrane. Locate the **superior, middle,** and **inferior turbinates** in the nasal cavity. The nasal cavity ends where two openings, the **internal (posterior) nasal apertures (nares),** or **choanae** (co-AH-nee), lead to the **pharynx** (FAIR-inks). These are seen in figure 39.3.

Pharynx

The pharynx can be divided into three regions based on location. The uppermost area is the **nasopharynx** (NAZE-oh-FAIR-inks), directly posterior to the nasal cavity. The nasopharynx has two openings on the lateral walls, which are the openings of the **auditory,** or **eustachian, tubes.** As the pharynx descends behind the oral cavity it becomes the **oropharynx.** The oropharynx is a common passageway for food, liquid, and air. The **uvula** is a small, pendulous structure that partially separates the oral cavity from the oropharynx. The uvula flips upwards during swallowing, helping prevent fluids from entering the nasopharynx. The most inferior portion of the pharynx is the **laryngopharynx** (la-RING-go-FAIR-inks), located posterior to the larynx. Find the regions of the pharynx in figure 39.3.

Larynx

The **larynx** (LAIR-inks) is commonly known as the "voice box" because it is an important organ for sound production in humans. It controls the pitch of the voice, while the shape of the oral cavity and the placement and size of the paranasal sinuses are responsible for the sonority of the voice. The larynx occurs at about the level of the fourth through sixth cervical vertebrae and consists of a number of cartilages. The most prominent cartilage in the larynx is the **thyroid cartilage,** a shield-shaped structure made of hyaline cartilage. Examine models and charts in the lab and compare the structures of the larynx to figure 39.4.

The thyroid cartilage is more prominent in males because it increases in size under the influence of testosterone. Inferior to the thyroid cartilage is the **cricoid** (CRY-coyd) **cartilage.** The cricoid cartilage is also composed of hyaline cartilage, and it is relatively narrow when seen from the anterior but increases in size at its posterior surface. Superior to the cricoid cartilage in the posterior wall of the larynx are the paired **arytenoid** (AR-ih-TEE-noyd) **cartilages.** These cartilages attach to the posterior end of the **true vocal folds (vocal cords).** Movement of the arytenoid cartilages pulls on the true vocal folds, causing them to stretch and increasing the pitch of the voice. This occurs by the contraction of intrinsic muscles attached to the arytenoid cartilages from the back, while the vocal cords are held stationary by the thyroid cartilage in the front. Above the true vocal folds are the **vestibular folds (false vocal folds)** (see figure 39.4).

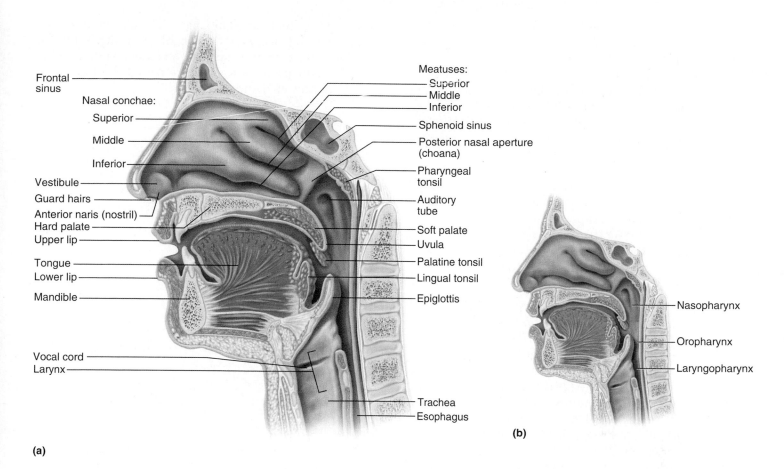

Frontal sinus

Nasal conchae:
Superior
Middle
Inferior

Vestibule
Guard hairs
Anterior naris (nostril)
Hard palate
Upper lip
Tongue
Lower lip
Mandible

Vocal cord
Larynx

Meatuses:
Superior
Middle
Inferior

Sphenoid sinus
Posterior nasal aperture (choana)
Pharyngeal tonsil
Auditory tube
Soft palate
Uvula
Palatine tonsil
Lingual tonsil
Epiglottis

Trachea
Esophagus

Nasopharynx
Oropharynx
Laryngopharynx

(a)

(b)

FIGURE 39.3 Upper Respiratory Tract (a) Midsagittal section of the head with nasal septum removed; (b) three regions of the pharynx.

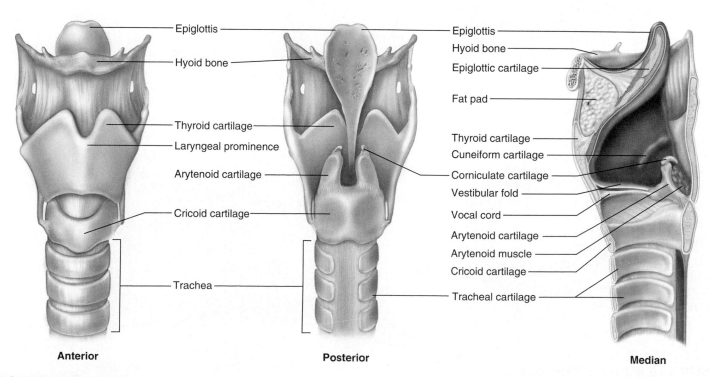

Epiglottis
Hyoid bone
Thyroid cartilage
Laryngeal prominence
Arytenoid cartilage
Cricoid cartilage
Trachea

Epiglottis
Hyoid bone
Epiglottic cartilage
Fat pad
Thyroid cartilage
Cuneiform cartilage
Corniculate cartilage
Vestibular fold
Vocal cord
Arytenoid cartilage
Arytenoid muscle
Cricoid cartilage
Tracheal cartilage

Anterior   **Posterior**   **Median**

FIGURE 39.4 Larynx

At the very posterior, superior edge of the larynx are the **corniculate** and **cuneiform** (cue-NEE-ih-form) **cartilages.** These are also made of hyaline cartilage. The last cartilage of the larynx is the **epiglottis**, composed of elastic cartilage. During swallowing the epiglottis protects the opening of the larynx, known as the **glottis** (see figure 39.3). In the swallowing reflex, muscles pull the epiglottis down over the glottis. This is not a perfect system, as anyone knows who has started swallowing a liquid and responded by laughing at a joke. Inhalation at the beginning of the laugh causes fluid to move into the larynx and trachea, irritating the respiratory lining. This causes another reflex called the **cough reflex,** which propels the liquid out of the respiratory system. Locate the structures of the larynx on models or charts in the lab and in figure 39.4.

Trachea and Bronchi

The **trachea** (TRAY-kee-uh) is commonly known as the "windpipe" because it conducts air from the larynx to the lungs. The trachea is a straight tube whose lumen is kept open by **tracheal** (C-shaped) **cartilages.** Examine these cartilages by running your fingers gently down the outside of your throat. Palpate the cartilage rings below the larynx. The tracheal cartilages are composed of hyaline cartilage. At the most inferior portion of the trachea is a center point known as the **carina** (ca-RY-na) (keel). Locate the features of the trachea in figures 39.5 and 39.6.

The trachea is also lined with respiratory epithelium. Obtain a prepared slide of the trachea and find the tracheal cartilage, respiratory epithelium, and **posterior tracheal membrane** (with the trachealis muscle) (see figure 39.6).

The trachea splits into two tubes, which enter the lungs. These tubes are the **main bronchi** (BRONK-eye). Each lung receives air from a main bronchus. These main bronchi contain hyaline cartilage and are lined with respiratory epithelium. The main bronchi of the lung divide into the **secondary bronchi,** and these further divide to form tertiary bronchi. The extensive branching of the bronchi produce a structure called the **bronchial tree** (see figures 39.5 and 39.7).

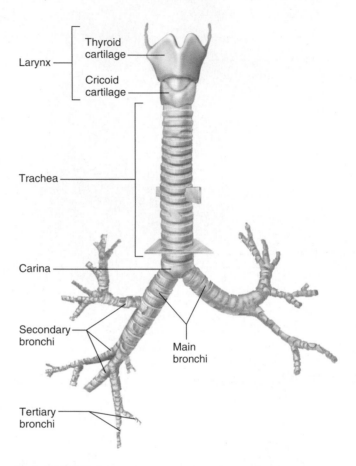

FIGURE 39.5 Larynx, Trachea, and Bronchi, Anterior View

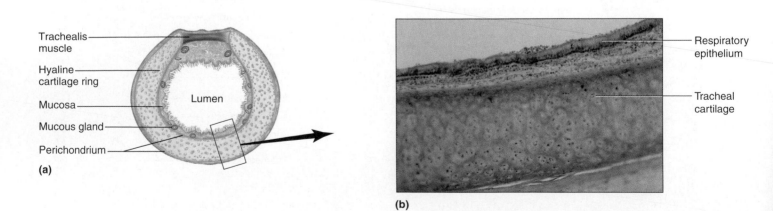

FIGURE 39.6 Trachea, Cross Section (a) Diagram; (b) photomicrograph (100×).

Trachea

Main bronchus

Secondary bronchus

Tertiary bronchus

FIGURE 39.7 Cast of Bronchial Tree

Lungs

Of the two lungs in humans, the right lung has three lobes: a **superior lobe,** a **middle lobe,** and an **inferior lobe.** The left lung has two lobes: a **superior lobe** and an **inferior lobe.** The left lung also has an indentation, called the **cardiac notch,** where the apex of the heart rests. Look at models or charts in the lab and identify the major features, as shown in figure 39.8.

The lungs are located in the **pleural cavities** on each side of the **mediastinum.** The **parietal pleura** is the outer membrane on the chest cavity wall, and the membrane that adheres to the surface of the lungs is the **visceral pleura.** The space between the membranes is known as the **pleural cavity.** These are shown in figure 39.9.

Histology of the Lung

The bronchi continue to divide until they become **bronchioles,** small respiratory tubules with smooth muscle in their walls and an inner lining of respiratory epithelium. Obtain a prepared slide of lung and scan it first under low power and then under higher powers. The

bronchioles in the lung further divide into **respiratory bronchioles,** so named due to small structures called alveoli attached to their walls. The respiratory bronchioles lead to passageways known as **alveolar ducts,** which branch into **alveoli.** Alveoli are air sacs in the lung that exchange oxygen and carbon dioxide with the blood capillaries of the lungs. From the bronchioles to the alveoli, the respiratory epithelium progressively decreases in height, eventually becoming simple squamous epithelium. Examine your slide and locate the bronchi, bronchioles, respiratory bronchioles, alveolar ducts, and alveoli. Alveoli clustered around an alveolar duct are collectively known as an **alveolar sac** (see figure 39.10).

The alveoli are composed of simple squamous epithelium **(type I alveolar cells)** and are in close proximity to the vascular endothelium of the capillaries surrounding the alveoli. The division of the lung into numerous alveoli tremendously increases the surface area of the lung. This increase is vital for the rapid and extensive *diffusion* of oxygen across the respiratory membranes.

Other cell shapes that occur in the prepared lung sections are **type II alveolar cells** (septal cells), and they decrease the surface tension of the lung by secreting surfactant. The lung also contains **alveolar macrophages,** which phagocytize dust particles entering the lungs.

Examine a prepared slide of smoker's lung. Note the dark material in the lung tissue and the general destruction of the alveoli. Breakdown of the alveoli leads to a disease known as **emphysema.**

Cat Dissection

Prepare your cat for dissection and remember to place all excess tissue in the appropriate waste container and not in a standard wastebasket or down the sink!

1. If you have not opened the chest cavity in your study of the cat, then you should do so now. Removal of the skin is discussed in Exercise 13, and the procedure for opening the thoracic cavity is described in Exercise 34.
2. Locate the **larynx** of the cat above the **trachea.** Notice the broad, wedge-shaped structure in the front. This is the **thyroid cartilage.**
3. Make a midsagittal incision through the thyroid cartilage and continue carefully cutting until you have cut completely through the larynx. To examine the larynx more completely you may wish to continue your midsagittal cut partway through the trachea.
4. Open the larynx and find the **epiglottis** of the cat. Notice how the elastic cartilage is lighter in color than that of the thyroid cartilage. This is because the epiglottis is made of elastic cartilage, while the thyroid cartilage is composed of hyaline cartilage.
5. Find the **cricoid cartilage,** the **arytenoid cartilages,** and the **true** and **false vocal folds** (see figure 39.11).

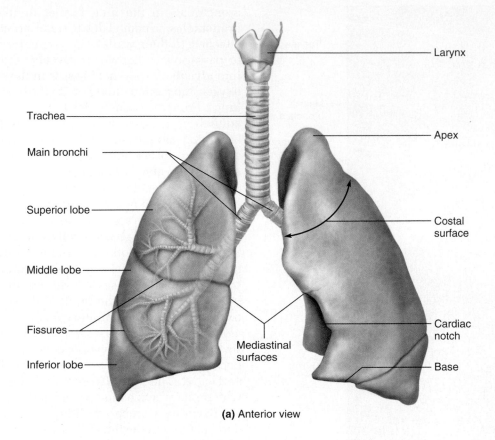

Larynx

Trachea

Main bronchi

Superior lobe

Middle lobe

Fissures

Inferior lobe

Apex

Costal surface

Cardiac notch

Mediastinal surfaces

Base

(a) Anterior view

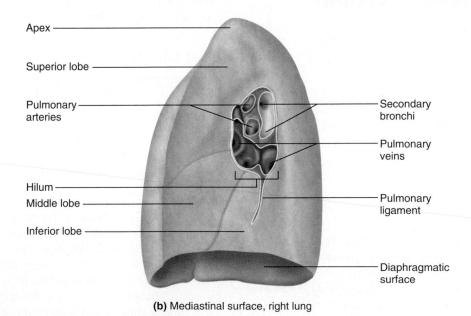

Apex

Superior lobe

Pulmonary arteries

Hilum

Middle lobe

Inferior lobe

Secondary bronchi

Pulmonary veins

Pulmonary ligament

Diaphragmatic surface

(b) Mediastinal surface, right lung

FIGURE 39.8 Lungs (a) Right lung showing bronchial tree, left lung surface view; (b) medial view.

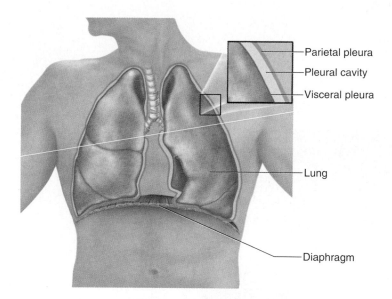

FIGURE 39.9 Pleural Membranes and Cavity

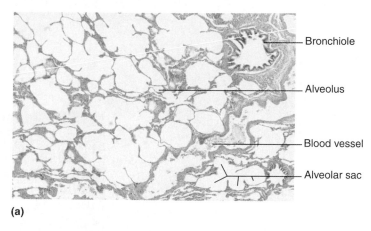

(a)

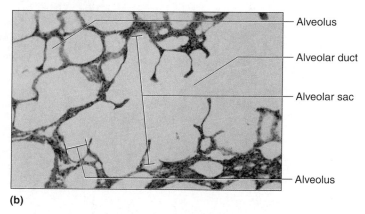

(b)

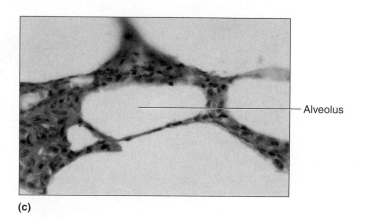

(c)

FIGURE 39.10 Histology of the Lung (a) Photomicrograph of lung (40×); (b) photomicrograph of alveolar duct (100×); (c) photomicrograph of alveolus (400×).

6. Examine the trachea as it passes from the larynx and into the thoracic cavity. Ask your instructor for permission before you cut the trachea in cross section. If you do, you should see the **tracheal cartilages** and the **posterior tracheal membrane.** The posterior portion of the trachea is located ventrally to the esophagus. Notice how the trachea splits into the two **bronchi,** which then enter the lungs. The lobes of the lungs are different in the cat than in the human. The right lung in the cat has four lobes, while the left lung has three lobes. How does this compare to the pattern in humans?

7. Examine the structure of the lungs in your specimen and compare it to figure 39.12. The lungs are covered with a thin serous membrane called the visceral pleura, while the outer covering (on the deep surface of the ribs and intercostal muscles) is called the parietal pleura. The space between these two membranes is the pleural cavity.

8. Cut into one of the lungs of the cat and examine the lung tissue. Note how the lungs appear like a very fine mesh sponge. The alveoli of the lungs are microscopic.

9. At the inferior portion of the thoracic cavity is the diaphragm. As it contracts the pressure in the thoracic cavity decreases and air fills the lungs. Examine the diaphragm in your specimen and compare it to figure 39.12.

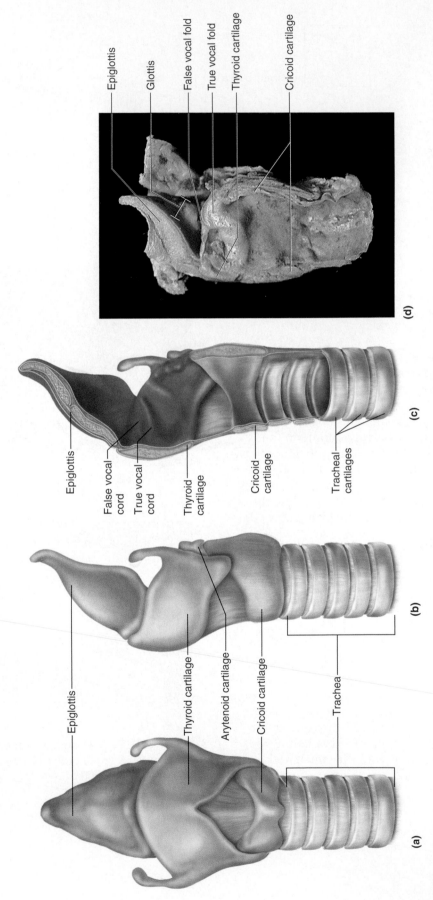

FIGURE 39.11 Larynx of the Cat Diagram (a) anterior view; (b) left lateral view; (c) midsagittal view; (d) photograph, midsagittal view.

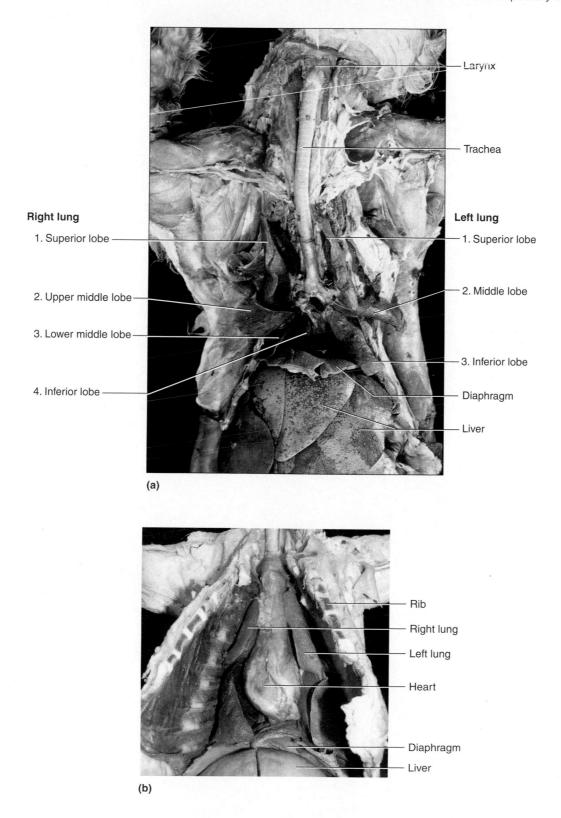

Right lung

1. Superior lobe

2. Upper middle lobe

3. Lower middle lobe

4. Inferior lobe

Larynx

Trachea

Left lung

1. Superior lobe

2. Middle lobe

3. Inferior lobe

Diaphragm

Liver

(a)

Rib

Right lung

Left lung

Heart

Diaphragm

Liver

(b)

FIGURE 39.12 Respiratory Structures of the Cat (a) Larynx, trachea, and lungs; (b) lungs.

Notes

REVIEW SECTION

Structure of the Respiratory System

Name _____ Date _____

Lab Section _____ Time _____

Review Questions

1. What is the common name for the external nares?

2. Some of the nasal cartilages are made of hyaline cartilage. What functional adaptation does cartilage have over bone in making up the external framework of the nose?

3. The nasal cavities are separated from each other by what structure?

4. In lung cancer there are frequently tumors in the lymphatic tissue. These may reduce the flow in the pulmonary arteries. What impact would this have on the respiratory system?

5. What is the function of respiratory epithelium and the superficial blood vessels in the nasal cavity?

6. Name the openings between the nasal cavity and the pharynx.

7. What is the name of the space behind the oral cavity and above the laryngopharynx?

8. What is the name of the large cartilage of the anterior larynx?

9. What membrane attaches directly to the lungs?

10. The trachea branches into two tubes that go to the lungs. What are these tubes called?

11. Where is the bronchial tree found?

12. What small structure in the lung is the site of exchange of oxygen with the blood capillaries?

13. The surface area of the lungs in humans is about 70 square meters. How can this be so if the lungs are located in the small space of the thoracic cavity? What role do alveoli play in the nature of surface area?

14. Emphysema is a destruction of the alveoli of the lungs. What effect does this have on the surface area of the lungs?

15. Fill in the following illustration of the human respiratory system using the terms provided.

main bronchus   internal nares
nasal cavity   middle lobe
epiglottis   trachea
superior lobe   inferior lobe
cricoid cartilage

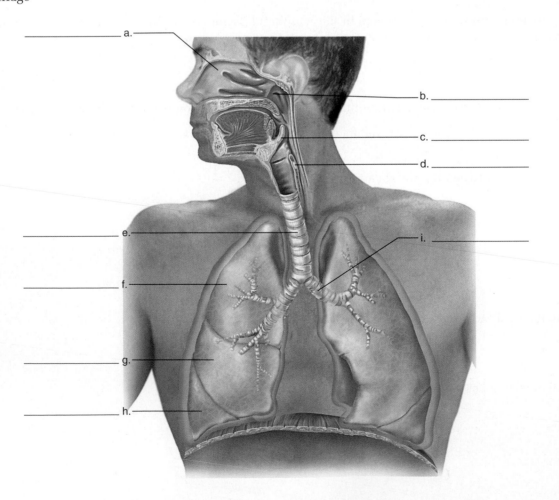

LABORATORY

Respiratory Function, Breathing, Respiration

INTRODUCTION

The function of the respiratory system is to exchange gases between the external air and the blood. Oxygen diffuses from the air into the blood of the lungs, and carbon dioxide diffuses from the blood into the air. The process can be divided into **pulmonary ventilation,** the mechanical process of moving air into and out of the lungs, and the **exchange of gases** across the respiratory membrane. Oxygen requirements vary with the body's need. Variations in the loading of oxygen by the lungs can be accomplished by increasing/decreasing the volume of air with each inhalation, the breathing rate, or both. Removal of carbon dioxide can also be accomplished by an increase in volume or breathing rate. This is important in maintaining an appropriate acid-base balance of the blood. These topics are discussed in the Saladin text in chapter 22, "The Respiratory System." In this exercise you measure breathing rates, lung volumes and capacities, and the effects of carbon dioxide on the acid-base balance of a solution.

OBJECTIVES

At the end of this exercise you should be able to

1. measure and understand the relationship of pulmonary volumes and capacities, including vital capacity, tidal volume, inspiratory reserve volume, and expiratory reserve volume;
2. describe the relationship between oxygen and carbon dioxide levels in the blood and breathing rates;
3. describe the mechanical process of breathing;
4. describe the use of the spirometer or airflow transducer available in your lab;
5. demonstrate the use of the stethoscope in obtaining respiratory sounds;
6. predict the flow of air in an individual, with changes in air pressure between the lungs and the external air or with changes in the resistance in the respiratory passages;
7. identify the tidal volume, expiratory reserve volume, and vital capacity on a spirogram.

MATERIALS

Respiration Model

Bell jar respiration model

Pulmonary Volume Setup

BIOPAC air transducer, wet spirometer, or handheld spirometers

Disposable mouthpieces to fit respirometer or spirometers

Watch or clock with accuracy in seconds

Biohazard bag

AFT6—600 mL calibration syringe

AFT1—disposable bacteriological filter

AFT2 disposable mouthpiece

SS11LA airflow transducer

Breathing Sounds and Breathing Rate Setup

Stethoscope

Alcohol wipes

Acid-Base Setup (One Setup per Table)

Litmus solution (2 g litmus powder in 600 mL water)

NaOH solution (1 N) in dropper bottles

Straws

100 mL Erlenmeyer flasks

Safety glasses

PROCEDURE

Mechanics of Breathing

Air moves from regions of higher pressure to lower pressure. The lungs fill with air or deflate due to changes in air pressure. Normal atmospheric air pressure is measured in millimeters of mercury (mmHg) and standard air pressure at sea level is 760 mmHg. When there is no movement of air into or out of the lungs, the pressure in the lungs is equal to that of the surrounding air, as illustrated in figure 40.1(1). During inspiration, the diaphragm

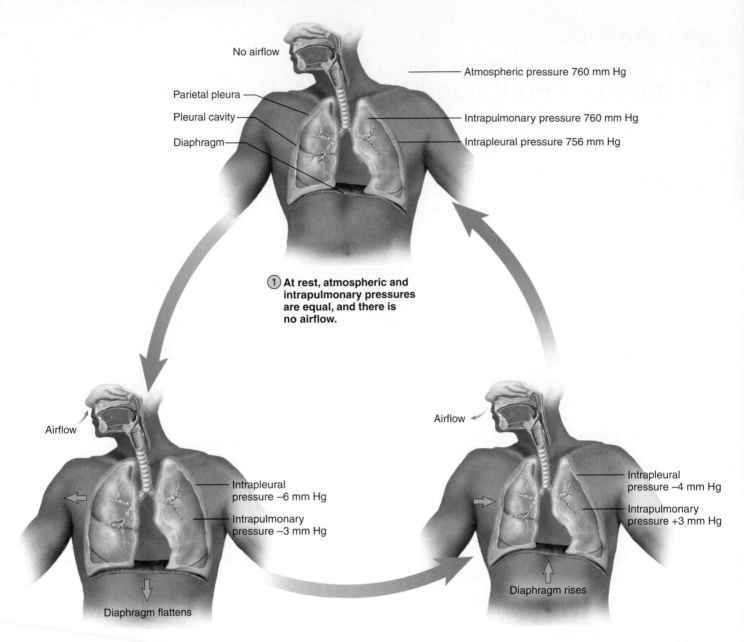

No airflow

Parietal pleura

Pleural cavity

Diaphragm

Atmospheric pressure 760 mm Hg

Intrapulmonary pressure 760 mm Hg

Intrapleural pressure 756 mm Hg

(1) **At rest, atmospheric and intrapulmonary pressures are equal, and there is no airflow.**

Airflow

Intrapleural pressure –6 mm Hg

Intrapulmonary pressure –3 mm Hg

Diaphragm flattens

Airflow

Intrapleural pressure –4 mm Hg

Intrapulmonary pressure +3 mm Hg

Diaphragm rises

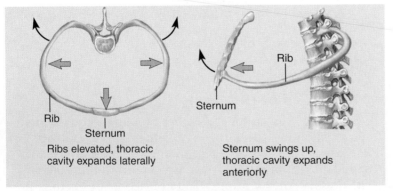

Rib

Sternum

Rib

Sternum

Ribs elevated, thoracic cavity expands laterally

Sternum swings up, thoracic cavity expands anteriorly

(2) **In inspiration, the thoracic cavity expands laterally, vertically, and anteroposteriorly; intrapulmonary pressure drops 3 mm Hg below atmospheric pressure, and air flows into the lungs.**

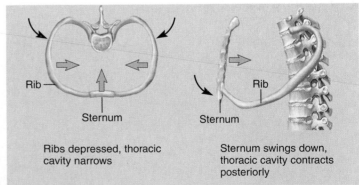

Rib

Sternum

Rib

Sternum

Ribs depressed, thoracic cavity narrows

Sternum swings down, thoracic cavity contracts posteriorly

(3) **In expiration, the thoracic cavity contracts in all three directions; intrapulmonary pressure rises 3 mm Hg above atmospheric pressure, and air flows out of the lungs.**

FIGURE 40.1 Mechanics of Breathing

contracts, increasing the volume in the thoracic cavity. This leads to a decrease in the pressure in the lungs and air moves from the atmosphere into the lungs. This is illustrated in figure 40.1(2). If the diaphragm relaxes, then the abdominal pressure forces the diaphragm upwards and increases the pressure in the thoracic cavity beyond that of the atmospheric pressure and the air moves out of the lungs, as illustrated in figure 40.1(3). You can use a bell jar model to demonstrate this in lab.

This model illustrates the movement of air into or out of the lungs, depending on air pressure differentials between the lungs and the external environment. The glass housing represents the thoracic cage; the balloons represent the lungs; the latex sheeting represents the diaphragm; and the rubber stopper represents the nose. Pull *gently* on the latex diaphragm. As you do this the volume in the pleural cavity (the space between the balloons and the glass jar) increases, causing a decrease in the pressure of the pleural cavity.

When the pressure in the pleural cavity decreases, air in the external environment is at a higher pressure than that in the jar and moves into the bell jar, filling the balloon lungs. As you release the diaphragm, the volume decreases and the pressure increases in the pleural cavity bell jar, exceeding the external air pressure. Air moves out of the lungs and through the nose (rubber stopper). Although the mechanics of human breathing differ somewhat from this model, the bell jar model is valuable to demonstrate the basic principle of air moving from a region of high pressure to a region of low pressure.

Measurement of Relaxed Breathing Rate

It is important that your lab partner be distracted from thinking about breathing.

1. Therefore, your lab partner should read the remainder of the laboratory exercise while you count the number of breaths he or she takes for a total of 2 minutes.
2. Divide the number by 2 to calculate the average number of breaths per minute.
3. Record that number for your lab partner in the space provided and in the Chapter Summary Data section at the end of the exercise.

? Breaths per minute: _____ 1

Breathing rate varies with oxygen demand and/or carbon dioxide levels in the blood. Before doing any exercise *estimate* the breaths per minute your lab partner might take after completing 2 minutes of strenuous exercise. Record your estimation.

? Estimation: _____ 2

Caution! If you have a heart condition, a family history of heart failure, or another medical condition that prevents you from doing strenuous exercise, then do not do the exercise sections of this lab.

Have your lab partner do strenuous exercise for 2 minutes (jumping jacks, running in place or outside of lab, running up and down stairs, etc.). As soon as your lab partner is finished, record the number of breaths per minute for the *first minute* after exercise and write this number in the space below and in the Chapter Summary Data section.

? Number of breaths per minute after 2 minutes of

exercise_____ 3

? How does this compare with your estimation?

_____ 4

Measurement of Pulmonary Volumes and Capacities

Pulmonary volumes are the amount of air that flows into or out of the lungs during a particular event. **Capacities** are the summation of volumes. There are many different ways to record pulmonary volumes. An affordable way to measure pulmonary volume is to use a handheld spirometer in which the exhaled air spins the vanes of the spirometer, estimating the volume of air exhaled. This is like the anemometer used by weather stations to measure wind speed. These are relatively inexpensive pieces of equipment and are good for measuring vital capacity but they are not as accurate as other equipment. The wet spirometer measures the volume of air exhaled into a chamber, so it measures the actual amount of air exhaled. Some digital recorders, such as those made by BIOPAC, use a flow meter to estimate volume.

Caution!   You should never inhale using a spirometer.

Tidal Volume with Handheld Spirometer

1. Place a disposable mouthpiece on the spirometer tube.
2. Make sure you set the indicator dial to zero by twisting the knurled ring on the top of the spirometer (zero may be the same as the maximum volume, 7,000 in some spirometers) (see figure 40.2).

FIGURE 40.2 Handheld Spirometer

3. As you exhale, estimate what a normal breath volume will be and exhale this amount rather forcefully. Do not exhale more than what you would for a normal breath. If you breathe gently into a handheld spirometer, you may not cause the vanes to spin enough to get a significant recording.
4. Exhale five total breaths (without resetting the spirometer to zero between breaths) and record your results.

Total of five breaths recorded by spirometer: _____

Divide the total number by 5 and enter the average tidal volume.

❓ Average tidal volume: _____ 5

You may find that the tidal volume is variable among members of your class. This is due to the difficulty of measuring tidal volumes with a standard lab apparatus. The average tidal volume is about 500 mL.

Tidial Volume with a Wet Spirometer

The tidal volume can be measured if your wet spirometer is accurate. Place a disposable mouthpiece into the flexible hose and breathe a normal breath into the mouthpiece. Record your data in the following space. Throw the disposable mouthpiece into the biohazard container when you are finished.

Tidal volume: _____ 5

Expiratory Reserve Volume

You can determine your **expiratory reserve volume (ERV)** with the wet spirometer or the handheld spirometer. The expiratory reserve volume is the maximal amount of air you can exhale after a *normal exhalation.* This is typically around 1,000 mL.

1. Make sure to close off your nostrils and, after a normal exhalation, forcibly expel the remainder of your breath through the mouthpiece into the spirometer.
2. Repeat this 2 more times for a total of 3 exhalations, and calculate the average expiratory reserve by dividing the sum of the volumes by 3.
3. Record your results.

Trial 1: _____

Trial 2: _____

Trial 3: _____

❓ Average expiratory reserve volume: _____ 6

Vital Capacity (VC)

The vital capacity (VC) is the total volume of air that can be forcefully expelled from the lungs after a maximum inhalation. Measuring vital capacity is like participating in the national championship of exhalation.

1. Measure the vital capacity with the use of a wet spirometer or handheld spirometer by first breathing in as deeply as you possibly can.
2. Close your nostrils, and then exhale through the mouthpiece completely until you cannot exhale anymore.
3. Force as much air from your lungs as you can.
4. Record your vital capacity.

❓ Vital capacity: _____ 7

Lung volumes are illustrated in figure 40.3.

Percent of Normal Vital Capacity

You can compare your vital capacity to those of other individuals of your sex, height, and age by using one of the two charts (one for females, one for males) in table 40.1. As you examine the chart for a person of your height, notice that the vital capacity decreases with age.

Calculate your percent of normal vital capacity by dividing your data by the normal data for a person of your age, sex, and height and multiplying the result by 100.

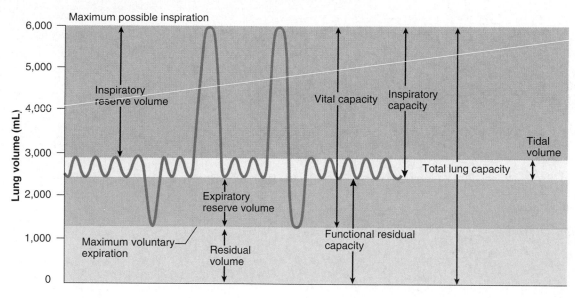

FIGURE 40.3 Lung Volumes

$$\frac{\text{Your vital capacity}}{\text{Normal vital capacity}} \times 100 = \text{Percent of normal vital capacity}$$

? Your percent of normal vital capacity: _____ 8

If you come out at 100% of normal, then you have an average vital capacity of individuals in your bracket. If your value is less than 100%, then you have a smaller vital capacity than normal for your bracket. A percent vital capacity of 80% of average, or better, is in the normal range. If your value is larger than 100%, then you have a larger vital capacity than average.

Calculation of the Inspiratory Reserve Volume (IRV)

The vital capacity consists of the expiratory reserve volume, the tidal volume, and the inspiratory reserve volume. You can *indirectly* determine the **inspiratory reserve volume (IRV)** by subtracting the expiratory reserve volume (ERV) and the tidal volume (TV) from the vital capacity (VC). This is indicated in figure 40.3. You cannot measure the inspiratory reserve volume directly with the use of a handheld spirometer (it records exhalations only). An average IRV is around 3,000 mL. Calculate your inspiratory reserve.

$$\text{IRV} = \text{VC} - (\text{ERV} + \text{TV})$$

? Your IRV: _____ 9

A decrease in lung capacity may be due to a decrease in lung elasticity (which reduces the pulmonary compliance), such as tuberculosis or pulmonary fibrosis. The obstructive disorders, such as asthma and emphysema, decrease the volume of the lung capacity as well.

BIOPAC Respiration Lab

Pulmonary volumes and capacities can be measured directly with the use of a respirometer or spirometer, or they can be measured indirectly with the use of a flow meter. One such device is an airflow transducer made by BIOPAC. You should have all the necessary material for measuring pulmonary flow, including an MP30 unit, transducer hardware, mouthpieces, bacteriological filters, a compatible computer, and software (BIOPAC Pro or BIOPAC Student Lessons). Every student in lab should record a spirogram. Make sure that everyone is able to have access to the equipment prior to doing any data analysis.

Caution!   When you change subjects always use a new disposable bacteriological filter.

Calibrating the Machine
1. Make sure that the computer is turned on, that the MP30 unit is connected to the computer and turned on, and that the transducer is plugged into the MP30 unit.

TABLE 40.1   Predicted Vital Capacities for Females and Males

Females

Height in Centimeters

Age	152	154	156	158	160	162	164	166	168	170	172	174	176	178	180	182	184	186	188
16	3,070	3,110	3,150	3,190	3,230	3,270	3,310	3,350	3,390	3,430	3,470	3,510	3,550	3,590	3,630	3,670	3,715	3,755	3,800
18	3,040	3,080	3,120	3,160	3,200	3,240	3,280	3,320	3,360	3,400	3,440	3,480	3,520	3,560	3,600	3,640	3,680	3,720	3,760
20	3,010	3,050	3,090	3,130	3,170	3,210	3,250	3,290	3,330	3,370	3,410	3,450	3,490	3,525	3,565	3,605	3,645	3,695	3,720
22	2,980	3,020	3,060	3,095	3,135	3,175	3,215	3,255	3,290	3,330	3,370	3,410	3,450	3,490	3,530	3,570	3,610	3,650	3,685
24	2,950	2,985	3,025	3,065	3,100	3,140	3,180	3,220	3,260	3,300	3,335	3,375	3,415	3,455	3,490	3,530	3,570	3,610	3,650
26	2,920	2,960	3,000	3,035	3,070	3,110	3,150	3,190	3,230	3,265	3,300	3,340	3,380	3,420	3,455	3,495	3,530	3,570	3,610
28	2,890	2,930	2,965	3,000	3,040	3,070	3,115	3,155	3,190	3,230	3,270	3,305	3,345	3,380	3,420	3,460	3,495	3,535	3,570
30	2,860	2,895	2,935	2,970	3,010	3,045	3,085	3,120	3,160	3,195	3,235	3,270	3,310	3,345	3,385	3,420	3,460	3,495	3,535
32	2,825	2,865	2,900	2,940	2,975	3,015	3,050	3,090	3,125	3,160	3,200	3,235	3,275	3,310	3,350	3,385	3,425	3,460	3,495
34	2,795	2,835	2,870	2,910	2,945	2,980	3,020	3,055	3,090	3,130	3,165	3,200	3,240	3,275	3,310	3,350	3,385	3,425	3,460
36	2,765	2,805	2,840	2,875	2,910	2,950	2,985	3,020	3,060	3,095	3,130	3,165	3,205	3,240	3,275	3,310	3,350	3,385	3,420
38	2,735	2,770	2,810	2,845	2,880	2,915	2,950	2,990	3,025	3,060	3,095	3,130	3,170	3,205	3,240	3,275	3,310	3,350	3,385
40	2,705	2,740	2,775	2,810	2,850	2,885	2,920	2,955	2,990	3,025	3,060	3,095	3,135	3,170	3,205	3,240	3,275	3,310	3,345
42	2,675	2,710	2,745	2,780	2,815	2,850	2,885	2,920	2,955	2,990	3,025	3,060	3,100	3,135	3,170	3,205	3,240	3,275	3,310
44	2,645	2,680	2,715	2,750	2,785	2,820	2,855	2,890	2,925	2,960	2,995	3,030	3,060	3,095	3,130	3,165	3,200	3,235	3,270
46	2,615	2,650	2,685	2,715	2,750	2,785	2,820	2,855	2,890	2,925	2,960	2,995	3,030	3,060	3,095	3,130	3,165	3,200	3,235
48	2,585	2,620	2,650	2,685	2,715	2,750	2,785	2,820	2,855	2,890	2,925	2,960	2,995	3,030	3,060	3,095	3,130	3,160	3,195
50	2,555	2,590	2,625	2,655	2,690	2,720	2,755	2,785	2,820	2,855	2,890	2,925	2,955	2,990	3,025	3,060	3,090	3,125	3,155
52	2,525	2,555	2,590	2,625	2,655	2,690	2,720	2,755	2,790	2,820	2,855	2,890	2,925	2,995	2,990	3,020	3,055	3,090	3,125
54	2,495	2,530	2,560	2,590	2,625	2,655	2,690	2,720	2,755	2,790	2,820	2,855	2,885	2,920	2,950	2,985	3,020	3,055	3,085
56	2,460	2,495	2,525	2,560	2,590	2,625	2,655	2,690	2,720	2,755	2,790	2,820	2,855	2,885	2,920	2,950	2,980	3,015	3,045
58	2,430	2,460	2,495	2,525	2,560	2,590	2,625	2,655	2,690	2,720	2,750	2,785	2,815	2,850	2,880	2,920	2,945	2,975	3,010
60	2,400	2,430	2,460	2,495	2,525	2,560	2,590	2,625	2,655	2,685	2,720	2,750	2,780	2,810	2,845	2,875	2,915	2,940	2,970
62	2,370	2,405	2,435	2,465	2,495	2,525	2,560	2,590	2,620	2,655	2,685	2,715	2,745	2,775	2,810	2,840	2,870	2,900	2,935
64	2,340	2,370	2,400	2,430	2,465	2,495	2,525	2,555	2,585	2,620	2,650	2,680	2,710	2,740	2,770	2,805	2,835	2,865	2,895
66	2,310	2,340	2,370	2,400	2,430	2,460	2,495	2,525	2,555	2,585	2,615	2,645	2,675	2,705	2,735	2,765	2,800	2,825	2,860
68	2,280	2,310	2,340	2,370	2,400	2,430	2,460	2,490	2,520	2,550	2,580	2,610	2,640	2,670	2,700	2,730	2,760	2,795	2,820
70	2,250	2,280	2,310	2,340	2,370	2,400	2,425	2,455	2,485	2,515	2,545	2,575	2,605	2,635	2,665	2,695	2,725	2,755	2,780
72	2,220	2,250	2,280	2,310	2,335	2,365	2,395	2,425	2,455	2,480	2,510	2,540	2,570	2,600	2,630	2,660	2,685	2,715	2,745
74	2,190	2,220	2,245	2,275	2,305	2,335	2,360	2,390	2,420	2,450	2,475	2,505	2,535	2,565	2,590	2,620	2,650	2,680	2,710

"Predicted Vital Capacities for Females and Males (tables)" by E. A. Gaensler, M.D. and G. W. Wright, M.D., AEH, Vol. 12, pp. 146–189, February 1966. Reprinted with permission of the Helen Dwight Reid Educational Foundation. Published by Heldref Publications, 1319 18th Street NW, Washington, DC 20036-1802. Copyright © 1966.

TABLE 40.1 Continued

Males

Height in Centimeters

Age	152	154	156	158	160	162	164	166	168	170	172	174	176	178	180	182	184	186	188
16	3,920	3,975	4,025	4,075	4,130	4,180	4,230	4,285	4,335	4,385	4,440	4,490	4,540	4,590	4,645	4,695	4,745	4,800	4,850
18	3,890	3,940	3,995	4,045	4,095	4,145	4,200	4,250	4,300	4,350	4,405	4,455	4,505	4,555	4,610	4,660	4,710	4,760	4,815
20	3,860	3,910	3,960	4,015	4,065	4,115	4,165	4,215	4,265	4,320	4,370	4,420	4,470	4,520	4,570	4,625	4,675	4,725	4,775
22	3,830	3,880	3,930	3,980	4,030	4,080	4,135	4,185	4,235	4,285	4,335	4,385	4,435	4,485	4,535	4,585	4,635	4,685	4,735
24	3,785	3,835	3,885	3,935	3,985	4,035	4,085	4,135	4,185	4,235	4,285	4,330	4,380	4,430	4,480	4,530	4,580	4,630	4,680
26	3,755	3,805	3,855	3,905	3,955	4,000	4,050	4,100	4,150	4,200	4,250	4,300	4,350	4,395	4,445	4,495	4,545	4,595	4,645
28	3,725	3,775	3,820	3,870	3,920	3,970	4,020	4,070	4,115	4,165	4,215	4,265	4,310	4,360	4,410	4,460	4,510	4,555	4,605
30	3,695	3,740	3,790	3,840	3,890	3,935	3,985	4,035	4,080	4,130	4,180	4,230	4,275	4,325	4,375	4,425	4,470	4,520	4,570
32	3,665	3,710	3,760	3,810	3,855	3,905	3,950	4,000	4,050	4,095	4,145	4,195	4,240	4,290	4,340	4,385	4,435	4,485	4,530
34	3,620	3,665	3,715	3,760	3,810	3,855	3,905	3,950	4,000	4,045	4,095	4,140	4,190	4,225	4,285	4,330	4,380	4,425	4,475
36	3,585	3,635	3,680	3,730	3,775	3,825	3,870	3,920	3,965	4,010	4,060	4,105	4,155	4,200	4,250	4,295	4,340	4,390	4,435
38	3,555	3,605	3,650	3,695	3,745	3,790	3,840	3,885	3,930	3,980	4,025	4,070	4,120	4,165	4,210	4,260	4,305	4,350	4,400
40	3,525	3,575	3,620	3,665	3,710	3,760	3,805	3,850	3,900	3,945	3,990	4,035	4,085	4,130	4,175	4,220	4,270	4,315	4,360
42	3,495	3,540	3,590	3,635	3,680	3,725	3,770	3,820	3,865	3,910	3,955	4,000	4,050	4,095	4,140	4,185	4,230	4,280	4,325
44	3,450	3,495	3,540	3,585	3,630	3,675	3,725	3,770	3,815	3,860	3,905	3,950	3,995	4,040	4,085	4,130	4,175	4,220	4,270
46	3,420	3,465	3,510	3,555	3,600	3,645	3,690	3,735	3,780	3,825	3,870	3,915	3,960	4,005	4,050	4,095	4,140	4,185	4,230
48	3,390	3,435	3,480	3,525	3,570	3,615	3,655	3,700	3,745	3,790	3,835	3,880	3,925	3,970	4,015	4,060	4,105	4,150	4,190
50	3,345	3,390	3,430	3,475	3,520	3,565	3,610	3,650	3,695	3,740	3,785	3,830	3,870	3,915	3,960	4,005	4,050	4,090	4,135
52	3,315	3,353	3,400	3,445	3,490	3,530	3,575	3,620	3,660	3,705	3,750	3,795	3,835	3,880	3,925	3,970	4,010	4,055	4,100
54	3,285	3,325	3,370	3,415	3,455	3,500	3,540	3,585	3,630	3,670	3,715	3,760	3,800	3,845	3,890	3,930	3,975	4,020	4,060
56	3,255	3,295	3,340	3,380	3,425	3,465	3,510	3,550	3,595	3,640	3,680	3,725	3,765	3,810	3,850	3,895	3,940	3,980	4,025
58	3,210	3,250	3,290	3,335	3,375	3,420	3,460	3,500	3,545	3,585	3,630	3,670	3,715	3,755	3,800	3,840	3,880	3,925	3,965
60	3,175	3,220	3,260	3,300	3,345	3,385	3,430	3,470	3,500	3,555	3,595	3,635	3,680	3,720	3,760	3,805	3,845	3,885	3,930
62	3,150	3,190	3,230	3,270	3,310	3,350	3,390	3,440	3,480	3,520	3,560	3,600	3,640	3,680	3,730	3,770	3,810	3,850	3,890
64	3,120	3,160	3,200	3,240	3,280	3,320	3,360	3,400	3,440	3,490	3,530	3,570	3,610	3,650	3,690	3,730	3,770	3,810	3,850
66	3,070	3,110	3,150	3,190	3,230	3,270	3,310	3,350	3,390	3,430	3,470	3,510	3,550	3,600	3,640	3,680	3,720	3,760	3,800
68	3,040	3,080	3,120	3,160	3,200	3,240	3,280	3,320	3,360	3,400	3,440	3,480	3,520	3,560	3,600	3,640	3,680	3,720	3,760
70	3,010	3,050	3,090	3,130	3,170	3,210	3,250	3,290	3,330	3,370	3,410	3,450	3,480	3,520	3,560	3,600	3,640	3,680	3,720
72	2,980	3,020	3,060	3,100	3,140	3,180	3,210	3,250	3,290	3,330	3,370	3,410	3,450	3,490	3,530	3,570	3,610	3,650	3,680
74	2,930	2,970	3,010	3,050	3,090	3,130	3,170	3,200	3,240	3,280	3,320	3,360	3,400	3,440	3,470	3,510	3,550	3,590	3,630

2. Select **Lesson 12: Pulmonary 1—Volumes & Capacities (L012-Lung-1)** and type your name where appropriate. If you have a folder with your name already on the computer, then select "Use it" if prompted.
3. Place a new bacteriological filter into the end with the calibration syringe. Insert the syringe into the side of the transducer labeled "Inlet."
4. Pull the plunger of the calibration syringe all the way out by holding the syringe by the body. Do not hold the unit by the transducer.
5. Click "Calibrate," read all the directions, and click "OK."
6. Push the plunger in and out for a total of 5 cycles (5 in and 5 out). This should take about 30 seconds.
7. Click on "End Calibration."
8. The peaks should be even. If the calibration looks abnormal on the screen, then click "Redo." If it looks OK, then begin recording the data.

Recording Data

1. Place a new mouthpiece into the transducer where the calibration syringe was.
2. Pinch off your nose with a nose clip.
3. Click on "Record" and begin breathing easily into the transducer.
4. Take three normal breaths **(tidal volume)** and then inhale maximally and exhale normally **(inspiratory reserve volume)**, as seen in figure 40.3.
5. Take a few normal breaths again and then exhale maximally **(expiratory reserve volume)**.
6. Take a few normal breaths; then inhale maximally and exhale maximally **(vital capacity)**.
7. Click on "Stop."

If you are not satisfied with your recording (compare what you see with figure 40.3), click on "Redo." If you are happy with your recording, then click on "Done" and remove the bacteriological filter from the apparatus.

You can select "Review Saved Data" from the menu.

Data Analysis

Select **CH2** (which is the volume) and select **p–p** (peak to peak) which selects the range of minimum and maximum values. Select CH2 and click on "max;" then select CH2 and click on "min." Select CH2 again and click on "delta." Compare your data with figure 40.3.

Select the I-beam and highlight one of the tidal volume measurements from the peak to the depth of a breath. Save this data as "Tidal Volume."

Select the area from the peak of a normal inhalation to the maximum air inhaled and save the data as "Inspiratory Reserve Volume."

Find the point of a normal exhalation (after a tidal volume) to the complete exhalation and save the data as "Expiratory Reserve Volume."

Find the point of the maximum inhalation to the maximum exhalation and save the data as "Vital Capacity."

Enter your data from the BIOPAC Respiration Lab in the following section.

Tidal volume: _____ 5

Inspiratory reserve volume: _____ 9

Expiratory reserve volume: _____ 6

Vital capacity: _____ 7

Your percent normal vital capacity can be calculated as it was using wet spirometry or handheld spirometers. Record your percent normal vital capacity in the following space.

Percent normal vital capacity: _____ 8

Residual Volume

There is still air left in the lungs after a maximal exhalation. This is known as the residual volume (RV) and it is approximately 1,000 mL. The total lung capacity (TLC) is the sum of the vital capacity and the residual volume.

Minute Ventilation

The total amount of air inspired and expired in 1 minute of normal (tidal volume) breathing is known as the minute ventilation. This is calculated by multiplying the number of breaths per minute by the tidal volume. Record your minute volume in the following space.

Minute volume: _____ 10

Construct a bar graph using your data in chart 40.1. The vital capacity should be composed of the ERV, TV, and IRV.

Flow and Resistance

The flow of air is proportional to the pressure of the air and inversely proportional to the resistance in the airways. As pressure difference between the outside air and the air inside the lungs increases, the flow of air increases. If the resistance in the air passageways increases, then the flow

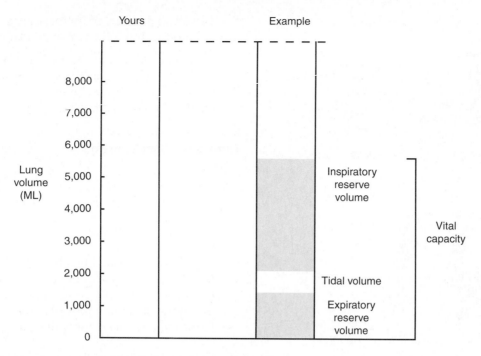

CHART 40.1 Fill in the chart with your own data. In the example, the IRV is 3,500mL, the TV is 500 mL, and the ERV is 1,500 mL, for a VC of 5,500 mL.

rate decreases. You can demonstrate the effects of changing the resistance of the airways in a simple demonstration. Do not do this experiment if you have a respiratory condition, such as asthma or severe sinus allergies.

1. Using a stopwatch or a watch with a second hand record how long it takes to forcibly inhale the maximum amount of air that your lungs can hold. Do this by breathing through both your mouth and your nose. Record the time in the following space.

? Time for maximum inhalation: _____ seconds 11

2. Close your mouth and one nostril and try the experiment again. Breathe only through one nostril and record the time that it takes to maximally inhale. Record this in the following space.

? Time for inhalation through one nostril: _____

_____ seconds 12

Closing off the respiratory passages (mouth and one nostril) increases the resistance in the respiratory system. How does this change the rate of airflow?

Asthma occurs due to a restriction of the bronchioles in the lung. If this occurs, is there a change in the pressure between the alveoli and the external air or an increase in the resistance to airflow into the lungs?

Respiratory Sounds

In this section, you listen to breathing sounds, which can be heard with a stethoscope.

1. Before you begin with the experiment, clean the earpieces of the stethoscope with an alcohol wipe and let them dry. The stethoscope earpieces should point toward the anterior as you insert them into your ears.
2. Locate the larynx of your lab partner and place the diaphragm of the stethoscope just inferior to it.
3. Listen for the sound as your lab partner inhales and exhales. These are the tracheal and bronchial sounds.
4. Locate the triangle of auscultation (see figure 40.4), an area just medial to the inferior angle of the scapula. This is an ideal area for listening to sounds because the thoracic cage is not covered by muscles in this location.
5. Have your lab partner inhale and exhale deeply several times.

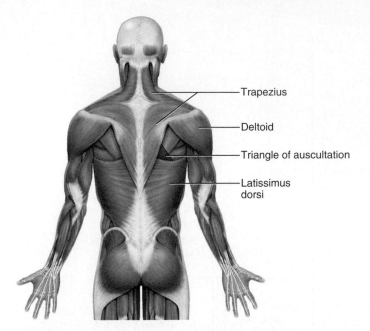

- Trapezius
- Deltoid
- Triangle of auscultation
- Latissimus dorsi

FIGURE 40.4 Triangle of Auscultation, Located in the Shaded Region

6. Listen for a smooth flow of air into and out of the lungs. Wheezing and rattling are indicators of congestion in the lungs.
7. Record the sounds you hear in the space below and indicate the condition of your lab partner.

Acid-Base Effects of the Respiratory Gases

Carbon dioxide combines with water to form carbonic acid, which subsequently dissociates into bicarbonate and hydrogen ions, as represented in the following equation:

$$CO_2 \; + \; H_2O \; \longrightarrow \; H_2CO_3 \; \longrightarrow \; HCO_3^- \qquad + \; H^+$$

| Carbon dioxide | Water | Carbonic acid | Bicarbonate ion | Hydrogen ion |

You can see the effects of increasing carbon dioxide levels in the blood in the following experiment.

1. Add about 50 mL of litmus solution to a small (100 mL) Erlenmeyer flask. If the solution is red, add the sodium hydroxide solution (NaOH) drop by drop until the color just begins to turn blue.

Caution! Wear safety goggles if you add NaOH to the flask.

---

2. Insert a drinking straw into the flask and gently blow air into the flask.
3. Bubble your exhaled breath into the flask and look for a color change. A blue color indicates an alkaline condition, and a red color indicates an acid condition.
4. Using the preceding formula, determine how the exhalation into the flask alters the acid-base conditions.

Cardiopulmonary Resuscitation

Cardiopulmonary resuscitation, or CPR, is typically used for people suffering from myocardial infarcts (heart attacks), drug overdoses, drowning or trauma, and obstruction of the airways, among other things. This is a technique that combines the use of chest compression of about 100 times per minute on the body of the sternum with mouth-to-mouth ventilation. The ratio is about 30 chest compressions followed by 2 ventilations. It is important to check to make sure that the airway is clear prior to ventilating the lungs. When the lungs are temporarily nonfunctional, CPR may keep a person alive until medical help arrives.

REVIEW SECTION

Respiratory Function, Breathing, Respiration

Name _____ Date _____

Lab Section _____ Time _____

Review Questions

1. What is pulmonary ventilation?

2. Why would carbon monoxide (it binds to hemoglobin) and carbon dioxide in cigarette smoke cause smokers to generally have a higher than average vital capacity?

3. What was your measured breathing rate in breaths per minute?

4. What is the tidal volume in liters for an average adult?

5. Define tidal volume.

6. What instrument do you use to measure breathing volumes?

7. If you inhale maximally, what is the name of the volume of air that you completely exhale?

8. Examine the predicted vital capacity chart. What is the approximate percent decrease of vital capacity in the same individual from age 25 to age 75?

9. How does the decrease in vital capacity potentially influence an individual's athletic performance or aerobic condition as aging occurs?

10. What is the pressure difference between the external air and the pleural cavity when inhalation just begins?

11. Calculate the IRV of an individual with a vital capacity of 4,400 mL, an expiratory reserve volume of 1,300 mL, and a tidal volume of 500 mL.

12. How does excess carbon dioxide change the acid-base condition of a solution?

? Chapter Summary Data

Use this section to record your results from questions within the exercise.

1. _____   7. _____

2. _____   8. _____

3. _____   9. _____

4. _____   10. _____

5. _____   11. _____

6. _____   12. _____

LABORATORY

Physiology of Exercise and Pulmonary Health

INTRODUCTION

An understanding of exercise physiology is important not only for athletic training but also for wellness and general health. Significant changes occur during exercise that affect many organ systems of the body. The changes vary, depending on the type of exercise. Exercise is broadly classified as either **aerobic exercise** or **anaerobic exercise.** Aerobic exercises increase heart rate and breathing rates at moderate levels for extended periods of time. Aerobic exercises do not deplete the level of oxygen consumption. Anaerobic exercises result in the consumption of available oxygen faster than it can be supplied to the muscle tissue. The tissue uses glucose anaerobically and produces lactic acid. Exercise has a profound effect on the muscular, skeletal, cardiovascular, and respiratory systems. Consider the increased metabolic demands placed on skeletal muscles during repeated contractions. Skeletal muscle is more metabolically active when contracting than when at rest; it uses additional oxygen and nutrients. Oxygen diffuses from the blood to the muscle tissue due to the concentration gradient of oxygen between these two areas. In addition to the diffusion of oxygen, the precapillary sphincters of the circulatory system open and the arterioles dilate, providing greater blood flow to the muscles. This additional blood flow requires an increase in the volume of blood passing through the heart with each contraction. Heart size increases and pulse rate decreases with long-term rigorous exercise. The respiratory system responds to this greater oxygen demand with increased volume per breath and a greater number of breaths per minute. This increases the minute volume. The lung capillaries expand as well, and a greater diffusion of oxygen occurs between the alveoli and the blood capillaries. This respiratory response is covered in the Saladin text in chapter 22, "The Respiratory System." In this exercise you examine respiratory health, the effects of exercise on the body, and the body's comparative responses to aerobic exercise.

OBJECTIVES

At the end of this exercise you should be able to

1. list the major organ systems directly involved in fitness;
2. describe basic physiological differences between a person who is physically active and one who is physically inactive;
3. determine the forced expiratory volume exhaled in 1 second;
4. calculate the personal fitness index of a subject who performs the Harvard step test or Cooper's 12-minute run test.

MATERIALS

BIOPAC Setup

AFT6—6000 mL calibration syringe

AFT1—disposable bacteriological filter

AFT2—disposable mouthpiece

SS11L or SS11LA—airflow transducer

MP30—data acquisition unit

Compatible computer

Noseclips

16-inch step

20-inch step

Metronome or clock with second hand

Treadmill

PROCEDURE

Pulmonary Health

Forced Expiratory Vital Capacity (FEV)

Indications of health can be roughly correlated with the amount of air expelled from the lungs in 1 second. This is usually expressed as a percent when compared to the person's vital capacity (VC) as $FEV_1/VC\%$. This should be approximately 75% of the VC in healthy adults. In this exercise you will use the BIOPAC setup, which is similar to the one used in Exercise 40. Decreases in forced expiratory vital capacity may be caused by asthma, emphysema, or other pulmonary conditions.

BIOPAC Lesson 13—Pulmonary Function II

1. Setup

Make sure that the computer is on but the MP30 unit is off. Plug in the airflow transducer to the MP30 unit and turn the unit on. Open the BIOPAC Student Lab (BSL) Software and select BIOPAC Lesson 13—Pulmonary Function II,. You will collect data for your respiratory volumes by breathing into a flow meter that measures the force of air passing through it, which it then will convert it to a volume. The resulting "spirogram" is displayed on the computer monitor and saved as a file in your BIOPAC folder. The following are the steps you will take.

Click on the BIOPAC icon on the computer desktop to start BIOPAC. A menu of lessons will appear. Select **BIOPAC Lesson 13—Pulmonary Function II,.** Type in your (folder) **name.** If you have a folder on this computer station, a window should appear with the message "A folder with this name already exists. Would you like to use it or create a new folder?" Choose **Use it.**

Place the bacterial filter onto the end of the calibration syringe. Insert the calibration syringe/filter assembly into the side of the airflow transducer labeled **"Inlet."**

2. Calibration

Pull the calibration syringe plunger all the way out and hold the calibration syringe horizontal, so that the airflow transducer is upright. It must remain vertical for the calibration and experimentation, as the membrane is sensitive to changes in deflection due to gravity. Hold the calibration syringe by the barrel (body), not by the airflow transducer (connecting duct). Click on **Calibrate.** After the first stage of the calibration is recorded a second dialogue box will appear, prompting you to read all the directions. Do so and click on **Yes.** Cycle the syringe plunger in and out 5 times (10 strokes). Use a rhythm of about 1 second per stroke, with 2 seconds between each stroke. Click on **End Calibration.** If the calibration looks good, then detach the calibration syringe and proceed to Step 3 **(Record).** (You should be able to see 5 downward and 5 upward deflections on the screen.) If the calibration peaks are uneven or significantly vary from one another, click **Redo.**

3. Record Data

Insert a clean mouthpiece (where the calibration syringe had been attached). Place a clean noseclip on your nose. Make sure that you hold the transducer upright at all times. Click on **Record** and breathe as follows (be sure to start the recording before taking the first normal breath):

1. Take three normal breaths (three inhales and three exhales).
2. Inhale as deeply as you can (take in as much air as you can). Hold it for an instant.
3. Exhale as fast and as much as you can.
4. Click on **Stop.**

If the recording is not good, click on **Redo** to repeat it. If the recording looks good, as seen in figure 41.1, click on **Done.** Select **Record from Another Subject** or, if you

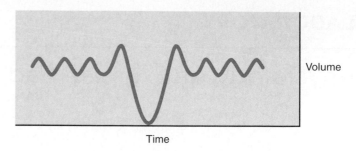

FIGURE 41.1 Spirogram

have recorded everyone's data, select **Analyze Current Data File.**

4. Data Analysis for FEV

Open your folder in the **Data Files** folder (probably listed as *"your name FEV-L13"*) and then open the data file. Select **Display Preferences** from the **File** menu, choose **Grids,** and click on **Show Grids** and then **OK.** Use the **I-beam** to highlight your graph. The **p–p** (peak to peak) will represent your vital capacity, as represented in figure 41.2.

Use the **I-beam** to select the volume of air you exhaled for the first second. This will be the FEV_1, as seen in figure 41.3.

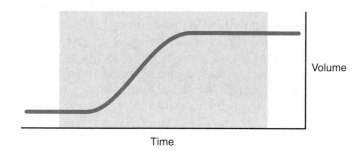

FIGURE 41.2 Peak to Peak Determination of Vital Capacity

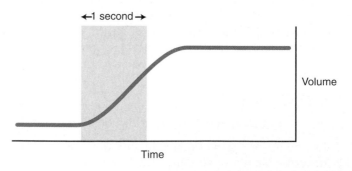

FIGURE 41.3 FEV_1 Highlight 1 Second of Time from Beginning of Expiration

Calculate your $FEV_1/VC\%$ and enter the value here:

_____

Save and print the entire spirogram you record to include in your lab report.

5. Interpretation

Spirometry is an important measurement tool to assist in diagnosing asthma, chronic obstructive pulmonary disease (COPD), and cystic fibrosis.

Pulmonary obstruction occurs when the respiratory passages are narrowed. This can be due to conditions such as asthma, excess mucus, and inflammation, such as bronchitis. In these cases, the vital capacity of the individual is normal but the $FEV_1/VC\%$ is low. It takes the subject longer to exhale completely.

Pulmonary restriction occurs when the lungs cannot fully inspire or expire the full volume of air. In these cases, the vital capacity of the individual is reduced. This can be due to fibrosis of the lungs (cystic fibrosis or fibrosis due to asbestos or silica); scarring of the lung tissue, as when a subject has had chronic lung infections; adhesions of the lung to the chest wall due to extreme emphysema; or removal of a section of lung. It may also be due to damage to the phrenic nerve, which stimulates the diaphragm. In these cases, the vital capacity of the individual is low but the $FEV_1/VC\%$ is normal. Most COPD is caused by smoking. Clinical values for FEV_1 compared to vital capacity are listed in table 41.1.

Measurement of Heart Rate

The measurement of fitness in this part of the exercise is on a voluntary basis. It is best if the class can obtain data from a person who regularly participates in aerobic exercises (three to six times per week) and from a person who does not exercise. The Harvard step test and the Cooper's 12-minute run test are two reliable measurements of fitness.

 Caution!   Do not do these exercises if you are at risk for heart disease or have a family history of heart disease or another condition, such as asthma, for which exercise is harmful. Stop if you feel exhausted or faint. If you develop chest pain or pain radiating down the left arm, seek medical attention immediately.

A correlation exists between heart rate and fitness level. The resting heart rate of people involved in regular, active aerobic exercise is lower than the rate of sedentary people. In addition, the heart recovers faster in people who have a regular exercise program than in those who do

TABLE 41.1	
$FEV_1/VC\%$	**Status**
100–75	Normal
74–60	Mild COPD
59–50	Moderate COPD
<50	Severe COPD

not exercise. In this experiment record the measurements of at least two volunteers from the class. The greater the number of students who participate, the better the data. Read the entire exercise before beginning.

Harvard Step Test

The Harvard step test was developed during World War II at Harvard University to determine a person's physical fitness. In this exercise students should select either the 20-inch step for people 5´8˝ or taller or the 16-inch step for people under 5´8˝. With one person acting as an observer, the subject should step up on the step in 1 second and down on the floor in another second, thus completing 30 complete cycles in 1 minute. The subject should keep the body upright and keep pace with the observer's count or with a metronome set at 60 beats per minute. The subject should exercise for at least 3 minutes but no more than 5 minutes. If the subject stops due to exhaustion, then the observer should note the time of exercise. After the period of exercise, the subject should sit and rest for 1 minute. The pulse should be taken from 1 minute to 1 minute 30 seconds and recorded in the following space. (See "A" on figure 41.4.)

Pulse from 1 minute to 1 minute 30 seconds: _____

The subject should rest for another 30 seconds before the pulse is taken from 2 minutes to 2 minutes 30 seconds after exercise and recorded here. (See "B" on figure 41.4.)

Pulse from 2 minutes to 2 minutes 30 seconds: _____

The subject should remain seated for another 30 seconds before the pulse is taken from 3 minutes to 3 minutes 30 seconds and recorded here. (See "C" on figure 41.4.)

Pulse from 3 minutes to 3 minutes 30 seconds: _____

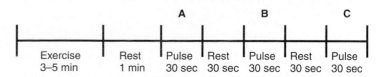

FIGURE 41.4 Harvard Step Test Periods

Add the three pulse counts recorded.

Sum of three pulse counts: _____

To determine the personal fitness index (PFI), use the following formula:

$$PFI = \frac{\text{Number of seconds of exercise}}{2 \text{ (sum of three pulse counts)}} \times 100$$

Therefore, if you exercised for 4 minutes and your pulse counts were 50, 48, and 45, then the PFI was

$$PFI = \frac{(4 \times 60) \times 100}{2(50 + 48 = 45)} = \frac{240 \times 100}{2(143)} = \frac{24,000}{286} = 84$$

Determine the fitness evaluation of the subject using the following data:

Below 55: poor physical condition

55–64: low average physical condition

65–79: high average physical condition

80–89: good physical condition

90 and above: excellent physical condition

Record the PFI of the subject: _____

What is the fitness evaluation for this person? _____

Cooper's 12-Minute Run Test

Another fitness test involves determining how far a person can run and/or walk in 12 minutes.

Caution! As with the Harvard step test, do not do these exercises if you are at risk for heart disease or have a family history of heart disease or another condition, such as asthma, for which exercise is harmful.

Procedure Warm up for 15 minutes prior to doing this test. Find a suitable running track or use a treadmill that is set to no incline and run or walk as fast and as far as you can in 12 minutes. You will not be able to sprint for the entire time, so pace yourself. If you cannot run anymore during the test, you should finish the test by walking or jogging or a combination of running and walking.

Record the distance you traveled in 12 minutes: _____

The results of your test and general fitness can be judged by the following table.

Under Age 40

Distance in 12 Minutes	Physical Condition
Over 2,700 m	Excellent
2,300 m to 2,700 m	Good
1,900 m to 2,299 m	Average
1,500 m to 1,899 m	Below average
Less than 1,500 m	Poor

Over Age 40

Distance in 12 Minutes	Physical Condition
Over 2,500 m	Excellent
2,100 m to 2,500 m	Good
1,700 m to 2,099 m	Average
1,500 m to 1,699 m	Below average
Less than 1,400 m	Poor

Target Heart Rate Zone for Exercise

The target heart rate zone for exercise is 60–80% of the maximum heart rate (MHR) for healthy adults. The maximum heart rate for an individual is his or her age subtracted from 220. You can calculate a basic target heart rate zone with the following procedure:

1. To calculate your maximum heart rate, subtract your age from 220. Enter the MHR value here: _____.
2. For 60% of the MHR, multiply the MHR by .6 and enter it in the space: _____.
3. For 80% of the MHR, multiply the MHR by .8 and enter it in the space: _____.
4. Your target heart rate zone for exercise should be between these two values.

Guidelines for Exercise

The American Heart Association recommends that healthy adults from the ages of 18 to 65 need moderate exercise (such as brisk walking) 30 minutes a day, 5 days a week, or vigorous exercise (such as jogging or running) for 20 minutes 3 days a week. Physical inactivity is a major public health issue.

> ### NOTE
>
> During the 2005 Tour de France, Lance Armstrong had a resting heart rate of 32-34 bpm. This is less than half the rate of the average resting heart rate of 72 bpm. He had 4% body fat, whereas the average person watching the race had body fat of 15-20%. The maximum amount of oxygen he could use (VO_2 max) was 84 mL/kg/min (6 liters of oxygen per minute). A physically active person has a VO_2 max of 40-50 mL/kg/min, whereas a physically inactive person has a VO_2 max of 30-40 mL/kg/min. The normal oxygen use at rest is about 3.5 mL/kg/min.

Body Mass Index

The body mass index (BMI) is a general guide to fitness that makes a couple of assumptions. One is that the person is of average build. Fitness level, gender, muscle mass, bone structure, and ethnicity can all influence the BMI. One way to get a general idea of the BMI is to use this calculation:

$$BMI = \frac{\text{Weight in pounds}}{(\text{Height in inches})^2} \times 703$$

Record your BMI here: _____

According to the National Heart, Lung, and Blood Institute, the BMI for average adults can be interpreted this way:

- Underweight = <18.5
- Normal weight = 18.5–24.9
- Overweight = 25–29.9
- Obesity = BMI of 30 or greater

If a person has a high muscle mass the BMI values may be overestimated, as muscle will show a person with higher weight than predicted. In older individuals who have less muscle mass the BMI may be underestimated.

Notes

REVIEW SECTION

Physiology of Exercise and Pulmonary Health

Name _____ *Date* _____

Lab Section _____ *Time* _____

Review Questions

1. Define $FEV_1/VC\%$.

2. Record your $FEV_1/VC\%$ or the one you measured in lab.

3. Does your $FEV_1/VC\%$ value fall within normal limits?

4. How does the $FEV_1/VC\%$ compare in a person with a pulmonary obstructive condition, such as asthma? Why?

5. How does the the $FEV_1/VC\%$ compare in a person with a pulmonary restrictive condition, such as asbestosis? Why?

6. If a person had a smaller body and therefore a smaller vital capacity, would the the $FEV_1/VC\%$ necessarily change?

7. What is your personal fitness index or the one measured in lab?

8. If the heart rate after 5 minutes of exercise was 70, 68, and 66 beats in the consecutive 30-second trials, what was the personal fitness index and what condition does that represent?

9. Record the results of the Harvard step test or the Cooper's 12-minute run test for the members in the class who performed them. Next to the Personal Fitness Index (Harvard) or Physical Condition (Cooper's) for your own results, write an *S* if you smoke cigarettes or an *N* if you are a nonsmoker.

PFI (Personal Fitness Index)
Physical Condition

1. _____

2. _____

3. _____

4. _____

5. _____

6. _____

7. _____

8. _____

9. _____

10. _____

11. _____

12. _____

13. _____

14. _____

15. _____

16. _____

17. _____

18. _____

19. _____

20. _____

21. _____

22. _____

23. _____

24. _____

LABORATORY

Anatomy of the Digestive System

INTRODUCTION

The digestive system can be divided into two major parts, the **alimentary canal** and the **accessory organs.** The alimentary canal is a long tube that runs from the mouth to the anus and comes into contact with food or the breakdown products of digestion. Some of the organs of the alimentary canal are the esophagus, stomach, intestines, rectum, and anus. The accessory organs are important in that they secrete many important substances necessary for digestion, yet these organs do not come into direct contact with food. Examples of accessory organs are the salivary glands, liver, gallbladder, and pancreas.

The functions of the digestive system are many and include ingestion of food, physical breakdown of food, chemical breakdown of food, food storage, water absorption, vitamin synthesis, food absorption, and elimination of indigestible material. These topics are covered in the Saladin text in chapter 25, "The Digestive System."

In this exercise you examine the anatomy of the digestive system in both human and cat, correlating the structure of the digestive organs with their functions.

OBJECTIVES

At the end of this exercise you should be able to

1. list, in sequence, the major organs of the alimentary canal;
2. describe the basic function of the accessory digestive organs;
3. note the specific anatomical features of each major digestive organ;
4. describe the layers of the wall of the gastrointestinal tract;
5. describe the major functions of the stomach and small and large intestines;
6. distinguish among different regions of the alimentary canal by their histology.

MATERIALS

Models, charts, or illustrations of the digestive system

Mirror

Materials for Cat Dissection

    Cats

    Dissection trays

    Scalpels or razor blades

    Protective gloves

    Waste container

Skull, human teeth, or cast of teeth

Cadaver (if available)

Microscopes

Microscope Slides

    Esophagus

    Stomach

    Small intestine

    Large intestine

    Liver

PROCEDURE

Overview of the Digestive Organs

Begin this exercise by examining a torso model or charts in the lab and compare them to figure 42.1. Locate the major digestive organs and place a check mark in the appropriate space.

_____ Mouth

_____ Teeth

_____ Pharynx

_____ Esophagus

_____ Stomach

_____ Small intestine

_____ Large intestine (colon)

    _____ Ascending colon

    _____ Transverse colon

    _____ Descending colon

    _____ Sigmoid colon

    _____ Rectum

_____ Anal canal

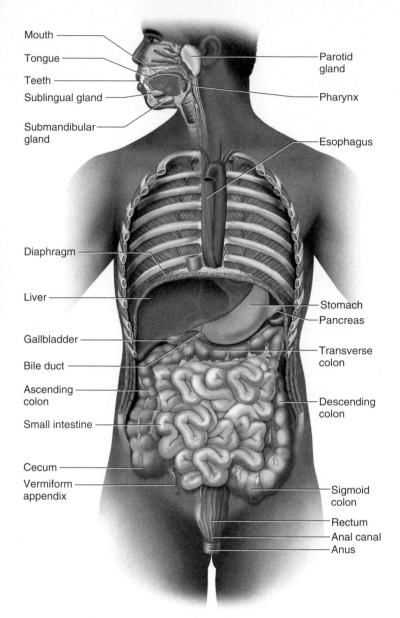

FIGURE 42.1 Overview of the Digestive System

_____ Anus

_____ Liver

_____ Vermiform appendix

_____ Pancreas

_____ Gallbladder

_____ Salivary glands

Alimentary Canal

Begin your study of the alimentary canal with the mouth. Examine a midsagittal section of the head, as represented in figure 42.2, and locate the major anatomical features.

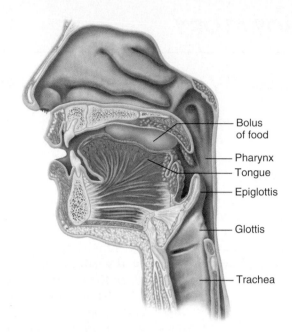

FIGURE 42.2 Oral Cavity, Midsagittal Section

Mouth

At the beginning of the alimentary canal is the **mouth.** The opening of the mouth is surrounded by **lips,** or **labia.** The **labial frenulum** is a membranous structure that keeps the lip adhered to the gums, or **gingivae.** The mouth is a space bordered in front by the lips, behind by the oropharynx, and on the sides by the inner wall of the cheeks. The hard and soft palates form the roof of the mouth, and the floor of the chin is the inferior border. The **hard palate** is composed of the palatine processes of the maxillae and palatine bones. The **soft palate** is composed of connective tissue and a mucous membrane. At the posterior portion of the mouth is the **uvula,** a small, grapelike structure suspended from the posterior edge of the soft palate. The uvula helps prevent food or liquid from moving into the nasal cavity during swallowing. The mouth is lined with **nonkeratinized stratified squamous epithelium,** which protects the underlying tissue from abrasion. The **tongue** is made of skeletal muscle. One of the major muscles of the tongue is the **genioglossus.** The tongue is important in speech, taste, the movement of food toward the teeth for chewing, and swallowing. The tongue acts as a piston to propel food to the **oropharynx,** the space behind the oral cavity. The tongue is held down to the floor of the mouth by a thin mucous membrane called the **lingual frenulum.** Use a mirror to examine the three types of **papillae,** or raised areas, on the tongue. These are **fungiform, filiform,** and **vallate** (circumvallate) papillae. Papillae increase the frictional surface of the tongue. **Taste buds** occur on the tongue along the sides of the papillae. The sense of taste is covered in Laboratory Exercise 25.

The mouth is important in digestion for the physical breakdown of food. This process is driven by powerful muscles called the **muscles of mastication.** The **masseter** and the **temporalis muscles** are involved in the closing of the jaws, and the **pterygoid muscles** are important in the sideways grinding action of the molar and premolar teeth.

Teeth Examine models of teeth, dental casts, or real teeth on display in the lab and compare them to figure 42.3. A tooth consists of a **crown, neck,** and **root.** The crown is the exposed part of the tooth; the neck is a constricted portion of the tooth that normally occurs at the surface of the gingivae; and the root is embedded in the jaw. Examine a model or an illustration of a longitudinal section of a tooth and find the outer **enamel,** an extremely hard material. Inside this layer is the **dentine,** which is made of bonelike material. The innermost portion of the tooth consists of the **pulp cavity,** which leads to the **root canal,** a passageway for nerves and blood vessels into the tooth. The nerves and blood vessels enter the tooth through the **apical foramen** at the tip of the root of the tooth. The teeth occur in depressions in the mandible or maxilla called **alveolar sockets** and are anchored into the bone by the periodontal ligament.

There are four different types of teeth in the adult mouth:

- **Incisors** are flat, bladelike front teeth that nip food. There are eight incisors in the adult mouth.
- **Canines,** or **cuspids,** are the pointed teeth just lateral to the incisors that function in shearing food. There are four canines in the adult, and they can be identified as the teeth with just one cusp, or point.
- **Premolars,** or **bicuspids,** are lateral to the canines; they grind food. There are typically eight premolars in adults, and they can be identified by their two cusps.
- **Molars** are found closest to the oropharynx. These grind food, and there are 12 molar teeth in the adult mouth (including the wisdom teeth). Molars typically have three to five cusps.

Examine a human skull or dental cast and identify the characteristics of the four different types of teeth (figure 42.4).

Humans have two sets of teeth: the **deciduous,** or "milk," **teeth** appear first, and these are replaced by the **permanent teeth.** There are 20 deciduous teeth. In children there are no deciduous premolar teeth, and there are only 8 deciduous molar teeth. In adults there are 8 premolar teeth and 12 molar teeth. Compare figure 42.4, the adult pattern, to figure 42.5, the deciduous teeth.

Frequently, the pattern of tooth structure is represented by a **dental formula,** which describes the teeth by quadrants. The dental formula for the deciduous teeth is

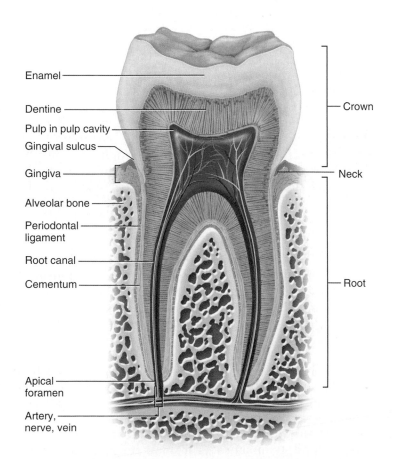

FIGURE 42.3 Tooth, Longitudinal Section

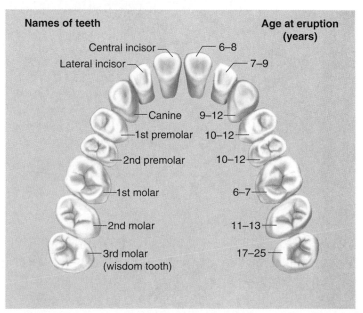

FIGURE 42.4 Upper Teeth of an Adult

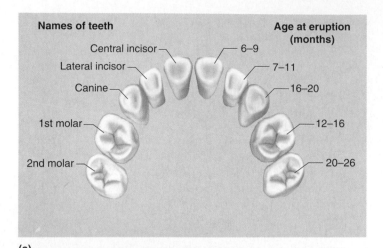

Names of teeth

Central incisor — 6–9
Lateral incisor — 7–11
Canine — 16–20
1st molar — 12–16
2nd molar — 20–26

Age at eruption (months)

(a)

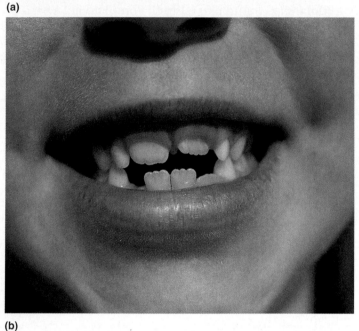

(b)

FIGURE 42.5 Deciduous Teeth (a) Upper teeth; (b) replacement of teeth.

illustrated here. I = incisor, C = canine, P = premolar, and M = molar.

Deciduous teeth (20 total)

dental formula	I	C	P	M	
	2	1	0	2	One side (quadrant) Top
	2	1	0	2	Bottom

The upper numbers refer to the number of teeth in the maxilla on one side of the head. The lower numbers refer to the number of teeth in the mandible on one side of the head. The adult dental formula is as follows:

Adult (permanent) teeth (32 total)

I	C	P	M
2	1	2	3
2	1	2	3

Oropharynx

The **oropharynx** is the space behind the mouth that serves as a common passageway for air, food, and liquid. Above the oropharynx is the **nasopharynx,** which leads to the nasal cavity, and below the oropharynx is the **laryngopharynx,** which leads to the larynx and the esophagus. The oropharynx is composed of nonkeratinized stratified squamous epithelium. Muscles around the wall of the oropharynx are the **pharyngeal constrictor muscles** and are involved in swallowing. Food is moved by the tongue to the region of the pharynx, where it is propelled into the esophagus. Locate the oropharynx and the esophagus in figure 42.2.

Esophagus

The esophagus conducts food and liquid from the oropharynx, through the diaphragm, and into the stomach. Normally, the esophagus is a closed tube that begins at about the level of the sixth cervical vertebra. As a lump of food, or **bolus,** enters the esophagus, skeletal muscle begins to move it toward the stomach. The middle portion of the esophagus is composed of both **skeletal** and **smooth muscle,** while the lower portion of the esophagus is made of smooth muscle. In the lower region of the esophagus, the smooth muscle contracts, moving the bolus by a process known as **peristalsis.** The esophagus has an inner epithelial lining of **stratified squamous epithelium** and an outer connective tissue layer called the **adventitia.** Identify the four layers of the esophagus—mucosa, submucosa, muscularis, and adventitia—under the microscope. The space inside the esophagus where food passes through is called the **lumen,** which continues through the gastrointestinal tract. The lower portion of the esophagus has an **esophageal (cardiac) sphincter,** which prevents the backflow of stomach acids. **Heartburn** occurs if the stomach contents pass through the esophageal sphincter and irritate the esophageal lining.

Abdominal Portions of the Alimentary Canal

The inner structure of the body has been referred to as a "tube within a tube." The body wall forms the outer tube; the gastrointestinal tract, including the stomach, small intestine, and large intestine, forms the inner tube. Specialized serous membranes cover the various organs and line the inner wall of the **coelom** (body cavity). The membrane lining the outer surface of the gastrointestinal tract is called the **visceral peritoneum (serosa)** and continues as a double-folded membrane called the **mesentery,** which attaches the tract to the back of the body wall. In between the linings of the mesentery are arteries, veins, nerves, and lymphatics. The mesentery is continuous with the membrane on the inner side of the body wall, where it is called the **parietal peritoneum.** Locate these three membranes in figure 42.6.

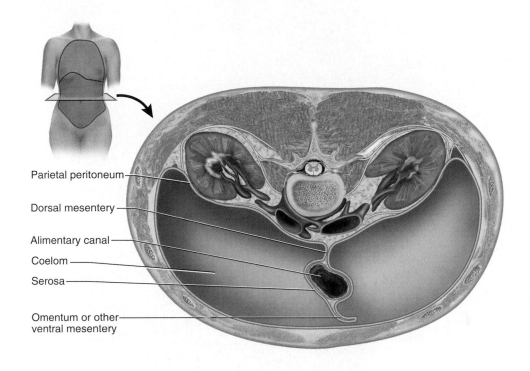

Parietal peritoneum

Dorsal mesentery

Alimentary canal

Coelom

Serosa

Omentum or other
ventral mesentery

Anterior

FIGURE 42.6 Membranes of the Gastrointestinal Tract (Idealized Drawing)

The gastrointestinal tract has a series of layers seen in microscopic sections. In general, the stomach, small intestine, and large intestine have the same layers from the lumen to the coelom. The innermost layer is the **mucosa,** which consists of a mucous membrane closest to the lumen, a connective tissue layer called the **lamina propria,** and an outer, muscular layer called the **muscularis mucosae.**

The next layer is called the **submucosa,** mostly made of connective tissue and containing numerous blood vessels. The next layer is the **muscularis (or muscularis externa),** typically made of two or three layers of smooth muscle. The muscularis propels material through the gastrointestinal tract and mixes ingested material with digestive juices. The outermost layer is called the **serosa,** or **visceral peritoneum,** and this layer is closest to the coelom. Examine a microscope slide of the gastrointestinal tract (small intestine) and locate these three membranes in figure 42.7.

Stomach

The **stomach** is located on the left side of the body and receives its contents from the esophagus. The food that enters the stomach is stored and mixed with the enzyme pepsin and hydrochloric acid to form a soupy material called **chyme.** The stomach can have a pH as low as 1 or 2. Chyme remains in the stomach as the acids denature proteins and pepsin reduces proteins to shorter

fragments. The acid of the stomach also has an antibacterial action, as most microbes do not grow well in conditions of low pH.

Examine a model of the stomach or charts in the lab and compare them to figure 42.8. Locate the upper portion of the stomach called the **cardia,** or **cardiac region.** A part of the cardia extends superiorly as a domed section called the **fundic region,** or **fundus.** The main part of the stomach is called the **body** and the terminal portion of the stomach, closest to the small intestine, is called the **pyloric region.** The pyloric region has an expanded area called the **antrum** and a narrowed region called the **pyloric canal.** The pylorus leads to the duodenum, and this opening is controlled by the **pyloric sphincter.** The left side of the stomach is arched and forms the **greater curvature,** while the right side of the stomach is a smaller arch, forming the **lesser curvature.** The inner surface of the stomach has a series of folds called **rugae,** which allow for expansion of the stomach. The contents are held in the stomach by two sphincters. The proximal **esophageal sphincter** prevents stomach contents from moving into the esophagus. The distal pyloric sphincter prevents the premature release of stomach contents into the small intestine. Locate the pyloric sphincter at the terminal portion of the stomach.

Stomach Histology Examine a prepared slide of stomach. Identify the four primary layers: **mucosa, submucosa,**

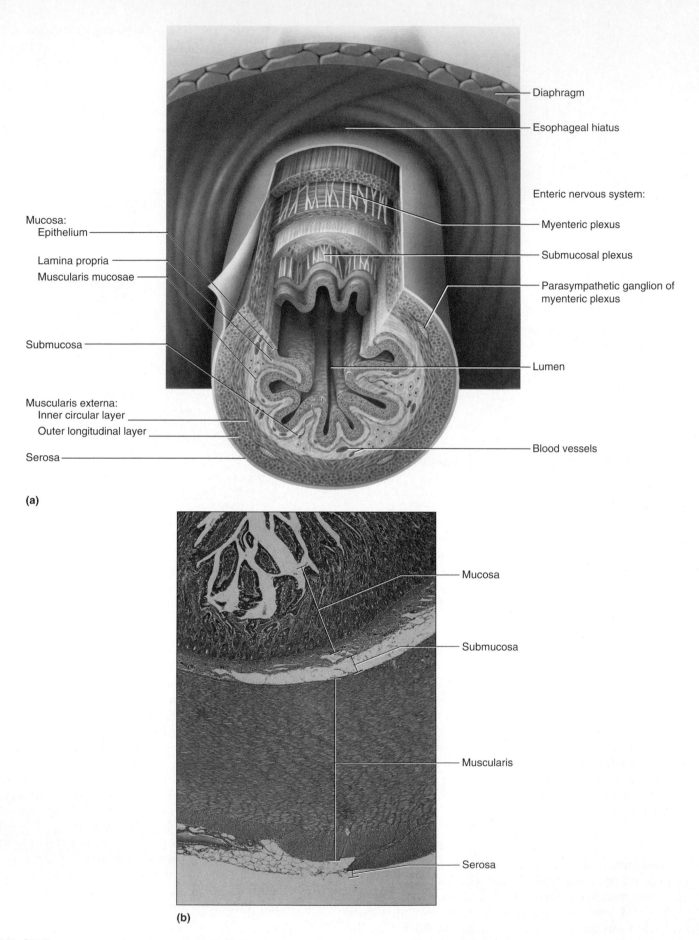

Mucosa:
   Epithelium

Lamina propria
Muscularis mucosae

Submucosa

Muscularis externa:
   Inner circular layer
   Outer longitudinal layer

Serosa

Diaphragm

Esophageal hiatus

Enteric nervous system:

Myenteric plexus

Submucosal plexus

Parasympathetic ganglion of
myenteric plexus

Lumen

Blood vessels

(a)

Mucosa

Submucosa

Muscularis

Serosa

(b)

FIGURE 42.7 **Gastrointestinal Tract, Cross Section** (a) Diagram; (b) photomicrograph (100×).

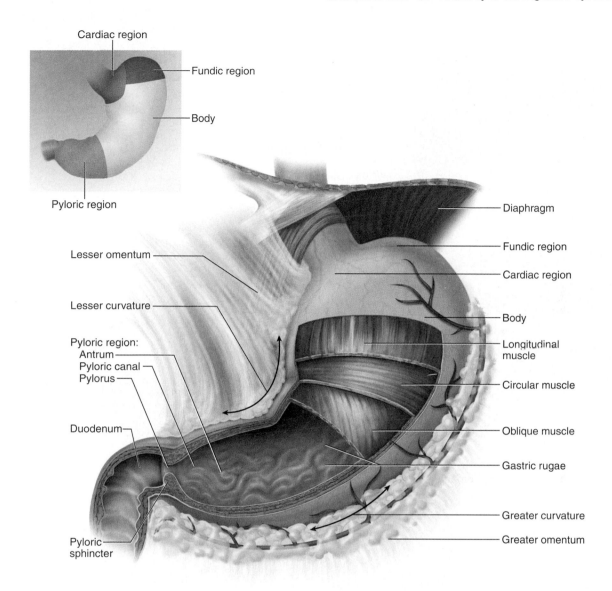

FIGURE 42.8 Gross Anatomy of the Stomach

muscularis, and **serosa.** Notice how the mucosa in the prepared slide has a series of indentations. These depressions are **gastric pits,** at the bottom of which are the **gastric glands.** These occur in the inner lining of the stomach. The mucous membrane is composed predominantly of **simple columnar epithelium. Surface mucous cells** occur in the membrane and secrete **mucus,** which protects the stomach lining from erosion by stomach acid and the **proteolytic** (protein-digesting) enzyme **pepsin.** Other specialized cells that you might find in the mucosa are **chief cells,** which secrete **pepsinogen** (the inactive state of pepsin). Chief cells contain blue-staining granules in some prepared slides. Other cells are **parietal cells,** which secrete HCl. They typically contain orange-staining granules. When pepsinogen comes into contact with HCl it is activated as pepsin.

Deeper to the mucous membrane locate the **lamina propria,** usually lighter in color. The **muscularis mucosae** is even farther away from the lumen and moves the mucous membrane. The **submucosa** is the next layer and is typically lighter in color in prepared slides.

The next layer of the stomach is the **muscularis.** In some parts of the stomach there are three layers of the muscularis—an **inner oblique layer,** a **middle circular layer,** and **an outer longitudinal layer.** The muscularis moves chyme from the stomach through the pyloric sphincter and into the small intestine. It also mixes the chyme.

The outermost layer is the **serosa,** and it is composed of a thin layer of connective tissue and **simple squamous epithelium.** Locate these structures and compare them to figure 42.9.

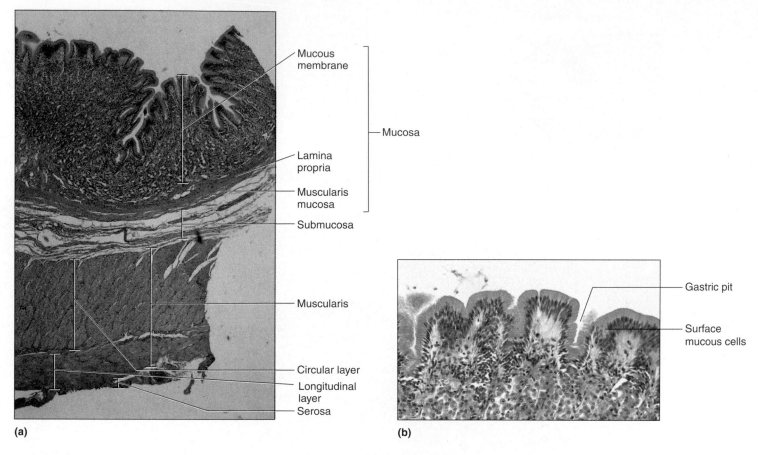

(a)

(b)

FIGURE 42.9 **Histology of the Stomach** (a) Overview (40×); (b) mucosa (100×).

Small Intestine

The **small intestine,** named because it is small in diameter, is approximately 5 m (17 ft) long. It is typically 3 to 4 cm (1.5 in.) in diameter when empty. Movement through the small intestine occurs by peristalsis, smooth muscle contraction. The primary function of the small intestine is nutrient absorption.

Locate the three major regions of the small intestine on a model or chart and compare them to figure 42.10. The first part of the small intestine is the **duodenum,** a C-shaped structure attached to the pyloric region of the stomach. The duodenum is partly retroperitoneal in that a portion of it is posterior to the parietal peritoneum. The duodenum is approximately 25 cm (10 in.) long. It receives fluid from both the **pancreas** and the **gallbladder.** The gallbladder releases **bile,** which emulsifies lipids, into the duodenum. The lipids break into smaller droplets, which increase the surface area for digestion. The bile is transported to the duodenum by the common bile duct. The pancreas secretes many digestive enzymes (proteases, amylases, and lipases) and bicarbonate, which neutralize the stomach acids. These are secreted into the duodenum by the **pancreatic duct.** The pancreatic duct often joins

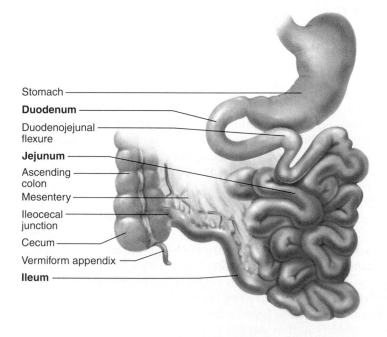

FIGURE 42.10 Gross Anatomy of the Small Intestine

with the common bile duct to form the **hepatopancreatic ampulla** (ampulla of Vater).

The second portion of the small intestine is the **jejunum,** approximately 2 m (6.5 ft) in length. The junction between the jejunum and the duodenum is known as the **duodeno-jejunal flexure.** The terminal portion of the small intestine is the **ileum,** approximately 3 m (10 ft) in length. A closure between the small intestine and large intestine is called the **ileocecal valve.** This valve keeps material in the large intestine from reentering the small intestine. Locate the small intestine and associated structures in figure 42.10.

Histology of the Small Intestine

Villi distinguish the small intestine from both the stomach and the large intestine. Villi are fingerlike projections that increase the surface area of the mucosa. Each villus contains **blood vessels,** which transport sugars and amino acids from the intestine to the liver. In addition to this, the villi contain **lacteals,** which transport triglycerides as chylomicrons via lymphatics to the venous system. When seen with the naked eye, villi give the lining of the small intestine a velvety appearance. The inner lining of the small intestine consists of **simple columnar epithelium** with **goblet cells.** You can distinguish the three sections of the small intestine by noting that the duodenum has **duodenal (Brunner's) glands** in the submucosa. The jejunum and the ileum lack these glands. The ileum is distinguished by the presence of **aggregated lymph nodules,** or **Peyer's patches,** present in the submucosa. These lymph nodules contain lymphocytes, which are activated and protect the body from the bacterial flora in the lumen of the small intestine. Compare the prepared slides of the small intestine to figure 42.11.

Large Intestine

The **large intestine** is so named because it is large in diameter. The large intestine is approximately 7 cm (3 in.) in diameter and 1.4 m (4.5 ft) in length. The function of the large intestine is the absorption of water, some vitamins, and solutes and the formation of feces. The mucosa of the large intestine is made of **simple columnar epithelium** with a large number of **goblet cells.** There are no villi present, nor aggregated lymph nodules, yet the wall of the large intestine has solitary lymph nodules.

Look at a model or chart of the large intestine and note the major regions. Compare them to figure 42.12.

- **Cecum:** first part of the large intestine. The cecum is a pouchlike area that articulates with the small intestine at the level of the ileocecal valve.
- **Ascending colon:** found on the right side of the body. It becomes the transverse colon at the right colic (hepatic) flexure.
- **Transverse colon:** traverses the body from right to left. It leads to the descending colon at the left colic (splenic) flexure.

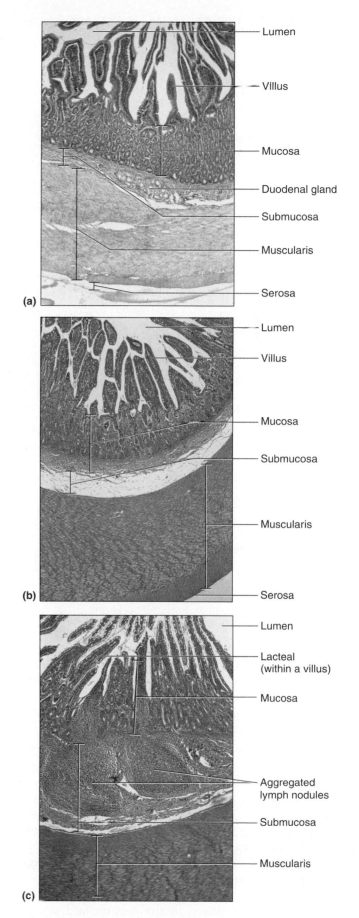

FIGURE 42.11 Three Sections of Small Intestine (40×)
(a) Duodenum; (b) jejunum; (c) ileum.

527

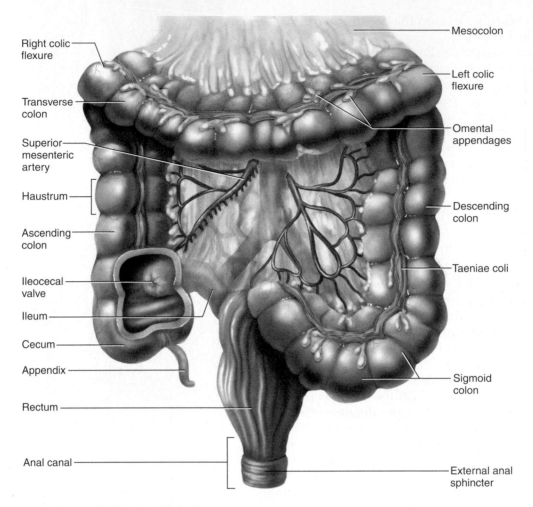

FIGURE 42.12 Gross Anatomy of the Large Intestine

- **Descending colon:** passes inferiorly on the left side of the body and joins with the sigmoid colon.
- **Sigmoid colon:** an S-shaped segment of the large intestine in the left inguinal region.
- **Rectum:** a straight section of colon in the pelvic cavity. The rectum has superficial veins in its wall called **hemorrhoidal veins.** These may enlarge and cause the uncomfortable condition known as **hemorrhoids.**

The large intestine has some unique structures. The longitudinal layer of the muscularis of the large intestine is not found as a continuous sheet but occurs along the length of the large intestine as three bands called **taeniae coli.** These muscles contract and form pouches or puckers in the intestinal tract called **haustra** (singular, *haustrum*). Another unique feature of the outer wall of the large intestine are fat lobules called, **omental (epiploic) appendages.** Locate these structures in figure 42.12.

Fecal material passes through the large intestine by peristalsis and is stored in the rectum and sigmoid colon. **Defecation** occurs as **mass peristalsis** causes a bowel movement.

Histology of the Large Intestine   Examine a prepared slide of the large intestine and compare it to figure 42.13. The large intestine is distinguished from the small intestine by the absence of villi and from the stomach by the presence of large numbers of goblet cells. Examine your slide for these characteristics.

Anal Canal

The **anal canal** is not part of the large intestine but is a short tube that leads to an external opening, the anus. Locate the anal canal in figure 42.12.

Accessory Organs
Salivary Glands

The **salivary glands,** located in the head, secrete **saliva** into the mouth. Saliva is a watery secretion that contains **mucus** (a protein lubricant) and **salivary amylase,** a starch-digesting enzyme. The average adult secretes about 1.5 liters of saliva per day.

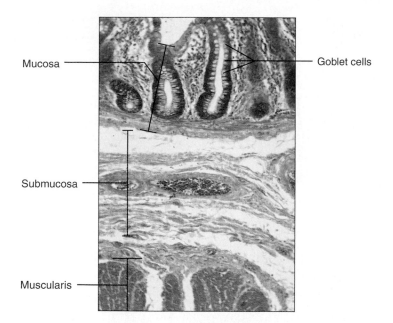

FIGURE 42.13 Histology of the Large Intestine (100×)

There are three pairs of salivary glands. The first of these, the **parotid glands,** are located just anterior to the ears. Each gland secretes saliva through a **parotid duct,** a tube that traverses the buccal (cheek) region and enters the mouth just posterior to the upper second molar. The second pair, the **submandibular glands,** are located just medial to the mandible on each side of the face. The submandibular glands secrete saliva into the mouth by a single duct on each side of the mouth inferior to the tongue. The third pair, the **sublingual glands,** are located inferior to the tongue, which open into the mouth by several ducts. Locate these salivary glands on a model of the head and in figure 42.14.

Vermiform Appendix

The **vermiform appendix** is about the size of your little finger and is located near the junction of the small and large intestines (at the region of the ileocecal valve). Locate the appendix on a torso model or chart in the lab and compare it to figure 42.15. The appendix is absent in the cat.

Omenta

The **lesser omentum** is an extension of the peritoneum that forms a double fold of tissue between the stomach and the liver. The **greater omentum** is a section of peritoneum that originates between the stomach and transverse colon and drapes over the intestines as a fatty apron. Locate the lesser and greater omenta in figure 42.16.

Liver

The **liver** is a complex organ with numerous functions; some are digestive but many are not. The liver processes

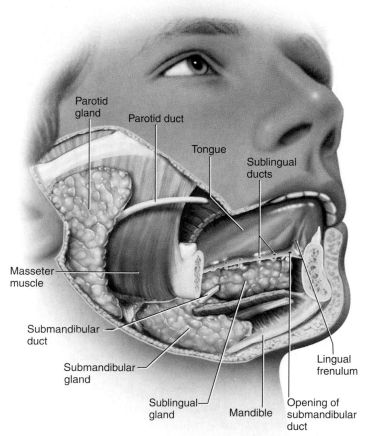

FIGURE 42.14 Salivary Glands

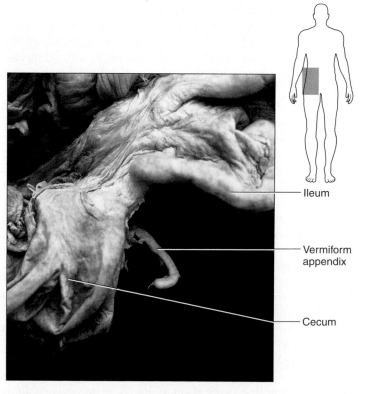

FIGURE 42.15 Vermiform Appendix

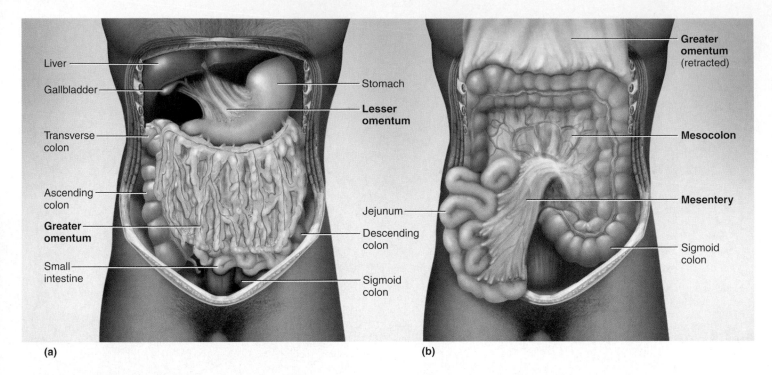

Liver

Gallbladder

Transverse colon

Ascending colon

Greater omentum

Small intestine

Stomach

Lesser omentum

Jejunum

Descending colon

Sigmoid colon

(a)

Greater omentum (retracted)

Mesocolon

Mesentery

Sigmoid colon

(b)

FIGURE 42.16 Omenta, Mesentery, and Mesocolon (a) Superficial view of greater omentum; (b) view of deeper layers.

digestive products from the vessels returning blood from the intestines and has a role in either moving nutrients into the bloodstream or storing them in the liver tissue. The liver also produces blood plasma proteins; detoxifies harmful material that has been produced by, or introduced into, the body; and produces bile.

Examine a model or chart of the liver and note its relatively large size. The liver is located on the right side of the body and is divided into four lobes, the **right, left, quadrate,** and **caudate lobes.** Only two lobes of the liver can be seen from the anterior side, and these are the large right lobe and the smaller left lobe. These two lobes are separated by a slip of mesentery called the falciform ligament, which suspends the liver from the diaphragm. In an inferior view of the liver all four lobes can be seen. The quadrate lobe is a small, rectangular lobe adjacent to the gallbladder, and the caudate lobe is at the posterior edge of the liver near the inferior vena cava. Locate the **gallbladder,** on the inferior aspect of the liver, and compare the gallbladder and liver structures to figure 42.17.

Liver Histology   Examine a prepared slide of liver tissue. Note the hexagonal structures in the specimen. These are known as **liver lobules.** Each lobule has a blood vessel in the middle called the **central vein.** Vessels that carry blood to the central vein are the liver **sinusoids,** and these are lined with a double row of cells called **hepatocytes.** Hepatocytes carry out the various functions of the liver as just described. The tissue of the liver is extremely vascular and functions as a sponge. Fresh, oxygenated blood

from the hepatic artery and deoxygenated blood from the hepatic portal vein mix in the liver. **Kupffer cells** are macrophages that occur throughout the liver tissue and function as phagocytic cells. Locate the lobules, central vein, sinusoids, bile ductule, and hepatocytes in a prepared slide, using figure 42.18 to aid you.

Pancreas

The **pancreas** is located inferior to the stomach and on the left side of the body. It has both endocrine and exocrine functions. The hormonal function of the pancreas is covered in Laboratory Exercise 28. The exocrine function of the pancreas is studied in Laboratory Exercise 43. The pancreas consists of a **tail,** found near the spleen; an elongated **body;** and a rounded **head,** located near the duodenum. Enzymes and buffers pass from the tissue of the pancreas into the **pancreatic duct** and then into the duodenum. Locate the pancreatic structures, as represented in figure 42.19.

Gallbladder

The **gallbladder** is located just inferior to the liver. The liver is the site of bile production, and the bile flows from the liver through **left** and **right hepatic ducts** to enter the **common hepatic duct.** Once in the common hepatic duct the bile flows into the **cystic duct** and then is stored in the gallbladder where it is concentrated. As the stomach begins to empty its contents into the duodenum the gallbladder constricts, and bile flows from

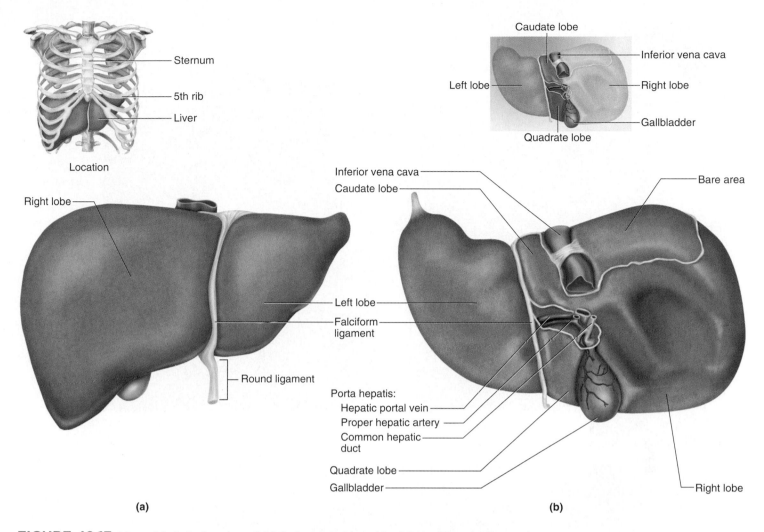

FIGURE 42.17 Liver (a) Anterior view; (b) inferior view (dorsal is at top of liver in diagram).

the gallbladder back into the cystic duct and into the **common bile duct,** which empties into the duodenum. Locate these structures on a model and compare them to figure 42.19.

Cat Dissection

Prepare for the dissection by obtaining a cat and dissection equipment. Remember to place all excess tissue in the appropriate waste container and not in a standard wastebasket or down the sink!

Begin your study of the digestive system of the cat by locating the large **salivary glands** around the face. You must first remove the skin from in front of the ear. If you have dissected the face for the musculature you may have already removed the **parotid gland,** a spongy, cream-colored pad anterior to the ear. You can also locate the **submandibular gland** as a pad of tissue slightly anterior to the angle of the mandible and lateral to the digastric muscle. The **sublingual gland** is just anterior

to the submandibular gland and is an elongated gland that parallels the mandible. Use figure 42.20 to help you locate these glands.

Dissection of the head of the cat should be done only if your instructor gives you permission to do so. To dissect the head you should first use a scalpel and make a cut in the midsagittal plane from the forehead of the cat to the region of the occipital bone. Cut through any overlying muscle in the cat. You should then use a small saw and gently cut through the cranial region of the skull, being careful to stay in the midsagittal plane. Once you make an initial cut through the dorsal side of the head, you should cut the mental symphysis of the cat, thus separating the mandible.

You can then use a long knife (or scalpel) to cut through the softer regions of the skull, brain, and tongue. Use a scalpel to cut through the floor of the oral cavity to the hyoid bone. It is best to use a scissors or small bone cutter to cut through the hyoid bone. This should allow you to open the head and examine the structures seen

Hepatocytes

Stroma

Central vein

Hepatic triad:

Branch of hepatic portal vein

Branch of proper hepatic artery

Bile ductule

Bile canaliculi

Hepatic sinusoid

Stroma

(a)

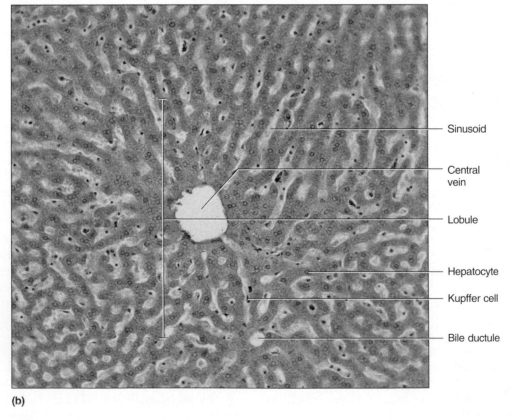

Sinusoid

Central vein

Lobule

Hepatocyte

Kupffer cell

Bile ductule

(b)

FIGURE 42.18 Histology of the Liver (a) Diagram; (b) photomicrograph (100×).

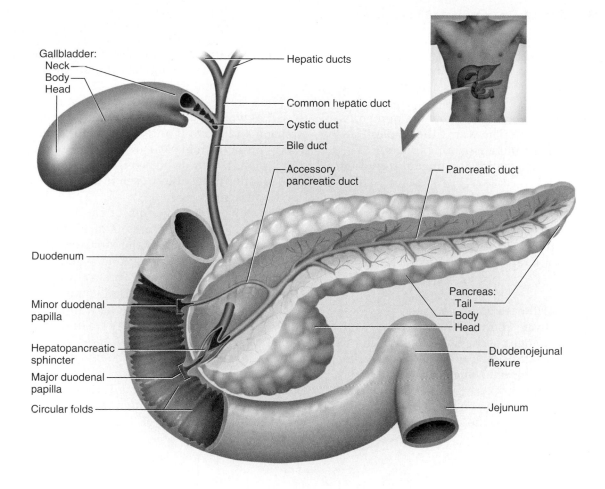

FIGURE 42.19 Gallbladder, Pancreas, and Duodenum

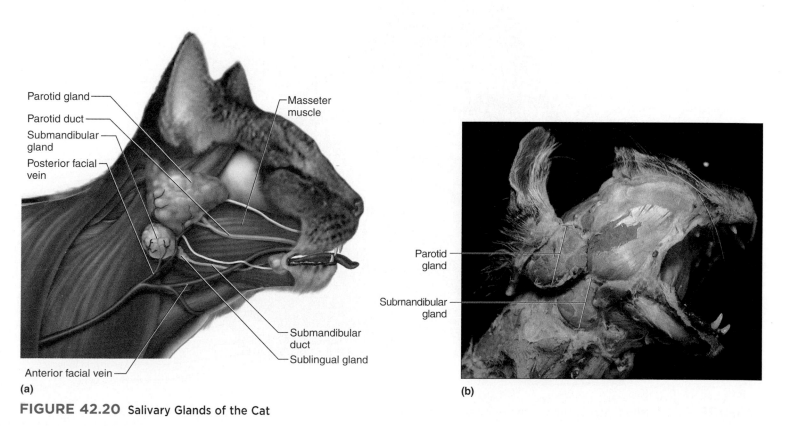

(a)

(b)

FIGURE 42.20 Salivary Glands of the Cat

in a midsagittal section, as seen in figure 42.21. Locate the hard palate, tongue, oral cavity, and oropharynx. Note how the larynx in the cat is closer to the tongue than in humans.

Examine the **tongue** of the cat and note the numerous papillae on the dorsal surface. These are **filiform papillae.** They have both a digestive and a grooming function. You can lift up the tongue and examine the **lingual frenulum** on the ventral surface.

Look for the muscular tube of the esophagus by carefully lifting the trachea ventrally away from the neck. You should be able to insert your blunt dissection probe into the oral cavity and gently wiggle it into the esophagus. The tongue and esophagus are represented in figure 42.21.

Abdominal Organs

To get a better view of the abdominal organs it is best to cut the **diaphragm** away from the anterior body wall. Carefully cut the lateral edges of the diaphragm from the ribs. You should see a fatty drape of material covering the intestines. This is the **greater omentum.** Just caudal to the diaphragm note the dark brown, multilobed **liver.** In the middle of the liver is the green **gallbladder.** To the left of the liver (in reference to the cat) is the J-shaped

stomach. Lift the liver and locate the **lesser omentum,** a fold of tissue that connects the stomach to the liver. Locate these structures in figure 42.22.

Make an incision into the stomach and locate the folds known as rugae. Place a blunt probe inside the stomach and move it anteriorly. Notice how the esophageal sphincter makes it difficult to pass through the diaphragm. If you do move the probe into the esophagus you should be able to see it as you look anterior to the diaphragm.

The stomach has an anterior cardia, a domed fundus, greater and lesser curvatures, and a pyloric region. Cut into the pyloric region of the stomach and through the duodenum. Locate a tight sphincter muscle between the stomach and the duodenum. This is the pyloric sphincter. As you move into the duodenum you may have to scrape some of the chyme away from the wall of the small intestine to be able to see the fuzzy texture of the intestinal wall. This texture is due to the presence of villi. As in the human, the small intestine is composed of three regions, the duodenum, the jejunum, and a terminal ileum. Compare the stomach and small intestine to figure 42.22.

Elevate the greater omentum to see the pancreas, along the caudal side of the stomach. The pancreas

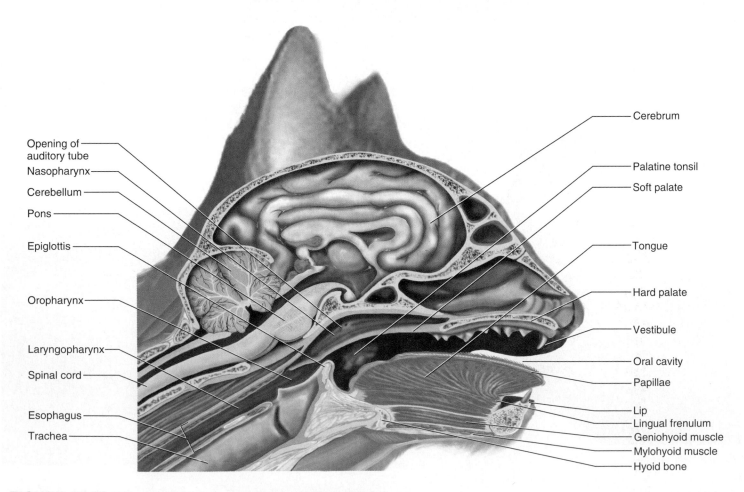

FIGURE 42.21 Midsagittal Section of the Head and Neck of a Cat

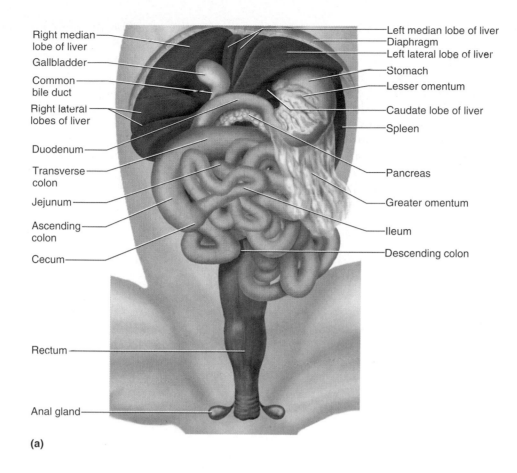

Right median lobe of liver
Gallbladder
Common bile duct
Right lateral lobes of liver
Duodenum
Transverse colon
Jejunum
Ascending colon
Cecum
Rectum
Anal gland

Left median lobe of liver
Diaphragm
Left lateral lobe of liver
Stomach
Lesser omentum
Caudate lobe of liver
Spleen
Pancreas
Greater omentum
Ileum
Descending colon

(a)

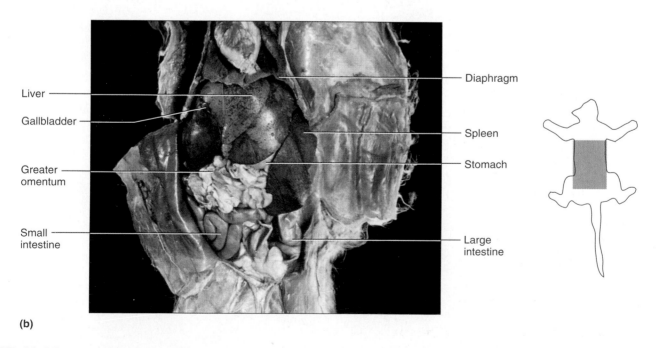

Liver
Gallbladder
Greater omentum
Small intestine

Diaphragm
Spleen
Stomach
Large intestine

(b)

FIGURE 42.22 Abdominal Organs of the Cat (a) Diagram; (b) photograph.

appears granular and brown. The tail of the pancreas is near the spleen, a brown, elongated organ on the left side of the body.

The small intestine is an elongated, coiled tube about the diameter of a wooden pencil. Note the **mesentery,** which holds the small intestine to the posterior body wall near the vertebrae. The small intestine is extensive in the cat and rapidly expands into the large intestine. There is no vermiform appendix in the cat. In humans it is found at the ileocecal junction. The large intestine in the cat is a fairly short tube, with a diameter slightly larger than your thumb. The first part of the large intestine is a pouch called the cecum. The remainder of the large intestine can be further divided into the ascending, transverse, and descending colon and the rectum. Compare these to figure 42.22. Examine the **parietal peritoneum** along the inner surface of the body wall and the visceral peritoneum that envelops the intestines.

Clean Up   When you are done with your dissection carefully place your cat back in the plastic bag.

REVIEW SECTION

Anatomy of the Digestive System

Name _____ Date _____

Lab Section _____ Time _____

Review Questions
Matching

Match the terms on the left with those on the right. A term may be used more than once or not at all.

Set One

1. pancreas
2. descending colon
3. part of the stomach closest to the small intestine
4. middle portion of the small intestine
5. distal portion of the small intestine

a. jejunum
b. pyloric region
c. ileum
d. alimentary canal
e. accessory organ

Set Two

6. outer surface of the stomach
7. layer adjacent to the lumen of intestine
8. cell type in the muscularis
9. location of the villi

a. submucosa
b. serosa
c. mucosa
d. smooth muscle

10. What is semidigested food in the stomach called?

11. Where do you find lacteals in the digestive tract?

12. What membrane holds the tongue to the floor of the mouth?

13. What part of the tooth is found above the neck?

14. What is the layer of a tooth superficial to the dentine?

15. What are the adult teeth directly posterior to the canine teeth called?

16. The segments or pouches of the large intestine have what particular name?

17. What are the names of the salivary glands anterior to the ear?

18. Where is the lesser omentum found?

19. Where does the cystic duct take bile for storage?

20. Stomach acidity is in the range of pH 1 to 2. How might this inhibit the growth of ingested bacteria?

21. Trace the flow of bile from the liver to the duodenum, listing all the structures that come into contact with the bile on its journey.

22. How does the large intestine differ from the small intestine in length?

23. How does the large intestine differ from the small intestine in diameter?

24. Name two functions of the pancreas.

25. Label the following illustration using the terms provided

ascending colon	vermiform appendix	sigmoid colon
rectum	tongue	small intestine
mouth	duodenum	esophagus
stomach	parotid gland	liver

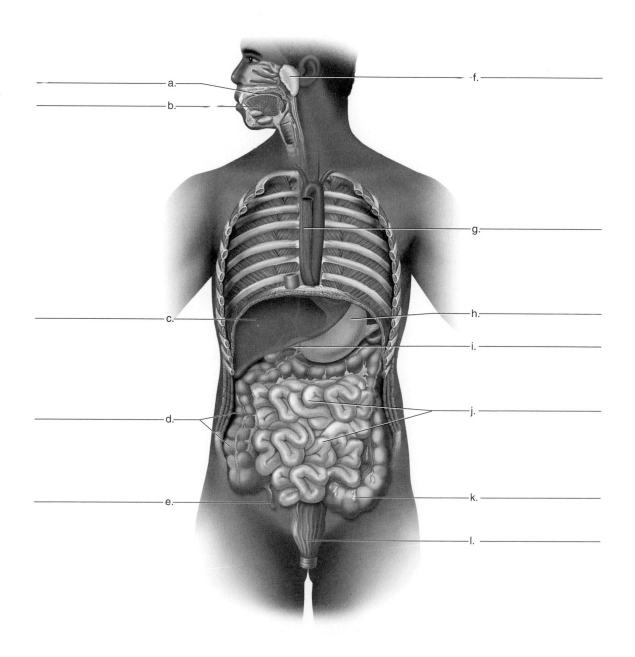

Notes

LABORATORY

Digestive Physiology

INTRODUCTION

An understanding of the digestive process is fundamental to the study of human physiology. We obtain almost all the raw materials the body uses for growth, development, life-sustaining energy, and other metabolic functions from digestion. Digestive physiology involves the physical and chemical breakdown of material, absorption of nutrients, coordination between sections of the digestive tract and accessory organs (liver secretions and processing of absorbed foods), and elimination of fecal material. In this exercise you look at the physical (mechanical) and chemical digestion of food. Digestive physiology is covered in the Saladin text in chapter 25, "The Digestive System" and chapter 26, "Nutrition and Metabolism."

In this experiment you analyze several enzymes important in digestion and their effectiveness in digesting four materials commonly found in plant or animal tissue: starch, lipid, cellulose, and protein. The goal of these experiments is to determine if various enzymes in the human digestive tract can break down the four substances into smaller products. If the enzymes are effective, then starch should be broken down into sugars, lipids broken down into monoglycerides and fatty acids, cellulose into sugars, and proteins into amino acids.

Furthermore, you examine the mechanical breakdown of food that precedes most of the chemical processes. By increasing the surface area of solid food, you will determine if the rate of digestion increases.

OBJECTIVES

At the end of this exercise you should be able to

1. discuss the importance of catabolism in the digestive process;
2. describe how an enzyme is important in chemical digestion;
3. list possible human digestive enzymes, as determined by experiment;
4. demonstrate the effectiveness of enzymes in food digestion;
5. describe the mechanisms involved in the mechanical digestion of food;
6. state one of the limitations of human digestive enzymes.

MATERIALS

Microscopes

General Supplies

80 test tubes (15 mL each)

12 test tube holders

6 test tube racks

12 test tube brushes and soap

Warm water bath, set at 35° C with test tube racks and thermometer

Hot water bath or hot plates and 400 mL beakers for hot bath (100° C) and thermometer

6 permanent markers

18 pipettes of 5 mL each

6 pipette pumps

7 100 mL beakers to pour from stock bottle for distribution

1 250 mL bottle of tap water

6 small spatulas

100 mL 1% pancreatin solution, fresh

300 mL 1% alpha amylase solution, fresh

100 mL 0.1% alpha amylase solution, fresh

300 mL Benedict's reagent (6 50 mL bottles)

Parafilm® squares (10 per table)

Biohazard container or container with 10% bleach solution

Starch Digestion

6 dropper bottles of iodine solution

250 mL of 0.5% potato starch solution

6 boxes of microscope slides (1 per table)

6 boxes of coverslips (1 per table)

6 eyedroppers

Sugar Test

100 mL of 1% maltose solution

6 10 mL graduated cylinders

Cellulose Digestion

Cellulose—2 cotton balls cut into small pieces

6 small dropper bottles of water

Lipid Digestion

200 mL litmus cream

6 dropper bottles of "acid solution" (lemon juice or vinegar)

6 dropper bottles of 1% sodium hydroxide solution (NaOH)

12 pairs of goggles (during use of NaOH)

12 pairs of latex gloves

Protein Digestion

25 mL of 1% BAPNA solution; BAPNA (benzoyl-DL-arginine-p-nitroanilide) is available from Sigma B4875 or Aldrich 85,711-4

Surface Area and Digestion

2 large potatoes per class

2 breadboards

2 #4 cork borers

2 small rods (applicator sticks) to push through the borers

12 rulers (15 cm)

6 razor blades

2 mortars and pestles

PROCEDURE

Read all the experiment before beginning! Your instructor may want you to do these experiments in student pairs or as members of a larger group. If you do these experiments as part of a larger group, make sure that you *witness* and *record* the results of your group. **Do not discard any material until everyone in your group has seen the result** or until your instructor directs you to do so. While you are waiting for one set of materials to incubate you can begin the next set of experiments.

Label all test tubes with your *group name* and the *test tube number.* This is important because removal of the wrong tube will give you erroneous results, and the removal of a tube belonging to another group will produce the wrong result not only for your group but for the other group as well.

Tests in this exercise are divided into **reagent tests** and **experimental tests.** The reagent tests allow you to see the results of an experiment when using known samples. The experimental tests determine if a digestive reaction has taken place.

The Nature of Enzymes

The function of a digestive enzyme is to break down **macromolecules** in the stomach and small intestine into smaller molecules in a process known as **catabolism.** Enzymes allow chemical changes to occur at body temperatures by lowering the activation energy of chemical reactions. These reactions occur faster because of the presence of enzymes.

The material acted on by enzymes is called the **substrate** and the result of the reaction produces the **product.** In catabolic reactions a substrate is broken down into products but there are **synthesis reactions** where substrates are joined together to form a product. This occurs in anabolic reactions. Digestion utilizes catabolic reactions. This is shown in figure 43.1.

Starch Digestion

Starch digestion occurs by the breakdown of a large molecule of starch into smaller molecules of sugar by enzymes called **amylases.** Amylases catabolize starches as represented by figure 43.2. One way to analyze the effectiveness

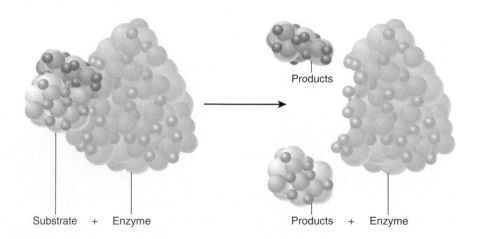

FIGURE 43.1 Catabolic Reactions with Enzymes

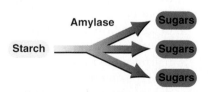

FIGURE 43.2 Starch Digestion

of amylases is to see if the enzyme removes starch from solution. The decrease in or absence of starch from solution after the reaction indicates that amylases are present.

Experiment 1: Reagent Test—Iodine

You will use a *reagent test* to determine the presence of starch. This test uses **iodine.** If starch is present, iodine produces a *blue-black color,* a **positive result.** If starch is absent, then the solution remains *yellow,* and this is a **negative result.**

1. Label a test tube with your group name and the number 1.
2. Pour a small amount of starch solution from the stock bottle *into* a small beaker. *Never return excess solution to the stock bottle!*
3. Take a pipette pump and pipette and place 1 to 2 mL of starch solution from the beaker in the test tube labeled #1.

Caution!   Never pipette material by mouth—use a pipette pump or bulb.

---

4. Add 4 or 5 drops of iodine solution to the starch solution in test tube #1. The solution should turn blue or black if starch is present. Save this test result for future comparison and record the outcome of your test.

Color of reaction (blue/black or yellow): _____

Test result (positive/negative for starch): _____

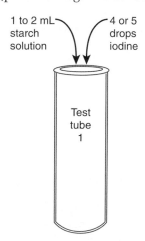

Experiment 2: Determination of Starch Digestion by Amylase

1. Using a pipette, remove 5 mL of starch solution from the beaker that received the stock solution, and place it in a small test tube that you have labeled with your group name and the number 2a.

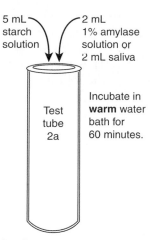

2. Into a small beaker pour a small amount of 1% amylase from the stock bottle. With a new pipette, withdraw 2 mL of amylase solution from the beaker and add it to the starch solution. Discard any remaining solution in your pipette in the sink. Do not return it to the beaker. You can substitute 2 mL of saliva for the amylase if directed by your instructor. Do not use your saliva if you recently had sugar in your mouth. Remember to treat bodily fluids with standard precautions. Stir the solutions in the test tube by flicking the bottom of the tube gently with your finger.
3. Incubate the mixture in test tube #2a in a *warm* water bath (35° C) for 60 minutes.
4. After 60 minutes take 1 to 2 mL of the solution from test tube #2a and place it in a new test tube labeled 2b.

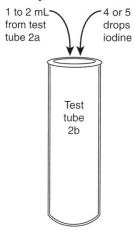

5. Test the solution in #2b with 4 to 5 drops of iodine solution and record your results.
6. Compare the results of your experiment with test tube #1.

Color of the solution with iodine: _____

Test results (positive/negative): _____

7. From test tube #2b, withdraw a small sample with an eye dropper and place a drop on a microscope slide.

8. Examine the slide under low power and look for starch grains. Compare this slide to one that you made from a drop of starch from the *undigested* starch solution (from test tube #1). Is there a difference? Record your observations in the following space.

Observations:

Do **not** discard the contents of tubes #2a and #2b at this time.

Sugar Production

The preceding experiment tested for the *decrease* in starch as a way to determine if a catabolic reaction had taken place. Another way to look at the effectiveness of amylase is to determine the *presence* of sugar after the exposure of starch to the enzyme. In this experiment, instead of looking for the absence of the **substrate** (starch) to determine enzyme effectiveness, you can test for the formation of the **product** (sugar). If amylase is effective, then the presence of sugar in solution, after exposure to the enzyme, indicates the conversion of starch to sugar.

Experiment 3: Reagent Test—Benedict's Test

The first test in this section involves the detection of sugar in a solution. This is done with the Benedict's test.

Caution! Benedict's reagent is poisonous. Use caution when mixing the solutions in the test tube.

1. Label a test tube with the number 3 and your group name and then pipette 1 to 2 mL of 1% maltose solution into the test tube. Maltose is a sugar.
2. Using a pipette and a pipette pump (*never* by mouth), add 1 mL of Benedict's reagent to the maltose solution in the test tube. Hold the test tube by the top and gently flick the bottom of the tube with the pad of your index finger to mix the contents.
3. Place this mixture in a *hot* (100° C) water bath for about 10 minutes. If the mixture turns green, yellow, orange, or brick red, then sugar is present.

The sequence of colors indicates the presence of sugar in increasing amounts:

Blue = 0

Green = trace

Yellow = moderate

Orange, red = large amounts

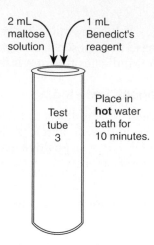

If there is no color change (for example, if the solution stays blue) after 10 minutes, then remove the tube from the bath and note the absence of sugar (a negative result). Record your results.

Color of test solution: _____

Test result (positive/negative for sugar): _____

Experiment 4: Determination of Sugar (Maltose) Production by Amylase

1. Place 4 mL from test tube #2a (the starch and amylase reaction) and put it in a test tube labeled with your group name and the number 4.
2. To the solution in test tube #4, add 4 mL of Benedict's reagent.
3. Place the tube in a hot water bath (100° C) for 10 minutes. Record your results.

Color of solution with Benedict's reagent:

Presence of sugar: _____

Effectiveness of digestion: _____

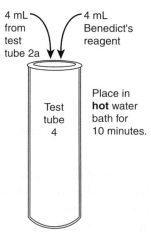

Cellulose Digestion

Cellulose is composed of repeating **glucose** units. Starch is also composed of repeating glucose units. If amylase converts cellulose to sugar, then the presence of sugar in solution indicates that the enzyme amylase is effective in breaking down cellulose into glucose. This is the same process as experiment 4, except that the substrate is changed from starch to cellulose. To test for the effectiveness of amylase in digesting cellulose, incubate a cellulose mixture with amylase or saliva in a warm water bath. A positive reaction indicates the presence of sugar after incubation.

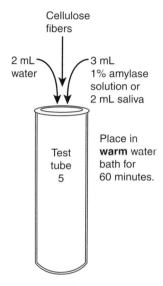

Experiment 5: Determination of Sugar Production from Cellulose

1. Place a small pinch of cellulose fibers (the size of a green pea) in a test tube and add 2 mL of water and either 3 mL of 1% amylase solution or 2 mL of saliva. You should have a total volume of 4–5 mL. Make sure the majority of cellulose fibers are not stuck to the sides of the test tube. You can use a small spatula to push the fibers into the liquid, if necessary. The test tube should be labeled with your group name and the number 5.
2. Incubate for 60 minutes in a *warm* water bath.
3. After 60 minutes remove the solution from the bath and add 4 mL of Benedict's reagent.
4. Place the tube in a *hot* water bath (100° C) for 10 minutes. Record the presence or absence of sugar. A positive reaction to the Benedict's test indicates the presence of sugar after incubation.

Color of Benedict's test: _____

Presence of sugar: _____

Effectiveness of digestion: _____

Lipid Digestion

In the digestive process some **lipids (triglycerides)** are broken down into **monoglycerides** and **free fatty acids.** The fatty acids make the solution more **acidic,** thus lowering the **pH.** See figure 43.3. The decrease in pH can be used as a measure of digestion. The greater the acidity of the solution, the greater the digestion.

Experiment 6: Reagent Test—Litmus Test

1. To determine the effectiveness of the litmus reagent pour 3 mL of **litmus cream** into a test tube labeled with your group name and the number 6. Litmus cream consists of dairy cream (containing fat) and litmus powder (a pH indicator).
2. Add the acid solution (lemon juice or vinegar) drop by drop while flicking the bottom of the test tube until the color changes from *blue to pink.* Do not add too much acid but just enough to change the color. A pink color indicates an acidic condition.
3. Now add 1% sodium hydroxide solution (NaOH) a drop at a time until you reverse the color. This indicates that the solution is alkaline.

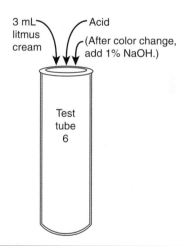

 Caution! Sodium hydroxide is a caustic solution that can dissolve your skin, so be careful! Wear protective gloves and goggles as you add the sodium hydroxide solution.

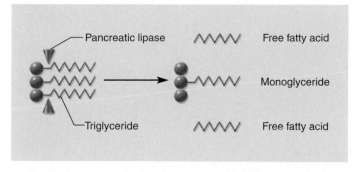

FIGURE 43.3 Breakdown of a Triglyceride into Two Free Fatty Acids and a Monoglyceride

Experiment 7: Determination of Lipid Digestion by Pancreatin

1. Pipette 3 mL of litmus cream into a small test tube labeled with your group name and the number 7.
2. Add 1 mL of pancreatin solution (pancreatin contains many digestive enzymes, including enzymes that digest lipids, called lipases) to the litmus cream and stir well with a spatula.
3. Incubate for 60 minutes in the *warm* (35° C) water bath.
4. After 60 minutes look for a color change. If the cream turns pink, then digestion has occurred. If the color remains blue, then no digestion has occurred. Record your results.

Color of solution after incubation (pink/blue): _____

Condition of the litmus cream (acidic/alkaline): _____

Extent of digestion: _____

Change in smell from normal cream: _____

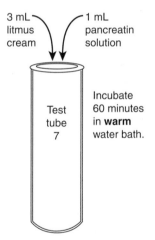

3 mL litmus cream → 1 mL pancreatin solution

Test tube 7

Incubate 60 minutes in **warm** water bath.

Protein Digestion

Proteins are made of **amino acids.** These amino acids are linked by **peptide bonds** to form **polypeptide chains.** If the chains are long enough and branch, they form **proteins.** Proteins are split into amino acids by a number of digestive enzymes called proteases. Aminopeptidases cleave amino acids from the amine end of polypeptide chains. Carboxypeptidases cleave amino acids from the carboxyl end. Pepsin cleaves polypeptides between tyrosine (Tyr) and phenylalanine (Phe) amino acids (see figure 43.4). The effectiveness of proteases can be studied with the use of a chromogenic (color-producing) substance known as BAPNA (benzoyl-DL-arginine-p-nitroanilide). If proteases are present they release a yellow aniline dye from the larger BAPNA molecule.

Experiment 8: Determination of Protein Digestion by Pancreatin

1. Label a test tube with your group name and the number 8.
2. Pipette 1 mL of BAPNA solution into the tube. To this add 1 mL of pancreatin solution and stir by flicking the bottom of the test tube.
3. Place the tube in the *warm* water bath (35° C) for 15 minutes.
4. After 15 minutes, remove the tube from the bath and examine the test tube to see if it has turned yellow. Yellow indicates protein-digesting abilities of the enzyme. Record your results.

Color of the tube (yellow/clear): _____

Digestion capability (yes/no): _____

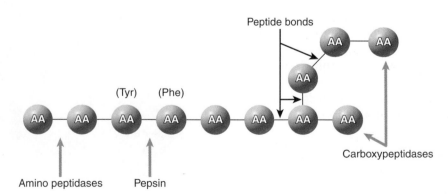

FIGURE 43.4 Enzymatic Catabolism of Proteins "AA" represents amino acids. Carboxypeptidase removes amino acids on the carboxyl end and amino peptidase removes amino acids on the amino end.

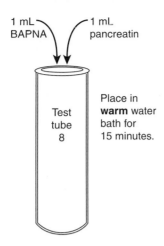

1 mL BAPNA   1 mL pancreatin

Test tube 8

Place in **warm** water bath for 15 minutes.

The amount of digestion can be estimated by the color of the test. As described in experiment 4, blue indicates no sugar, with green, yellow, orange, and brick red indicating increasing amounts of sugar. In which tube did the greatest amount of digestion take place? In which tube did the least amount of digestion take place? Record your results.

Color of tube (entire potato): _____

Color of tube (mashed potato): _____

Relative digestion of intact piece: _____

Relative digestion of mashed piece: _____

Experiment 9: Surface Area and Digestion

As food is broken down physically, the surface area increases relative to the volume of the food. This experiment examines the effectiveness of an increase in surface area for digestion.

1. Label two test tubes, each having your group name and the number 9. Write "entire" on one tube and "mashed" on the other.
2. Carefully plunge a #4 cork borer into the center of a raw potato. *Make sure your hand is not on the receiving end of the cork borer! Use a breadboard.*
3. Remove the cork borer from the potato and, using a small rod, push the potato cylinder from the borer.
4. Using a ruler and a razor blade, cut the potato cylinder into two pieces, each 1 cm in length. *It is important that the two pieces are the same length!*
5. Thoroughly rinse both pieces, and place one piece of potato in the test tube labeled "entire."
6. Chop the other piece into several little pieces and mash these with a mortar and pestle until they are well pulverized.
7. Carefully place all the mashed potato in the other test tube with the "mashed" label.
8. Pipette exactly 4 mL of 0.1% amylase solution* (or 0.5 mL of saliva and 4 mL water) into each test tube and incubate them for 20 minutes in the *warm* (35° C) water bath.
9. After incubation, place exactly 2 mL of Benedict's reagent into each tube and place these two tubes in the hot water bath for 10 minutes.

*You can use a solution of 0.1% alpha amylase or dilute the 1% stock solution by adding 1 mL of 1% amylase solution to 9 mL of water.

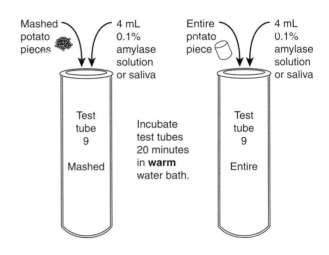

Mashed potato pieces   4 mL 0.1% amylase solution or saliva

Test tube 9 Mashed

Incubate test tubes 20 minutes in **warm** water bath.

Entire potato piece   4 mL 0.1% amylase solution or saliva

Test tube 9 Entire

When you clean up, place all the test tubes, pipettes, or other material in either a biohazard container or a 10% bleach solution.

Notes

REVIEW SECTION

Digestive Physiology

Name _____ Date _____

Lab Section _____ Time _____

Review Questions

Matching

Set One

Match the enzyme in the left column with the product produced in the right column.

1. amylase   a. amino acid

2. protease   b. monoglycerides and free fatty acids

3. lipase   c. sugar

Set Two

Match the test result in the left column with the color in the right column.

4. positive starch test with iodine   a. orange-red

5. positive sugar test with Benedict's solution   b. pink

6. positive lipid test with litmus cream   c. blue-black

7. positive protein test with BAPNA   d. yellow

8. negative starch test with iodine

9. Substrates are converted into what substances by enzymes?

10. What is catabolism?

11. What are the bonds that hold amino acids together?

12. Explain why a negative iodine test for starch would indicate a positive result for the enzymatic degradation of starch by amylase.

13. What would a positive iodine test indicate in the preceding reaction?

14. Record all your negative results. Determine if these negative results indicate digestion or no digestion.

15. Explain why cellulose could or could not be digested by amylase.

16. What effect does chewing your food have on digestion? What experiment did you perform to evaluate the effectiveness of chewing for digestion?

17. Amylase is now frequently prepared with lactose or other sugars as extenders. Determine which experiments would give erroneous results if contaminated amylase were used.

18. How does amino peptidase differ from carboxypeptidase in terms of protein digestion?

LABORATORY

Urinary System

INTRODUCTION

The urinary system filters dissolved material from the blood, regulates electrolytes and fluid volume, concentrates and releases waste products, and reabsorbs metabolically important substances back into the circulatory system. **Filtration** occurs when one or more substances pass through a selectively permeable membrane while others do not. Filtration in the kidney involves both metabolic waste products (urea) and material beneficial to the body. Not all filtered material is desirable to have **excreted** from the body. Glucose and other materials, such as sodium and potassium ions, are **reabsorbed** from the kidney back into the circulatory system. The kidney **secretes** urea; some drugs; hydrogen and hydroxyl ions; and hormones, such as rennin and erythropoietin. Finally the kidneys excrete metabolic wastes, hydrogen ions, toxins, water, and salts. The system consists of two kidneys, two ureters, a single urinary bladder, and a single urethra. These topics are discussed in the Saladin text in chapter 23, "The Urinary System." In this exercise you examine the gross and microscopic anatomy of the urinary system, study the major organs as represented in humans, and dissect a mammal kidney if available.

OBJECTIVES

At the end of this exercise you should be able to

1. list the major organs of the urinary system;
2. describe the blood flow through the kidney;
3. describe the flow of filtrate through the kidney;
4. name the major parts of the nephron;
5. trace the flow of urine from the kidney to the exterior of the body;
6. distinguish between the parts of the nephron in histological sections.

MATERIALS

Models and charts of the urinary system

Models and illustrations of the kidney and nephron system

Microscopes

Microscope slides of kidney and bladder

Samples of renal calculi (if available)

Preserved specimens of sheep or other mammal kidney

Dissection Trays and Materials

Scalpels or razor blades

Forceps

Blunt (mall) probes

Latex gloves

Waste container

PROCEDURE

Overview

You can begin the study of the urinary system by locating its principal organs. Look at figure 44.1 and compare it to the material available in the lab. Find the **kidneys,** the **ureters,** the **urinary bladder,** and the **urethra.**

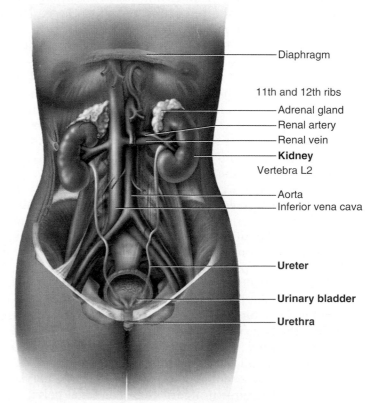

FIGURE 44.1 Major Organs of the Urinary System

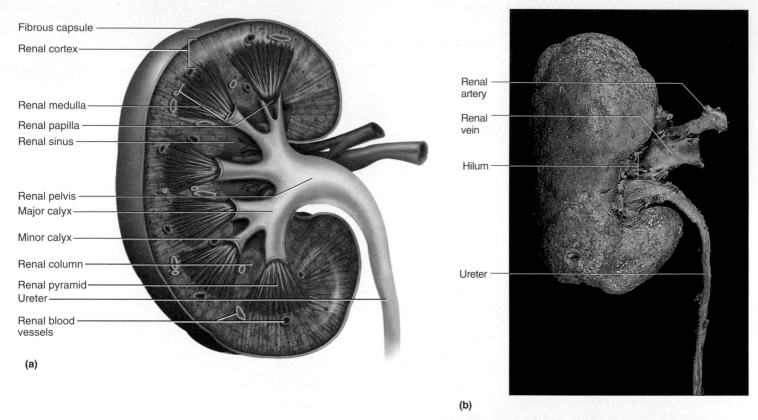

(a)

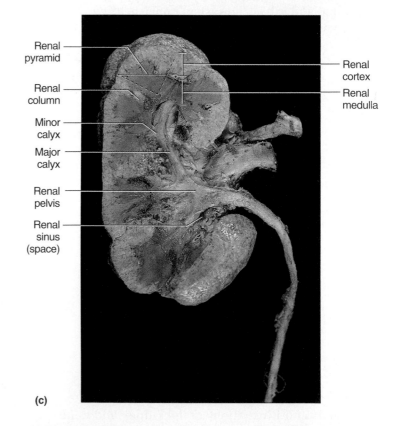

(b)

Kidneys

The kidneys are **retroperitoneal** (posterior to the parietal peritoneum) and are embedded in the **perirenal fat capsule.** The capsule cushions the kidneys, which are found mostly below the protection of the rib cage. The kidneys are located adjacent to the vertebral column about at the level of T12 to L3. The right kidney is slightly more inferior than the left.

Examine a model of the kidney and compare it to figure 44.2. Locate the outer **fibrous capsule,** a tough connective tissue layer; the outer **cortex;** and the inner **medulla** of the kidney. The kidney has a depression on the medial side where the renal artery enters the kidney and the renal vein and the ureter exit the kidney. This depression is called the **hilum.**

If you examine a coronal section of the kidney, you can find the **renal (medullary) pyramids,** separated by the **renal columns.** Each renal pyramid ends in a blunt point called the **renal papilla** (see figure 44.2). Urine drips from many papillae toward the middle of the kidney.

The urine drips into the **minor calyces** (singular, *calyx*), which enclose the renal papillae. Minor calyces are somewhat like funnels that collect fluid. Minor calyces lead to the **major calyces** and these, in turn, conduct urine into the large **renal pelvis.** The renal pelvis is found in the **renal sinus.** The renal pelvis is like a glove in a coat pocket. The pocket is the renal sinus, and the membranous glove that occupies the space is the renal pelvis. Locate these structures in figure 44.2. The renal pelvis is connected to the ureter at the medial side of the kidney.

(c)

FIGURE 44.2 **Gross Anatomy of the Kidney, Entire and Coronal Section** (a) Diagram of coronal section of the kidney and details of renal pyramid; (b) photograph of entire kidney; (c) coronal section.

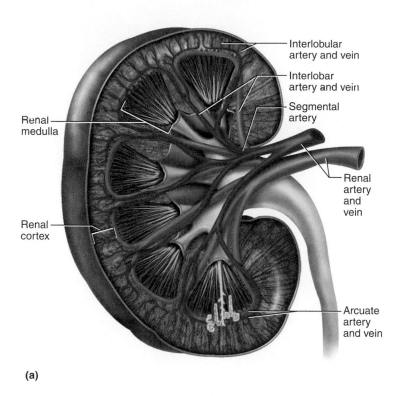

(a)

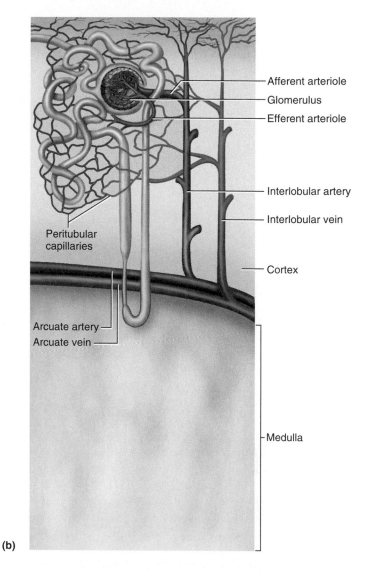

(b)

FIGURE 44.3 Blood Flow Through the Kidney (a) Major arteries and veins of the kidney; (b) diagram of blood flow.

Blood Flow Through the Kidney

There are two fluids that flow through the kidney. One is filtrate; the other is blood. The **renal arteries** take blood to the kidneys and soon divide inside the renal sinus to form the **segmental arteries.** Each segmental artery divides into several **interlobar arteries,** which take blood through the renal columns. The interlobar arteries make a sharp curve and form the **arcuate arteries,** found between the renal cortex and the renal medulla. From the arcuate artery the **interlobular arteries** penentrate into the renal cortex. These blood vessels can be seen in figure 44.3. From an interlobular artery, blood flows into an **afferent arteriole,** through the **glomerulus,** and to the **efferent arteriole.** The blood then enters the **peritubular capillaries** and returns via the **interlobular veins,** the **arcuate veins,** and the **interlobar veins** to the **renal vein.** The separation between the cortex and the medulla occurs at the level of the arcuate veins.

There are vessels that branch off of the efferent arterioles and form a network of blood vessels known as the **vasa recta.** The vasa recta vessels are found primarily in the medullary area, and though they represent only a small fraction of the capillary bed of the kidney, they are important in producing a concentrated urine. The vasa recta are found in association with **juxtamedullary nephrons.**

The blood flow in the kidney forms a portal system, in which blood flows from one capillary bed (the glomerulus) to another capillary bed (the peritubular capillaries) prior to returning to the heart. The capillaries of the glomerulus are the site of filtration, whereas the peritubular capillaries are the primary site of reabsorption.

Ultrastructure of the Kidney

Before you examine the sections of kidney under the microscope, first become familiar with the structure of the **nephron.** Examine models and charts in the laboratory and compare them to figure 44.4. Locate the **glomerular capsule (Bowman's capsule),** the **proximal convoluted tubule,** the **nephron loop (loop of Henle),** and the **distal convoluted tubule.** The glomerulus along with the glomerular capsule are known as the **renal corpuscle.** The nephron consists of the renal corpuscle, proximal convoluted tubule, nephron loop, and distal convoluted tubule. Blood travels to the nephron via the afferent arteriole. When the blood reaches the glomerulus (a cluster of capillaries in

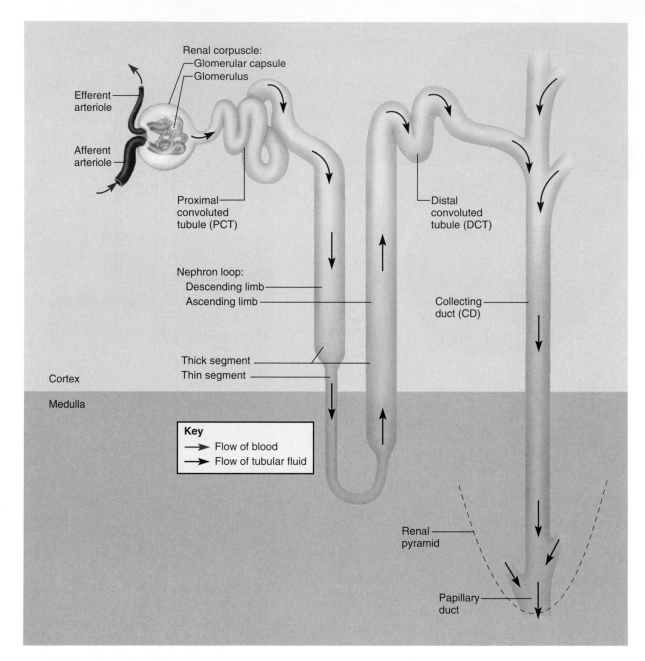

FIGURE 44.4 Nephron

the renal cortex) the plasma is filtered by blood pressure forcing fluid across the capillary membranes. Blood pressure is the driving force for renal filtration and this pressure is opposed by hydrostatic pressure in the Bowman's capsule space outside of the glomerular capillaries and the osmotic pressure that occurs due to the proteins left in the blood plasma. This filtered fluid, present in the nephron, is called **filtrate.**

As the filtrate flows through the nephron, water, glucose, and many electrolytes are returned to the blood. The urea is concentrated as it passes through the nephron and the **collecting duct,** a tube that receives the fluid from the nephrons. The amount of reabsorption in the kidney varies with the material that flows through it.

In normal conditions, glucose is completely reabsorbed into the blood. About 99% of the water in the filtrate is reabsorbed. Only 50% of the urea is excreted into the urine. The remainder of the urea returns to the blood and is filtered again. Urea is concentrated in the collecting duct, and some of it diffuses into the medulla, increasing the osmolarity in the medulla. This increases the flow of water out of the nephron, thus concentrating the urine.

There are two types of nephrons in the kidney. The majority of the nephrons are found in the cortex of the kidney and are thus called **cortical nephrons.** Those found in the medulla are called **juxtamedullary nephrons** and are far fewer in number, some extending to near the tip of the renal pyramids.

In summary, urine is produced from filtered blood. Some of the liquid portion of blood flows from the glomerulus into the nephron. Materials valuable to the body, such as glucose and other solutes, are reabsorbed by the nephron and return to the cardiovascular system by the peritubular capillaries. The main metabolic byproduct, urea, is removed from the kidney and passes as urine from the collecting ducts to the minor calyces. The volume of urine and some of the constituents found in urine are controlled by hormones such as aldosterone and antidiuretic hormone (ADH). Aldosterone increases the reabsorption of sodium and reduces water volume.

Antidiuretic hormone causes the distal convoluted tubules to reabsorb water, also decreasing the urine output.

Microscopic Examination of the Kidney

Examine a kidney slide under low power. You should see the cortex of the kidney, which has a number of round structures scattered throughout. These are the **glomeruli,** and they are composed of capillary tufts. The medullary region of the kidney slide has open, parallel spaces called **collecting ducts.** Compare the slide to figure 44.5(a).

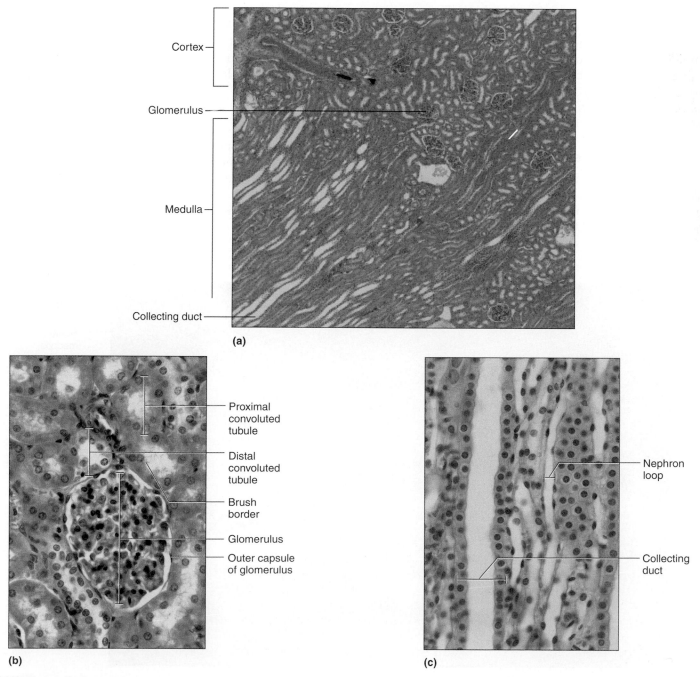

FIGURE 44.5 Photomicrographs of a Kidney (a) Overview (40×); (b) cortex (400×); (c) medulla (400×).

Examine the slide under higher magnification and locate the glomerulus and the **glomerular capsule.** The capsule is composed of simple squamous epithelium and specialized cells called podocytes. Examine the outer edge of the capsule around the glomerulus. If you move the slide around in the cortex, you should find the proximal convoluted tubules with the **brush border** or microvilli on the inner edge of the tubule. The inner surface of the tubule appears fuzzy. The microvilli increase the surface area of the proximal convoluted tubule. What functional value could be ascribed to having such a surface area?

The distal convoluted tubules do not have brush borders; therefore, the inner surface of a tubule does not appear fuzzy. The cells of the distal convoluted tubules generally have darker nuclei and relatively clear cytoplasm when compared to the cells of the proximal convoluted tubules, as seen in figure 44.5b. Examine the medulla of the kidney under high magnification and locate the thin-walled nephron loop and the larger-diameter collecting ducts, as seen in figure 44.5c.

Dissection of the Sheep Kidney

Place a sheep kidney and dissection equipment on your table. Examine the outer **capsule** of the kidney. You may see some tubes coming from a dent in the kidney. The dent is the **hilum,** and the tubes are the **renal artery, renal vein,** and **ureter.** The renal artery is smaller in diameter and has a thicker wall than the renal vein. The ureter has an expanded portion near the hilum. Make an incision in the sheep kidney a little off center in the coronal plane (see figure 44.6). This section allows you to better see the interior structures of the kidney. Locate the **renal cortex,** the outer layer of the kidney. You should also locate the **renal medulla,** which has triangular regions known as the **renal pyramids.** At the tip of each pyramid is a **papilla.** Urine from the papillae drips into the **minor calyx.** Many minor calyces lead to a **major calyx.** The major calyces take fluid to the **renal pelvis.**

Lift the renal pelvis somewhat to pull it away from the **renal sinus.** The sinus is the space in the kidney, which may be filled with adipose tissue. You should also examine the exit of the renal pelvis as it becomes the **ureter.** When you are finished with the dissection, place the material in the proper waste container provided by your instructor.

Ureters

The ureters are long, thin tubes that conduct urine from the kidneys to the urinary bladder. The ureters have **transitional epithelium** as an inner lining and smooth muscle in their outer wall. Urine is expressed by peristalsis from the kidney to the urinary bladder. Examine the models in the lab and compare them to figures 44.1 and 44.6.

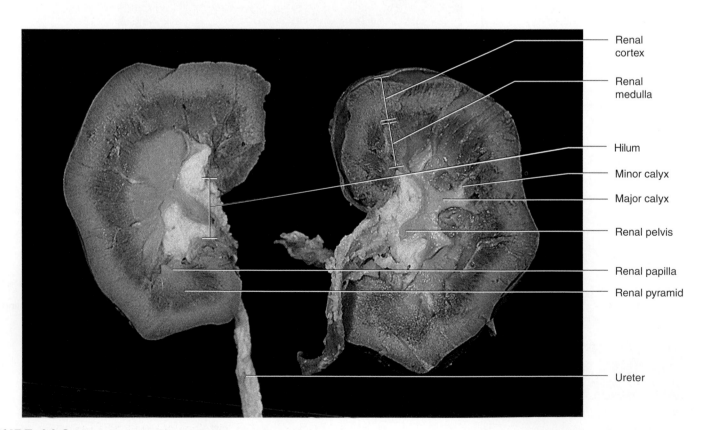

FIGURE 44.6 Dissection of a Sheep Kidney

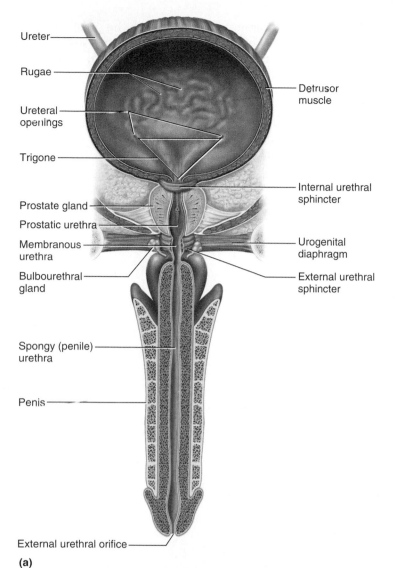

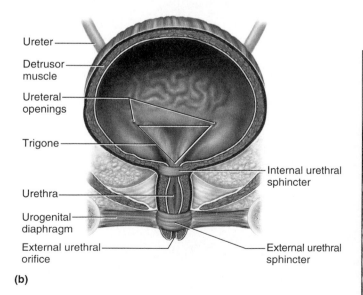

FIGURE 44.7 Bladder and Urethra, Coronal Sections
(a) Male; (b) female.

Urinary Bladder

The **urinary bladder** is located anterior to the parietal peritoneum and is thus described as being **anteperitoneal.** Locate the urinary bladder in a torso model in lab. It is found just posterior to the symphysis pubis. On the posterior wall of the urinary bladder is a triangular region known as the **trigone.** The trigone is defined by the superior entrances of the ureters and the inferior exit of the urethra. Transitional epithelium lines the inner surface of the bladder, while layers of smooth muscle known as **detrusor muscles** are found in the wall of the bladder. Compare models in lab to figure 44.7.

Histology of the Bladder

Transitional epithelium has a special role in the urinary bladder. This epithelium can withstand a significant amount of stretching (distention) when the bladder fills with urine. Examine a slide of transitional epithelium under the microscope and compare it to figure 44.8. Transitional epithelium is also discussed and illustrated in Laboratory Exercise 6. Look at the inner surface of the section for the epithelial layer. The cells are shaped somewhat like teardrops. Transitional epithelium can be distinguished from stratified squamous epithelium in that the cells of transitional epithelium from an empty bladder do not flatten at the surface of the tissue. Below the transitional epithelium is an underlying layer known as the **lamina propria.** You should also examine the smooth muscle layers of the urinary bladder that make up the detrusor muscle.

FIGURE 44.8 Histology of the Bladder (100×)

Urethra

The terminal organ of the urinary system is the **urethra.** The urethra is approximately 3 to 4 cm long in females. It passes from the urinary bladder to the **external urethral orifice** located anterior to the vagina and posterior to the clitoris, as seen in figures 44.7b and 44.9b. In figure 44.9 you can also see the position of the urinary bladder in relationship to the symphysis pubis. The urethra is about 20 cm long in males (figures 44.7a and 44.9a). It begins at the urinary bladder and passes through the prostate gland as the **prostatic urethra.** It continues and passes through the body wall as the **membranous urethra,** then exits through the penis to the external urethral orifice at the tip of the glans penis. This terminal portion of the urethra is known as the **penile,** or **spongy, urethra.** Urinary bladder infections are more common in females than in males (figures 44.7a and 44.9a). How can you explain this in terms of microorganism movement into the bladder through the urethra?

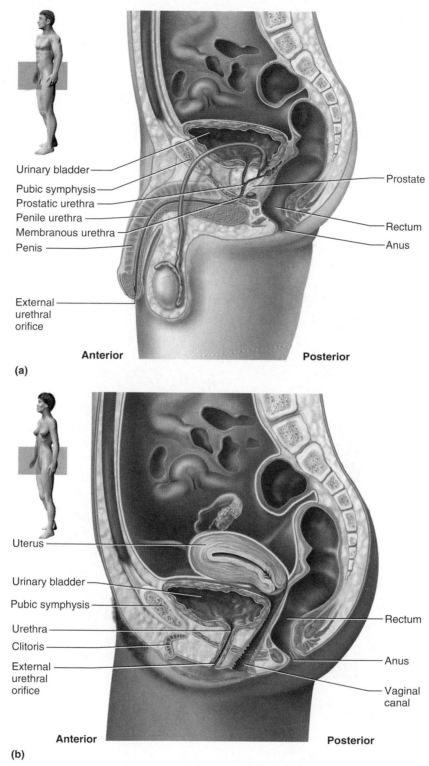

Urinary bladder
Pubic symphysis
Prostatic urethra
Penile urethra
Membranous urethra
Penis
Prostate
Rectum
Anus
External urethral orifice
Anterior
Posterior
(a)

Uterus
Urinary bladder
Pubic symphysis
Urethra
Clitoris
External urethral orifice
Rectum
Anus
Vaginal canal
Anterior
Posterior
(b)

FIGURE 44.9 Midsagittal Section of the Male and Female Pelves (a) Male pelvis; (b) female pelvis.

Cat Dissection

You should open the abdominal region of the cat, if you have not done so already. The kidneys are on the dorsal body wall of the cat and are located dorsal to the parietal peritoneum. Examine the kidneys and the structures that lead to and from the hilum of the kidney. You should locate the renal veins that take blood from the kidney. The veins are larger in diameter than the renal arteries and they are attached to the posterior vena cava of the cat, which runs along the ventral, right side of the vertebral column. Find the renal arteries that take blood from the aorta to the kidneys. The aorta lies to the left of the posterior vena cava. You should be able to locate the ureters as they run posteriorly from the kidney to the urinary bladder. Do not dissect the urethra at this time. You can locate the urethra during the dissection of the reproductive structures in Exercises 46 and 47. You should compare your dissection to figure 44.10.

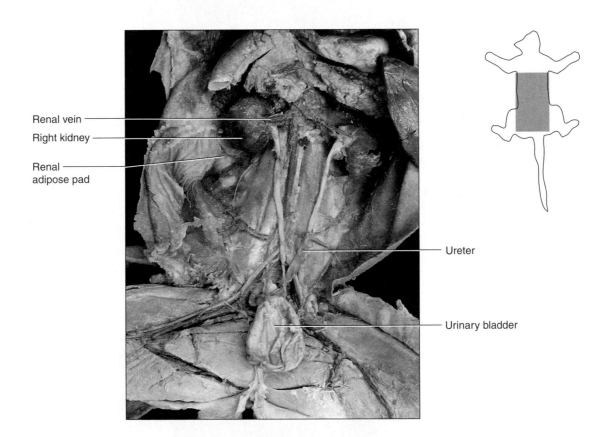

FIGURE 44.10 Urinary System in the Cat

REVIEW SECTION

Urinary System

Name _____ *Date* _____

Lab Section _____ *Time* _____

Review Questions
Matching
Match the descriptions in the left column with the terms in the right column.

1. outermost part of the kidney

2. storage organ of the urinary system

3. takes blood from the kidney

4. separates the renal cortex from the medulla

5. receives urine from the renal papilla

6. leads directly to the renal pelvis

   a. renal vein

   b. major calyx

   c. minor calyx

   d. urinary bladder

   e. fibrous capsule

   f. arcuate arteries

Fill in

7. The _____ is found between the kidney and the urinary bladder.

8. The _____ is the terminal part of the nephron.

9. Filtration occurs at the _____ part of the nephron.

10. What is the outer region of the kidney called that contains glomeruli?

11. Describe the kidneys with regard to their position to the parietal peritoneum.

12. What takes urine from the bladder to the exterior of the body?

13. What blood vessel takes blood to the kidney?

14. On the posterior bladder is a triangular region. What is it called?

15. Distal convoluted tubules flow directly into what structures?

16. What is a renal papilla?

17. Name the protective capsule that covers a kidney.

18. Blood in the glomerulus next flows to what arteriole?

19. Which shows greater anatomical difference between the sexes: ureters, urinary bladder, or urethra?

20. Blood in an arcuate vein would next flow into what structure?

21. What histological feature distinguishes a proximal convoluted tubule from a distal convoluted tubule?

22. Name the parts of the nephron.

23. What type of cell lines the bladder?

24. Fill in the following illustration using the terms provided.

    renal artery   renal vein   renal pelvis

    minor calyx   major calyx   ureter

    fibrous capsule   renal pyramid   renal cortex

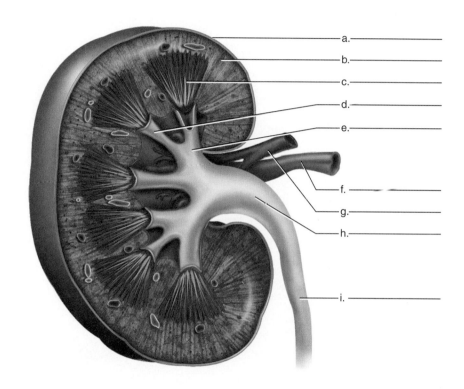

Notes

LABORATORY

Urinalysis

INTRODUCTION

Urinalysis is important in clinical assessment of an individual's physical condition. Traces of blood in the urine can indicate the presence of kidney stones or renal damage; elevated levels of glucose can indicate diabetes mellitus; high bilirubin levels can be indicative of liver problems; and high bacterial counts can indicate bladder or kidney infections. The analysis of urine provides a broad diagnostic indicator of general health. You can read more about the urinary system in the Saladin text in chapter 23, "The Urinary System."

Normal values of urine vary based on water intake and diet. A person who drinks 3 quarts of water per day will have a different urine composition than one who drinks 3 cups of water per day. Not only will the volume of water in the urine vary, but the concentration of other urinary solutes will vary as well. Some cations, such as potassium and sodium, are also diet-dependent. In this exercise you will analyze a sample of urine for materials dissolved in the urine or suspended in it. The suspended material may include cells.

OBJECTIVES

At the end of this exercise you should be able to

1. determine the specific gravity of a urine sample and compare it to normal values;
2. list the sediments commonly found in urine;
3. discuss the importance of urinalysis as a general diagnostic tool;
4. distinguish among casts, crystals, and microbes in a urine sample;
5. prepare a stained sediment slide and identify major components of the sediment.

MATERIALS

Sterile urine collection containers

Permanent marker or wax pencil

Microscope slides

Coverslips

Urine sediment stain (Sedistain, Volusol, etc.)

Chemstrip or Multistix 10SG urine test strips

Tapered centrifuge tubes

Test tube racks

Pasteur pipettes and bulbs

Centrifuge

Protective gloves

Protective eyewear

Biohazard bag

Microscopes

Urine specimen (yours or synthetic/sterilized urine)

10% bleach solution

Caution! *Urine is potentially contaminated with pathogens.* **Wear protective barrier gloves and protective eyewear during the entire exercise.** Place all disposable material that comes into contact with urine in the biohazard bag. Work only with your own urine and avoid any contact between urine and an open wound or cut. If you spill your sample, notify your instructor and wipe up the spill and swab the countertop with a 10% bleach solution. When you are finished with the exercise place reusable glassware in a 10% bleach solution. Read all procedures before beginning this exercise.

PROCEDURE

You can use either your own urine for urinalysis or simulated urine. If you use simulated urine you can test for color, pH, specific gravity, glucose, and protein. Some tests are provided in kit form from biological supply houses. If you use your own urine follow the directions given next.

Using a marker or wax pencil, write your name on a sterile collection container. Proceed to the restroom and obtain a urine sample for testing. If you collect your urine at home, do so in the morning and store the sample in the refrigerator. The sample should be a **midstream** collection because the urethra is initially full of microbes and the first volume of urine will contain abnormally elevated levels of microorganisms. Take the specimen back to the lab and note the color of the sample.

Urine color: _____

The color of urine should normally be a pale yellow. This is due to the pigment **urochrome,** a metabolic product of hemoglobin breakdown. Higher levels of B vitamins may artificially color the urine brighter yellow, and a low fluid intake may also cause the urine to be a deeper yellow. Urine is normally clear, but the sample may be cloudy due to bacterial infection or other contaminants. Record the color and turbidity of your urine sample.

Color: _____

Turbidity: _____

Smell the urine sample and record the odor.

Odor: _____

Urine should have a faint but characteristic odor. Consumption of certain foods, such as asparagus, may produce sulfur compounds in urine, which produce stronger odors.

Urine Characteristics

Gently swirl the urine before testing and pour a 10 mL sample into a clean test tube. Follow your instructor's directions for determining standard values in urine. You can test for specific compounds in urine individually, or you can use a Chemstrip or Multistix to run a battery of tests in a few minutes. The Multistix procedure is outlined next. You should have the urine sample, the test strip, a pencil or pen, and a container (so that you can read the values) handy before you proceed with the test. The test covers 10 specific exams in 2 minutes, so you have to be organized to record the results. Consider any result read after 2 minutes invalid. Multistix test strips are composed of sections of paper with test reagents embedded into the fibers. They react with urine components if present. Examine figure 45.1 and note the way the stick is read. Read the test strip first from the region near where you hold the strip and continue away from the area to the end of the strip.

Multistix/Chemstrip Procedure

If you are using a Multistix or Chemstrip, be sure to pay attention to the time limits provided on the container for reading the results. Wear protective gloves to dip the strip

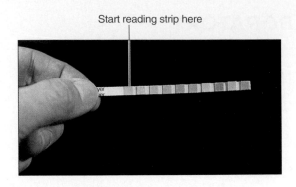

Start reading strip here

FIGURE 45.1 Reading a Urine Test Strip Read the strip near the end where you hold onto the strip with your fingers and progress to the other end of the strip.

into the urine sample and leave the test strip in the urine for no longer than 1 second. Follow directions precisely on the test container, and record the results from your test strip in the spaces provided. Use the bottle the test strips came in as a reader to determine your results.

Glucose: _____

Although glucose is present in blood plasma there is normally no glucose in urine. **Glycosuria** is the condition of having glucose in the urine. Fasting levels of glucose should be minimal (less than 40 mg/dL) or none. Trace amounts of glucose can be found after meals high in carbohydrates, especially sugar, and significant levels can be found in individuals with diabetes mellitus.

Bilirubin: _____

The bilirubin test should be negative, since bilirubin is not normally present in urine. Bilirubin is a metabolic waste product from the destruction of erythrocytes. In the condition of **bilirubinuria** there is the presence of the bile pigments and bilirubin in urine. This can be due to erythrocyte destruction (hemolytic anemia), blockage of the bile duct, or liver damage, such as hepatitis or cirrhosis.

Ketones: _____

Normally, there should be a trace (5 mg/mL) or no ketones (acetone) in the urine. Ketone bodies in urine **(ketonuria)** represent the general body condition known as **ketosis.** This is due to mobilization and use of fat stores in people during periods of starvation, in individuals with diabetes mellitus, or in people with an abnormally high fat diet.

Specific gravity: _____

The specific gravity of pure water is 1.000. The specific gravity of urine varies from a dilute 1.001 to 1.035 in concentrated urine. When urine has a high specific gravity there are more dissolved solutes in the urine. A high specific gravity may be due to lower water intake or such diseases as diabetes mellitus.

Hematuria: _____

There should normally be no blood present in urine. The presence of erythrocytes or hemoglobin in urine may be due to some forms of anemia, erythrocyte destruction after incompatible blood transfusions, renal disease or infection, kidney stones, or urinary contamination during the menses (bleeding period) of a female's menstrual cycle. If the indicator strip shows a spotted pattern, then the erythrocytes are intact (nonhemolyzed). If the red blood cells have ruptured, then the presence of **hematuria** shows up as uniform color changes on the indicator strip.

pH: _____

The pH of urine can vary from around 4.5 to 8.2. This represents over a 1,000-fold change in hydrogen ion concentration. The dramatic fluctuation in pH is seen as the kidneys regulate blood pH by removing ions from the blood. Urine is normally slightly acidic. In a diet high in protein, the urine is more acidic, while a diet high in vegetable material yields a urine that is more alkaline.

Protein: _____

Normally, there should be no protein in urine. The most common blood protein is albumin, and its presence in urine is known as **albuminuria.** This may be due to diabetes mellitus, renal damage (kidney disease), extreme physical activity, or hypertension.

Urobilinogen: _____

Normal ranges of urobilinogen are 0.2 to 1 mg/dL. Increases in the secretion of urobilinogen indicate significant **hemolysis** of erythrocytes to the point that the liver cannot process the bilirubin. The bilirubin increases in the plasma, and the formation of urobilinogen in the intestines increases. The urobilinogen diffuses into the blood, where it is filtered by the kidneys.

Nitrites: _____

The normal blood value of nitrites should be negative. A positive value for nitrites indicates the presence of large numbers of bacteria in the urinary tract.

Leukocytes: _____

The normal value for leukocytes is negative or in trace amounts. Elevated levels of leukocytes in urine is known as **pyuria.** This indicates a potential urinary tract infection, typically bladder or kidney infection.

Some drugs and vitamins can give false positive results in urinalysis tests. Any test that indicates renal disease should be further validated by professionals prior to any course of action.

After you have finished your results, complete table 45.1 in the review section.

Sediment Study

1. Obtain a centrifuge tube from the supply area and write your name on the tube for identification.
2. Pour 10 mL of urine into the tube and place it in the centrifuge opposite another tube containing an equal volume.
3. Balance the centrifuge. If there is an uneven number of tubes, make sure to fill a tube with water to the same level as the unpaired tube and place it opposite that tube in the centrifuge. This balances the centrifuge while it is operating.
4. Spin the tubes for about 4 minutes at a slow speed (1,500 rpm) and let the rotor of the centrifuge slow down over time. If you brake the spinning you may resuspend the sediments.
5. Carefully remove your tube and place it in a test tube rack on your desk.
6. Examine the tube. You should have an upper fluid layer and a small amount of urine sediment at the bottom of the tube.
7. Place one drop of urine sediment stain (Sedistain or other urine stain) on a clean microscope slide.
8. Withdraw a small sample from the bottom of the centrifuge tube with a Pasteur pipette and place a drop of the sediment on the slide. **Do not put the urine on the slide before you put on the stain because you may contaminate the stain if the stain bottle comes in contact with the sediment.**
9. Place a coverslip on the sample and examine the slide under the microscope. Examine the slide under low power first and look at the free edge of the coverslip. Red-orange, elongated crystals may appear here later as the stain evaporates. Do not confuse the stain crystals with sediment crystals.
10. Now examine the slide under high power and compare your sample to the common urine sediments in figures 45.2, 45.3, and 45.4. These are not to scale.

11. Record the material you find in urine sediment in number 17 in the review section at the end of this laboratory exercise. Some of the common items found in urine are as follows:

Organisms (see figure 45.2)

Yeasts (other than *Candida*) Many yeasts are a normal constituent.

Trichomonas vaginalis Common protozoan infection

Candida albicans causes vaginal yeast infections

Cells (see figure 45.2)

Epithelial cells

Cells of urethra (squamous cells)—normal constituent

Cells of bladder, ureter (transitional cells)—normal constituent

Bacterial cells in urine indicate a condition known as bacteriuria.

Erythrocytes (RBCs)—menstrual blood in females or blood from the passage of kidney stones

Leukocytes (WBCs)—indicative of urinary tract (bladder/kidney) infections

Crystals (see figure 45.3) (usually indicative of reduced water intake)

Struvite (magnesium ammonium phosphate)

Calcium carbonate

Calcium oxalate—small, green crystals

Cystine—oxidation product of an amino acid

Ureates

Cholesterol

Tyrosine—amino acid

Casts and Artifacts (see figure 45.4)

Casts are conglomerations of cells and other materials from the kidneys

Hyaline—from sclerosis of arteries or glomerulus

Leukocytes (indicate pyelonephritis, inflammation of the kidney)

Erythrocytes (indicate glomerulonephritis, inflammation of glomeruli)

Artifacts

Fibers—clothing fibers, mucous threads

skin oil, powders

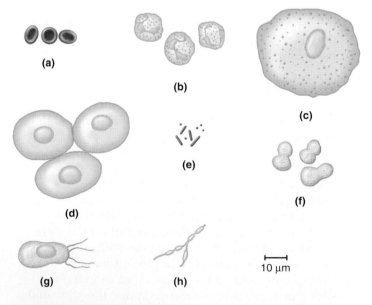

FIGURE 45.2 Urine Sediments—Cells (a) Red blood cells; (b) white blood cells; (c) squamous epithelial cells; (d) transitional epithelial cells; (e) bacteria; (f) yeast; (g) *Trichomonas*; (h) *Candida*.

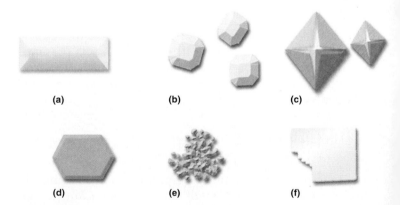

FIGURE 45.3 Urine Sediments—Crystals Sizes are variable: (a) struvite; (b) calcium carbonate; (c) calcium oxalate; (d) cystine; (e) ureate; (f) cholesterol.

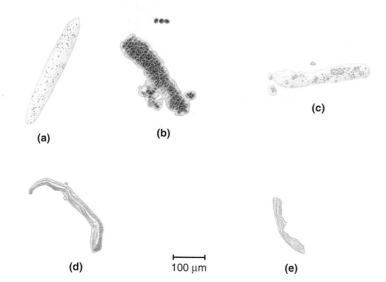

(a) **(b)** **(c)**

(d) 100 μm **(e)**

FIGURE 45.4 Urine Sediments—Casts and Fibers (a) Hyaline cast; (b) erythrocyte cast; (c) leukocyte cast; (d) vegetable fiber; (e) mucous thread.

Renal Calculi

Kidney stones, or **renal calculi,** are due to mineral build-up in the kidneys, amino acid deposition, or amino acid oxidative products, such as tyrosine or cystine crystals. Calculi may occur from not drinking enough fluids, excessive mineral uptake, or metabolic disorders, such as an inability or a decreased ability to metabolize some amino acids. The calculi are masses of crystals that fuse into a larger form and can block the ureter, causing a retention of fluid in the kidney. Calculi can be exceptionally painful. Examine renal calculi in the lab, if available.

Clean Up Place urine-contaminated material to be disposed of in the biohazard container and all the urine-contaminated material to be reused in the 10% bleach container. Swab the countertops with 10% bleach, which destroys infectious agents.

Notes

REVIEW SECTION

Urinalysis

Name _____ Date _____

Lab Section _____ Time _____

Review Questions

1. What metabolic by-product from hemoglobin colors the urine yellow?

2. How can adequate water intake be judged by the color of urine?

3. What is hematuria?

4. Which urine sediment material is probably due to reduced water intake?
   a. vegetable fibers   b. crystals   c. epithelial cells   d. bacterial cells

5. There are about 200 grams of protein in the blood plasma and normally none in urine. What mechanism keeps protein out of the urine, and what structure would be potentially damaged (review Laboratory Exercise 44 if necessary) if protein was found in significant amounts in urine?

6. The kidneys are efficient at balancing the blood pH. It is critical that blood acidity remain constant. If excess hydrogen ions are present in the blood and increase blood acidity, the kidneys secrete the hydrogen ions. What effect would this have on the pH of urine?

7. If you process 180 liters of water through the kidney each day yet produce only 1.8 liters of urine, what is the percent reabsorption of water?

8. Assume that a person did not collect a midstream sample of urine but collected a sample from the beginning of urination. What additional materials might be in greater numbers in this sample?

9. How much water do you normally drink each day? Does this correspond to the specific gravity of your urine?

10. What is the name of the condition of having measurable amounts of glucose in urine?

11. What must be present in urine to have the condition of ketonuria?

12. Elevated levels of white blood cells produce what condition in urine?

13. What cells would be found in urine that originally came from the urethra?

14. What cells would be found in urine that came from the bladder?

15. If you had large numbers of calcium crystals in the urine, what could this tell you about the amount of water you drink?

16. List your results from urinalyis here. Indicate if they are within the normal range or, if not, what this may indicate.

TABLE 45.1

Characteristic	Normal Value or Range	Your Results	Indication
Appearance	Almost clear to deep amber		
Odor	Faint odor		
Glucose	None to trace		
Bilirubin	None		
Ketones	None to trace (5 mg/mL)		
Specific gravity	1.001–1.035		
Free hemoglobin	None		
pH	4.5–8.2		
Albumin	None to trace		
Urobilinogen	Trace (0.2–1 mg/dL)		
Nitrites	None		
Leukocytes	None to trace		

17. List the sediments that you found in your urine sample.

Notes

LABORATORY

Male Reproductive System

INTRODUCTION

The male reproductive system produces **male gametes (spermatozoa);** transports the gametes to the female reproductive tract; and secretes the male reproductive hormone, **testosterone.** The gonad- or gamete-producing structure of the male reproductive system is the testis (plural, *testes*). The structure and function of the male reproductive system are covered in the Saladin text in chapter 27,"The Male Reproductive System." In this exercise, you examine the gross anatomy of the male reproductive system, the histology of the system, and the male reproductive system of the cat.

OBJECTIVES

At the end of this exercise you should be able to

1. describe the gamete-producing organ of the male reproductive system;
2. describe the formation of sperm cells in the testis;
3. list the pathway that sperm cells follow from production to expulsion;
4. describe the anatomy of the spermatic cord;
5. list the four components of semen and where they are produced;
6. name the three cylinders of erectile tissue in the penis.

MATERIALS

Charts, models, and illustrations of the male reproductive system

Microscopes

Prepared slides of a cross section of testis and sperm smear

Cat

Materials for Cat Dissection

Dissection trays

Scalpel and two or three extra blades

Barrier gloves

Blunt (mall) probe

Forceps and sharp scissors

First aid kit in lab or prep area

Sharps container

Animal waste disposal container

PROCEDURE

Overview of the Gross Anatomy of the Male Reproductive System

Examine the models and charts of the male reproductive system available in the lab and locate the following structures in figure 46.1:

testis

epididymis

scrotal sac (scrotum)

ductus (vas) deferens

seminal vesicle

prostate gland

bulbourethral gland

penis

urethra

Testes

Examine charts and models of the testes and compare them to figures 46.1 and 46.2. The **testes** are paired organs wrapped in a tough connective tissue sheath called the **tunica albuginea** (see figure 46.2). They are surrounded by the **scrotal sac (scrotum),** which keeps the testes on the exterior of the body cavity, where the temperature is somewhat cooler. The testes are the site of **spermatozoa** (sperm cell) production, and this process must occur at about 35° C. The scrotal sac is lined with a layer of smooth muscle called the **dartos muscle.** These muscle fibers contract when the testes are cold, tightening the scrotal sac and bringing the testes closer to the body. When the environment around the testes is warm, the dartos muscles relax and the testes descend from the body, thus becoming cooler.

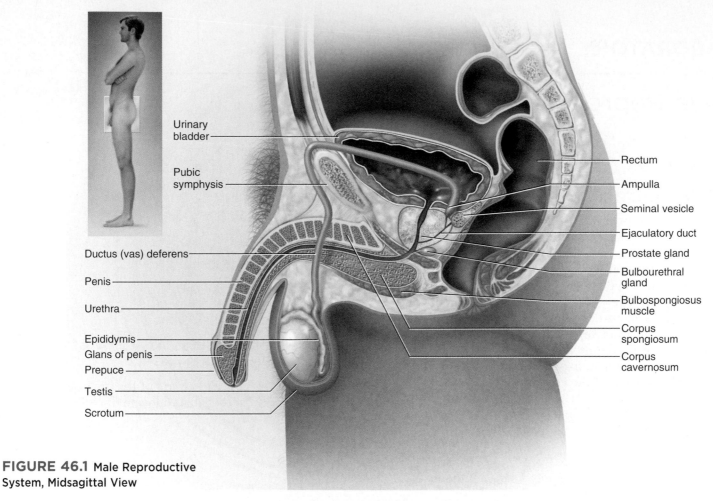

FIGURE 46.1 Male Reproductive System, Midsagittal View

Urinary bladder

Pubic symphysis

Ductus (vas) deferens

Penis

Urethra

Epididymis

Glans of penis

Prepuce

Testis

Scrotum

Rectum

Ampulla

Seminal vesicle

Ejaculatory duct

Prostate gland

Bulbourethral gland

Bulbospongiosus muscle

Corpus spongiosum

Corpus cavernosum

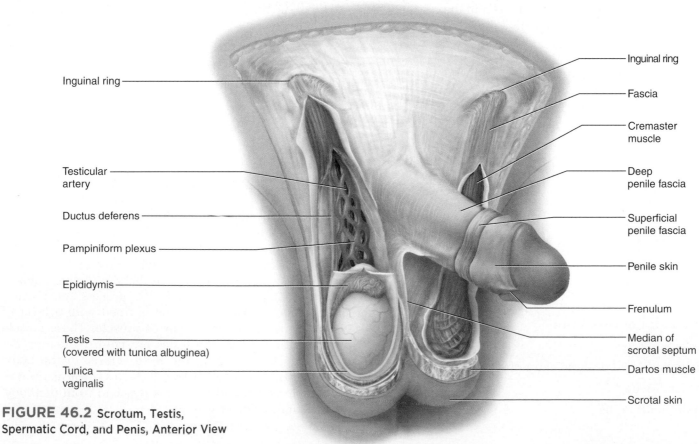

FIGURE 46.2 Scrotum, Testis, Spermatic Cord, and Penis, Anterior View

Inguinal ring

Testicular artery

Ductus deferens

Pampiniform plexus

Epididymis

Testis (covered with tunica albuginea)

Tunica vaginalis

Inguinal ring

Fascia

Cremaster muscle

Deep penile fascia

Superficial penile fascia

Penile skin

Frenulum

Median of scrotal septum

Dartos muscle

Scrotal skin

Cremaster muscles, which consist of skeletal muscle, also control the temperature of the testes. When they contract they bring the testes closer to the body, which increases their temperature. When they relax the testes descend.

Histology of the Testis

Examine the testis in a prepared slide. Numerous tubules are seen in cross section. These are the **seminiferous tubules.** The gametes, or spermatozoa, are produced in seminiferous tubules in the testis (see figure 46.3).

Find the clusters of cells that frequently appear as triangles in between the tubules. These are the **interstitial (Leydig) cells.** They produce the male sex hormone, **testosterone.** Examine the seminiferous tubules under high magnification. You should be able to see the outer row of cells called the **spermatogonia.** These cells reproduce by mitosis to produce **primary spermatocytes.** The primary spermatocytes undergo meiosis, or reduction division, to eventually produce the sex cells (spermatozoa). The primary spermatocytes divide to form **secondary spermatocytes,** found closer to the lumen. The secondary spermatocytes become **spermatids.** Spermatids lose their remaining cytoplasm and mature into **spermatozoa.** Compare this process to figures 46.3 and 46.4 and locate the spermatogonia, primary and secondary spermatocytes, spermatids, and spermatozoa. **Sustentacular (Sertoli) cells** assist in the movement of the primary spermatocytes.

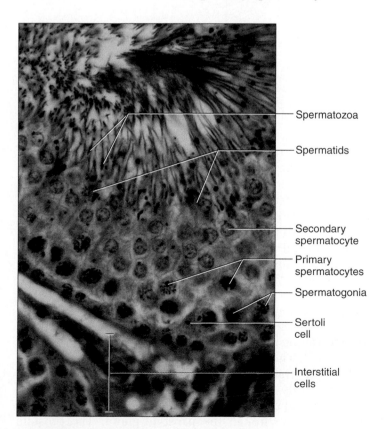

FIGURE 46.3 Histology of Testis (400×)

After you have examined your prepared slide of a section of testis, you should draw what you see in the following space. You should include the interstitial cells, seminiferous tubules, spermatogonia, spermatocytes, spermatids, and spermatozoa. You may need to look at several sections to see all these items. Your drawing may be a composite of many tubules.

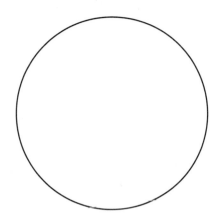

Sperm

The structure of an individual sperm cell consists of a **head, midpiece,** and **tail.** The head contains the genetic information (DNA) as well as a cap known as the **acrosome.** The acrosome contains digestive enzymes that digest the exterior covering of the female gamete. The midpiece of the sperm contains mitochondria that provide ATP to the sperm cell tail. The tail of the sperm is a flagellum that propels the sperm forward. Examine a prepared slide of sperm and compare it to figure 46.5.

Epididymis

Spermatozoa (or sperm cells) from each testis travel from tubules in the testis to the **rete testis** and into the **epididymis,** where they are stored and mature. Each epididymis has a blunt, rounded **head;** an elongated **body;** and a tapering **tail** that leads to the **ductus deferens.** Sperm maturation, or **capacitation,** occurs in the epididymis. If spermatozoa are removed from the testis proper, they are not capable of fertilizing the female oocyte (egg). Spermatozoa move slowly through coiled tubules of the epididymis. Examine a model or chart of the longitudinal section of a testis and epididymis and locate the structures by comparing them to figure 46.6.

Spermatic Cord

Spermatozoa travel from the epididymis into the ductus deferens. The ductus deferens is enclosed in the **spermatic cord,** a complex cable consisting of the ductus deferens, the **testicular artery** and **vein,** and the testicular **nerves.** The cremaster muscle is a cluster of skeletal

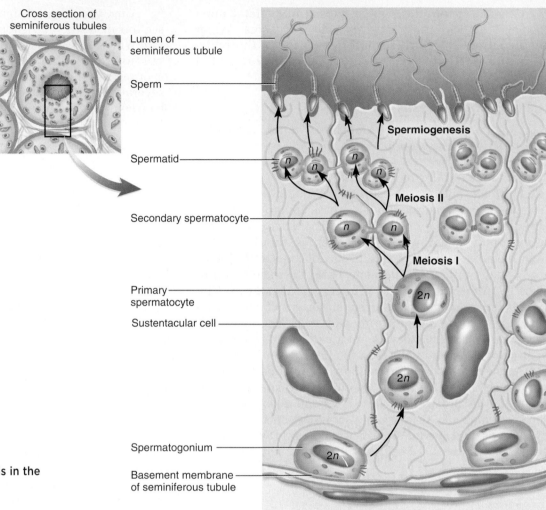

Cross section of
seminiferous tubules

Lumen of
seminiferous tubule

Sperm

Spermiogenesis

Spermatid

Meiosis II

Secondary spermatocyte

Meiosis I

Primary
spermatocyte

Sustentacular cell

Spermatogonium

Basement membrane
of seminiferous tubule

FIGURE 46.4 Spermatogenesis in the
Seminiferous Tubule

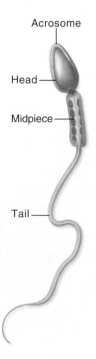

Acrosome

Head

Midpiece

Tail

FIGURE 46.5 Sperm Cell

muscle fibers. The spermatic cord is longer on the left side than on the right; therefore, the left testis is lower than the right. Locate the structures of the spermatic cord in figures 46.2 and 46.6.

As the spermatic cord reaches the **inguinal ring,** the ductus deferens travels around the posterior surface of the urinary bladder. You can trace the course of the ductus deferens until it reaches the inferior portion of the bladder. The ductus deferens enlarges somewhat here to form the **ampulla** of the ductus deferens. Each ductus deferens joins with a **seminal vesicle,** a gland that adds fluid to the spermatozoa. The union of the ductus deferens and the seminal vesicle produces the **ejaculatory duct.** Locate the seminal vesicle and the ejaculatory duct in figure 46.7. The seminal fluid adds about 60% to the final volume of semen.

From this location the ejaculatory duct leads to the inferior portion of the bladder and joins with the urethra, which passes through the **prostate gland.** The prostate gland is located just inferior to the urinary bladder, and

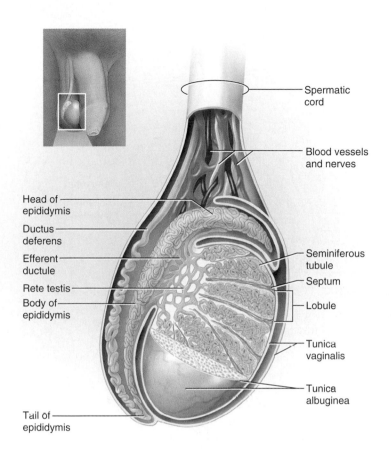

FIGURE 46.6 Testis and Epididymis, Lateral View

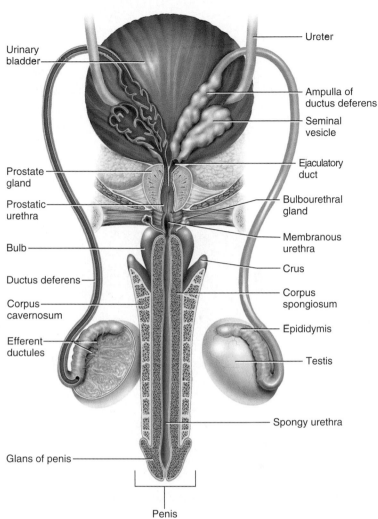

FIGURE 46.7 Urinary Bladder with Seminal Vesicles, Posterior View

the urethra that passes through the gland is known as the **prostatic urethra.** The prostate gland adds a buffering fluid to the secretions of the testes and seminal vesicles. The prostate fluid makes up slightly less than 30% of the final semen volume. As the prostatic urethra exits the prostate gland it becomes the **membranous urethra** and passes through the body wall. Here the paired **bulboure-thral (Cowper's) glands** are found, which add a lubricant to the seminal fluid. **Seminal fluid** consists of secretions from the seminal vesicles, prostate gland, and bulboure-thral glands. **Semen** consists of seminal fluid plus sper-matozoa from the testes. Spermatozoa make up much less than 1% of the total volume of semen. The urethra passes out of the body cavity and becomes the **spongy (penile) urethra** of the penis. Locate the portions of the urethra and the accessory glands in figure 46.7.

External Genitalia
Penis

The **penis** consists of an elongated **shaft** and a distally expanded **glans penis.** The glans is covered with the **prepuce,** or **foreskin,** removed in some males by a proce-dure called a **circumcision.** At the inferior portion of the

glans is a region richly supplied with nerve endings called the **frenulum.** The glans penis is an expanded region that stimulates the genitalia of the female. The erect penis is, on average, about 16 cm in length. The anatomy of the penis can be seen in figures 46.2 and 46.8.

The penis contains three cylinders of **erectile tissue.** The **corpus spongiosum** is the cylinder of erectile tissue that contains the **spongy urethra.** The two **corpora cav-ernosa** are located anterior to the corpus spongiosum. Examine a model or chart of a cross section of penis and compare it to figure 46.8. The proximal parts of the cylinders of erectile tissue are anchored to the body. Locate the **crus,** an expansion of the corpora cavernosa. The **bulb** of the penis is an extension of the corpus spongiosum. The crus and the bulb form the **root** of the penis, as illustrated in figure 46.7. The corpus spon-giosum expands distally to form the glans penis. Note the **dorsal arteries** and **deep arteries** of the penis that take blood to the penis. Locate the **dorsal vein** and the

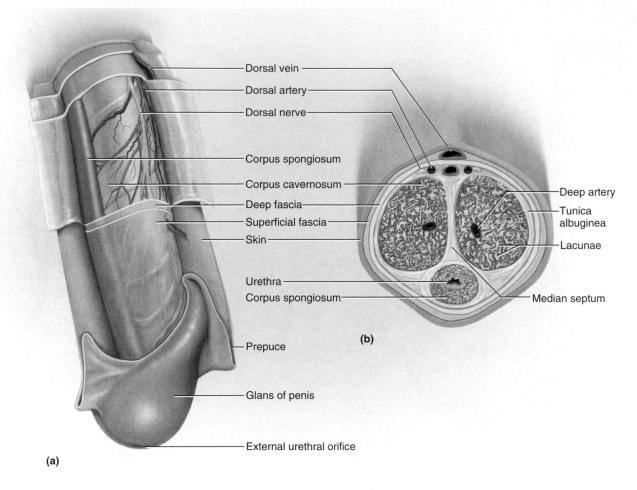

FIGURE 46.8 Penis (a) Three-quarter view; (b) cross section.

deep dorsal vein of the penis. When the arteries of the penis dilate, the erectile tissues engorge with blood and the penis becomes erect. The erection subsides as the arteries constrict, decreasing blood flow into the penis. Examine a model of the penis and find the features listed in figure 46.8.

Perineum

The floor of the pelvis as seen from the outside is referred to as the **perineum.** It is a diamond-shaped structure defined by the pubic symphysis at the anterior point, the lateral points being the ischial tuberosities and the posterior point being the coccyx. It can be divided into a posterior **anal triangle** and an anterior **urogenital triangle.** The anal triangle surrounds the anus and the urogenital triangle encloses the penis and scrotum (see figure 46.9).

Male Contraception

The main goal in male contraception is the prevention of spermatozoa from reaching the female gametes.

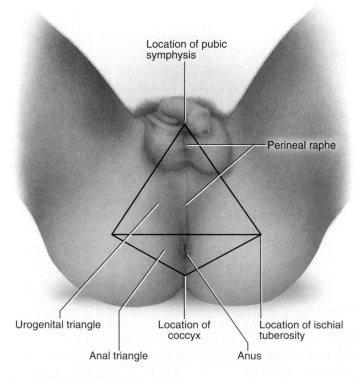

FIGURE 46.9 Male Perineum

Abstention, or refraining from sexual intercourse, is the most effective method. The next most effective method is male sterilization, commonly performed by a procedure called a **vasectomy** (the cutting and tying of the two ductus deferens). In a vasectomy an incision is made on each side of the scrotal sac and both ductus deferens are cut. The free ends of the cut ductus deferens are tied, preventing sperm from traveling from the testes to the spermatic cords (see figure 46.10). Use of a barrier, such as a condom, is relatively effective if used properly, because it prevents ejaculated sperm from entering the female reproductive tract. Coitus interruptus, or preejaculatory withdrawal, is not an effective method because sperm may be present in seminal fluid prior to ejaculation and pregnancy can result.

Cat Dissection

Prepare for the cat dissection by obtaining a cat and dissection equipment. Check to see whether your cat is male or female. The penis may be retracted in your specimen, so look for the opening of the penile urethra and the

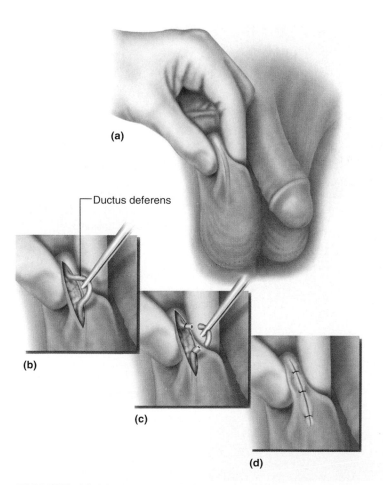

FIGURE 46.10 Vasectomy Procedure (a) Location of ductus deferens; (b) isolation of ductus deferens; (c) cutting and tying; (d) external sutures.

scrotal sac. Team up with a lab partner or group that has a cat of a different sex than your specimen, so that you can learn both male and female reproductive systems. Once you are sure you have a male cat, locate the **scrotal sac** and paired **testes**. Make an incision on the lateral side of the scrotal sac and locate the testis. If your cat was neutered, you will not be able to locate the testis. If the testes are present, cut through the connective tissue of the scrotal sac (the **tunica vaginalis**) and observe both the testis and the **epididymis.** The testis is covered by a tough connective tissue membrane called the **tunica albuginea.** Use figure 46.11 as a guide. Spermatozoa move from the testis and into the epididymis, where the sperm cells mature.

Once you have located the epididymis, proceed in an anterior direction and trace the thin **ductus deferens** from the epididymis into the spermatic cord. The spermatic cord traverses the body wall on the exterior and enters the body of the cat at the inguinal ring (an opening through the inguinal ligament). You may want to gently insert a blunt probe into the inguinal ring so that you can locate the ductus deferens as it passes into the coelom. Locate the ductus deferens as it enters the body cavity and notice how it arches around the ureter on the dorsal side of the urinary bladder. You may have cut the ductus deferens in an earlier exercise, so if you cannot find it on one side look for it on the other side.

You may want to look for the accessory organs of the male reproductive system, but this takes some significant dissection. *Check with your instructor before cutting through the pelvis of your cat.* If your instructor directs you to do so, then begin by cutting through the musculature of the cat at the level of the symphysis pubis. Make a midsagittal incision through the groin muscles, and carefully cut through the cartilage of the symphysis pubis. You should now be able to open the pelvic cavity and locate the single **prostate gland** and the paired **bulbourethral glands** (see figure 46.11). Much of the anatomy of the cat is similar to the human except that there are no seminal vesicles in the cat. Trace the ductus deferens from the posterior surface of the bladder through the prostate gland and the penis. You can make either a longitudinal section through the penis to trace the urethra in the erectile tissue or a cross section of the penis to see the three cylinders of erectile tissue—the corpus spongiosum and the two corpora cavernosa.

 Clean Up   Remember to place all excess tissue in the appropriate waste container and not in a standard wastebasket or down the sink!

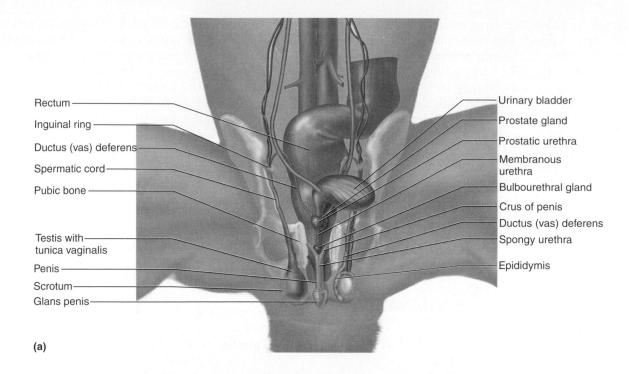

Rectum

Inguinal ring

Ductus (vas) deferens

Spermatic cord

Pubic bone

Testis with tunica vaginalis

Penis

Scrotum

Glans penis

Urinary bladder

Prostate gland

Prostatic urethra

Membranous urethra

Bulbourethral gland

Crus of penis

Ductus (vas) deferens

Spongy urethra

Epididymis

(a)

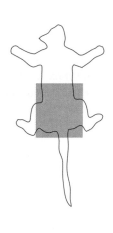

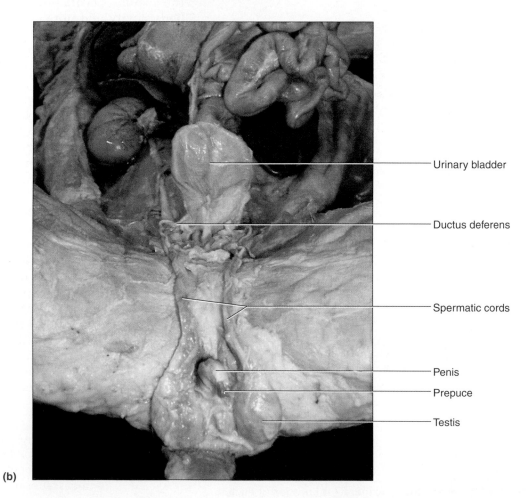

Urinary bladder

Ductus deferens

Spermatic cords

Penis

Prepuce

Testis

(b)

FIGURE 46.11 Male Reproductive Organs of the Cat (a) Diagram; (b) photograph.

Male Reproductive System

Name _____ Date _____

Lab Section _____ Time _____

Review Questions

1. What is the gonad in the male reproductive system?

2. The testes have both endocrine and exocrine functions. Describe the endocrine and exocrine products that come from the testes.

3. Proper sperm production must occur at what temperature?

4. Male sterility can result from excessively high temperatures around the testes. What mechanisms counteract the effects of high temperature?

5. Name the lining of the scrotal sac consisting of a smooth muscle layer.

6. What structure of the testis produces spermatozoa?

7. What cells initiate spermatozoa production?

8. Where does sperm move after being in the epididymis?

9. Where is the cremaster muscle found?

10. List all the structures involved in producing semen.

11. How does spermatozoa differ from seminal fluid?

12. A vasectomy is the cutting and tying of the two ductus deferens at the level of the spermatic cords. Review the percent of spermatozoa that composes semen and determine what effect a vasectomy has on semen volume.

13. Which one of the seminal fluid glands is not a paired gland?

14. Name the three sections of the urethra and where they occur.

15. Where is the glans penis located?

16. What is the cylinder of erectile tissue below the corpora cavernosa?

17. What male reproductive gland is missing in the cat but present in the human?

18. Label the following illustration using the terms provided.

seminal vesicle	prostate gland	bulbourethral gland	glans penis
prepuce	epididymis	bulb of penis	urinary bladder
corpus cavernosum	ductus (vas) deferens	testis	scrotum

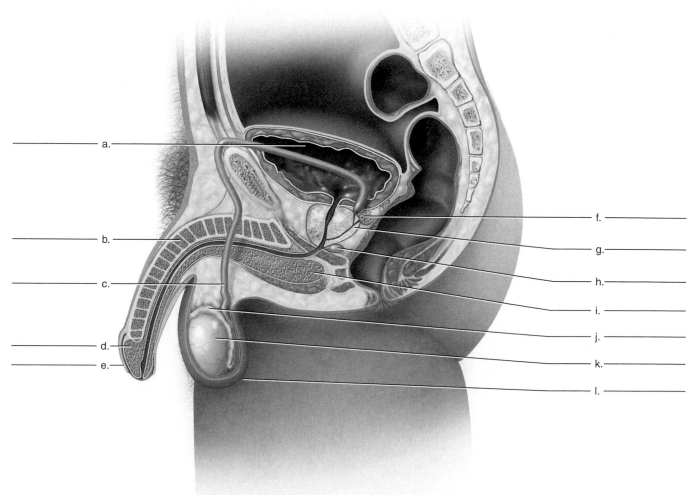

Notes

LABORATORY

Female Reproductive System

INTRODUCTION

The female reproductive system is functionally more complex than the male reproductive system. In the male, the reproductive system produces gametes and delivers them to the female reproductive system. The female reproductive system not only produces gametes and receives the gametes from the male but also, under hormonal influence of HCG from the early cell mass and later from the placenta, provides space and maternal nutrients for the developing **conceptus.** Finally, the female reproductive system delivers the child into the outer environment. The female reproductive system is covered in the Saladin text in chapter 28, "The Female Reproductive System." The ovaries are the gamete-producing organs of the female reproductive system. They produce oocytes and the female sex hormones, estrogen (estradiol) and progesterone. In this exercise you learn about the structure and function of the female reproductive system.

OBJECTIVES

At the end of this exercise you should be able to

1. identify the gamete-producing organ of the female reproductive system;
2. trace the pathway of a gamete from the ovary to the usual site of implantation;
3. list the structures of the vulva;
4. describe the function of each organ in the female reproductive system;
5. name the layers of the uterus from superficial to deep;
6. compare and contrast the anatomy of the cat reproductive system with that of the human.

MATERIALS

Charts, models, and illustrations of the female

reproductive system

Microscopes

Prepared slides of ovary and uterus

Cat

Materials for Cat Dissection

    Dissection trays

    Scalpel and two or three extra blades

    Gloves (household latex gloves work well for repeated use)

    Blunt (mall) probe

    Forceps and sharp scissors

    First aid kit in lab or prep area

    Sharps container

    Animal waste disposal container

PROCEDURE

Overview of the Gross Anatomy of the Female Reproductive System

Examine a model or chart of the female reproductive system and locate the following major reproductive organs there and in figure 47.1.

    ovary

    uterine (fallopian) tube

    uterus

    vaginal canal

    clitoris

    labia minora (singular, *labium minus*)

    labia majora (singular, *labium majus*)

Ovaries

Each **ovary** is approximately 3 to 4 cm long and oblong. The ovaries produce **oocytes,** shed from the outer surface of the ovary during **ovulation.** From here the oocytes move into the **uterine (fallopian) tube.** The ovaries are not directly attached to the uterine tube, and the oocytes must move from the surface of the ovary into the uterine tube.

Histology of the Ovary

Examine a prepared slide of the ovary under the microscope on low power. Locate the background substance of the ovary, known as the **stroma.** Look for circular structures

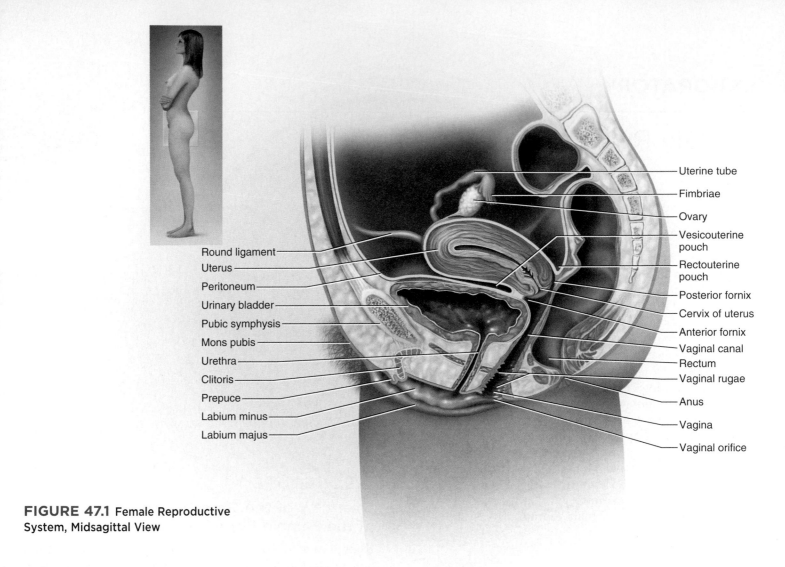

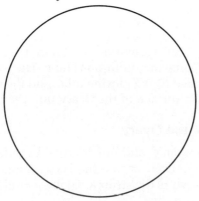

Round ligament
Uterus
Peritoneum
Urinary bladder
Pubic symphysis
Mons pubis
Urethra
Clitoris
Prepuce
Labium minus
Labium majus

Uterine tube
Fimbriae
Ovary
Vesicouterine pouch
Rectouterine pouch
Posterior fornix
Cervix of uterus
Anterior fornix
Vaginal canal
Rectum
Vaginal rugae
Anus
Vagina
Vaginal orifice

FIGURE 47.1 Female Reproductive System, Midsagittal View

in the ovary. These are the **ovarian follicles.** Locate the **primordial follicles** in your slide and compare them to the follicles in figure 47.2.

You should also locate the **primary** and **secondary follicles.** Some of the follicles may contain **oocytes.** Primary follicles contain **primary oocytes,** secondary follicles contain **secondary oocytes.** The largest follicles in the ovary are the **mature ovarian follicles (Graafian follicles),** and you may be able to see one if it is present in your slide. During **ovulation** in humans usually one secondary oocyte is shed from the ovary. In cats many secondary oocytes may be shed. Examine your slide and compare it to figures 47.2 and 47.3. Draw what you see in the following space.

After ovulation the remains of a mature ovarian follicle become a **corpus luteum,** which primarily secretes **progesterone.** If pregnancy does not occur, the corpus luteum decreases in size and becomes the **corpus albicans.** Examine a prepared slide of the ovary with a corpus luteum or corpus albicans and compare it to figures 47.2 and 47.4.

Uterine Tubes

The uterine tube has a small fringe on the distal region known as the **fimbriae.** These are small, fingerlike projections attached to an expanded region known as the **infundibulum.** The uterine tube also has an enlarged region known as the **ampulla** and a narrower portion toward the uterus. Examine a model or chart of the female reproductive system and compare it to figure 47.5.

Uterus

The **uterus** is a pear-shaped organ with a domed **fundus;** a **body;** and a circular, inferior end called the **cervix.** The uterine tubes enter the uterus at about the junction of the fundus with the uterine body. The uterine wall is composed of three layers. The outer surface of the uterus is called the **perimetrium.** This is located near the body cavity. The

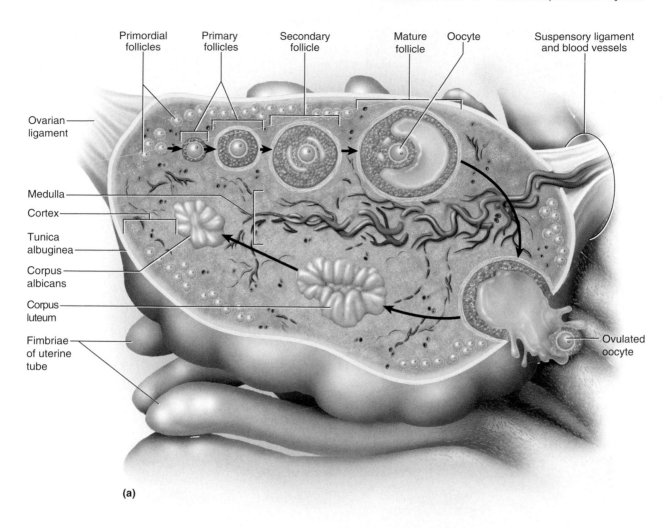

(a)

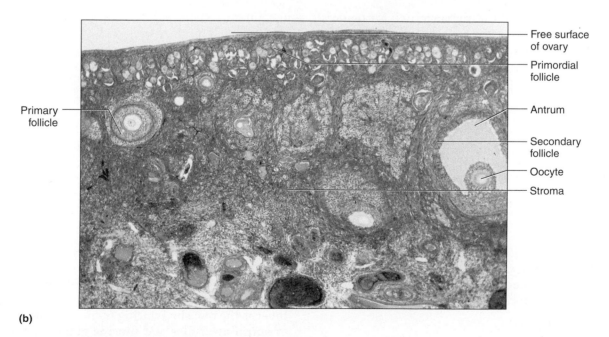

(b)

FIGURE 47.2 Histology of the Ovary (a) Diagram—arrows indicate a timeline of development from primordial follicles to the corpus albicans; (b) photomicrograph of cat ovary (40×).

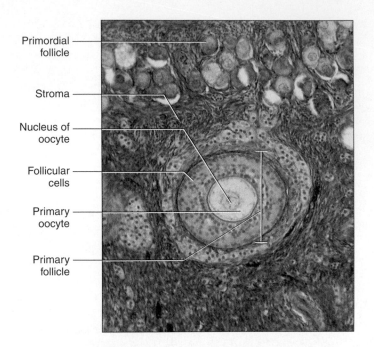

Primordial follicle

Stroma

Nucleus of oocyte

Follicular cells

Primary oocyte

Primary follicle

FIGURE 47.3 Primary Oocyte in Follicle (100×)

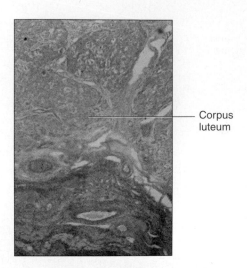

Corpus luteum

FIGURE 47.4 Post-Ovulatory Ovary (40×)

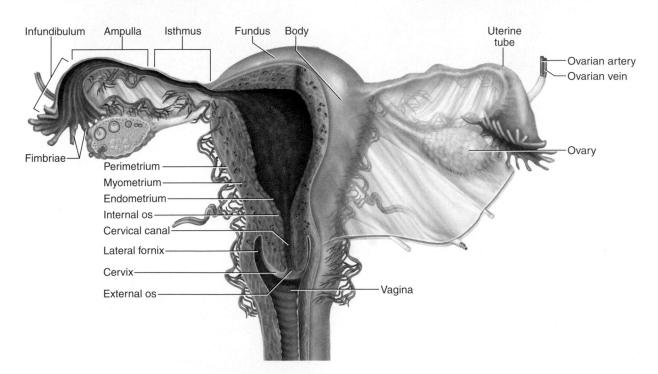

Infundibulum   Ampulla   Isthmus   Fundus   Body

Uterine tube

Ovarian artery
Ovarian vein

Fimbriae

Perimetrium
Myometrium
Endometrium
Internal os
Cervical canal
Lateral fornix
Cervix
External os

Ovary

Vagina

FIGURE 47.5 Female Reproductive System, Posterior View

majority of the uterine wall consists of the **myometrium,** a thick layer of smooth muscle, and the innermost (deepest) layer of the uterus is the **endometrium.** Examine the models or charts in the lab and locate the structures in figure 47.5.

Histology of the Uterus

Examine a prepared slide of the **uterus** and locate its three layers. Locate the outer **perimetrium,** the smooth

muscle of the **myometrium,** and the inner **endometrium.** Compare these to figures 47.5 and 47.6.

Now examine the two layers of the endometrium of the uterus under higher magnification. The **functional layer** is the one shed during menstruation. It is composed of **spiral arterioles** and **uterine glands,** which appear as wavy lines toward the edge of the tissue. The **basal layer** is deeper and contains **straight arterioles.** Deep to the

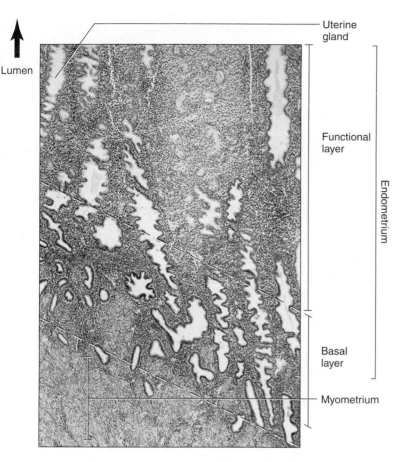

Lumen

Uterine
gland

Functional
layer

Endometrium

Basal
layer

Myometrium

FIGURE 47.6 Histology of the Uterus, Secretory Phase (40×)

endometrium is the myometrium, which can be distinguished by the smooth muscle found in it. As you study the layers of the endometrium, refer to the cyclic changes that occur with the endometrium, as illustrated in the bottom part of figure 47.7. Pap smears are scrapings of the cervix to check for the presence of abnormal cells. Normal epithelial cells of the cervix have a relatively large cytoplasm-to-nucleus ratio. When the nucleus occupies a larger portion of the cell, there is concern for cervical cancer. Regular pap smears are important because early detection may significantly reduce the threat of cervical cancer.

Ovarian and Menstrual Cycles

The endometrium undergoes dynamic changes during the **menstrual cycle.** The effects of **luteinizing hormone (LH)** and **follicle-stimulating hormone (FSH),** from the anterior pituitary, and subsequently **estrogen** and **progesterone** from the ovary, have a significant effect on the endometrium. FSH stimulates the ovarian follicles to secrete estrogens and some progesterone. Elevated levels of these hormones promote the thickening of the endometrium, as indicated in figure 47.7. The **menstrual phase** of the

cycle occurs typically from days 1–5 on a 28-day cycle. The **proliferative phase** follows from days 6–14 and the endometrium increases in thickness. The **secretory phase** occurs from days 15–26 in an average cycle when glycogen is produced in the endometrium. After ovulation, progesterone levels increase. Toward the end of a woman's menstrual cycle, estrogen and progesterone levels drop and this causes the functional layer of the endometrium to slough off. The loss of the endometrial layer is the beginning of a woman's period, or **menstruation.** One method of birth control is to alter the levels of hormones in females. This is known as **hormonal contraception.** Pills that use higher levels of estrogen prevent a woman from ovulating. The stimulus for ovulation is a spike in estrogen levels; therefore, if estrogen is already elevated, no ovulation occurs. Pills that have high levels of progesterone increase the thickness of the cervical mucous plug, decreasing the ability of the sperm to pass into the uterus. Examine figure 47.7 and note how the endothelial layer begins to decrease when estrogen and progesterone levels fall (about day 25 in figure 47.7).

Ligaments

The uterus and ovaries are suspended in the pelvic cavity by a number of connective tissue sheaths called ligaments. The **broad ligament** anchors the uterus to the lateral pelvic wall. The **round ligament** attaches the uterus to the anterior body wall at about the region of the inguinal canal. The **ovarian ligament** directly attaches the ovary to the uterus, and the **suspensory ligament** attaches the ovaries to the lumbar region. Locate these structures in models or charts in the lab and in figure 47.8.

Vagina

The vagina consists of the **vaginal canal** and the **vaginal orifice.** The uterus joins with the vaginal canal at the cervix. The vaginal canal is a tough, muscular tube with a recessed region around the cervix known as the **fornix.** The vagina is about 8–10 cm long, although it can stretch considerably during intercourse and delivery. It is located between the urethra, on the anterior side, and the rectum, which is posterior. The outer layer of the vaginal wall is the adventitia. There is a middle muscularis layer, and the layer of the vagina near the lumen is the mucosa. The mucosa is composed of stratified squamous epithelium in adult women. The vaginal canal is poorly supplied with nerves, and the wall of the vagina has cross ridges called **rugae.** The vaginal canal and orifice can expand greatly during the delivery of a child. Locate the vaginal canal, vaginal orifice, fornix, and rugae in figures 47.1 and 47.5.

External Genitalia

As in the male, the floor of the pelvis as seen from the outside is referred to as the **perineum.** It can be divided into a posterior **anal** (or **rectal**) **triangle,** a region enclosing the

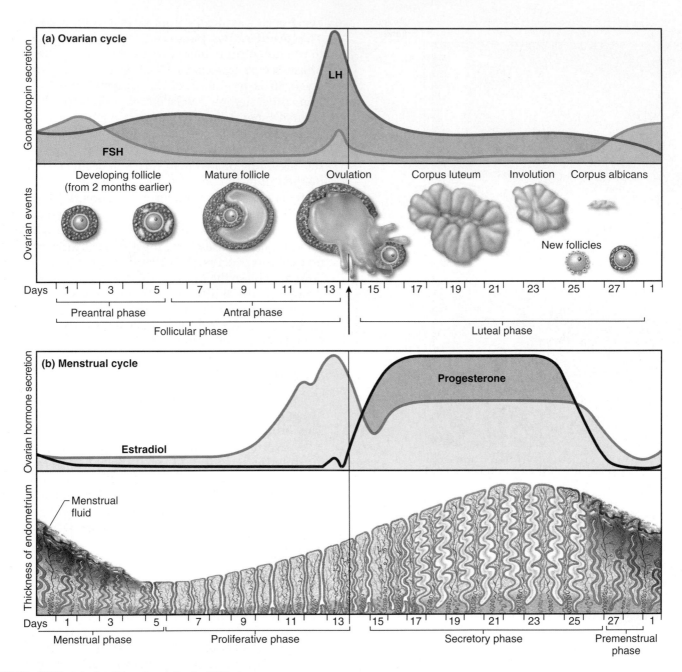

FIGURE 47.7 Ovarian/Menstrual Cycle (100×)

anus, and an anterior **urogenital triangle** consisting of the reproductive and urinary structures (see figure 47.9).

The external female reproductive structures are called the outer genitalia and are collectively referred to as the **vulva.** The **mons pubis** is the anterior-most structure of the vulva and is an adipose pad that overlies the symphysis pubis. Posterior to the mons pubis is the **clitoris,** a cylinder of erectile tissue embedded in the body wall that terminates anterior to the urethral orifice as the **glans clitoris.** The clitoris has the same embryonic origin as the penis in males, and like the penis it is richly supplied with nerve endings. The body of the clitoris is a curved structure illustrated in figure 47.1, and the

glans clitoris is the terminal portion. The glans clitoris is enclosed by the **prepuce,** an extension of the **labia minora.** Posterior to the clitoris is the external **urethral orifice** and posterior to this is the **vaginal orifice.** The vaginal orifice is partially enclosed by a mucous membrane structure known as the **hymen.** The hymen is variable anatomically and has historically (and sometimes incorrectly) been used as an indicator of virginity. Lateral to the vaginal orifice are the labia minora (singular, *labium minus*). The space between the labia minora is known as the **vestibule,** and located laterally and posteriorly to the vestibule are the **greater vestibular glands** (or **Bartholin's glands**). Lateral to the labia minora are the paired **labia**

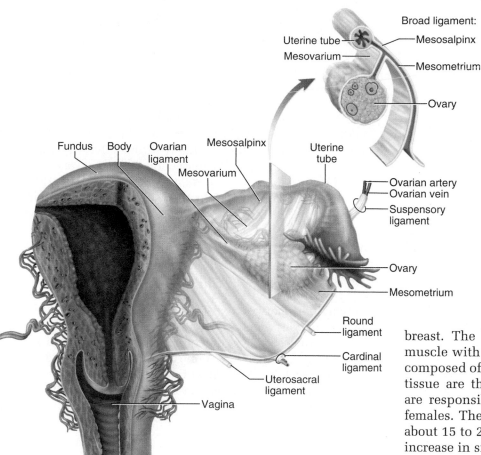

FIGURE 47.8 Ligaments of the Female Reproductive System

majora. Locate these structures on models or charts in the lab and in figure 47.9.

Anatomy of the Breast

The structure of the breast derives from the integumentary system, yet the role of the female breast in reproduction is important as a source of nourishment for the offspring. The major structures of the external breast are the pigmented **areola,** the protruding **nipple,** the **body** of the breast, and the **axillary tail (tail of Spence).** The axillary tail is of clinical importance in that breast tumors frequently occur there. Examine the surface features of the breast in figure 47.10 and locate the structures listed.

Compare models or charts in the lab and locate the internal structures of the breast. The breast is anchored to the pectoralis major muscle with **suspensory ligaments.** Much of the breast is composed of **adipose tissue,** and embedded in the adipose tissue are the **mammary glands.** The mammary glands are responsible for the production of milk in lactating females. The glands are clustered in **lobes,** and there are about 15 to 20 lobes in each breast. The mammary glands increase in size in nursing women and lead to **lactiferous ducts,** which subsequently lead to **lactiferous sinuses (ampullae)** that exit via the nipple. Humans have several ampullae leading to each nipple. The mammary glands in females begin to undergo changes prior to puberty and

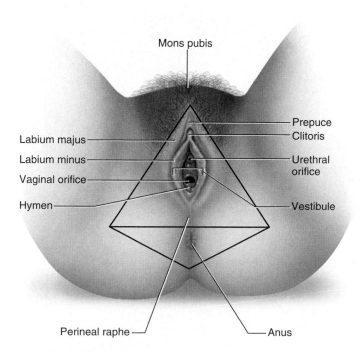

FIGURE 47.9 Female Perineum and External Genitalia

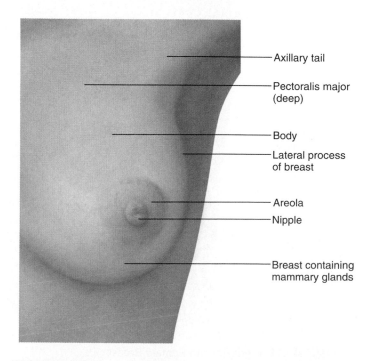

FIGURE 47.10 Surface Features of the Female Breast

become functional glands after delivery of a child. Note the features illustrated in figure 47.11.

Female Contraception

As with male contraception the prevention of male and female gametes from uniting is the goal in female contraception. Abstention, or refraining from intercourse, is the most effective method of birth control. Other methods involve oral contraceptives or hormonal implants, tubal ligation (cutting and tying of the uterine tubes), use of barrier methods (such as the condom or diaphragm), and use of spermicidal foams. The use of oral contraceptives, synthetic estrogens and progesterones, decreases the levels of LH and FSH in the pituitary (by negative-feedback mechanisms), thus preventing ovulation from occurring. Progesterone implants, which also prevent ovulation, elevate hormone levels. Examine the levels of progesterone in figure 47.7. When these levels are elevated note how the levels of FSH and LH are low.

Cat Dissection

Prepare for the cat dissection by obtaining a dissection tray, scalpel, scissors or bone cutter, string, forceps, and plastic bag with label.

You probably already removed the multiple mammary glands of the female cat during the removal of the skin. If this is the case, then examine the external genitalia for the **urogenital orifice.** *Check with your instructor before cutting through the pelvis of your cat.* If your instructor directs you to do so, then begin by cutting through the musculature of the cat at the level of the **symphysis pubis.** Continue to cut in the midsagittal plane through the cartilage of the symphysis pubis and then cranially through the abdominal muscles, as described in the dissection of the male cat. This should expose the reproductive organs (see figure 47.12).

Unlike the human female, the cat has a **horned (bipartite) uterus.** The uterus of a cat has an appearance of a Y, with the upper two branches being the **uterine horns** and the stem of the Y the **body** of the uterus. Humans normally have single births from a pregnancy, while the expanded uterus in cats facilitates multiple births. If your cat is pregnant the uterus will be greatly enlarged, and you may find many fetuses inside. If your cat was spayed then the ovaries were removed.

The **ovaries** in cats are caudal to the kidneys and are relatively small organs. Examine the paired ovaries and the short **uterine tubes (oviducts)** in the cat. If you have difficulty locating them, trace the uterus toward the uterine horns and locate the ovaries. The uterine tubes in the human

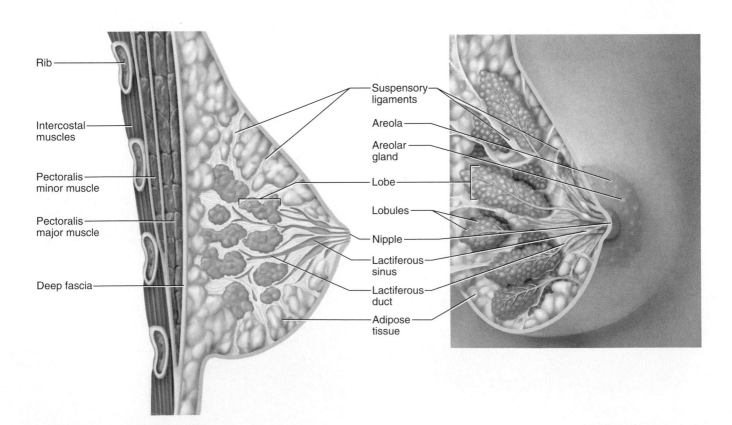

Rib
Intercostal muscles
Pectoralis minor muscle
Pectoralis major muscle
Deep fascia

Suspensory ligaments
Areola
Areolar gland
Lobe
Lobules
Nipple
Lactiferous sinus
Lactiferous duct
Adipose tissue

FIGURE 47.11 Interior of the Female Breast

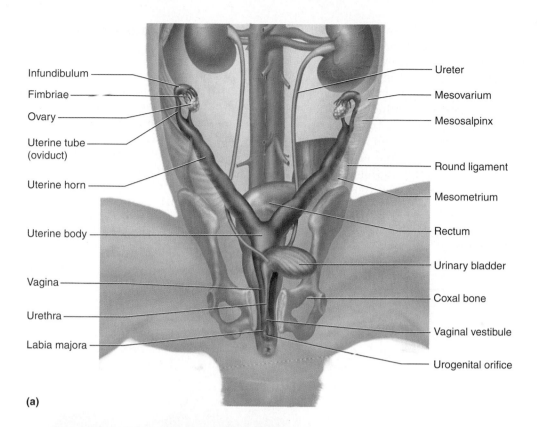

Infundibulum

Fimbriae

Ovary

Uterine tube
(oviduct)

Uterine horn

Uterine body

Vagina

Urethra

Labia majora

Ureter

Mesovarium

Mesosalpinx

Round ligament

Mesometrium

Rectum

Urinary bladder

Coxal bone

Vaginal vestibule

Urogenital orifice

(a)

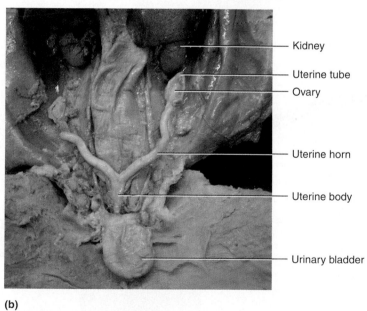

Kidney

Uterine tube

Ovary

Uterine horn

Uterine body

Urinary bladder

(b)

FIGURE 47.12 Female Reproductive Organs of the Cat
(a) Diagram; (b) photograph.

female are proportionally longer than those in the cat. The opening of the uterine tube near the ovary is called the **ostium,** and it receives the **oocytes** during **ovulation.**

At the termination of the uterus is the **cervix,** which leads to the **vaginal canal.** The vagina in cats is different than in humans in that the **urethral opening** is internally enclosed in the vaginal canal. This region where the vagina and the urethral opening occurs is called the **vaginal vestibule.** Thus, the opening to the external environment is a common urinary and reproductive outlet called the **urogenital orifice.** Locate these structures in figure 47.12.

Clean Up   Remember to place all excess tissue in the appropriate waste container and not in a standard wastebasket or down the sink!

Stages of Development
Early Development

The union of the sperm and egg in a process known as **fertilization** initiates a remarkable phenomenon of growth and differentiation from the single-celled **zygote** to the adult human. Fertilization usually occurs in the uterine tube, and the zygote divides into 2 cells, then 4, 8, 16, and so on, until a solid cluster of cells called a **morula** is formed. The morula continues to divide until it becomes a hollow ball of cells known as the **blastocyst.** The covering of cells on the outside of the blastocyst is called the **trophoblast,** and the cluster of cells on the inside is known as the **inner cell mass.** Review these stages in figure 47.13. The placenta is important during development, as it produces hormones that maintain pregnancy.

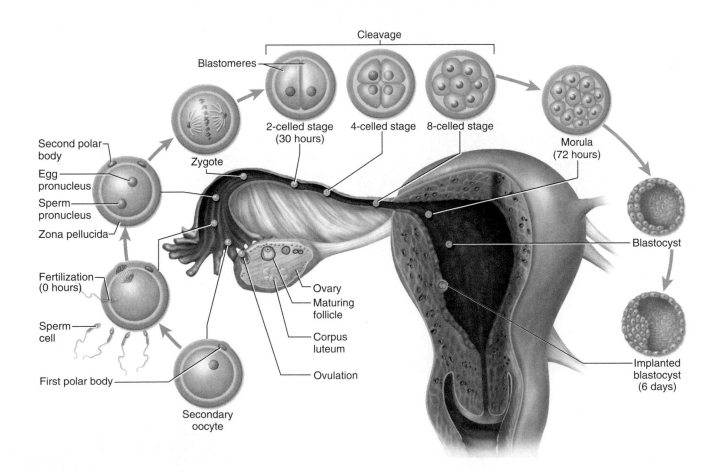

FIGURE 47.13 Early Stages of Human Development

Embryonic Tissues

The early stage of development continues to progress and the formation of three embryonic tissues occurs. These are the **ectoderm,** the **mesoderm,** and the **endoderm.** The ectoderm gives rise to the outer layer of skin and the nervous tissue, the mesoderm gives rise to bones and muscles, and the endoderm gives rise to many internal organs, such as digestive and respiratory organs. The early development of these layers is illustrated in figure 47.14.

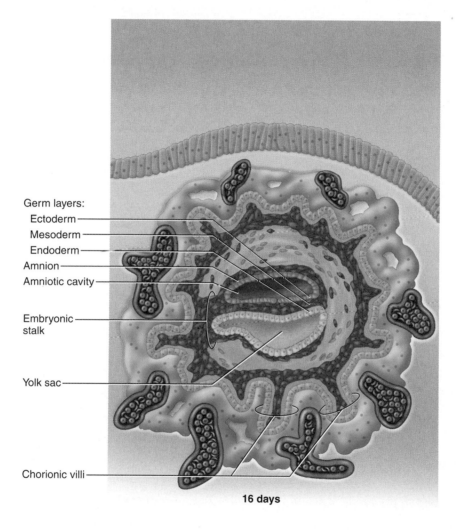

Germ layers:
Ectoderm
Mesoderm
Endoderm
Amnion
Amniotic cavity
Embryonic stalk
Yolk sac
Chorionic villi

16 days

FIGURE 47.14 Embryonic Development

Notes

REVIEW SECTION

Female Reproductive System

Name _____ Date _____

Lab Section _____ Time _____

Review Questions

1. What is the gonad in the female reproductive system?

2. What is the inner layer of the uterus called?

3. What happens to estrogen and progesterone levels just prior to menstruation?

4. The ovaries attach to the uterus by what structure?

5. What three hormones are at elevated levels just prior to ovulation?

6. Where is the fornix in the female reproductive system?

7. Which is more anterior, the urethral opening or the clitoris?

8. What is the background substance of the ovary called?

9. What is the name for the expulsion of the secondary oocyte from the ovary?

10. What is the deepest layer of the endometrium called?

11. What is the name of the part of the breast that is near the shoulder?

12. What are the milk-producing glands of the breast called?

13. A zygote is formed from the fusion of what two cells?

14. Embryonic tissue consists of three layers. What are these called?

15. What is a morula?

16. Trace the pathway of milk from the mammary glands to expulsion.

17. Ectopic pregnancies are those that occur outside of the endometrial layer of the uterus. Provide an explanation for how pregnancies may occur in the uterine tube (thus a tubal pregnancy) or in the abdominopelvic cavity.

18. How does the uterus of the human female differ from that of the cat?

19. How does the structure of the uterus in the cat correlate to multiple births from each pregnancy?

20. Label the following illustration using the terms provided.

cervix   clitoris   vaginal orifice   fundus

urinary bladder   fornix   labium minus   rugae

vaginal canal

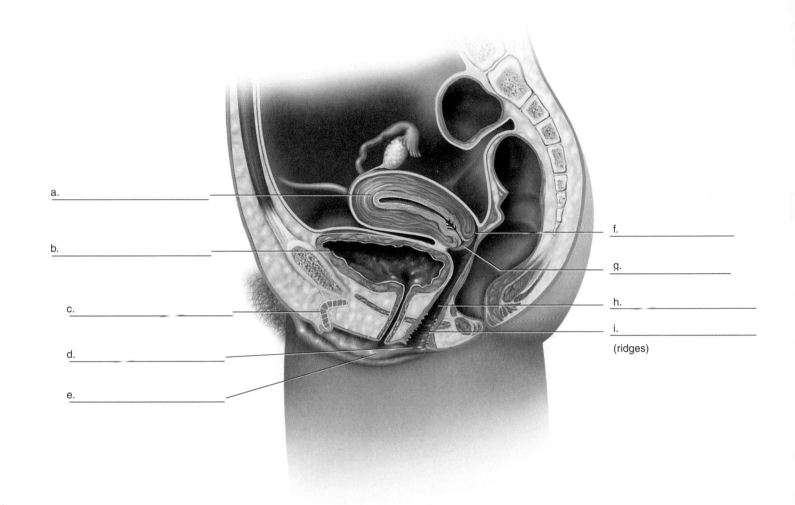

a. _____

b. _____

c. _____

d. _____

e. _____

f. _____

g. _____

h. _____

i. _____
(ridges)

Notes

MEASUREMENT CONVERSIONS

The **metric system** is universally used in science to measure certain **values** or **quantities.** These values are **length, volume, mass** (weight), **time,** and **temperature.** The metric system is based in units of 10, and conversion to higher or lower values is relatively easy when compared to using the U.S. customary system. In the United States, length is typically measured in inches, feet, yards, or miles. The **base unit** for length in the metric system is the meter. As shown in table A.1, 1 meter is equivalent to 1.09 yards, or 39.4 inches. When the measurement of great lengths makes the use of meters impractical, then the kilometer (1,000 meters) is used. When the measurement of extremely small lengths is needed, then the centimeter (1/100 of a meter), millimeter, micrometer, nanometer, and angstrom are used. The common multiples or fractions for the values are outlined in table A.1.

TABLE A.1	Conversions of Standard and Metric Systems				
Value	**Base Unit**	**×1,000**	**1/100**	**1/1,000**	**1/1,000,000**
Length	Meter (m)	Kilometer (km)	Centimeter (cm)	Millimeter	Micrometer
	1 m = 39.4 in.	1 km = $\frac{5}{8}$ mi	2.54 cm = 1 in.		(1 nanometer = 10^{-9} m)
	1 m = 1.09 yd				(1 angstrom = 10^{-10} m)
Mass	Gram (g)	Kilogram (kg)	Centigram	Milligram	Microgram
	454 g = 1 lb	1 kg = 2.2 lb			
	1 g = .036 oz				
Volume	Liter (L)	Kiloliter	Centiliter	Milliliter (mL)	Microliter
	1 liter = 1.06 qt			1 mL = 1 cc	
	1 gal = 3.78 liters				
Time	Second (sec)	Kilosecond	Centisecond	Millisecond	Microsecond
Temperature	Degrees/Celsius				
	0° C = freezing				
	100° C = boiling				
	°F = $\frac{9}{5}$° C + 32				

Appendix B

PREPARATION OF MATERIALS AND SOLUTIONS

The following solutions are designed for a lab of 24 students. It is best to estimate how many milliliters of stock solutions are required for all the students in a lab and then double that amount. The preparations are listed alphabetically and the solutions are the approximate volume for the needs of a class of 24. The number of the lab exercise follows the solution description for cross-referencing.

Acetylcholine Chloride Solution (0.1%)

Add 0.1 g of acetylcholine chloride to 100 mL frog Ringer's solution. Pour into a small, labeled dropper bottle. (Laboratory Exercise 33)

Acid Solution

Pour 200 mL lemon juice or vinegar into dropper bottles labeled "Acid Solution." (Laboratory Exercise 43)

Agar Plates

Add 20 g agar to enough water to make 1 liter of solution. Boil and stir the agar until it all dissolves. Pour into Petri dishes for three dishes per table. (Laboratory Exercise 5)

Alpha Amylase Solution (0.1%)

Add 0.1 g of alpha amylase to a graduated cylinder and fill to 100 mL with water or make a 0.1 serial dilution by taking 10 mL of the following solution and adding 90 mL of water. Pour into a small, labeled bottle or beaker. (Laboratory Exercise 43)

Alpha Amylase Solution (1%)

Add 3 g of alpha amylase to a graduated cylinder and add water to the 300 mL mark. Pour into a small, labeled bottle or beaker. (Laboratory Exercise 43)

Ammonia

Small bottle of ammonia. (Laboratory Exercise 25)

BAPNA Solution (1%)

N-alpha-benzoyl-DL-arginine-p-nitroanilide hydrochloride. Add 0.5 g of BAPNA powder in 50 mL of water. Pour into two bottles of 25 mL each. BAPNA is an expensive material but you do not need much of it for the lab. Available from Sigma B4875 or Aldrich 85,711-4. (Laboratory Exercise 43)

Benedict's Reagent

A copper sulfate solution that turns color if reducing sugars are present and remains blue in the absence of reducing sugars. Add 35 g sodium citrate and 20 g sodium carbonate (Na_2CO_3) to 160 mL water. Filter through paper into a glass beaker. Dissolve 3.5 g copper sulfate ($CuSO_4$) in 40 mL water. Pour the copper sulfate solution into the 160 mL, stirring constantly. Add to six 50 mL bottles. Label "Benedict's Reagent." (Laboratory Exercise 43)

Bleach Solution (10%)

Mix 100 mL household bleach (sodium hypochlorite) with 900 mL tap water. (Laboratory Exercises 29, 30, and 45)

Caffeine Solution, Saturated

Add small amounts of caffeine to 50 mL water until no more will dissolve. Decant the solution into small dropper bottles. (Laboratory Exercise 33)

Calcium Chloride Solution (2%)

Weigh 5 g calcium chloride and place in a graduated cylinder. Add frog Ringer's solution to make 250 mL. Pour in dropper bottles. (Laboratory Exercise 33)

Cat Wetting Solution

Numerous formulations are available for keeping preserved specimens moist. Some commercial preparations are available that reduce the exposure of students to formalin or phenol. You may not need any wetting solution if the cats are kept in a securely closed plastic bag. You can make a wetting solution by putting 75 mL formalin, 100 mL glycerol, and 825 mL distilled water in a 1-liter squeeze bottle. Another mixture consists of equal parts Lysol and water. (Laboratory Exercises 13–18, 34–37, and 46–47)

Cellulose

Cut several (3–4) g pure cotton (cotton wool, cotton balls) into fine pieces (0.5 cm or less). Label "Cellulose." (Laboratory Exercise 43)

Epinephrine Solution (0.1%)

Add 0.1 g of adrenalin chloride in 100 mL frog Ringer's solution. Label and pour the solution into small dropper bottles. (Laboratory Exercise 33)

Essential Oils Preparation

Fill several small screw-top vials with peppermint, almond, wintergreen, and camphor oils (available from local drug stores). Label "Peppermint," "Almond," "Wintergreen," and "Camphor," respectively, and keep vials in separate wide-mouthed jars to prevent cross-contamination of scent. (Laboratory Exercise 25)

Fill four small vials halfway to the top with cotton and color them red with food coloring. Label the vials "Wild Cherry" and add benzaldehyde solution until the cotton is moist. Fill four small vials halfway to the top with cotton. Add benzaldehyde solution until the cotton is moist. Label "Almond." (Laboratory Exercise 25)

Filtration Solution (1% Starch, Charcoal, and Copper Sulfate Solution)

Place 5 g starch, 5 g powdered charcoal, and 5 g copper sulfate ($CUSO_4$) in a 1-liter beaker. Add enough water to make 500 mL. Stir well and pour into a 500 mL bottle. Label "Filtration Solution." (Laboratory Exercise 5)

Frog Ringer's Solution

Weigh and place the following materials in a 1-liter graduated cylinder:

> 6.5 g NaCl (sodium chloride)
>
> 0.2 g $NaHCO_3$ (sodium bicarbonate)
>
> 0.1 g $CaCl_2$ (calcium chloride)
>
> 0.1 g KCl (potassium chloride)

To these add enough water to make 1,000 mL. This solution should be prepared fresh and used within a few weeks. (Laboratory Exercises 23 and 33) (In Laboratory Exercise 33, three solutions are needed—one at room temperature, one at 37° C, and one in an ice bath.)

Hydrochloric Acid Solution (0.1%)

Add 1 mL concentrated HCl to 1 liter of water. Remember "AAA"—Always Add Acid to water. (Laboratory Exercise 23)

India Ink

Dropper bottles of India ink. (Laboratory Exercise 5)

Iodine Solution

Prepare by adding 10 g I_2 (iodine) and 20 g KI (potassium iodide) to 1 liter of distilled water. Store in small, dark dropper bottles. Label "Iodine Solution." (Laboratory Exercises 5 and 43)

Litmus Cream

Use approximately 250 mL heavy cream. To this add powdered litmus until the cream is a light blue. Pour into two separate bottles and label "Litmus Cream." (Laboratory Exercise 43)

Litmus Solution

Weigh 2 g litmus powder and dissolve in 600 mL water. Titrate HCl into the solution until it begins to turn red; then add NaOH solution by drops until it just turns back to blue. Pour into two bottles. (Laboratory Exercise 40)

Maltose Solution (1%)

Add 1 g maltose in enough water to make 100 mL. Stir until dissolved and pour into two clean bottles. Label "1% Maltose Solution." (Laboratory Exercise 43)

Methylene Blue (1%)

Add 5 g methylene blue powder to 500 mL distilled water. Pour into dropper bottles. (Laboratory Exercise 3)

Methylene Blue Solution (0.01 M)

Add 3.2 g methylene blue (MW 320) to distilled water to make 1 liter of solution. Place in dropper bottles. (Laboratory Exercise 5)

Molasses or Concentrated Sucrose Solution (20%)

Use undiluted molasses or a 20% sugar solution. To make the sugar solution add 100 g table sugar (sucrose) to water to make 500 mL of solution. Make sure the sucrose is completely dissolved. (Laboratory Exercise 5)

Nitric Acid (1 N) (for Decalcifying Bones)

Add 64 mL concentrated nitric acid (70%) slowly to water to make 1 liter of solution. Remember "AAA"—Always Add Acid to water. (Laboratory Exercise 8)

Pancreatin Solution (1%)

Place 1 g pancreatin powder in a graduated cylinder and add water to make 100 mL. Stir well. Adjust the pH with 0.05 M sodium bicarbonate until neutral (pH 7). Pour into different stock bottles and label "1% Pancreatin Solution." Preparation note: Use fresh pancreatin. Pancreatin may be stored frozen (not in a frost-free freezer that regularly cycles between freezing and defrosting). Commercially prepared pancreatin has an optimum pH. If the pH is too low, the reaction will be slowed or stopped. (Laboratory Exercise 43)

Perfume, Dilute Solution

Add 10 mL inexpensive perfume to 50 mL isopropyl alcohol. (Laboratory Exercise 25)

Phosphate Buffer Solution

Add 3.3 g potassium phosphate (monobasic) and 1.3 g sodium phosphate (dibasic) to 500 mL water. Place in squeeze bottles. (Laboratory Exercise 29)

Potassium Dichromate Solution (0.01 M)

Add 2.94 g potassium dichromate (MW 294) crystals to water to make 1 liter of solution. Label and pour into dropper bottles. (Laboratory Exercise 5)

Potassium Permanganate Solution (0.01 M)

Add 1.58 g potassium permanganate (MW 158) crystals to water to make 1 liter of solution. Label and pour into dark brown dropper bottles. (Laboratory Exercise 5)

Procaine Hydrochloride

Place 1 g procaine hydrochloride solution in 1 mL water. Add to this 30 mL pure ethanol. Place in small screw-capped bottle. (Laboratory Exercise 23)

Salt Solution (3%)

Add 15 g NaCl crystals (food-grade table salt) to water to make 500 mL solution. Fill clean, food-grade dropper bottles labeled "Salty." (Laboratory Exercise 25)

Sodium Chloride Solution (0.9%) (Physiological Saline)

Put 9.0 g NaCl in 1,000 mL water and pour into small dropper bottles. (Laboratory Exercise 5)

Sodium Chloride Solution (5%)

Add 25 g NaCl crystals to water to make 500 mL of solution. Label the solution and place in small dropper bottles. (Laboratory Exercise 5)

Sodium Chloride Solution (5%)

Add 2.5 g NaCl crystals to water to make 50 mL solution. Label the solution and place in small beaker. (Laboratory Exercise 23)

Sodium Hydroxide Solution (1%)

Add water to 2 g NaOH to make 200 mL solution. Pour into 6 small dropper bottles. (Laboratory Exercise 43)

Caution!   NaOH is caustic. Wear gloves and goggles. Pour into dropper bottles and label as "1% NaOH Solution."

Sodium Hydroxide Solution (1 Normal)

Add 40 g NaOH crystals or powder to water to make 1 liter of solution. Pour into small dropper bottles with a caution label. (Laboratory Exercise 40)

Caution!   The reaction is exothermic and generates heat. Wear protective gloves and eyewear. If you spill this on your skin, make sure you flush your skin immediately with cold water.

Starch Solution (0.5%)

A potato starch solution is made by first boiling 500 mL water. Remove the water from the hot plate and add 2.5 g potato starch powder. Stir and cool the mixture. Do not boil the starch and water mixture because this will lead to some hydrolysis of starch to sugar. Test for the presence of sugar by using Benedict's reagent. There should be no sugar present. Place into 250 mL bottles and label "0.5% Starch Solution." (Laboratory Exercise 43)

Starch Solution (1%)

Boil 1 liter of distilled water. Remove the water from the heat and add 10 g cornstarch (or 10 g potato starch). Filter the mixture through cheesecloth into bottles. (Laboratory Exercise 5)

Sugar Solution (3%)

Using a clean container, for food use, dissolve 15 g table sugar in enough water to make 500 mL of solution. Fill clean, food-grade dropper bottles labeled "Sweet." (Laboratory Exercise 25)

Sugar Solutions

Four table sugar solutions of 2 liters each. (Laboratory Exercise 5)

   0%—2 liters of water

   5%—dissolve 100 g sugar in enough water to make 2 liters of solution

   15%—dissolve 450 g sugar in enough water to make 3 liters of solution; place 2 liters in one bottle and 1 liter in a bottle labeled "15% sucrose solution"

   30%—dissolve 600 g sugar in enough water to make 2 liters of solution

Tonic water

Select a clean, food-grade dropper bottle that holds 100mL. Label "Bitter," Fill with commercial tonic water. (Laboratory Exercise 25)

Umami Solution

Add 15 g monosodium glutamate (MSG) to 500 ml water in a clean, food-grade container. Fill clean, food-grade dropper bottles labeled "Umami." (Laboratory Exercise 25)

Vinegar Solution

Use household vinegar or make a 5% food-grade acetic acid solution by adding 5 mL concentrated acetic acid to about 50 mL water and then adding additional water to make 100 mL. Fill clean, food-grade dropper bottles labeled "Sour." (Laboratory Exercise 25)

Wright's Stain

Wright's stain is available as a commercially prepared solution from a number of biological supply houses. (Laboratory Exercise 29)

LAB REPORTS

Part of working in science involves writing lab reports. You should write lab reports in a certain style and follow basic guidelines that are generally accepted in the field. The general format for the lab reports falls into four categories: **introduction, materials and methods, results,** and **conclusion** sections. Each of these parts is important in the write-up, and each of these four sections must be included in the lab report. Your report should have a title, your name, your instructor's name, the course name and semester, and your lab section.

The purpose of a scientific report is to explain an investigation. Your instructor is your primary audience for your report in this class, so you should communicate that you understand the basic information. Part of a grade in a lab report is dependent on doing the experiment correctly and demonstrating that you are aware of the outcome of the experiment and its relationship to theory or applications you have learned in lecture.

If your experiment did not come out as you thought it might, you still can do well in your lab report. Results from your experiment may have come out satisfactorily and be fine even if you think that the data should have been different from what you obtained.

Introduction

The introduction section consists of a description of the problem and the subject of your study. In professional journals the introduction often includes a history of past experimentation or current knowledge in the field, but you will probably not have this in your lab report. You should pose the question or hypothesis that your experiment is trying to resolve in the introduction. You may be conducting an experiment to determine whether an enzyme functions on a substrate, whether a particular effect occurs when you perform an action, or the like.

Materials and Methods

The materials and methods section consists of a clear description of what equipment, animals, chemicals, and so on were used and the experimental procedure followed. You must write this section in clear and precise terms. From this section you should expect that a person could repeat your experiment and produce the same results.

In one way this is the "recipe" for your experiment. As in baking a cake it is not good enough simply to list the materials. You must include quantity, what sequence the materials were added, how long things were stirred, how long the cake baked, and so on. You must provide specific details to your procedure. Make sure that you state how much material was added to a sample. For example, you should write: "We added 5 mL." This is a known quantity that people can duplicate, whereas "We added a little" is vague. You must make sure that you include all steps in your procedure. If you leave something out in your description, a person following your directions might get different results.

Results

The results section is where the outcome from your experiment is listed in a clear and defined way. Your data must be clearly presented. This may consist of the tabulation of data that you acquired from your experiment in the form of a line graph, bar graph, or table. Remember that any graph should have a complete description. If you study the effect of exercise on heart rate, then exercise is the independent variable and is listed on the horizontal axis, while heart rate is the dependent variable and it is listed on the vertical axis, as illustrated in the following graph.

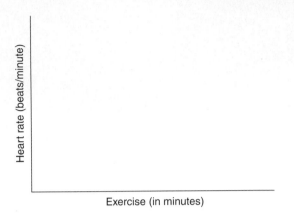

Exercise (in minutes)

The data for your lab result section may be in raw form (direct measurements from your experiment), or you may want to organize the data by calculating the mean (average) data, range (high and low), and so on, as determined from your experiment. There should be no interpretation of the results at this point. You should not try to explain your results but simply list what the outcome was. Save the interpretation of the data for the conclusion.

Conclusion

In the conclusion you analyze and interpret your data and determine whether the results of your experiment proved your initial question (make sure to answer your original question). This is the most important part of your lab report in terms of finding out if you understood the lab. Did you prove your hypothesis, did you disprove your hypothesis, or are the results inconclusive? Where should you go from here?

Sometimes experiments go awry and the results do not support the original hypothesis. This can be due to many things, such as experimental error, a big factor in lab experiments, especially at the undergraduate level. Someone may have not followed the experimental procedure correctly and added too much; too little; or, in some cases, none of the materials that should have been part of the procedure. Sometimes it is a matter of timing; the process did not go on as long as it should or went on too long. In some cases the data are taken down incorrectly or read wrong. A "3" in the data might be mistaken for an "8," or the data are entered in the wrong location. There are hundreds of possibilities as to why error plays a part in experiments and your instructor is probably familiar with many of them.

Other factors are faulty equipment, supplies, and setup. Some problems are caused by the variation in living organisms. Humans and other animals have idiosyncratic physiological responses. Not all people and not all experimental animals respond in the same way. For example, almost all product information sheets that come with pharmaceutical drugs list adverse reactions seen in some people. If everyone responded the same way there would be no need for warnings on drug products.

If you obtain variances from the results you expect to get, then you should write this in your report. Sometimes experimental procedures are included that produce results other than what you expect to see. You should be honest in your recording of data and resist the temptation to "fudge" the data so that you can get a better result and, therefore, a better grade on your report.

Sometimes people in industry change their data to produce more desirable results. If this is discovered by independent investigations the corporation that allowed this to happen usually pays significant fines. Students who change data and are discovered are also subject to penalty far more severe than what they would receive due to the "bad" data they think they might have gotten. In addition to being honest in the reporting of data you also want to make sure that your words are your own. Even though you may have performed an experiment word for word from this lab manual or another source, make sure that you do not plagiarize the material. You should paraphrase material if you want to have the same meaning but not the same words as someone else.

Your lab report should follow the format described here unless your instructor decides to make significant changes in the write-up. Lab reports tend to take a long time to write, but the analysis in the lab report is where you work toward understanding the experiment.

COMMON PREFIXES, SUFFIXES, AND ROOT WORDS IN ANATOMY AND PHYSIOLOGY

Anatomy and physiology have words that may seem long and confusing. Much of the vocabulary has Latin or Greek origin. The **root word** is the core portion of the term. In the word *histology* the *histo-* part is the root word, which means "tissue." **Prefixes** are parts that come before the root word. These are indicated in the following list with a hypen after the term. In the term *cytoskeleton* the prefix is written *cyto-* to show that it occurs first. **Suffixes** are parts that come after the root word and they are indicated with a hyphen before the term. In the term *fungicide* the suffix is *-cide,* indicating that it occurs after the root word. Words that have hyphens before and after can be either prefixes or suffixes.

Term	Meaning	Example
a-	Without	Acellular (without cells)
ab-	To take away	Abduct (to take away)
acoust-	Hearing	Acoustic meatus (ear canal)
acro-	Tip, peak	Acromion (tip of shoulder)
ad-	Next to	Adduct (to bring a limb next to the body)
adipo-	Fat	Adipose (fat tissue)
-al	Relating to	Costal (relating to the ribs [costa])
alb-	White	Tunica albuginea (white layer)
-algia	Pain	Neuralgia (nerve pain)
an-	Without	Anaphylaxis (without protection—a systemic reaction)
ana-	Up	Anatomy (to cut up)
andros-	Male	Androgens (male sex hormones)
angio-	Vessel	Angiogram (radiograph of the blood vessels)
ante-	In front of	Antebrachial (forearm)
anti-	Against	Antiviral (against viruses)
arthro-	Joint	Arthrology (study of joints)
-ase	An enzyme	Lipase (lipid-digesting enzyme)
-aur-	Ear	Auricle (part of the ear)
auto-	Self	Autoimmune (against one's own immune system)
basi-	Base or bottom	Basicranial (base of the skull)
bi-	Two	Bipolar neuron (neuron with two poles)
bio-	Life	Biology (study of life)
brady-	Long, slow	Bradycardia (slow heart rate)
bucco-	Cheek	Buccinator (muscle of the cheek)

carcin-	Cancer	Carcinoma (cancerous tumor)
cardio-	Heart	Cardiology (study of the heart)
cata-	Downward	Catabolism (to break down)
-cephal-	Head	Hydrocephaly (water on the brain)
-cele	Space	Blastocele (space in a blastocyst)
celi-	Abdomen	Celiac artery (artery in the abdomen)
cerebro-	Brain	Cerebrum (large structure of the brain)
chondros-	Cartilage	Chondrocyte (cartilage cell)
-cide	To kill	Fungicide (something that kills fungus)
circum-	Around	Circumcise (to cut around, such as the foreskin of the penis)
-clast-	To break	Osteoclast (cell that dissolves bone)
co-	Together	Coenzyme (molecule that functions with an enzyme)
colpo-	Vagina	Colposcope (scope used to see the vagina)
com-	Join together	Gray commissure (part that joins two halves of the spinal cord)
con-	Join together	Conduct (to go with)
contra-	Opposite	Contralateral (on the opposite side)
corp-	Body	Corpus callosum (callous [tough] body in the brain)
cort-	Bark	Renal cortex (outer part of the kidney)
crypto-	Hidden	Cryptorchidism (hidden testes)
cyano-	Blue	Cyanosis (low oxygen level, causing blue color)
cyst-	Bladder	Cystitis (inflammation of the bladder)
-cyte- (cyto-)	Cell	Leukocyte (white blood cell)
de-	Without	Deoxyribonucleic acid (RNA without oxygen)
demi-	Half	Demifacet (half of one face)
derma-	Skin	Dermatology (study of the skin)
di-	Two	Diploe (spongy bone between two hard sections of bone)
dia-	Through, across	Diapedesis (move through cell spaces)
dis-	Apart	Distend (to move apart from resting condition)
-duct-	Lead	Conduct (to lead to)
dys-	Bad, painful	Dysentery (pain in the intestines)
e-	Away from	Evaporate (to take vapor away from water)
ec-	Out	Eccrine (to take out, as in a sweat gland takes out liquid from the body)
ecto-	Outside	Ectoderm (outer germ layer)
-ectomy	Surgical removal	Appendectomy (to remove the appendix)
-edem-	Swell	Edemia (swelling of tissue)
-emia	Blood	Anemia (not enough blood)
en-	Inside	Encephalitis (inflammation in the brain)
endo-	Within	Endocytosis (within the cell)
-entero-	Intestine	Gastroenterology (study of the stomach and intestines)
epi-	On, above	Epidermis (on the skin)
erythro-	Red	Erythrocyte (red blood cell)
eu-	True, good	Eukaryotic (cells with a true nucleus)
ex-	Out, away from	Expiration (to breathe out)
exo-	Outside	Exocytosis (moving outside the cell)
extra-	Outside	Extracellular (outside of the cell)
-facet	Face	Costal facet (flat surface where a rib attaches)
-fere	Carry	Efferent (to carry away)

-ferous	Carry	Calciferous (producing calcium)
-form	To look like	Fungiform (looks like a fungus or mushroom)
gastro-	Stomach	Gastritis (inflammation of the stomach)
-genesis	To make	Spermatogenesis (formation of sperm)
glossus-	Tongue	Genioglossus (muscle of the tongue)
glyc-	Sugar	Hypoglycemic (low in blood sugar)
-gram	A recording	Electocardiogram (recording of electrical activity of the heart—ECG)
-graph	Recording instrument	Electrocardiograph (instrument measuring ECG)
gyno-	Female	Gynecology (study of the female reproductive system)
hemo-	Blood	Hemolytic (that which destroys blood)
hemi-	Half	Hemisphere (half of a sphere)
hepato-	Liver	Hepatocyte (liver cell)
hex-	Six	Hexagonal (six-sided)
hist-	Tissue	Histology (study of tissues)
hydr-	Wet, water	Hydrate (to give water to)
hyper-	More, above	Hyperactive (excessive activity)
hypo-	Less, below	Hypodermis (under the skin)
hyster-	Uterus	Hysterectomy (surgical removal of the uterus)
-id	State of being	Putrid (state of being rotten)
in-	Into	Infected (to have disease move into the body)
infra-	Below	Infraspinous (below a spine)
inter-	Between	Intercellular (between cells)
intra-	Within	Intravenous (within the veins)
ipsi-	Same	Ipsilateral (on the same side)
-ism	State or condition	Hypothyroidism (having low thyroid hormone levels)
iso-	Equal	Isometric (same length)
-itis	Inflammation	Hepatitis (inflammation of the liver)
juxta-	Near	Juxtaglomerular (near the glomerulus)
kerato-	Hornlike	Keratinocyte (toughened skin cell)
-kin-	Move	Kinase (enzymes that cause movement)
lact-	Milk	Lactose (milk sugar)
-lacrima-	Tear	Lacrimal glands (tear gland)
leuko-	White	Leukocyte (white blood cell)
liga-	Bind	Ligand (chemical that binds to a receptor)
lipo-	Fat	Liposuction (surgical removal of fat)
litho-	Stone	Lithotripsy (breaking up of kidney stones)
-logos	Study	Endocrinology (study of the endocrine system)
-lysis	To break up, destroy	Hemolysis (destruction of red blood cells)
macro-	Big	Macrophage (large, phagocytic cell)
mal-	Bad	Malnourished (with bad nutrition)
-malacia	Softening	Osteomalacia (softening of bone)
mamma-	Breast	Mammogram (X-ray of the breast)
mast-	Breast	Mastitis (inflammation of the breast)
mega-	Big	Megacolon (enlarged colon)
melano-	Black	Melanin (black pigment)
meso-	Middle	Mesothelium (endothelium arising from the middle embryonic layer)

-metrium	Uterus	Endometrium (tissue in the uterus)
meta-	Change	Metastasis (change from original)
micro-	Small	Microsurgery (surgery on a small thing)
mono-	One, single	Mononucleosis (disease with many cells, having one nucleus each)
multi-	Many, much	Multipolar neuron (neuron with many poles)
myo-	Muscle	Myometrium (muscle of the uterus)
necro-	Death	Necrotic (dead material)
neo-	New	Neonate (newborn)
nephro-	Kidney	Nephritis (inflammation of the kidney)
neuro-	Nerve	Neurology (study of the nervous system)
oculo-	Eye	Oculomotor nerve (nerve that moves the eye)
odonto-	Tooth	Odontoid process (toothlike process)
-oid	Looks like	Sigmoid (S-shaped)
-ole	Small	Bronchiole (smaller than a bronchus)
oligo-	Few, medium	Oligosaccharide (carbohydrate of a few sugars)
-oma	Tumor	Lymphoma (tumor of the lymphatic system)
oo-	Egg	Oocyte (egg cell)
-opia-	To see	Hyperopia (farsighted)
ophthalmo-	Eye	Ophthalmologist (person who studies the eye)
-orchid-	Testis	Orchidectomy (removal of the testis)
-ory	Belonging to	Respiratory (belonging to the lungs, etc.)
-osis	A condition	Kyphosis (condition of having a bent spine)
osteo-	Bone	Osteology (study of bones)
oto-	Ear	Otoliths (bones of the ear)
-ous	Producing material	Mucous membrane (membrane that produces mucus)
pan-	All	Pandemic (worldwide)
para-	Near, next to	Paravertebral (near the vertebrae)
-pathy	Illness	Neuropathy (nerve disease)
ped-	Child	Pediatrician (doctor who treats children)
-penia	Lacking	Leukopenia (low white blood cell number)
penta-	Five	Pentagonal (with five sides)
per-	Through	Perfuse (to pass liquid through)
peri-	Around	Perichondrium (around the cartilage)
-phago-	Eat	Phagocytosis (process in which cells "eat" material)
-pharyn-	Throat	Glossopharyngeal (of the tongue and throat)
-phas	Speech	Aphasia (lack of speech)
-phil-	To love	Hydrophilic (water-loving)
phleb-	Vein	Phlebitis (inflammation of a vein)
-phobia	Fear	Hydrophobic (does not mix well with water)
physio-	Nature	Physiology (study of the nature of the body)
-plasm-	Cell, tissue	Plasma membrane (cell membrane)
-plegia	Paralysis	Quadriplegia (paralysis of the four limbs)
pneumo-	Air	Pneumonia (inflammation of the lungs)
pod-	Foot	Podiatrist (doctor who deals with conditions of the feet)
-poie-	Make	Erythropoiesis (make red blood cells)
poly-	Many	Polycythemia (too many red blood cells)
post-	Behind, after	Posterior (back side of something)

pre-	Before	Precentral gyrus (bump before the central sulcus)
pseudo-	False	Pseudostratified epithelium (epithelium with false layers)
psycho-	Mind, soul	Psychology (study of the mind)
py-	Pus	Pyuria (pus in urine)
quad -	Four	Corpora quadrigemina (brain structure with four parts)
re-	Return, back	Relaxation (return to being loose)
recto-	Straight	Rectum (straight tube of the colon)
ren-	Kidney	Renal vein (vein of the kidney)
retro-	Backward, behind	Retroperitoneal (in back of the peritoneum)
rhino-	Nose	Rhinovirus (cold in the nose)
-rrhagia	Outpouring	Hemorrhage (to bleed profusely)
-rrhea	Flow, discharge	Diarrhea (watery bowel movement)
sarco-	Flesh	Sarcoplasm (cytoplasm of muscle cells)
scler-	Hard	Sclera (hard, white covering of the eye)
-scopy	See, look	Microscopy (to see with a microscope)
semi-	Half	Semicircular ducts (ducts resembling half a circle)
sigma-	The letter *S*	Sigmoid colon (part of the large intestine that takes the shape of an *S*)
soma-	Body	Somatic nerves (nerves going to the body)
sperm-	Male sex cells	Spermatogeneis (formation of male sex cells)
sphygm-	Pulse	Sphygmomanometer (instrument that measures blood pressure)
-stasis	Stop, maintain	Homeostasis (steady state condition)
steno-	Narrow	Mitral stenosis (narrowing of the mitral valve)
-stoma	Mouth	Tracheostomy (to make a mouth [opening] in the trachea)
sub-	Beneath, under	Subdermal (under the skin)
super-	Above, more	Superficial (on top of)
supra-	Above	Supraspinous (above a spine)
sym-	Together	Sympathetic (with emotion)
syn-	Together	Synapse (where two nerves join together)
tachy-	Fast, quick	Tachycardia (fast heart rate)
tetra-	Four	Tetracycline (antibiotic with four hydrocarbon rings)
therm-	Heat	Thermogenesis (to produce heat)
-tomy	Cut	Anatomy (to cut up)
tox-	Poison	Toxicology (study of toxins)
trans-	Across	Transmembrane proteins (proteins that occur across membranes)
tri-	Three	Triceps brachii (three-headed muscle of the arm)
-troph-	Feed	Trophic hormone (one that "feeds" other hormones to produce hormones)
-tropic	To influence	Thyrotropic hormones (hormone that influences the thyroid gland)
uni-	One	Unipolar neuron (neuron with one pole)
-uria	Urine	Ketonuria (ketones in the urine)
vaso-	Vessel	Vasodilator (substance that opens blood vessels)
viscer-	Gut	Visceral peritoneum (lining of the viscera)

Credits

Photographs

Chapter 5
Figure 5.4 a-c: © David M. Phillips/Visuals Unlimited, Inc.

Chapter 14
Figure 14.9: © The McGraw-Hill Companies, Inc./Cordi Smith, photographer.

Chapter 16
Figure 16.4 top left: © The McGraw-Hill Companies, Inc./Cordi Smith, photographer.

Chapter 17
Figure 17.2b: © The McGraw-Hill Companies, Inc./Photo and dissection by Christine Eckel; **17.10**: © The McGraw-Hill Companies, Inc./Cordi Smith, photographer.

Chapter 18
18.4b: © The McGraw-Hill Companies, Inc./Rebecca Gray, photographer; **18.5b**: © The McGraw-Hill Companies, Inc./Photo and dissection by Christine Eckel.

Chapter 30
Figure 30.3: Courtesy Clay Adams, Inc.: **30.4**: Courtesy of Coulter International Corporation.

Chapter 37
Page 479: From E.K. Markell and M. Vogue, Medical Parasitology, 5th ed., W. B. Saunders.

Chapter 38
Figure 38.3: © Creatas/Punchstock/RF.

Chapter 39
Figure 39.12b: © The McGraw-Hill Companies, Inc./Cordi Smith, photographer.

Chapter 40
Figure 40.2: Courtesy of Propper, Inc.

Chapter 42
Figure 42.22b: © The McGraw-Hill Companies, Inc./ Cordi Smith, photographer.

Chapter 44
Figure 44.10a: © The McGraw-Hill Companies, Inc./Cordi Smith, photographer.

Chapter 46
46.11b: © The McGraw-Hill Companies, Inc./Cordi Smith, photographer.

Photos not listed are courtesy of the author.

Index